# Remote Sensing: Techniques and Applications

# Remote Sensing: Techniques and Applications

Edited by **Henry Collier**

SYRAWOOD
PUBLISHING HOUSE
New York

Published by Syrawood Publishing House,
750 Third Avenue, 9th Floor,
New York, NY 10017, USA
www.syrawoodpublishinghouse.com

**Remote Sensing: Techniques and Applications**
Edited by Henry Collier



International Standard Book Number: 978-1-68286-213-1 (Hardback)



Printed in China.

# Contents

# Preface

Remote sensing plays a prominent role in atmospheric measurements and physical observations of earth's topography. This book contains a detailed explanation of the various concepts and applications of remote sensing such as climate change, greenhouse emissions, various remote sensing instruments, remote sensing techniques for monitoring of agricultural productivity, etc. The case studies and researches provided in this book will help students, researchers and all associated with the field of remote sensing in gaining advanced information of the field.

This book unites the global concepts and researches in an organized manner for a comprehensive understanding of the subject. It is a ripe text for all researchers, students, scientists or anyone else who is interested in acquiring a better knowledge of this dynamic field.

I extend my sincere thanks to the contributors for such eloquent research chapters. Finally, I thank my family for being a source of support and help.

**Editor**

# 1

# LOCALIZATION AND RECOGNITION OF DYNAMIC HAND GESTURES BASED ON HIERARCHY OF MANIFOLD CLASSIFIERS

M. Favorskaya*, A. Nosov, A. Popov

Institute of Informatics and Telecommunications, Siberian State Aerospace University, 31 Krasnoyarsky Rabochy av., Krasnoyarsk, 660014 Russian Federation - favorskaya@sibsau.ru, alexander@nosov.org, vm_popov@sibsau.ru

**Commission WG V/5, WG III/3**

**KEY WORDS:** Dynamic Gesture, Skeleton Representation, Motion History, Gesture Recognition

**ABSTRACT:**

Generally, the dynamic hand gestures are captured in continuous video sequences, and a gesture recognition system ought to extract the robust features automatically. This task involves the highly challenging spatio-temporal variations of dynamic hand gestures. The proposed method is based on two-level manifold classifiers including the trajectory classifiers in any time instants and the posture classifiers of sub-gestures in selected time instants. The trajectory classifiers contain skin detector, normalized skeleton representation of one or two hands, and motion history representing by motion vectors normalized through predetermined directions (8 and 16 in our case). Each dynamic gesture is separated into a set of sub-gestures in order to predict a trajectory and remove those samples of gestures, which do not satisfy to current trajectory. The posture classifiers involve the normalized skeleton representation of palm and fingers and relative finger positions using fingertips. The min-max criterion is used for trajectory recognition, and the decision tree technique was applied for posture recognition of sub-gestures. For experiments, a dataset "Multi-modal Gesture Recognition Challenge 2013: Dataset and Results" including 393 dynamic hand-gestures was chosen. The proposed method yielded 84–91% recognition accuracy, in average, for restricted set of dynamic gestures.

## 1. INTRODUCTION

During three decades after appearance of graphical user interface with mouse and keyboard, the sensor and display technologies evolve persistent improving and expanding the novel devices ranging from very large displays for design and educational projects to small smartphones or smart watches for conventional consuming. All these devices keep pushing researchers to develop new interaction techniques based on natural human possibilities, first of all, using human gestures and body movements as more natural and intuitive communication between people and devices. Hand gestures have been studied for a long time since 1990s starting with indispensable attributes such as different color gloves in order to provide the simplified tracking of hands and fingers in 3D environment.

Design of formal visual languages with the goal of easy human–computer communication through the use of graphics, drawings, or icons meets the challenge of complicated technical implementation because of variety of locations, shapes, overlaps of hands and cluttered background in a scene. Human gestures are classified as the head, the hand, and the body gestures. Each group assumes special capturing, tracking, and recognition methods. The upper-body gestures are represented in two forms: as natural and artificial gestures. The natural gestures are uncertain, with cultural and local diversity while the artificial gestures are more comprehensible for predefined actions. Also existing gesture systems are classified as encumbered (when a user ought to hold an external device to make gestures (Lu et al., 2014)), touch-based (systems with the touch-screen and different commands), and vision-based (allow users to make gestures without contact). The detailed review of vision-based hand gesture recognition is presented by Rautaray and Agrawal (Rautaray and Agrawal, 2015).

Before recognition, a segmentation stage is necessary. Any gesture segmentation is a difficult process, especially for dynamic gestures. The segmentation process is characterized by ambiguities, when the start and end time instants of dynamic gesture are difficult identified in continuous sequence (Yang et al., 2007), and a spatio-temporal variability, which is caused by gesture variety in shape, location, duration, and trajectory, even for the same person at different time instants (Kosmopoulos and Maglogiannis, 2006). However, a simultaneous procedure of gesture segmentation and recognition is discussed in some researches (Kim et al., 2007).

Our contribution deals with building of hierarchy of manifold classifiers including two levels: the trajectory classifiers in any time instants and the posture classifiers in selected time instants. The trajectory classifiers include skin detector, normalized skeleton representation of hand/hands, and motion history representing by motion vectors normalized through predetermined directions (8 and 16 in our case). Each dynamic gesture is separated into a set of sub-gestures in order to predict a trajectory and remove those samples of gestures, which do not satisfy to current trajectory. The posture classifiers involve the normalized skeleton representation of palm and fingers and relative fingers position using fingertips.

The rest of this paper is organized as follows. Section 2 reviews the existing methods of dynamic gesture capturing and tracking as well as methods of gesture recognition. The proposed gesture

* Corresponding author

trajectory classifiers are described in Section 3. Section 4 provides a discussion about the posture classifiers. Experimental results are situated in Section 5. Finally, Section 6 concludes the paper.

## 2. RELATED WORK

In this section, two mutual issues will be briefly reviewed representing the dynamic gesture capturing and tracking methods and the recognition methods in Sections 2.1–2.2, respectively.

### 2.1 Methods of Gesture Capturing and Tracking

The Hidden Markov Models (HMMs) and the Dynamic Time Warping (DTW) are the most popular methods for dynamic gesture recognition. Many researchers employed the HMM for gesture capturing in the spatio-temporal framework successfully as well as for gesture recognition. The HMM and its modifications was especially popular in 1990s. The HMM-based threshold model was proposed by Lee and Kim (Lee and Kim, 1999). The start and end points were fixed by a likelihood threshold calculated for predefined gesture models. The DTW is a method for sequence comparison, initially used in various applications. For gestures with varying lengths, the DTW warps the tested trajectory in order to match with a predetermined template of an exemplary gesture trajectory. A precursor to the DTW was the Longest Common Subsequence (LCS) method of alignment. It was successfully applied by Stern et al. (Stern et al., 2010) for TV remote control. The LCS algorithm was developed for matching sub-word sequences in documents temporally, using feature distance costs. The LCS is more robust to noise and outliers the DTW in computational speed. Instead of a complete mapping between all points, the LCS algorithm ignores a point without good matching. The classifier uses the "spotted" gesture trajectories or moving window of trajectory points.

Alon et al. (Alon et al., 2005) developed the method with core of Dynamic Space-Time Warping (DSTW) algorithm. The DSTW algorithm as an extension of the DTW aligns a pair of query and model gestures in both space and time dimensions. A warping path in time is aligned with detection the best hand candidate region in every query frame by dynamic programming. The DSTW algorithm models the multiple candidate feature vectors (hypotheses) for hand location in each frame. The warping path in the spatial and the temporal dimensions has some constraints such as the boundary conditions, the temporal continuity, and the temporal monotonicity. The system worked in cluttered background, multiple moving objects, and multiple skin-colored image regions. However, the starting and ending frame of each gesture were pointed manually.

Krishnan et al. (Krishnan et al., 2010) proposed to form a gesture spotting network using the individual gesture models and the adaptive threshold model learnt from Adaptive Boosting algorithm. This technique was evaluated for hand gestures spotting from continuous accelerometer data streams and recognizing by the HMM based on the adaptive threshold model with precision of 78% and recall of 93%.

The attempts of simultaneous gesture spotting and recognition attract some researchers due to a good idea to reduce an unavoidable time delay between segmentation and recognition for on-line continuous gesture recognition. The forward scheme computing a Competitive Differential Observation Probability (CDOP) between a gesture and a non-gesture was proposed by Kim et al. (Kim et al., 2007). This scheme modelled a sequentially variant time-series of gestures using the HMM and Self Organizing Maps (SOMs). The authors computed observation probability of gesture or non-gesture by use a number of continuing observations within the sliding window for several observations. This helps to avoid an undesirable effect of an abrupt change of observations within a short interval. An association mapping technique was used to estimate a correlation measure between 2D shape data and 3D articulation data. Then the gestures were recognized by the trained HMM.

The Conditional Random Fields (CRFs) as an alternative approach to the HMM are studied by Elmezain et al. (Elmezain et al., 2010) for simultaneous spotting and recognition of digits. Such interpretation can be useful for dynamic gesture recognition. The CRFs are undirected graphical models for labeling sequential data, which overcome the weakness of maximum entropy Markov models. The authors conducted good idea about initially constructed the non-gesture pattern by the CRFs due to the CRFs uses a single model for the joint probability of the sequences. As a result, they obtained good reliability accuracy estimations of 96.51% and 90.49% for recognition of isolated and meaningful gestures, respectively.

Extraction and tracing of hand region using entropy analysis was utilized by Lee et al. (Lee et al., 2004). It was one of the first attempts for proposition of video-based non-contact interaction techniques. A hand region was extracted based on color information and entropy distribution. In common case, a gesture tracking can be based on various techniques, e.g., fingertips tracking for restricted set of gestures, hand's position tracking by optical flow method, motion energy analysis to estimate a dominant motion of hand, and adaptive skin color model and Motion History Image (MHI), splitting a feature trajectory into sub-trajectories.

### 2.2 Methods of Dynamic Gesture Recognition

In general, two types of recognition techniques, 2D shape data and 3D articulation data, are possible for dynamic gesture recognition. Bobick and Davis (Bobick and Davis, 2001) recognized human movements by 2D temporal templates including a motion energy image and the MHI. Hu moments were descriptors of the temporal templates. A human gesture recognition based on fitted quadratic curves as a place of centre points of skin region and 2D foreground silhouettes was proposed by Dong et al. (Dong et al., 2006). The principal disadvantage consists in the evident dependence of the obtained 20 features from a viewing angle between a human and cameras. Also a gesture may be classified in a conventional manner, when a sequence of postures is processed, using principal component analysis, and then it is recognized applying Finite State Machines (FSMs) also called as a finite-state automata (Hong et al., 2000).

Bhuyan (Bhuyan, 2012) proposed a concept of key frames detection, in which a hand changes its shape significantly. Next keyframes are extracted by measuring a shape similarity using a Hausdorff distance measure. Thus, such key frame selection eliminates redundant frames. Each FSM constructed through the training stage corresponds to a particular gesture. Recognition is executed by matching only a current keyframe state with the states of different FSMs obtained during the training stage. All the FSMs in the gesture vocabulary use an angular radial transformation shape descriptor that speeds up the recognition in many times.

Dynamic Bayesian Network (DBN) is a generalized class of probabilistic models including the HMM and Kalman filter or particle filter as special cases. Suk et al. (Suk et al., 2010) proposed the DBN model for recognition of hand gestures implementing a control of media players and slide presentations. Ten isolated one-hand and two-hand gestures are enough simple, and the authors received the recognition rate of 99.59%. More practical continuous gesture recognition was addressed based on a cyclic spotting network connecting with gesture DBN. A Viterbi dynamic programming method was used to recognize gesture and detect the start and end points of gesture simultaneously.

Al-Rousan et al. (Al-Rousan et al., 2010) studied the dynamic gestures of Arabic sign language using two-stage scheme, when, first, a group of gestures is recognized and, second, a gesture within a group is recognized. A set of spatial features was extracted including a hand region, coordinates of its centre, a direction angle of hand region, and a hand vector representing a shape of hand. The authors considered these features invariant to scale and transition. The HMM was used for hand recognition. The authors divided all features into two types: six simple features and seventeen complex (in vector representation) features. The recognition rate for the signer-dependent achieved 92.5% while for the signer-independent was 70.5%.

3D modelling of hand gesture is usually connected with multiple cameras shooting during training stage or Kinect application in order to obtain the depth images of a hand. For example, Keskin et al. (Keskin et al., 2011) proposed a 3D skinned mesh model with a hierarchical skeleton, consisting of 19 bones, 15 joints and 21 different parts for representation of American Sign Language. The methodology is based on fitting 3D skeleton to the hand. Random decision forests were trained on animated 3D hand model for pixel classification during the testing stage.

Holte and Moeslund (Holte and Moeslund, 2008) proposed to build and analyze the harmonic shape context as 3D motion primitives, which are received from motion characteristic instances of gestures. The authors used 3D image data from a range camera to achieve invariance to viewpoint. For this purpose, a family of orthogonal basis functions in a form of spherical harmonics was applied. 3D motion primitives representing as the strings of letters and/or digits with different lengths were compared by probabilistic edit distance method. An overall recognition rate of 82.9% was achieved under invariance of 3D location.

## 3. GESTURE TRAJECTORY CLASSIFIERS

Introduce assumptions, according to which a video sequence contains a hand gesture:

1. A hand gesture is performed on approximately uniform background.
2. A distance between a camera and a hand is nearly constant so that scale factor is non-significant.
3. Consider that a moving hand appears in video sequence, if a motion is continuous in a predetermined interval (1–2 s).
4. A hand gesture is performed in a priori known region of frame.
5. A moving hand is a dominant moving object.
6. A duration of moving is longer that $L_1$ frames but not more than $L_2$ frames with the temporal continuity and monotonicity.
7. All types of gestures are a priori known.
8. Images of hand gestures are captured with a single video camera and then processed by a single computer.

Dynamic gesture spotting is very difficult task without simplifying guesses, e.g., when the gesture trajectories are initiated by a button press or a long interval without motion. Li and Greenspan (Li and Greenspan, 2011) had solved the endpoint localization of dynamic gestures using a multi-scale gesture model representing as 3D spatio-temporal surfaces.

A motion history of dynamic gesture is a set of trajectory classifiers of sub-gestures, which model a realistic gesture by elements from special vocabulary. First, hand localization ought to be implemented (Section 3.1). Second, skeleton representation of gesture is built (Section 3.2). Third, trajectory classifiers are constructed (Section 3.3).

### 3.1 Hand Localization

For hand localization, it is reasonable to use two classifiers based on skin and motion detection. The skin-like color regions may be detected in a scene using various color spaces such as Red Green Blue (RGB), YUV (Y is a luminance component (brightness), U and V are the blue and red difference samples, respectively), YCbCr (Cb and Cr are the blue and red chromatic components, respectively), Hue Saturation Value (HSV), normalized RGB, and Log opponent (uses the base 10 logarithm to convert RGB values) color spaces (Favorskaya, 2013). The YCrCb color space was chosen for skin classifier. The linear equations, which determine the boundaries of skin color classifier, are provided by Equation 1:

$$\begin{cases} Cr \geq -2(Cb+24) \\ Cr \geq -4(Cb+32) \\ Cr \geq -(Cb+17) \\ Cr \geq 25(Cb+Q_1) \\ Cr \geq Q_3 \end{cases} ; \begin{cases} Cr \leq \dfrac{220-Cb}{6} \\ Cr \leq \dfrac{4}{3}(Q_2-Cb) \\ Cr \geq 0.5(Q_4-Cb) \end{cases} \quad (1)$$

where $Q_1$–$Q_4$ = additional parameters computed by Equations 2–3 in dependence of $Y$ value:

$$\begin{cases} \text{if } Y > 128 \text{ then } Q_1 = -2 + \dfrac{256-Y}{16}, \\ Q_2 = -20 - \dfrac{256-Y}{16},\ Q_3 = 6,\ Q_4 = -8 \end{cases} \quad (2)$$

$$\begin{cases} \text{if } Y \leq 128 \text{ then } Q_1 = 6\ \ Q_2 = 12, \\ Q_3 = 2 + Y/32,\ Q_4 = -16 + Y/16 \end{cases} \quad (3)$$

For reliable detection of hand/hands in frames, some authors proposed the approach based on a prior facial skin analysis as an individual skin tuning (Li et al., 2013). Such initial procedure is very useful for hand localization because the articulated user can be a single end-user of gesture recognition system, and this procedure can be considered as an adaptive step. Face detection based on Adaboost algorithm is fast and accuracy method proposed by Viola and Jones (Viola and Jones, 2001). As in any boosting algorithm, a cascade classifier is adopted during 2–3 iterations in order to detect the most possible regions of human faces. Additionally, eyes detection can be recommended in regions similar to skin color. Then hands are detected in enlarged surrounding region, which sizes are chosen empirically. After hand detection using skin classifiers, the morphological processing is recommended for improvement "broken" skin regions or skin regions with "holes" (Favorskaya and Nosov, 2014).

The use of Microsoft Kinect camera is in the area of interest for many segmentation tasks. The simple background subtraction can be applied under assumption of enough number of frames in static scene. First, a background accumulation is executed based on maximum values of depths in a set of frames. As a result, a background model can be constructed. Notice, an accurate background model is not required for hand capturing that means a possibility of on-line background model building. Second, the moving objects are extracted using a background subtraction model. Third, the object boundaries are computed based on detection and merger of contour components in order to receive the closed boundaries of moving objects. The depth map cannot be used only for body parts contour extraction but also for body silhouette building. The last possibility is useful for the body gesture recognition. For hand segmentation, the color distribution is analysed into segmented regions that increases the accuracy of palm and fingers localization in a cluttered background.

### 3.2 Skeleton Representation of Gesture

In this research, the idea of compact and informative description of dynamic gesture is conducted. For this purpose, a skeleton representation of hand including elbow, wrist, palm, and fingers is built. The normalized skeleton representation is invariant to shape but do not invariant to position in 3D space. However, in skeleton representation a "central" point as a centre point of circle, which is inscribed in a palm image region, can be defined in the most cases very fast.

A binary hand gesture image is an object with multi-linked polygonal shape. In this case, a skeleton representation is built using a term "maximum empty circle" (Mekhedov and Mestetskiy, 2010). For polygonal shape $F$, a maximum empty circle is any circle $B$, which is fully inscribed into a shape $F$ such that other circle $B'$ inscribing into a shape $F$ does not include a circle $B$. Thus, a skeleton of polygonal shape $F$ is a set of centres of maximum empty circles. A radial function $R(x, y)$ defining a radius value in any skeleton point $(x, y)$ is determined on a skeleton of an object.

Geometrically, a skeleton of polygonal shape is a graph including nodes (points in the XOY plane) and edges (lines joining some nodes pairs). A straight line or a parabolic arc will be the edges of such graph. A degree of any edge is equal 1, 2, or 3. The existing effective algorithms build a skeleton by time $O(n \log n)$, where $n$ is a number of nodes in a polygon shape. A skeleton building of 2D shape is described in details by Mekhedov and Mestetskiy (Mekhedov and Mestetskiy, 2010). An execution time of a skeleton building depends directly from a number of nodes in a polygon binary shape. The procedure of shape approximation based on Douglas-Peucker algorithm is represented and implemented in our previous research (Favorskaya and Nosov, 2014) with good speed/accuracy numerical results. After skeleton building, an additional procedure called pruning is implemented in order to remove noisy lines. Then a central point in palm image can be easily defined.

### 3.3 Trajectory Classifiers Based on Motion History

Let DG = {$DG_1$, $DG_2$, ..., $DG_n$} be a set of available dynamic gestures represented by skeleton vectors. Each dynamic gesture DG is composed from sub-gestures differing by directions and acceleration values from other sub-gestures. A set of sub-gestures generates a vocabulary, which helps to describe a realistic dynamic gesture by its model representation. A sequence of sub-gesture forms a gesture model. Due to a hand image is a compact set of mass points, a trajectory can be described by a single unique point, which can be a centroid of mass of a palm or a central point in skeleton representation (Section 3.2). A sub-gesture description SG is a set of vectors SG = {$\mathbf{SG}_1$, $\mathbf{SG}_2$, ..., $\mathbf{SG}_m$}. The beginning of each following vector $\mathbf{SG}_{j+1}$ is the ending of the previous vector $\mathbf{SG}_j$; thus, a connected set of directions performs a total direction of a sub-gesture. Each vector $\mathbf{SG}_j$ includes two components: a length $L_j$ and an angle $\alpha_j$ (between vector direction and the axis OX) in relative coordinates (Equation 4):

$$L_j = \sqrt{(x_{j+1} - x_j)^2 + (y_{j+1} - y_j)^2}; \quad \alpha_i = \arctan\frac{y_{j+1} - y_j}{x_{j+1} - x_j} \tag{4}$$

where $x_j$, $y_j$ = coordinates of current branch point
$x_{j+1}$, $y_{j+1}$ = coordinates of following branch point

Then a sub-gesture SG is recalculated by compression and normalization procedures in order to provide invariance to affine transform. The compression procedure rejects non-essential vectors with small value $L_j$ according to a predetermined threshold value. The normalization procedure provides a normalization of vectors lengths (total sum of vectors lengths is equal 1) by Equation 5:

$$L_j^{nr} = L_j \Big/ \sum_{j=1}^{n} L_j \tag{5}$$

and a normalization of vectors directions, when a current vector direction is replaced by one of normalized directions from a unit vectors set. Such unit vector $\mathbf{U}_j$ has a view of Equation 6:

$$\mathbf{U}_j = \left\{1, \alpha_j^{nr}\right\}; \quad \alpha_j^{nr} = j \cdot 2\pi/Z \tag{6}$$

where $Z$ = a number of directions (Z = 8 or Z = 16)

Let the temporal representation of all gesture samples be $\mathbf{G}_{c,k}$, $c = 1, 2, ...,$ where $C$ is a number of temporal classes of hand gestures and $k = 1, 2, ..., K$ is a number of samples into a temporal class. The reference temporal template $TT_c$ estimates dissimilarity $D(\cdot)$ by min-max criterion in the temporal dimension by Equation 7:

$$TT_c = \arg\min_k \arg\max_m D\left(\mathbf{G}_{c,k}, \mathbf{G}_{c,m}\right) \tag{7}$$

Template-based classification technique is announced as follows. The extracted and processed temporal representation is compared with reference temporal template for each class $c$

(Equation 7). Afterward, distances to all classes are computed. If a minimal distance is below a predetermined threshold value, then a gesture belongs to some class, otherwise it is concerned to a non-gesture class.

During the testing stage, the extracted sub-gestures help to predict a type of gesture. Several candidates may be determined. The posture classifiers of palm and fingers provide the final decision.

## 4. GESTURE POSTURE CLASSIFIERS

Each sub-gesture ought to include the detailed information about palm and fingers position. Consider that this analysis will be accomplished at the end of sub-gesture. Such information clarifies the class of dynamic hand gesture. Posture classifiers are discussed in Section 4.1. A procedure of rule-based recognition is situated in Section 4.2. A set of rules is applied to select the best candidate.

### 4.1 Posture Classifiers

The posture classifiers are constructed based on skeleton representation of hand. As one can see from Figure 1, the fingertips can be detected as endpoints. Then the relative locations of skeleton segments of each finger are analyzed, in particular lengths of fingers $S = \{s_1, s_2, s_3, s_4, s_5\}$ and angles $\Theta = \{\theta_1, \theta_2, \theta_3, \theta_4\}$ between the neighbouring fingers. A length value and a view of skeleton line of finger determine a position of finger. A set of angles may be restricted to one or two values. These values are graduated by threshold: if $TH_\theta < 3°$, then the fingers are grouped. Notice that not all gestures may have full description, some fingers may be invisible.

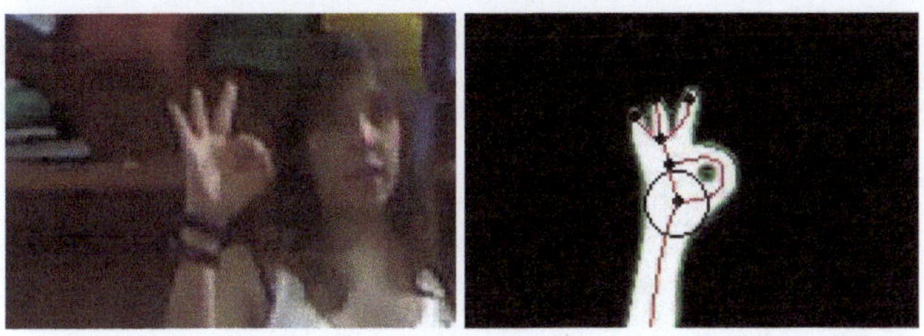

Figure 1. Original image (left), image with keypoints (right)

Figure 1 illustrates a skeleton building based on a binary image. Use of gray-scale image of hand permits to detect a skeleton for fingers overlapping a palm. Therefore, a posture classifier includes relative location of a centre point of a palm (a centre of inscribed circle), coordinates of fingertips, based on which it is possible to determine the fingers stretched, half-bend, half-closed, or closed and group, separate, cross, or loop. In the same manner, two hands can be analyzed, if a gesture is presented using both hands.

### 4.2 Rule-based Recognition

In spite of variety of 3D hand position, a hand model includes a fixed number of components, and each finger configuration is associated with a finger pose. Each hand posture is described by a rule capturing a hand configuration and involving the following parameters with available values:

1. The coordinates of centre point of a palm.
2. A set of lengths $S = \{s_1, s_2, s_3, s_4, s_5\}$. If ($s_i > TH_s$) and ($s_i$ is a straight line), then a finger is stretched. If ($s_i > TH_s$) and ($s_i$ is a curve), then a finger is half-bend or half-closed. If ($s_i > TH_s$) and ($s_i$, $_{sj}$ are closed curves), then two fingers form a loop. If ($s_i \leq TH_s$) and ($s_i$ is a straight line), then a finger is closed.
3. A set of angles $\Theta = \{\theta_1, \theta_2, \theta_3, \theta_4\}$. If $\theta_i > TH_\theta$, then a two fingers are separated. If ($\theta_i \leq TH_\theta$) and (a thickness of two fingers is less than the double thicknesses of one finger), then two fingers are crossed.
4. Relative coordinates of visible fingertips.
5. Detection a finger group based on a convexity analysis of a hand contour. For example, one can detect the groups from two, three, or four closed fingers estimating a thickness of group.

Items 1–5 mentioned above describe each sub-gesture as a set of finger orientations in the XOY plane (up, down, towards, or side), a set of finger inter-relations (grouped, separated, looped, or crossed), and a set of basic finger poses (stretched, half-bend, half-closed, or closed). The main positions of fingers based on depth performances and depicted in Table 1 were proposed by Mo and Neumann (Mo and Neumann, 2006).

| Category | Position | | | |
|---|---|---|---|---|
| Basic fingers orientation | 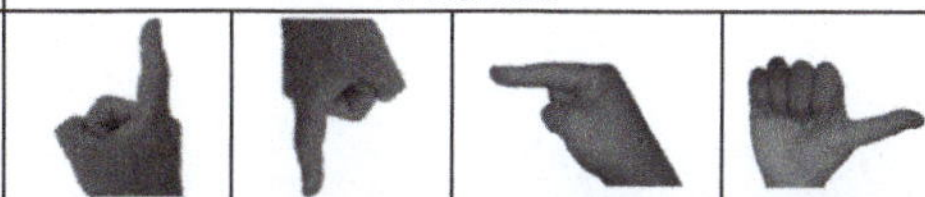 | | | |
| | Up | Down | Towards | Side |
| Basic finger inter-relations | 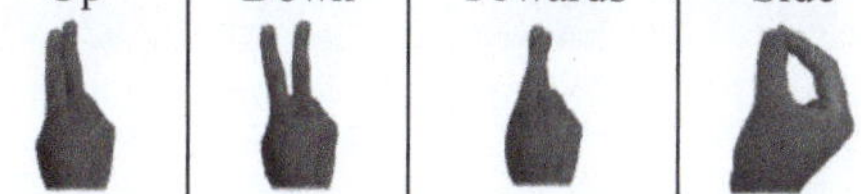 | | | |
| | Group | Separate | Cross | Loop |
| Basic finger poses | 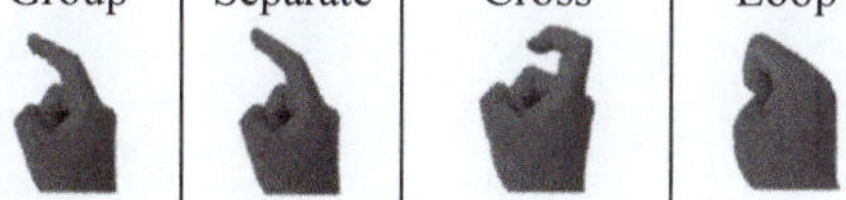 | | | |
| | Half-bend | Bend | Half-closed | Closed |

Table 1. Main positions of fingers

As a result, a sequence of sub-gesture descriptors can be received. If dominant singularity/singularities of a gesture are determined during the training stage, then the testing stage is simplified. The decision trees based on the proposed rules are used for recognition. A hierarchy description of sub-gestures including trajectory analysis finds a good mapping in decision trees. Due to rejection of non-suitable branches, the recognition is a fast procedure. When a dynamic gesture is finished, the descriptions of sub-gestures are finally analyzed in the temporal dimension, and the final decision is concluded. In spite of difficult analysis, a continuous sequence provides high data volume that objectively leads to better recognition results.

## 5. EXPERIMENTAL RESULTS

For experiments, a part of the dataset "Multi-modal Gesture Recognition Challenge 2013: Dataset and Results"[1] containing dynamic gestures was used. This large video dataset includes 13, 858 gestures from a lexicon of 20 Italian gesture categories recorded with a Kinect™ camera, providing the audio, skeletal model, user mask, RGB and depth images (Escalera et al., 2013). For better visibility in Figures 2–9, fragments of images with sizes 140 × 105 pixels were cut from the test video files.

[1] http://gesture.chalearn.org/2013-multi-modal-challenge

Figure 2. Sample00827, frames 1175-1231

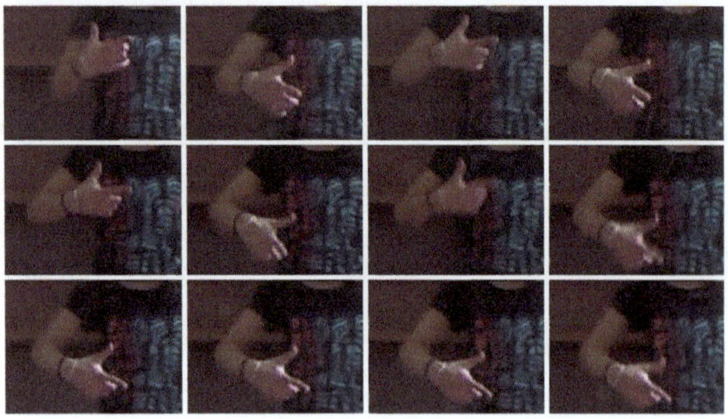

Figure 3. Sample00824, frames 164-211

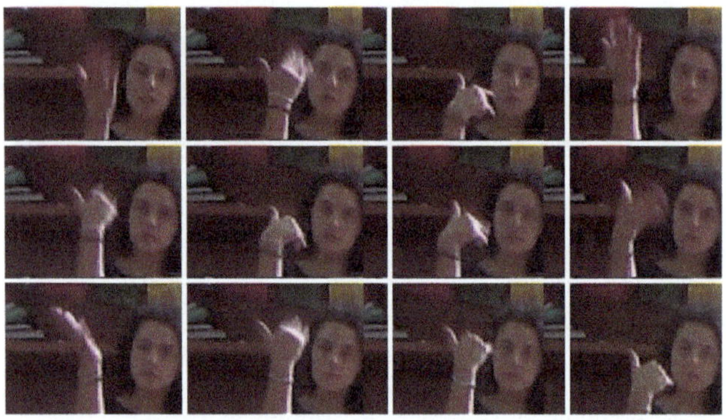

Figure 4. Sample00825, frames 189-232

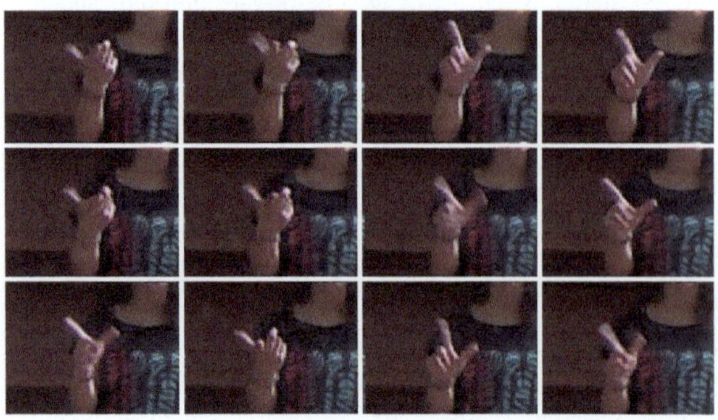

Figure 5. Sample00831, frames 921-1015

Figure 6. Sample00804, frames 216-251

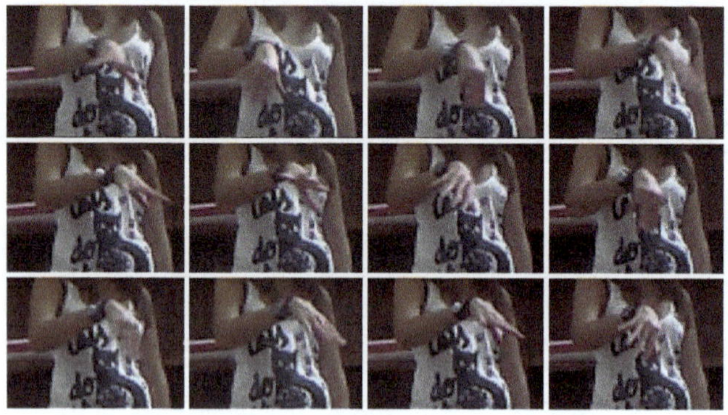

Figure 7. Sample00802, frames 913-970

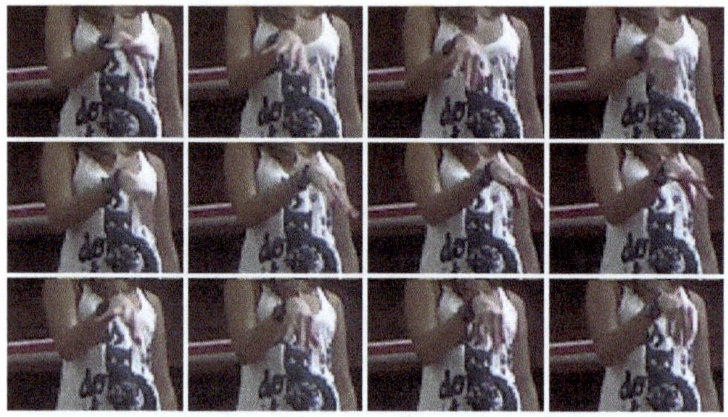

Figure 8. Sample00816, frames 1368-1408

Figure 9. Sample00803, frames 432-475

The software tool "DynGesture", v. 1.52 was developed using C# language in environment "Visual Studio 2012". The software includes two main algorithms. The segmentation algorithm contains a hand/hands localization based on skin classifiers, a topological skeleton building, a skeleton representation of palm and fingers with keypoints, a building of trajectory classifiers, and a vector description of sub-gestures. The recognition algorithm calculates the posture classifiers based on five proposed rules (Section 4.2) providing a recognition of main positions of fingers at the end of each sub-gesture. Then a sequence of sub-gesture descriptors is analyzed by decision tree procedure. Samples from each category involving 20–40 dynamic gestures were divided into the training and the testing sets in relation 20% and 80%, respectively. The results of gesture segmentation and gesture recognition are located in Tables 2 and 3, respectively. These estimations include True Recognition (TR), False Rejection Rate (FRR), False Acceptance Rate (FAR), and time costs of algorithms as average estimators for processing of a single frame. Experiments were executed using computer with the processor Intel(R) Core(TM) i5 750 2.67 GHz. The TR, FRR, and FAR estimators were received based on frames segmented manually and recognized by expert.

| Video file | TR (%) | FRR (%) | FAR (%) | Time (ms) |
|---|---|---|---|---|
| Sample00827 | 92.9 | 3.3 | 8.5 | 9.2 |
| Sample00824 | 86.7 | 3.1 | 6.7 | 8.6 |
| Sample00825 | 93.2 | 7.7 | 9.1 | 10.4 |
| Sample00831 | 90.4 | 2.3 | 4.6 | 7.7 |
| Sample00804 | 93.5 | 3.9 | 6.4 | 8.0 |
| Sample00802 | 83.3 | 6.6 | 10.2 | 9.3 |
| Sample00816 | 84.3 | 8.3 | 4.5 | 4.2 |
| Sample00803 | 90.8 | 5.2 | 9.1 | 8.5 |

Table 2. Gesture segmentation results and temporal estimators

As one can see from Table 2, all true and false segmentation results have close values in ranges of 84-93% and 3-10%, respectively. The deviation is explained by some blurred frames in video files, for example, "Sample00802" and "Sample00816". The temporal values of frame processing have the same order.

| Video file | TR (%) | FRR (%) | FAR (%) | Time (ms) |
|---|---|---|---|---|
| Sample00827 | 84.9 | 8.3 | 8.9 | 4.4 |
| Sample00824 | 78.6 | 7.1 | 5.2 | 5.6 |
| Sample00825 | 91.1 | 8.8 | 6.7 | 5.7 |
| Sample00831 | 81.6 | 5.3 | 4.5 | 4.2 |
| Sample00804 | 83.6 | 7.5 | 9.1 | 4.8 |
| Sample00802 | 93.4 | 5.8 | 5.1 | 5.3 |
| Sample00816 | 85.4 | 7.3 | 4.6 | 4.0 |
| Sample00803 | 89.4 | 8.7 | 5.2 | 6.1 |

Table 3. Gesture recognition results and temporal estimators

Gesture recognition results representing in Table 3 demonstrate the dependence from more simple or complex trajectory of dynamic gesture and visible shape of a hand. In spite of all difficulties, the final results are promising due to the analysis of a sequence of sub-gesture descriptors, which represent a single dynamic gesture. The temporal estimators show a possibility of real-time implementation of the designed software tool.

## 6. CONCLUSION

The proposed method for localization and recognition of dynamic gestures is based on two-level classifiers including the trajectory classifiers in any time instants and the posture classifiers for sub-gesture extraction in selected time instants. Our efforts were directed on building of such classifiers, which are invariant to scale, rotation, and shift in the XOY plane. Skeleton representation of wrist, palm, and fingers provides a description of special points (e.g., fingertips or centre point of a palm), based on which some rules were formulated. Motion history of trajectory and a set of posture descriptions of sub-gestures are the input information for decision trees. A dataset "Multi-modal Gesture Recognition Challenge 2013: Dataset and Results" was used for experiments. The numerical results were obtained for 393 dynamic gestures applicable in learning systems of sign languages as well as in human-computer interaction systems. The proposed approach yielded 84–91% recognition accuracy for restricted set of dynamic gestures.

## REFERENCES

Alon, J., Athitsos, V., Yuan, Q., Sclaroff, S., 2005. Simultaneous Localization and Recognition of Dynamic Hand Gestures. *7th IEEE Workshops on Application of Computer Vision*, 2, pp. 254-260.

Al-Rousan, M., Al-Jarrah, O., Al-Hammouri, M., 2010. Recognition of dynamic gestures in arabic sign language using two stages hierarchical scheme. *International Journal of Knowledge-based and Intelligent Engineering* 14(3), pp. 139-152.

Bhuyan, M.K., 2012. FSM-based recognition of dynamic hand gestures via gesture summarization using key video object planes. *World Academy of Science, Engineering and Technology* 6(8), pp. 724-735.

Bobick, A., Davis, J., 2001. The recognition of human movement using temporal templates. *IEEE Trans. Pattern Anal. Mach. Intell.* 23 (3), pp. 257-267.

Dong, Q., Wu, Y., Hu, Z., 2006. Gesture recognition using quadratic curves. In: *The 7th Asian Conference on Computer Vision*, pp. 817-825.

Elmezain, M., Al-Hamadi, A., Michaelis, B., 2010. A Robust Method for Hand Gesture Segmentation and Recognition Using Forward Spotting Scheme in Conditional Random Fields. In: *The 10th IEEE International Conference on Pattern Recognition*, pp. 3850-3853.

Escalera, S., Gonzàlez, J., Baró, X., Reyes, M., Lopes, O., Guyon, I., Athitsos, V., Escalante, H.J., (2013) Multi-modal Gesture Recognition Challenge 2013: Dataset and Results. In: *15th ACM International Conference on Multimodal Interaction*, pp. 445-452.

Favorskaya, M., 2013. Visual system of sign alphabet learning for poorly-hearing children. In: Kountchev, R., Iantovics, B. (eds) *Advances in Intelligent Analysis of Medical Data and Decision Support Systems*, SCI 473, pp. 23-39.

Favorskaya, M., Nosov, A., 2014. Hand-gesture Description Based on Skeleton Representation and Hu Moments. *Frontiers in Artificial Intelligence and Applications* 262, pp. 431-440.

Holte, M.B., Moeslund, T.B., 2008. View invariant gesture recognition using 3D motion primitives. *IEEE International Conference on Acoustics, Speech and Signal Processing*, pp. 797-800.

Hong, P., Turk, M., Huang, T.S., 2000. Gesture Modeling and Recognition Using Finite State Machines. In: *The IEEE Conference on Face and Gesture Recognition*, pp. 410-415.

Keskin, C., Kira, F., Kara, Y.E., Akarun, L. 2011. Real time hand pose estimation using depth sensors. In: *The IEEE International Conference on Computer Vision Workshops*, pp. 1228-1234.

Kim, D., Song, J., Kim D., 2007. Simultaneous gesture segmentation and recognition based on forward spotting accumulative HMMs. *Pattern Recognition* 40(11), pp. 3012-3026.

Kosmopoulos, D.I., Maglogiannis, I.G., 2006. Extraction of Mid-Level Semantics from Gesture Videos using a Bayesian Network. *International Journal of Intelligent Systems Technologies and Applications* 1(3-4), pp. 359-375.

Krishnanm, N.C., Lade, P., Panchanathan, S., 2010. Activity Gesture Spotting Using a Threshold Model Based on Adaptive Boosting. In: *The IEEE International Conference on Multimedia and Expo*, pp. 155-160.

Lee, H., Kim, J., 1999. An HMM-based threshold model approach for gesture recognition, *IEEE Trans. Pattern Anal. Mach. Intell.* 21(10), pp. 961-973.

Lee, J.S., Lee, Y.J., Lee, E.H., Hong, S.H., 2004. Hand region extraction and Gesture recognition from video stream with complex background through entropy analysis. *26th Annual International Conference of the IEEE Engineering in Medicine and Biology Society*, pp. 1513-1516.

Li, H., Greenspan, M., 2011. Model-based segmentation and recognition of dynamic gestures in continuous video streams. Pattern Recognition 44 (8), pp. 1614-1628.

Li, J., Zheng, L., Chen, Y., Zhang, Y., Lu, P., 2013. A Real Time Hand Gesture Recognition System Based on the Prior Facial Knowledge and SVM. *Journal of Convergence Information Technology* 8(11), pp. 185-193.

Lu, Z., Chen, X., Li, Q., Zhang, X., Zhou, P., 2014. A Hand Gesture Recognition Framework and Wearable Gesture-Based Interaction Prototype for Mobile Devices. *IEEE Transactions on Human-Machine Systems* 44(2), pp. 293-299.

Mekhedov, I., Mestetskiy, L., 2010. Skeleton of a Multi-ribbon Surface. In: Taniar, D., Gervasi, O., Murgante, B., Pardede, E., Apduhan, B.O. (eds) *Computational Science and Its Applications*, LNCS 6016, pp. 557-573.

Mo, Z., Neumann, U., 2006. Real-time Hand Pose Recognition Using Low-Resolution Depth Images. *IEEE Computer Society Conference on Computer Vision and Pattern Recognition* 2, pp. 1499-1505.

Rautaray, S.S., Agrawal, A., 2015. Vision based hand gesture recognition for human computer interaction: a survey. *Artificial Intelligence Review* 43(1), pp 1-54.

Stern, H., Frolova, D., Berman, S., 2010. Hand Gesture Recognition for TV Remote Control using Tree-Based Ensemble and LCS Classifiers. *International Conference on Image Processing, Computer Vision and Pattern recognition*, pp. 687-693.

Suk, H.I., Sin, B.K., Lee, S.W., 2010. Hand gesture recognition based on dynamic Bayesian network framework. *Pattern Recognition* 43(9), pp. 3059-3072.

Viola, P., Jones, M., 2001. Rapid Object Detection using a Boosted Cascade of Simple Features [C]. In: *The IEEE Computer Society Conference on Computer Vision and Pattern Recognition*, pp.511-518.

Yang, H.D., Park, A.Y., Lee, S.W., 2007. Gesture Spotting and Recognition for Human–Robot Interaction. *IEEE Trans. on Robotics* 23(2), pp. 256-270.

# Multi-level Indoor Path Planning Method

**Qing Xiong[a], Qing Zhu[a,b], Sisi Zlatanova[c],**

**Zhiqiang Du[a], Yeting Zhang[a], Li Yang Zeng[b]**

*a State Key Laboratory of Information Engineering in Surveying Mapping and Remote Sensing, Wuhan University, P.O. Box C310, 129 Luoyu Road, Wuhan, Hubei, 430079, P.R. China*

*b Faculty of Geosciences and Environmental Engineering, Southwest Jiaotong University, Chengdu, Sichuan, P.R. China*

*c GISt, OTB, Delft University of Technology, Jaffalaan 9, 2628 BX Delft, The Netherlands, Email: s.zlatanova@tudelft.nl*

**Key words**: indoor, path planning, semantic

**Abstract**: Indoor navigation is increasingly widespread in complex indoor environments, and indoor path planning is the most important part of indoor navigation. Path planning generally refers to finding the most suitable path connecting two locations, while avoiding collision with obstacles. However, it is a fundamental problem, especially for 3D complex building model. A common way to solve the issue in some applications has been approached in a number of relevant literature, which primarily operates on 2D drawings or building layouts, possibly with few attached attributes for obstacles. Although several digital building models in the format of 3D CAD have been used for path planning, they usually contain only geometric information while losing abundant semantic information of building components (e.g. types and attributes of building components and their simple relationships). Therefore, it becomes important to develop a reliable method that can enhance application of path planning by combining both geometric and semantic information of building components. This paper introduces a method that support 3D indoor path planning with semantic information.

## 1. Introduce

Indoor navigation is very useful in reality as people spend large parts of their lives in indoor spaces such as office buildings, commercial centers, conference venues, transportation facilities, and so on(Battles and Fu, 2014; Liu and Zlatanova, 2011; Mikkel et al., 2014). Furthermore, Indoor path planning is the core of indoor navigation. Many path planning algorithms have been developed in the fields of computer science and robotic in the past decades years. Today some studies have begun to exploit the path planning methods in various applications due to the significant improved in computer development. One direct application is indoor navigation (e.g. for the blind/wheelchair/pedestrian), in which a fast and efficient path planning algorithm often generates the collision-free paths for providing the navigational assistance. Another application is emergency evacuation simulation (e.g. for fire protection design) in the public buildings, where the planed paths provide the possible evacuation alternatives in different risky occurrence processes. In additional, path planning can help the architectural designer for improving automatic rule-based checking of building designs, for example, the shortest route between a public area and fire exits should be less than a minimum value in many fire protection design checking.

Most of the above applications require 2D drawing or building layouts as input, possibly with few attached attributes for obstacles, in path planning. However, they usually contain only geometric information while losing abundant semantic information. Successful path planning for 3D indoor spaces should depend on the accurate and updated geometry and semantics of building components, which include the correct geometrical positioning of indoor spaces, functions and properties of spaces and obstacles.

For solving this issue, this paper introduce an improved method to cope with path planning in 3D indoor space through complex 3D building models with semantic information. The input data conform to the definition of LOD4 in CityGML. Compared to the traditional CAD data, CityGML is capable of restoring both geometric and rich semantic information of building components, as well as their relationships.

This paper aims to study the particular problem of path planning for 3D indoor spaces, all relying on the building models with semantic information input. The application developed in this paper automatically extracts and manages geometric and semantic information on the building space and the inside components from input file, which enhances application of path planning. After extracting geometric and semantic information of building components for input data, the next step is to fine the most suitable path between two given location in the 3D indoor spaces. Various approaches can be used for the problem of path planning computation. In a general 3D indoor space, the problem becomes much harder, and it has been proved that computing a fine path connection two points considering the obstacles is another hard problem.

The traditional method of determining an optimal path to traverse from one point to another are often based on the graph exploration methods, such as the Dijkstra algorithm, Fast Marching Method, A* algorithm, and the ant colony algorithm. The Dijkstra algorithm could always find the global optimization solution, but it will consume a large amount of time to search. The Fast Marching Method is quite related to the Dijkstra algorithm, and it uses the upwind difference operators to approximate the gradient, which makes it capable of computing a large class of continuous problem. Being different from the Dijkstra algorithm, the A* algorithm does not have to carry out a global search, which uses a heuristic function to search directly towards the goal and find the possible shortest route, so its computational time is much shorter than the Dijkstra algorithm. The ant colony algorithm has a better probability in achieving the globally optimal solution, especially with a larger number of iterations, but this will lead to cost of a larger computation time. However, the traditional methods do have some limitations in 3D indoor environment. One is the execution efficiency and precision of the algorithm do not conform the real scene very well, especially at the part of corner. Second is they could not deal with multi-floor issues with foreseeing a wonderful effect.

This paper introduces an improved semantic-oriented path planning method based on A* algorithm. Firstly, the valuable geometric and semantic information is extracted from the input data, where the geometric information will traverse from vector data to grid data. Then, the method will find the most suitable path for a start location to goal destination by evaluating the weight factors of each criterion at grid nodes, which considering the vertical circulations like stairs or elevators between different floors. There will be an experiment for testing the method, and the input data includes two 3D building models containing some obstacles.

## 2. Coarse-to-fine indoor path planning method

This paper's method is to transform the input geometric data, which is triangularization, into voxelization data. The first process should inherit the semantic information from the input data. Then creating the semantic region based on the voxels. Additionally, we get the navigation meshes based on the contours of the regions. Finally, we use the A* algorithm based on the navigation meshes to find the path with input semantic conditions.

### 2.1 Voxelization process

Refer to classic voxelization methods(Jones and Satherley, 2000; Oomes et al., 1997; A et al., 1993), we use voxels to present the 3D model's surface information for preparing the data of the next steps. As show in Figure 1, the voxels inherit the triangle's semantic information.

Figure 1 Voxelization process

### 2.2 Regions generation

After the voxelization process, we now have the voxels set with unrelated relations. This step is to figure out which voxels form potential surfaces of contiguous regions. For this purpose, we introduce the voxel pillar in this paper. We set the voxel pillar contains the same semantic information and the same projected coordinates in x-y plane. As show in Figure 2, for each pillar, its axis-neighbour columns are searched for candidates. A pillar in the neighbour is considered a neighbour pillar if both of the following conditions are met: the up or down between the floors of the two pillars is less than the threshold value. This setting allows for surfaces such as stair steps to be detected as valid neighbours.

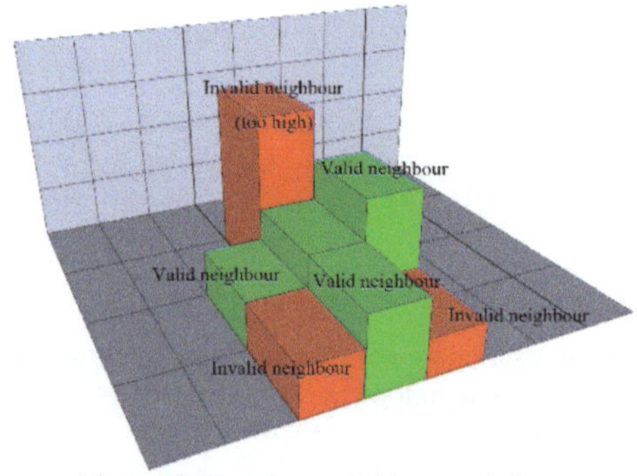


Figure 2 Creating neighbour relations

In this paper, the regions are groups of contiguous pillar that represent walkable surface area with same semantic information. The watershed algorithm is used for initial region creation(Haumont et al., 2003). Using the watershed analogy, the pillars which are furthest from a border represent the lowest points in the watershed. A border pillar represents the highest possible water level. The main loop iterates, starting at the lowest point in the watershed, then increments with each loop until the highest allowed water level is reached. This slowly "floods" the pillars starting at the lowest points. During each iteration of the loop, spans that are below the current water level are located and an attempt is made to either add them to exiting regions or create new regions. During the region expansion phase, if a newly flooded pillar borders an existing region, it is usually added to the region. Any newly flooded span that survives the region expansion phase is used as a seed for new region creation. Regions that are unnecessarily small, which may be caused by the size of input geometric data, are ignored for indoor path planning's presentation.

### 2.3 Contour generation

For creating contours, we care about the edges of the pillars. There are two types of edges: region and internal. Region edges are pillar edges across which the neighbour is in another region. Internal edges are pillar edges across which the neighbour is in the same region.

The first step is to categorize edges as either region edges or internal edges, as shown in Figure 3. For each pillar, we can check all axis-neighbours. If an axis-neighbour is not in the same region as the current span, the edge is flagged as a region edge.

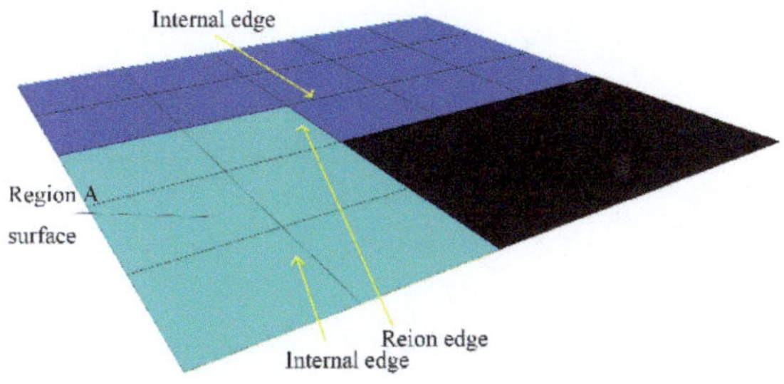


Figure 3 Searching for region edges

Once we have information on which pillar edges are region edges, we can build a contour by walking the edges, as shown in Figure 4. Once again we iterate the pillars. Also, the regions inherit the semantic information of pillars.

Figure 4 Building the contour of regions

### 2.4 Navigation meshes generation

The navigation meshed generation is the triangulation process of contours of regions(Oliva and Pelechano, 2013). Triangulation is performed by walking the edges of the contour and, for each group of three vertices, determining whether a valid internal triangles can be formed. From all potential candidates, select the one with the shortest new edge. The process continues with the remaining vertices until triangulation is complete.

Firstly, as shown in Figure 5, finding potential partitions of contour by walking the edges.

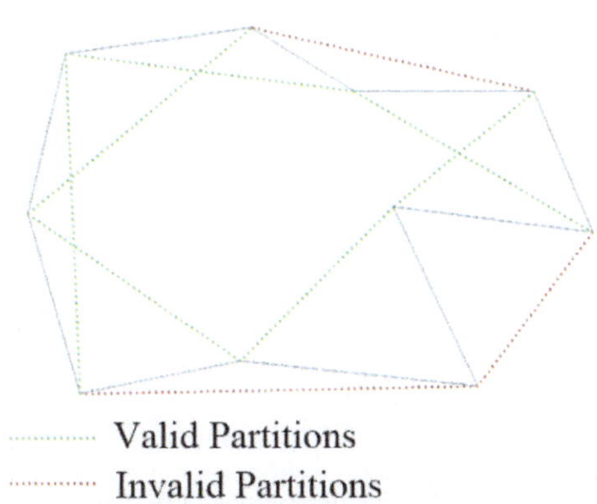


Figure 5 Finding potential partitions

Then selecting the shortest partition and form a triangle. Getting rid of invalidated partitions and generate new potential partitions, as shown in Figure 6.

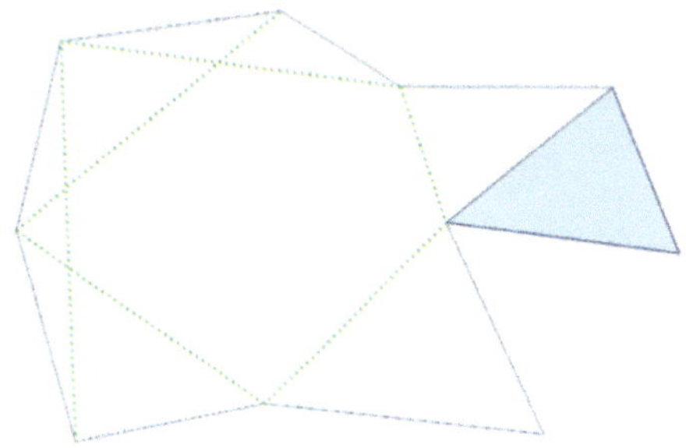

Figure 6 Triangulation of contour

Iterating the partitions, finally, we get the final triangulation, as shown in Figure 7.

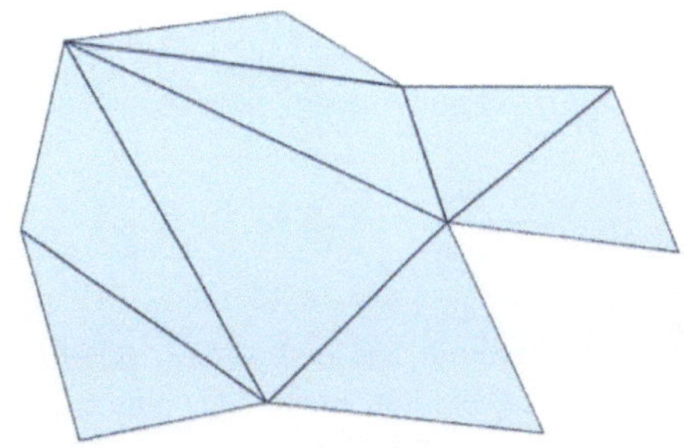

Figure 7 Final triangulation

### 2.5 Path planning with A* algorithm

After navigation meshes generation, each triangle has its own semantic information. This paper refer to the existing A* algorithm with triangulation (Cui and Shi, 2011; Zhongshan et al., 2011). As shown in , we set a start point A and an end point B, and finally generating a path from A to B.

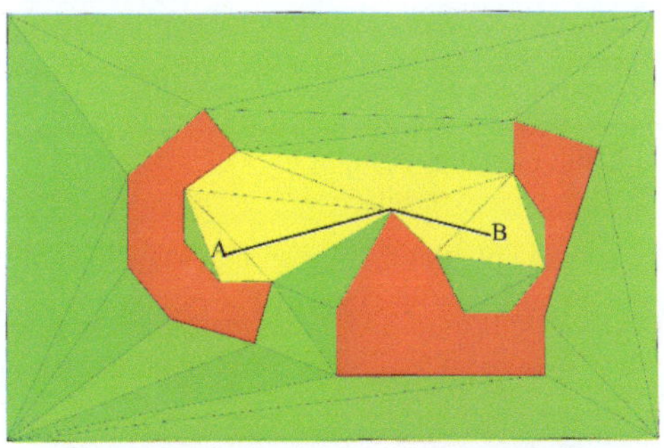


Figure 8 A* algorithm with triangulation

However, the triangulation result contains semantic information, which means any point in the navigation meshes should have a semantic name, or location name. Then, we could find a path from an indoor location to another indoor location.

## 3. Experiment

According to the indoor path planning method introduced above, this paper using a two floors building for experiment. The input data and voxelization result show in Figure 9.

Figure 9 The input data and voxelization result

It contains 2300 vertexes and 3600 triangles as input geometric data. And the voxelization result contains 35752 voxels as the voxel size is 0.3 * 0.3 * 0.3.

The region generation result shows in Figure 10. The reason of top-right region does not have connection relations with other regions is that the room's door is closed (as shown in Figure 9).

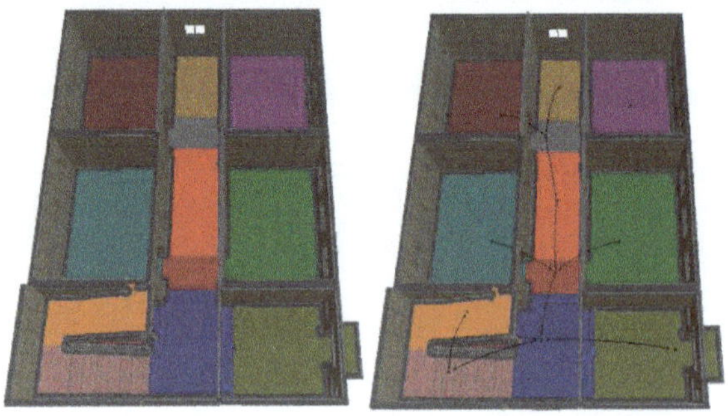

Figure 10 Regions generation result and connection between regions

The contour and navigation meshes generation results show in Figure 11.

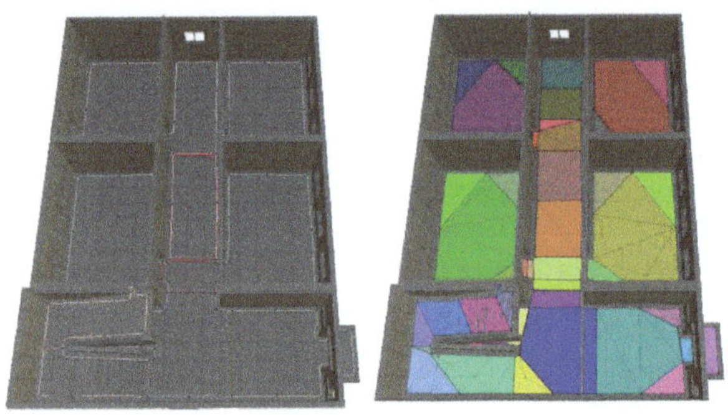

Figure 11 Contour and navigation meshes generation results

Then we set an example the finding an indoor path from "Hall" to "Room 101", the result with A* algorithm shows in Figure 12.

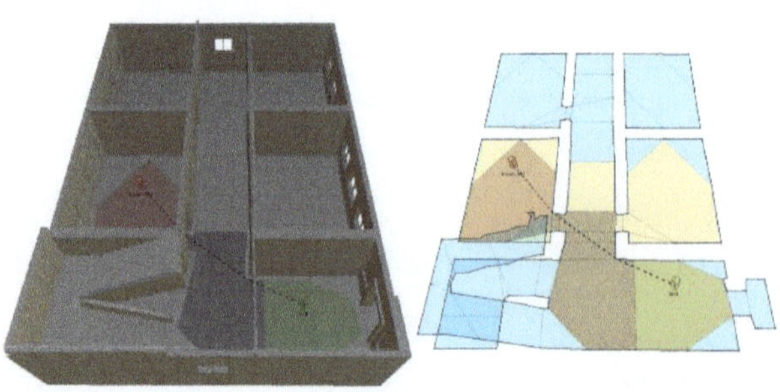

Figure 12 Indoor path planning from "Hall" to "Room 101"

As the same above, shows the path from "Room 102" to "Staircase".

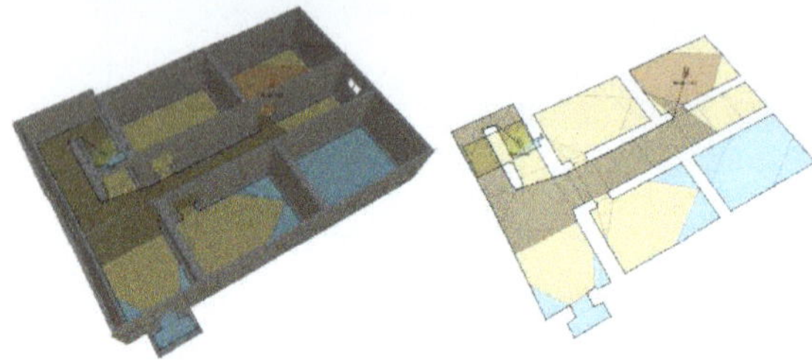

Figure 13 Indoor path planning from "Room 102" to "Staircase"

## 4. Conclusion

As the experiment shows, this paper's indoor path planning method could generate a fine rout from one semantic location to another semantic location. And this process does not break the integrity of original geometric data, and also inherit the semantic information. This should perform nice in indoor navigation application. However, the finally path did not conform to the human's walking habit in indoor, as some part of the path is too close to the wall or staircase. This defect will be improved in the next research.

## 5. Acknowledgement

This work is supported by National Science and Technology Support Program of China (No. 2012BAH35B03), and the National Natural Science Foundation of China (No. 41171311).

**References:**

A, K., D, C. and R, Y., 1993. Volume graphics. IEEE Computer, 51 - 64(26).

Battles, A. and Fu, W., 2014. Navigating Indoor with Maps: Representations and Processes.

Cui, X. and Shi, H., 2011. A*-based pathfinding in modern computer games. International Journal of Computer Science and Network Security, 11(1): 125--130.

Haumont, D., Debeir, O. and Sillion, F., 2003. Volumetric cell-and-portal generation. Blackwell Publishing, Inc, pp. 303--312.

Jones, M.W. and Satherley, R., 2000. Voxelisation: Modelling for volume graphics. Press, pp. 319--326.

Liu, L. and Zlatanova, S., 2011. Towards a 3D network model for indoor navigation. CRC Press, pp. 79-94.

Mikkel, B., de Haas, C., Lu, H., Xie, X. and Pilvinyte, A., 2014. Constructing indoor navigation systems from digital building information. Data Engineering (ICDE), 2014 IEEE 30th International Conference on, pp. 1194-1197.

Oliva, R. and Pelechano, N., 2013. NEOGEN: Near optimal generator of navigation meshes for 3D multi-layered environments. Computers & Graphics, 37(5): 403 - 412.

Oomes, S., Snoeren, P. and Dijkstra, T., 1997. 3D shape representation: Transforming polygons into voxels. In: B. ter Haar Romeny, L. Florack, J. Koenderink and M. Viergever (B. ter Haar Romeny, L. Florack, J. Koenderink and M. Viergever)^(B. ter Haar Romeny, L. Florack, J. Koenderink and M. Viergevers)|,*Scale-Space Theory in Computer Vision. Lecture Notes in Computer Science. Springer Berlin Heidelberg, pp. 349-352.

Zhongshan, X.X.D.O., Zhongshan, P.R.C.K. and China, P.R., 2011. Pathfinding of 3D Game with Navigation

Mesh, Pathfinding of 3D Game with Navigation Mesh. Proceedings of 2011 3rd IEEE International Conference on Information Management and Engineering(ICIME 2011) VOL.04, Zhengzhou,China, pp. 5.

# 3

# OPERATIONAL 333m BIOPHYSICAL PRODUCTS OF THE COPERNICUS GLOBAL LAND SERVICE FOR AGRICULTURE MONITORING

R. Lacaze [a, *], Bruno Smets [b], F. Baret [c], M. Weiss [c], D. Ramon [a], B. Montersleet [a], L. Wandrebeck [a], J.-C. Calvet [d], J.-L. Roujean [d], F. Camacho [e]

[a] HYGEOS, Euratechnologies, 165 avenue de Bretagne, 59000 Lille, France – rl@hygeos.com, dr@hgeos.com, bm@hygeos.com' lw@hygeos.com
[b] VITO, Boeretang 200, 2400 Mol, Belgium – bruno.smets@vito.be
[c] INRA, EMMAH UMR1114, site Agroparc, 84914 Avignon, France – frederic.baret@paca.inta.fr; marie.weiss@paca.inra.fr
[d] CNRM/Météo-France, 42 avenue G. Coriolis, 31057 Toulouse Cedex, France – jean-christophe.calvet@meteo.fr, jean-louis.roujean@meteo.fr
[e] EOLAB, G.V. Marques del Turia, 20-9, 46 005 Valencia, Spain – fernando.camacho@eolab.es

**KEY WORDS:** Copernicus, Service, bio-geophysical, agriculture, global, land, vegetation

**ABSTRACT:**

The Copernicus Global Land service provides continuously a set of bio-geophysical variables describing, over the whole globe, the vegetation dynamic, the energy budget at the continental surface and some components of the water cycle. These generic products serve numerous applications including agriculture and food security monitoring. The portfolio of the Copernicus Global Land service contains Essential Climate Variables like the Leaf Area Index (LAI), the Fraction of PAR absorbed by the vegetation (FAPAR), the surface albedo, the Land Surface Temperature, the soil moisture, the burnt areas, the areas of water bodies, and additional vegetation indices. They are generated every hour, every day or every 10 days on a reliable automatic basis from Earth Observation satellite data. Beside this timely production, the available historical archives have been processed, using the same innovative algorithms, to get consistent time series as long as possible. All products are accessible, free of charge after registration through FTP/HTTP (http://land.copernicus.eu/global/) and through the GEONETCast satellite distribution system. The evolution of the service towards the operations at 333m resolution is partly supported by the FP7/ImagineS project which focuses on the retrieval of LAI, FAPAR, fraction of vegetation cover and surface albedo from PROBA-V sensor data. The paper presents the innovations of the 333m biophysical products, make an overview of their current status, and introduce the next steps of the evolution of the Copernicus Global Land service.

## 1. INTRODUCTION

Since the 1st January 2013, the global component of the Copernicus Land service, known as Global Land Service, continuously monitors the state of continental ecosystems providing a set of bio-geophysical products describing the state and the dynamics of the vegetation, the energy budget at the surface and some components of the water cycle.

## 2. CURRENT STATIUS OF GLOBAL LAND SERVICE

The Global Land Service is a global systematic monitoring service which generates a number of biophysical variables over the globe, in Near Real Time (NRT) every hour, every day or every 10 days, depending on the users' requirements, and the satellite sensors acquisition capabilities. Beside this timely production, the available historical archives of satellite input date have been processed to get consistent time series as long as possible.

The aim of the Global Land Service is to support and consolidate the policies and commitments of the European Union at international level for a wide range of applications such as the crop monitoring and food security inside and outside Europe, the monitoring of biodiversity, protected areas, and forest cover, the drought assessment and the desertification, the carbon modelling, the land use and land cover change, and the support to the Earth observation activities in Africa. The Global Land Service is also part of the EU contribution to GEO/GEOSS.

### 2.1 The Portfolio

The products contained into the portfolio are gathered in 3 families describing (Table 1):

1. The vegetation state, dynamics and disturbances like the Leaf Area Index (LAI), the Fraction of photosynthethically active radiation absorbed by the vegetation (FAPAR), the fraction of green vegetation cover (FCover), the Dry Matter Productivity (DMP), the Normalized Difference Vegetation Index (NDVI), the Vegetation Condition Index (VCI), the Vegetation Productivity Index (VPI), and the maps of burnt areas

2. The energy budget at the interface between the surface and the atmosphere like the Top Of Canopy reflectance (TOC-ref), the Surface Albedo (SA) and the Land Surface Temperature (LST)

3. Some components of the water cycle like the Soil Water Index (SWI) and the maps of water bodies.

All products are generated over the globe. They are available within 3 days after the acquisition of the most recent

* Corresponding author. This is useful to know for communication with the appropriate person in cases with more than one author.

observation used to create the product, except the LST and the SWI which are available within 1 day.

| Variable | Temporal resolution | Spatial resolution | NRT | | Archive |
|---|---|---|---|---|---|
| | | | Sensor | Status | Sensor |
| LST | 1 hour | 0.05° | ΣGeo | Op | ΣGeo |
| SWI | 1 day | 0.1° | Metop / ASCAT | Op | Metop / ASCAT |
| NDVI / VCI / VPI | 10 days | 1/112° | PROBA-V | Demo | SPOT/VGT |
| DMP | 10 days | 1/112° | PROBA-V | Demo | SPOT/VGT |
| LAI / FAPAR / FCover | 10 days | 1/112° | PROBA-V | In dvlp | SPOT/VGT |
| Burnt Area | 1 day | 1/112° | PROBA-V | In dvlp | SPOT/VGT |
| TOC-ref | 10 days | 1/112° | PROBA-V | In dvlp | SPOT/VGT |
| Surface Albedo | 10 days | 1/112° | PROBA-V | In dvlp | SPOT/VGT |
| Water Bodies | 10 days | 1/112° | PROBA-V | In dvlp | SPOT/VGT |

Table 1: Portfolio of the Global Land Service. "Op" means "operational", "demo" means "Demonstration", "In dvlp" means "in development".

Till end of 2013, all 1km resolution products (1/112°) were generated from the SPOT/VEGETATION sensor data. The continuity of the NRT production is ensured with the PROBA-V sensor data. Note that, at the time this article is written, the PROBA-V products are still in development and not yet available to users. The Burnt Areas, the LAI, the FAPAR and the FCover should be made available in "Demonstration" mode (see §2.2) before mid-2015.

## 2.2 The Quality Control

The production is associated with a two-fold quality check: 1) a technical quality control is performed at each production center, and 2) an assessment of the scientific quality of products is carried out by expert teams.

During the production, automated tools monitor the individual processing steps to be check that the output files fit the expected characteristics, and their timeliness is compliant with the specifications (§2.1). Statistics are computed to verify that no unreliable values are present into the files.

A scientific quality assessment is performed per variable, for each new product or version, following the protocols and metrics defined by the Land Product Validation group of CEOS, when they exist. It mainly consists in indirect validation, i.e. inter-comparison with existing similar satellite-derived products, checking for the spatio-temporal consistency of products, and in direct validation, i.e. comparison with in-situ reference data when they exist and are available. In particular, the Global Land Service benefits from ground campaigns, conducted in the framework of the FP7 ImagineS project, which collect in-situ measurements, according to the sampling schemes and protocols defined by CEOS/LPV, and process them to retrieve reference high resolution maps. The resulting values are used to assess the accuracy of the satellite-derived biophysical products (Figure 1). The in-situ data and the reference maps are made available to the community through the ImagineS website at the address http://fp7-imagines.eu/pages/services/ground-data.php.

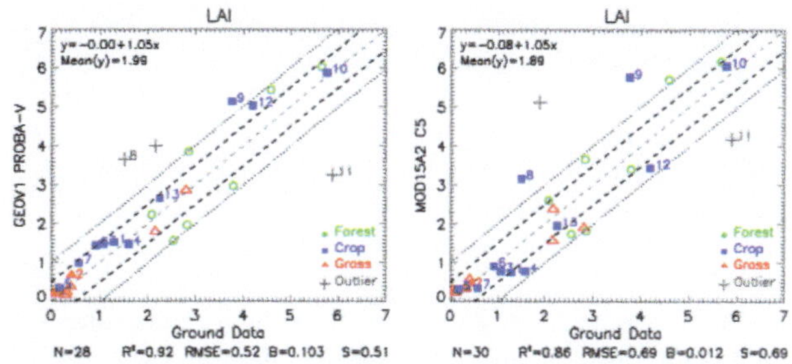


Figure 1: Comparison of satellite LAI product with the ground-based maps. Left side: PROBA-V GEOV1. Right side: MODIS C5. Filled symbols correspond to concomitant values of 2014 from ImagineS project and unfilled symbols to a different year.

Three levels of quality assessment are associated with three development stages of the products (Table 1):

1. "***Demonstration***": products are provided to users with a very limited commitment on the quality or availability of the service but they have been considered useful to be disseminated in order to enable users to test them and provide feedbacks. The demonstration phase ends with a limited quality assessment.

2. "***Pre-operational***": the limitations of the products are documented and can satisfy the majority of the applicable requirements and/or they have been considered useful for distribution to the users. The pre-operational phase ends with a full quality assessment.

3. "***Operational***": products have well documented non-relevant limitations and largely satisfy the requirements applicable and/or have been considered mature enough for distribution to users. The operational phase regularly performs quality control.

In addition to the quality assessment, a quality monitoring, applying the same approach and metrics, is also carried out, every 6 months, on the "operational" products to check that their quality keeps stable along time. Further, an across variables quality monitoring is also performed, every 6 months, integrating the variables into a land surface model using a data assimilation scheme. The SWI and the LAI are assimilated over France (8km x 8km) into the Land Data Assimilation system (LDAS) while a passive monitoring (i.e. comparison of model output resulting from the assimilation with the satellite products) of FAPAR, Surface Albedo and LST is also carried out. Here again, the Copernicus Global Land service benefits from the achievements of the ImagineS project which achieved upgrade and consolidation of the LDAS.

In parallel, an independent assessment of the service quality is organized through 2 bodies: 1) panel of external experts which give advices and recommendations on service evolution through half-yearly reviews; 2) a technical user committee which defines the product specifications.

## 2.3 The Distribution

The Global Land website, http://land.copernicus.eu/global, is the entry point for all information related to the service and the products. Individual product pages provide more details about the product status, its main characteristics in terms of algorithm, quality, technical properties, and applications. A set of 3 documents is associated with each product/version:

1. the Algorithm Theoretical Basis Document (ATBD) which describes the retrieval methodology,
2. the Quality Assessment Report which presents the results of the validation exercise,
3. the Product User Manual which summarizes the algorithm and the validation results and includes a detailed description of technical characteristics of the product.

The website gives access to the distribution data portal. The access to the products is free and open, whatever the user category (public body or private company) and whatever the application (research or commercial use). Anonymous query is possible to search the products and check their availability. It is only necessary to register, through an automated procedure, before ordering and downloading the products. Two data dissemination channels exist: via internet where both the NRT and the historic (archive) products are accessible; via EUMETCast towards Africa and South-America where only the NRT products are distributed.

## 3. EVOLUTION OF THE SERVICE

The major next evolution of the service is the introduction into the portfolio of the 333m products generated from the full resolution PROBA-V data. This evolution is partly supported by the ImagineS project which focuses on the LAI, FAPAR, FCover and surface albedo. ImagineS performs the research and development activities, i.e. defining innovative algorithms, assessing their performances and demonstrating their improvements comparing to previous methods, developing and qualifying the corresponding processing chains. The Global Land service integrates the qualified processing chain into its production unit, carries out the NRT processing associated with a continuous quality monitoring, and disseminates the products.

### 3.1 Production at 333m resolution

The retrieval algorithms of LAI/FAPAR/FCover, on one hand, of the surface albedo, on the other hand, were defined by INRA and Meteo-France, respectively, in the framework of the FP7/ImagineS project.

#### 3.1.1 LAI/FAPAR/FCover retrieval methodology

It is made of two steps as outlined in Figure 2.

The PROBA-V daily synthesis are the input data. First, they are translated into "SPOT/VGT-like" reflectances through a spectral correction. Then, they are corrected for atmospheric effects (ozone, water content, pressure) to get Top-of-Aerosols reflectances which are the input of neural networks (NNT) (one per variable). Only the reflectances inside a definition domain are used. Similarly, only the output of NNT inside the physical range and tolerance are kept. The NNT are trained with SPOT/VGT reflectances as input and the 'best' products derived from a combination of MODIS and CYCLOPES original products (Baret et al., 2007) similarly to what was achieved for SPOT/VGT LAI/FAPAR/FCover algorithm (Baret al al., 2013), which ensures the consistency between the 333m PROBA-V products and the 1km SPOT/VGT products.

Step B is a compositing and gap filling module. According to the pixel identification (Evergreen Broadleaf Forest (EBF) or not, based upon the time series of daily estimates or the Globcover land cover map (http://www.esa-landcover-cci.org ) when the time series does not provide a reliable identification of the vegetation class), the temporal smoothing is performed with specific parameters. Finally, a post-processing module is applied to fill the small gaps by linear interpolation.

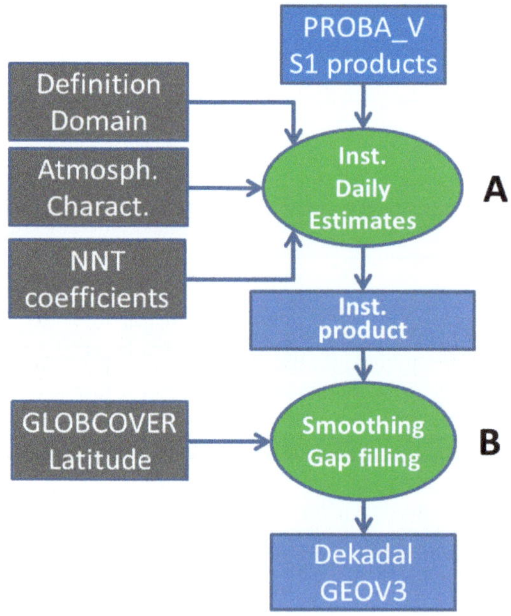


Figure 2: Flow chart of the LAI/FAPAR/FCover algorithm

HYGEOS has translated this algorithm into a software compliant with the coding specifications and the processing performances of the Global Land service. After its technical qualification and the thematic verification of the output products, the chain has been delivered to the Global Land service that should start the NRT production over Europe mid-2015. In parallel, in the framework of ImagineS, VITO will perform offline production over identified areas outside Europe where local expert teams will make a utility assessment of the products.

#### 3.1.2 Surface Albedo retrieval methodology

The outline of the methodology is depicted in Figure 3.

.

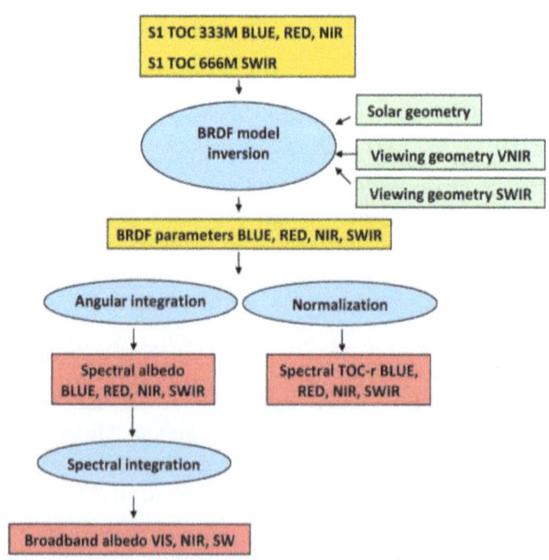


Figure 3: Flow chart of the surface albedo algorithm

The calculation of surface albedo and the normalized TOC-ref requires the knowledge of the complete bi-directional reflectance distribution function (BRDF) of the surface. A semi-empirical kernel-based reflectance model (Roujean et al., 1992) is adjusted to the measurements. This delivers an estimate of the complete angular dependence of the bi-directional reflectance factor in each spectral channel of the sensor. The proposed algorithm consists in calculating the "black-sky" and "white-sky" albedos in all instrument channels by using the coefficients provided by the directional reflectance model inversion. The narrow- to broadband conversion is performed with a linear regression formula.

This methodology is the same than the one used to generate the Surface Albedo products at 1km resolution (Table 1). The differences are related to the time compositing for the model inversion (20 days instead of 30 days) and in the temporal resolution of the products (5 days instead of 10 days). A shorter compositing period is possible because a finer spatial resolution allows a better cloud decontamination. Regarding the finer frequency, it will allow, in a further evolution step, to discriminate the albedo from bare soil and from vegetation.

Starting from a prototype delivered by Meteo-France, HYGEOS has developed the input and output modules able to ingest the PROBA-V Top-Of-Canopy daily synthesis and to create the albedo products according to the Global Land service specifications. It has been checked that the processing performances were compliant with the product timeliness. After verification of the output product consistency, the chain will be delivered to the Global Land service that should start the NRT production before the end of 2015.

### 3.1.3 Application for agriculture monitoring

The agricultural production is achieved per species since the markets are organized in such a way. This has strong implications on the remote sensing products required for agriculture monitoring: the acreage of each crop/fodder should be known accurately concurrently with the associated productivity. The use of 333m resolution observations will improve greatly the performance of global agricultural monitoring services, currently based on kilometric resolution images of mixed (several crops/fodder fields, natural areas) pixels.

### 3.2 Long-term evolution

At longer term, the evolution of the Global Land Service will ensure a sustainable global monitoring of the continental ecosystems with the continuity of 1km and 333m biophysical products generated from PROBA-V and Sentinel-3. A further step towards a production at 100m could be envisaged if users expressed such requirements. In parallel, a "worldwide" monitoring of hot-spots, like the major protected areas, will be added in the service, providing biophysical products derived from Sentinel-2 data at resolution between 10m and 10m, depending of the needs.

Further, the current biophysical products will benefit from the capabilities of the new sensors. For instance, a 1km resolution SWI product is in preparation, using jointly the Metop/ASCAT and Sentinel-1/SAR data.

Finally, new variables will be included into the service portfolio, according to the users' requirements. Some candidate variables are the phenology, the snow extent, or the water level of lakes and rivers.

## 4. CONCLUSION

The Copernicus Global Land service is operational providing sustainable NRT delivery of global products, associated with consistent historic time series. A continuous quality monitoring guaranties the product quality reached a level satisfying the users' requirements and keeps stable along time. The access to products is free and open through http://land.copernicus.eu/global/. Continuing the 1km products time series with the replacement of the SPOT/VEGETATION sensor by the PROBA-V sensor was challenging: some products are still in development. Full quality assessment of PROBA-V products is ongoing.

The next major evolution of the Global Land service is the production of a number of biophysical variables at 333m resolution from the PROBA-V sensor data. This is a first step before using jointly the data of PROBA-V and Sentinel-3 that should be launch before end 2015.

## 5. CONTACTS

For the Copernicus Global Land service:

- Website: http://land.copernicus.eu/global
- Coordinator: bruno.smets@vito.be
- S&T contact: rl@hygeos.com
- Helpdesk: helpdeskticket@vgt.vito.be
- Contract Management: copernicuslandproducts@jrc.ec.europa.eu
- Consortium:
    - Leader: VITO
    - Partners: HYGEOS, IPMA, Meteo-France, ZAMG
    - Associates: EOLAB, INRA, CREAF, TU Wien, University of Leicester

For the ImagineS project:

- Website: http://fp7-imagines.eu
- Coordinator: HYGEOS - rl@hygeos.com
- Consortium: ECMWF, EOLAB, INRA, Meteo-France, OMSZ, UCL, VITO

## 6. REFERENCES

Baret, F., M. Weiss, R. Lacaze, F. Camacho, H. Makhmara, P. Pacholcyzk, B. Smets, 2013. GEOV1: LAI and FAPAR essential climate variables and FCOVER global time series capitalizing over existing products. Part1: Principles of development and production. *Remote Sensing of Environment 137: 299–309.*

Baret , F., Hagolle, O., Geiger, B., Bicheron, P., Miras, B., Huc, M., Berthelot, B., Weiss, M., Samain, O., Roujean, J.L., Leroy, M., 2007. LAI, fAPAR and fCover CYCLOPES global products derived from VEGETATION. Part 1: Principles of the algorithm. *Remote Sensing of Enviroment 110, 275-286.*

Roujean J.-L., M. Leroy, and P.-Y. Deschamps, 1992, A bidirectional reflectance model of the Earth's surface for the correction of remote sensing data, *Journal of Geophysical Research, 97(D18), 20455-20468.*

# 4

# ESTIMATION OF GRASS YIELD IN LARGE REGION ON GEOGRAPHICALLY WEIGHTED REGRESSION MODEL

Luo Chengfeng [a, *], Yu Xiujuan [b], Liu Caijuan [c], Du Yingkun [c]

[a] Institute of Photogrammetry and Remote Sensing, Chinese Academy of Surveying and Mapping, Beijing 100830, P.R. China – chfluo@ casm.ac.cn

[b] National Calibration Center for Electro-optical Distance Meter, Chinese Academy of Surveying and Mapping, Beijing 100830, P.R. China –xjyrspxf@163.com

[c] Geomatics of College，Shandong University of Science and Technology，Qingdao 266590, P.R. China –liucaijuan15@163.com

[d] College of Information Science and Engineering，Shandong Agricultural University，Taian 271018, P.R. China - dyk9118@126.com

**Commission VI, WG VI/4**

**KEY WORDS:** Geographically Weighted Regression (GWR) Model；Grass Yield；QingHai Province；Remotely Sensed Data

**ABSTRACT:**

The grass yield embodies its productivity，and also is ground for developing animal husbandry production management. Now the remote sensing technology has been becoming an efficient and feasible mean to estimate the grass yield. In the study, the thought about Geographically Weighted Regression (GWR) was involved in estimating the grass yield. The special characteristics of samples measured on field were considered, and then each sample has a local function covering area around. And the parameters for the function are decided by the weighted function which is associated with the spatial distance between the sample and others around. GWR is a good solution to the model without spatial stationarity, as a consequence a significant model-fitting degree comes out. Based on GWR model an ideal production of grassland can be estimated. In this study, Qinghai province, about 0.72 million square kilometres, was taken as an example. The province is an important one on the Qinghai Tibet Plateau. Here the grassland not only closely relates with the local animal husbandry economy, but also directly affects the regional ecosystem security. Landsat TM data in 2013 and samples on field were used to estimate the production. As input parameters, OSAVI and FVC have high correlation coefficient more than 97% with grass yield. There were 201 samples involved in modelling, and the accuracy is 87.27%, above about 47% than that of multiple linear regression model, a widely used traditional statistic model. Another 220 samples were used to verify the results, and here the accuracy can reach 81.3%. Out results indicated that in 2013 the yield of grass in Qinghai province is $1.018*10^8$ ton. The difference between our data and that from professional sector is less than 10%.

## 1. INTRODUCTION

As the natural resources for livestock production forage, the productivity of grassland can reflect the capabilities of pasture carrying directly. There have many studies about estimating grass yield by remote sensing technology（Jin Y X，2011；Benie G.B.，2005；Liu X Y，2010）. Issues in focus include how to estimate the grass yield quickly and efficiently, monitor the productivity of grassland in a large region and know the current situation and development trend well. Such issues has becoming key topic in ecology and grassland academia field. It is also practical problem for grassland management urgently needed to be solved (Xu B, 2007). To estimate grass yield there are two models used widely, biological - physical model and statistical empirical model（Lv H Y，2010）. The former is always adopted to work for large scale region with coarse resolution data, like NOVAA and MODIS. And the result has low accuracy. When it comes to estimation with high accurate for local region, the modelling is becoming more complex and affected by the natural characteristics of the geographical closely（Chen J，2009）. Establishing the model based on ground truth data and remote sensing data, the statistical empirical model is relatively simple and practicable. Because the spatial heterogeneity in nature always is ignored in practice, to get high estimation accuracy is difficult（Tao W G，2007）. According to the first law of geography: the closer the distance between the features, the greater similarity（Miller H J，2004）. It is necessary to take the rule into modelling in order to get better result.

According to the first law of geography: the closer the distance between the features, the greater similarity（Miller H J，2004）. It is necessary to take the rule into modelling in order to get better result. Geographically Weighted Regression (GWR) model create the conditions for regression analysis of the relationship between spatial features. The grass yield has relation with geographical location, adjacent relationship between features with geographical proximity makes the grass yield spatial correlation. In the study the GWR was adopted to model, which can reflect spatial characteristics of grass yield. Another valuable side is the estimation was applied using imageries with30 m resolution in a large region, about 720,000 square kilometres.

* Corresponding author. chfluo@casm.ac.cn

## 2. DATA AND METHOD

### 2.1 Study Area

Qinghai Province, one of the four largest provinces in China, extending 8 degree of latitude. It also is one of the national five major pastoral areas, the grassland area here accounting for 50.46% of the total land area. The grassland resources here is very abundance, and available pasture area accounts for 87% of the province's natural grassland area, about 15% of the available grassland area in China (Shen Y C, 1991). Grassland mainly appeared in Southern Qinghai Plateau, Qilian Mountain and the south-eastern margin of the Qaidam basin mountain. There are 9 Grassland categories , including alpine meadow, alpine grassland, alpine meadow grassland, alpine desert, temperate steppe, temperate desert, temperate desert steppe, mountain meadow and lowland meadow. Here alpine meadow and alpine grassland present the main body of natural grassland, about 2948.16×104hm$^2$ and accounting for 80.88% of the province's total area of grassland. With remote sensing technology the grass yield can be monitored macro and comprehensively. It has great significance during balancing the livestock development and ecological construction (Yan D L, 2007).

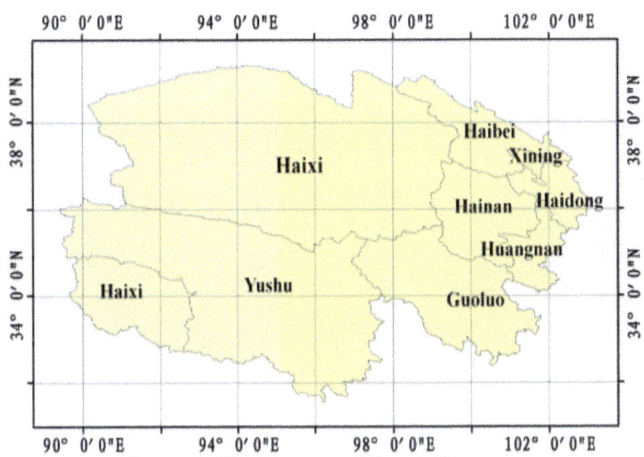


Figure 1. Sketch map of the study area

### 2.2 Data and Preprocessing

Landsat 8 TM data with resolution 30m was used in the study. According to the growth of grass is best in august each year, the images, receiving before and after a month from august 15th, were selected to work. There involves 46 scenes in all.

The field data comes from the ground field survey by the local grassland supervising department in 2013 august. There are about 600 typical and representative samples totally.

The preprocessing included radiometric calibration, atmospheric correction, terrain correction and removing cloud. Here the FLASH was used to finish atmospheric correction, C-correction model (Wang S, 2013) to terrain correction. The method referred from Li B X (2010) was used to remove the effects of cloud.

Figure 2. The mosaic image of study area after preprocessing

### 2.3 Method

GWR model is an expansion on the traditional model, and the spatial characteristics of the data are introduced the model. The geographic location of sample are involved into modelling based on multivariate linear regression. And the regression coefficient is assumed as the function of the location of sample. While modelling, a local function for every observation sample is calculated by GWR. And the coefficients are measured by weighting function among adjacent relationship of the samples. The non-stationary of the model is solved well by GWR. Thereby the goodness of fitting of the model is improved and better simulation results can be made.

GWR model can be expressed as following (Xuan H Y,2007):

$$y_i = \beta_0(u_i,v_i) + \sum_{k=1}^{p} \beta_k(u_i,v_i)x_{ik} + \varepsilon_i, i = 1,2,\cdots,n \quad (1)$$

Where ( $y_i; x_{i1}, x_{i2}, \cdots, x_{ip}$ ) is the observation value of dependent variable $y$ and independent variable $x_1, x_2, \cdots, x_p$ in the position $(u_i, v_i)$ ( $i = 1,2,\cdots,n$ ) . $\beta_j(u_i,v_j)(j = 0,1,\cdots,p)$ is the $ith$ observation point unknown parameter in the position $(u_i, v_i)$ ,and it is an arbitrary function of $(u_i, v_i)$ l $\varepsilon_i (i = 1,2,\cdots,n)$ are error independent and identically distributed, generally are assumed to obey $N(0,\sigma^2)$ . According to Tobler's first law of geography, the influence on the estimation of parameters for point $i$ from the observation value near point $i$ on is greater than influence from that far away from $i$ , GWR model based on linear regression model assumes that the regression coefficient is an arbitrary function of observation point location and it brings the spatial character into the model. The model calculates a local equation at each point, and observation value weight is no longer remain constant in the regression process, and the weight is related to the proximity position to $i$ :

$$\hat{\beta}(u_i, v_i) = (X^T W(u_i, v_i) X)^{-1} X^T W(u_i, v_i) Y \quad (2)$$

Where $\hat{\beta}$ is the estimation of $\beta$ , the first row elements of the independent variable matrix are 1; $X$ is the independent variable matrix about model factors; $Y$ is the vector of the values of the dependent variable and it is the matrix about the grass yield ground measured data; $W$ is a square matrix of weights relative to the position of $(u_i, v_i)$ in the study area; in practices the calculation method of spatial weights matrices are the Gauss distance, exponential distance and tribute distance. This study chose Gauss distance to determine the weight:

$$W(u_i, v_i) = e^{-\frac{1}{2}(\frac{d_{(u_i,v_i)}}{b})^2} \quad (3)$$

where $W(u_i, v_i)$ is the geographical weight of the $ith$ observation in the dataset relative to the location $(u_i, v_i)$, $d_{(u_i,v_i)}$ is some measurement of the distance between the $ith$ observation and the location $(u_i, v_i)$, $b$ is a quantity known as the bandwidth.

### 2.4 Input Parameter

To estimate grass yield with remote sensing data, vegetation index is adopted by many researches（Skinner R H，2011，JOSEM.PARCEIO，1997）. In the same time some weather elements, like temperature, rainfall, dryness also be used as input parameters（Li W J, 2012; Wei Y X, 2012）. Generally correlation coefficient（Xu H L，2013） and estimation accuracy are two important index to compare and choose the input parameters. In this study, five vegetation indices were compared, including normalized difference vegetation index (NDVI), enhanced vegetation index (EVI), soil adjusted vegetation index (SAVI), modified soil adjusted vegetation index (MSAVI) and optimal soil adjusted vegetation index (OSAVI)（Chen P F, 2007）. After several tests with five VIs, weather elements and fraction of vegetation coverage (FVC), finally EVI and FVC was adopted as input parameter to estimate grass yield.

The method mentioned by Brian Johnson（2012）was used to estimate the FVC, and the accuracy for the entire study area in August 2013 is 85.8%.

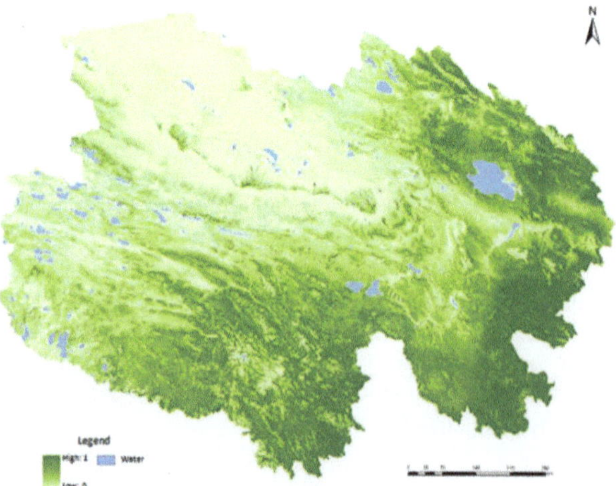


Figure 2. The FVC map of QingHai Province in 2013 August

### 2.5 Modelling with GWR

To model with GWR, the fresh weight of grass yield measured on the spot is chosen as the dependent variable and other factors include EVI and FVC are independent variables. After modelling, there is an equation for every samples calculated by GWR. While there is only one equation if the binary linear regression model is used.

A total of 201 field samples involved in grass yield estimation model building, the parameters of GWR model change in different geographical locations. As table 2，the parameters of GWR model change with different geographical locations.

| Variable | Minimum | Lwr Quartile | Upr Quartile |
|---|---|---|---|
| EVI | -22.94947 | -0.29945 | 1.9236 |
| VFC | -26.45646 | -0.4415 | 1.31991 |
| Constant | -2.71119 | 2.37819 | 3.22238 |

| Variable | Maximum | Median | Mean | Std |
|---|---|---|---|---|
| EVI | 52.29209 | 0.5347 | 1.0831 | 6.59432 |
| VFC | 10.59244 | 0.2882 | 0.27507 | 3.54877 |
| Constant | 11.58604 | 2.8394 | 2.77774 | 1.14701 |

Table 1. The range change of parameters from GWR model

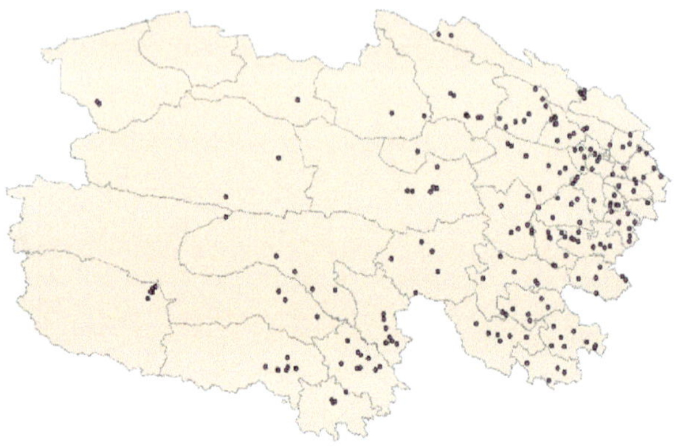

Figure 3. The distribution map of the samples for modelling

The multiple linear regression model is as following:

$$Y=0.684X_1- 0.006X_2+2.706 \quad (4)$$

Here Bandwidth has a great influence on the operation results of GWR model. the bigger bandwidth the bigger smoothing. As the bandwidth gets larger the weights approach unity and the local GWR model approaches the global OLS model. After several tests and analysis of test results, we use the bandwidth 230km and the corrected Akaike Information Criterion (Hurvich et al, 1998) extensively in GWR for as the measurement of goodness of fit Model establishment and Precision verification With the 201 samples, the modeling accuracy for GWR is 87.3%, while the accuracy for OLS is 40.35%. When it come to the statistical results of model parameters, it can be found the statistical index for GWR are better than that for OLS. Here, AICc is short for Akaike Information Criterion, r is correlation coefficient, and $r^2$ the coefficient of determination, $r^2$ Adj Adjusted r-square.

| Variable | Sigma | AICc | r | $r^2$ | $r^2$Adj |
|---|---|---|---|---|---|
| OLS | 0.193 | -86.437 | 0.815 | 0.665 | 0.663 |
| GWR | 0.012 | 749.849 | 0.987 | 0.973 | 0.891 |

Table 2. The Diagnostic statistics for parameters from GWR

### 2.6 Grass yield estimating

While estimating the grass yield for the whole region, the input parameters included raster data of the EVI and FVI. The value was calculated pixel by pixel. To find the nearest sample away the current pixel, and take the fitting equation of the nearest sample as basis to figure the grass yield for the pixel. For large region, there is huge amount of data, and the key to improve efficiency is to identify the nearest samples in shortest time. After repeated practice, it was found that building four forks tree index for all samples can improve the operation rate obviously.

The estimation result of grass yield in 2013 for Qinghai province is show as figure 4. Here from southeast to northwest the grass yield is becoming lower. The region with high grass yield distributes in southeast, including Haibei, Haidong, Huang Nan, Guoluo, and the south-eastern region of Yushu. And Haixi Autonomous region is desert and wilderness, so there has lowest grass yield. According the estimation, the total grass yield of Qinghai province is 1.018×108 Ton in 2013. Here the non-grass area was not taken into operate and such area were identified by the thematic map of second grassland survey in Qinghai Province.

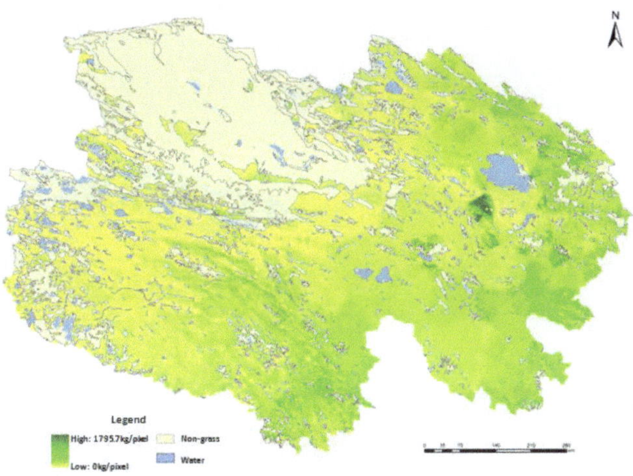


Figure 4. The grass yield map of QingHai Province in 2013

### 2.7 Precision verification

To verify the accuracy, root mean squared error (RMSE) (Jin Y X, 2011) was used. Theoretically, if RMSE=0, the model is error-free and it is considered an ideal state. In practical applications, the closer predictive value to the critical value, the better the model fitting results. The actual estimation accuracy of the model is 1-RMSE.

There are 210 ground truth samples involved into precision verification and the result value is 81.3%. The total grass yield estimated by the study were also compared with that from local Grassland professional department based on ground survey. The difference is less than 10%.

## 3. CONCLUSIONS

The location is introduced to building estimation model based on GWR. The parameters for different geographical samples are diverse, which reflect the spatial difference of these points. That means the same factors have different influence to the dependent variable because of deferent geographical position. The spatial difference of the grass yield in a large region can be reflected adequately by GWR. The experiments in the study have shown with GWR the goodness-of-fit of the model can be improved obviously, In the same time the estimation accuracy and verify accuracy are more than 80%. The difference between estimation with remote sensing data and ground survey is less than 10%. When it comes to long-term continuous monitoring in large area, technical method can be used.

To estimate grass yield with GWR model and remote sensing data, the estimation accuracy depends on ground samples. The accuracy is affected directly by spatial distribution characteristics of ground samples, as well as their representation of grassland production capacity. The high accuracy in the study benefited from the typical and representative samples. Now it takes grassland specialized department a lot of manpower and material resources every year to carry out the ground surveys. With the method mentioned above, only using RS data combined with the ground samples, it can be explored whether the ground samples can be reduced and how many samples is minimum requirements within an acceptable estimation error range. Such work will alleviate amount of field work for grassland specialized department.

Based on the grass yield estimated quickly with remote sensing technique, It is very meaningful to carry further research about expertly valuing the service function of the grassland ecosystem near real-time. Both from theoretical and practical view, such study is helpful for such work, like understanding the region ecological importance, scientific management and rational utilization of grass resources , maintaining grassland ecological balance, reasonable arrangements for livestock production, protection grass ecosystems and establish a comprehensive economic accounting system. The following work will focus on ecological evaluation.

## ACKNOWLEDGEMENTS

This research is supported by the basic scientific research fund of Chinese Academy of Surveying and Mapping (No. 7771512). The authors would like to thank Mr. Wang K and Mr. Wu C (Grassland supervision station of Qinghai Province) for their useful suggestion.

## REFERENCES

Benie G.B., Kabore S.S., Goita K., et al. Remote sensing-based spatio-temporal modeling to predict biomass in Sahelian grazing ecosystem. Ecological Modeling, 2005, 184:341-354.

Brian Johnson, Ryutaro Tateishi, Toshiyuki Kobayashi. Remote Sensing of Fractional Green Vegetation Cover Using Spatially-Interpolated Endmembers [J], Remote Sensing, 2012, 4: 2619-2634.

Chen J, Gu S, Shen M, et al, Estimating aboveground biomass of grassland having a high canopy cover: an exploratory analysis of in situ hyper spectral data[J].International Journal of Remote Sensing, 2009, 30（24）: 6497-6517.

Fotheringham A S. Chris Brunsdon, Martin Charlton. Geographically Weighted Regression: the Analysis of Spatially Varying Relationships [M]. U. K. West Sussex: John Wiley & Sons Ltd, 2002.

Hurvich CM, Simonoff JS and Tsai C9L 1998 Smoothing parameter selection in nonparametric regression using an improved Akaike information criterion Journal of Royal Statistical Societ y. B 60 2719293.

Jin Y X, Xu B, Yang X C, et al. Remote sensing dynamic estimation of grass production in Xilinguole, Inner Mongolia[J]. Science China, 2011, 41: 1185–1195.

Josem. Parceio, Howard E. Epstein, W .J. Iam K. Lauenroth, et al, A NPP Estimates From NDVI for the Central Grassland and Region of The United States[J], Ecology, 1997, 78(3): 953-958.

Li B X, Ma Z B, Qi Q W, et al. Cloud and shadow removal from Landsat TM data [J]. Journal of Remote Sensing, 2010, 14(3), 540-545.

Li W J, Jiu C L, Tan Z H, et al. Natural grassland productivity and the livestock-feeds balance in Qinghai Province [J]. Resources Science, 2012, 34(2):367-372.

Liu X Y, Feng Q SH, Liang T G, et al. Spatial-temporal dynamic balance between livestock carrying capacity and productivity of rangeland in Gannan of Gansu Province, China [J]. Chinese Journal of Grassland, 2010, 32(1):99-106.

Lv H Y. Study on the method of grass yield model [D].Beijing: Chinese Academy of Agricultural Sciences, 2010.

Miller H J. Tobler’ s first law and spatial analysis [J].Annals of the Association of American Geographers, 2004, 94(2):284-289.

Qu C P, Guan D X, Wang Z A, et al. Comparison of grassland biomass estimation models based on MODIS data [J]. Chinese Journal of Ecology, 2008, 27(11): 2028-2032.

Shen Y C, Xiang L P. The physical geography of Qinghai Province [M].Beijing: China Ocean Press, 1991.

Skinner R H, Wylie B K, Gilmanov T G. Using normalized difference vegetation index to estimate carbon fluxes from small rotationally grazed pastures[J].Agronomy Journal, 2011, 103: 972-979.

Tao W G. Remote sensing methods for monitoring grassland production in Inner Mongolia [D].Beijing: Chinese Academy of Agricultural Sciences, 2007.

Wang S, Li X, Zhang Q, et al. Analysis on the applicability of topographic correction models for Landsat images [J].Journal of Huazhong Normal University (Nat. Sci.)2013, 47(4):571-577.

Wei Y X, Wang L W, Shi Y C. Net Primary Productivity of Grassland Resources Monitoring Based on Remote Sensing Data in Qinghai Province, Scientia Geographica Sinica, 2012, 32(5):621-627.

Xu B, Yang X C, Tao W G, et al. remote sensing monitoring upon the grass production in China [J] Acta Ecologica Sinica, 2007, 2007(2): 405-413.

Xu H L, Li J P, Feng J. The pair-wise deletion correlation coefficient testing method and its applications [J]. Acta Meteorological Sinica, 2013, 71(5):901-912.

Xuan H Y, Li S P, Liu S Q. The geographically weighted regression model and its fit methods [J]. Journal of Gansu Sciences, 2007, 19(1):51-52.

Yan J W, Li Ch E , Yuan L, et al. Application Summary of EOS- MODIS data in the monitoring of grassland resources [J]. Pratacultural Science, 2008, 25(4):1-9.

Yan L D, Yin Q J, Zhang H Z.et al. The applied research of remote sensing material in Qinghai grassland resources monitoring and appraisal [J]. Journal of Natural Resources, 2007, 22(4): 640-648.

Zhang Y S, Zhao X Q, Zhou X M. Strategy and counter measure for sustainable development of animal husbandry in Qinghai [J]. Journal of Natural Resources, 2000, 15 (4): 328-334.

# MULTI-QUADRATIC DYNAMIC PROGRAMMING PROCEDURE OF EDGE–PRESERVING DENOISING FOR MEDICAL IMAGES

*Cong Thang Pham**, *Andrey V. Kopylov*

Institute of Applied Mathematics and Computer Science, Tula State University-Russia
(pacotha, And.Kopylov)@gmail.com

**Commission WG V/5, WG III/3**

**KEYWORDS**: Dynamic programming, Bayesian framework, Markov random fields (MRFs), Edges–preserving smoothing, Separable optimization, Potential functions.

**ABSTRACTS**: In this paper, we present a computationally efficient technique for edge preserving in medical image smoothing, which is developed on the basis of dynamic programming multi-quadratic procedure. Additionally, we propose a new non-convex type of pair-wise potential functions, allow more flexibility to set a priori preferences, using different penalties for various ranges of differences between the values of adjacent image elements. The procedure of image analysis, based on the new data models, significantly expands the class of applied problems, and can take into account the presence of heterogeneities and discontinuities in the source data, while retaining high computational efficiency of the dynamic programming procedure and Kalman filter-interpolator. Comparative study shows, that our algorithm has high accuracy to speed ratio, especially in the case of high-resolution medical images.

## 1. INTRODUCTION

Low–level image processing is a functional step in almost every medical image analysis system. It is a prerequisite for proper data interpretation, diagnosis and suggestion the corresponding treatments. Medical images (example: Magnetic Resonance, Ultrasound, Computed Tomography, X–Ray), may be corrupted by a disruptive noise during acquisition and transmission process and the essential requirement for every noise reduction procedure is to preserve local image features for an accurate and effective diagnosis.

Many edge-preserving denoising methods for medical images have been proposed in the literature. For example, Perona and Malik introduced nonlinear anisotropic diffusion (Perona and Malik, 1990; Gerig et al., 1992), which is an efficient method for noise reduction in MR images with local features preserving. Portilla (Portilla et al., 2003) have suggested alternative approach based on Wavelet transform. Furthermore, some other methods have good results of edge-preserving denoising: Wiener filter (Brown and Hwang, 1996), bilateral filter (Tomasi and Manduchi, 1998; Chaudhury et al., 2011), non – linear total variation (Rudin et al., 1992; Drapaca, 2009). Nevertheless, the monitoring of dynamic processes needs to improve performance of the low-level processing stage in both speed and accuracy.

In this paper, we develop new non–iterative parametric procedure for edge preserving in image smoothing. This procedure can effectively remove Gaussian noise as well as Rician noise (Dekker and Sijbers, 2014), typical for MR images, with high quality. We use here Bayesian framework as one of the most popular approaches to image processing. Under this approach, the problem of image analysis can be expressed as the problem of estimation of a hidden Markov component of a two-component random field, where the analyzed image plays the role of the observed component. An equivalent representation of Markov random fields in the form of Gibbs random fields, according to Hammersley-Clifford Theorem (Hammersley and Clifford, 1971), can be used to define prior probability properties of a hidden Markov field by means of so-called Gibbs potentials for cliques. In the case of a singular loss function, the Bayesian estimation of the hidden component can be found as a maximum a posteriori probability (MAP) estimation by the minimization of the objective function often called the Gibbs energy function. The quadratic form of the Gibbs potentials corresponds to the assumption of a normal priori distribution. In the paper (Kopylov, 2010), the more general case is considered, when pair-wise Gibbs potentials are selected as a minimum of a finite set of quadratic functions, and an efficient optimization procedure is proposed for the Blake and Zisserman function (Blake and Zisserman, 1987), which is often used in edge-preserving procedures. This highly effective global optimization procedure, is based on a recurrent decomposition of the initial problem of minimizing function of many variables into a succession of partial problems, each of which consists in minimizing a function of only one variable. Such a procedure is nothing else than the enhanced version of the famous dynamic programming procedure (Kalman and Bucy, 1961). In the case of a minimum of a finite set of quadratic functions pair-wise Gibbs potentials, the procedure breaks down at each step into several parallel procedures, according to the number of quadratic functions forming the intermediate optimization problems of one variable. The corresponding intermediate objective functions are called Bellman functions and the procedure itself - a multi quadratic dynamic programming procedure. It was proven, that in the case of signal processing the number of quadratic functions that are required for representation of a Bellman function, generally, does not increase by more than one at each step (Kopylov, 2010).

However, using this approach for processing the two-dimensional data on the basis of tree-like approximation of the

*Corresponding author

lattice neighborhood graph (Mottl, 1998), the number of quadratic functions in the Bellman functions may be too large and leads to a lack of effective implementation of the procedure. The basic idea of the proposed procedure is to find the groups of closest, in the appropriate sense, quadratic functions using k-means clustering algorithm (MacQueen, 1967), and to replace each of these groups by one quadratic function having the smallest minimun value.

During the experimental study, we compare the performance of MR images denoising algorithms by using the Mean Structure Similarity Index-MSSIM (Zhou Wang, 2004 ), Signal to Noise Ratio - SNR, and Peak to Signal Noise Ratio-PSNR in the presence of Gaussian and Rician noise. Wiener filtering (WF), non – linear Total Variation (TV), Anisotropic Diffusion Filter (ADF), Fast Bilateral Filter (FBF), the Bayesian least squares with Gaussians Scale Mixture (BLS-GSM) and proposed algorithms were tested. Results show, that our algorithm has the best accuracy to speed ratio, especially in the case of high-resolution images.

## 2. BAYESIAN FRAMEWORK FOR IMAGE DENOISING

Under this approach, the problem of image analysis can be expressed as the problem of estimating a hidden Markov component $X=(x_{\mathbf{t}}, \mathbf{t}\in T)$ $T=(\mathbf{t}=\{t_1,t_2\},$ $t_1=1..N_1,$ $t_2=1..N_2)$ of a two-component random field $(X,Y)$ , where the analyzed image $Y=(y_{\mathbf{t}}, \mathbf{t}\in T)$ plays the role of the observed component. An equivalent representation of Markov random fields in the form of Gibbs random fields, according to Hammersley Clifford Theorem (Hammersley JM, Clifford PE, 1971), can be used to define a priori probability properties of a hidden Markov field by means of so-called Gibbs potentials for cliques (Geman S., German D. 1984). In the case of a singular loss function, the Bayesian estimation of the hidden components can be found as Maximum a Posteriori Probability (MAP) by performing the minimization of the objective function, often called the Gibbs energy function (Mottl, 1998; Kopylov, 2010):

$$J(X\mid Y)=\sum_{\mathbf{t}\in T}\psi_{\mathbf{t}}(x_{\mathbf{t}}\mid Y)+\sum_{(\mathbf{t}',\mathbf{t}'')\in G}\gamma_{\mathbf{t}',\mathbf{t}''}(x_{\mathbf{t}'},x_{\mathbf{t}''}) \qquad (1)$$

The structure of the objective function (1) reflects the ordering property of analyzed data and determined by means of an undirected graph neighborhood $G\subset T\times T$ . In image analysis objective function can often be represented as the sum of the two types of potential functions for cliques, called node functions and edge functions. In image smoothing node functions $\psi_{\mathbf{t}}(x_{\mathbf{t}}\mid Y_{\mathbf{t}})$ play the role of a penalty on the difference between the values of input data $Y$ and the seeking function $X$ , and are usually chosen in quadratic form $\psi_{\mathbf{t}}(x_{\mathbf{t}}\mid Y_{\mathbf{t}})=(x_{\mathbf{t}}-y_{\mathbf{t}})^2$ . Each edge function $\gamma_{\mathbf{t}',\mathbf{t}''}(x_{\mathbf{t}'},x_{\mathbf{t}''})$ imposes penalty upon the difference of values in the corresponding pair of adjacent elements of the edge $(\mathbf{t}',\mathbf{t}'')$ of neighborhood graph $G$ , and can have various forms. The quadratic form of Gibbs potentials corresponds to the assumption of normal priori distribution $X$ . In a Bayesian framework, the edge functions are usually called a pair-wise potential functions.

There are different pair-wise potentials in the literature. A variety of non-convex pair-wise potential functions were considered for preserving large differences of values in the corresponding pairs of adjacent elements of the edges and smoothing the other differences (Nikolova, 2010). Convex edge-preserving pair-wise potential functions were proposed to avoid the numerical involutions arising with non-convex regularization (Xiaoju and Li, 2009). However, non-convex regularization offer the best possible quality of image reconstruction with neat and exact edges (Nikolova, 2010). One of the main problems in these approaches is high computational complexity of corresponding minimization procedures which can hardly be applied to high-resolution images.

In the paper (Kopylov, 2010), the different case is considered, when pair-wise Gibbs potentials are selected as a minimum of a finite set of quadratic functions, and an efficient optimization procedure for signal processing is proposed for the Blake and Zisserman function $\gamma(x_{\mathbf{t}'},x_{\mathbf{t}''})=u[(x_{\mathbf{t}'}-x_{\mathbf{t}''})^2,\Delta^2]$ , which is often used in edge-preserving procedures. This highly effective global optimization procedure, is based on a recurrent decomposition of the initial problem of minimizing function of many variables into a succession of partial problems, each of which consists in minimizing a function of only one variable. Such a procedure is nothing else than one of versions of the famous dynamic programming procedure. In the case of a minimum of a finite set of quadratic functions pair-wise Gibbs potentials, the procedure breaks down at each step into several parallel procedures, according to the number of quadratic functions forming the intermediate optimization problems of one variable. The corresponding intermediate objective functions are called Bellman functions and the procedure itself - a multi quadratic dynamic programming procedure.

According to the central idea of the variation approach to the analysis of ordered data sets, the result of analysis $\hat{X}(Y)$ is defined by the condition:

$$\hat{X}(Y)=\arg\min_{x_t\in \mathrm{X}} J(X\mid Y) \qquad (2)$$

The procedure of dynamic programming will search for optimization of the objective function (1) in the forward and backward recurrent relation. In the forward recurrent with upward recurrent relation defined by the Bellman function (Mottl, 1998):

$$\tilde{J}_{\mathbf{t}}(x_{\mathbf{t}})=\psi_{\mathbf{t}}(x_{\mathbf{t}})+\min_{x_{\mathbf{t}-1}}\left[\gamma_{\mathbf{t}}(x_{\mathbf{t}-1},x_{\mathbf{t}})+\tilde{J}_{\mathbf{t}-1}(x_{\mathbf{t}-1})\right] \qquad (3)$$

The global minimum of the Bellman function from the last variable $\min_{x_{\mathbf{N}}}\tilde{J}_{\mathbf{N}}(x_{\mathbf{N}})$ coincides with the global minimum of the total objective function in all variables, so the optimal value of the last variable minimize its function Bellman (Mottl, 1998):

$$\hat{x}_{\mathbf{N}}=\arg\min_{x_{\mathbf{N}}}\tilde{J}_{\mathbf{N}}(x_{\mathbf{N}}) \qquad (4)$$

The other variables can be found by applying backward recurrent relation ( $\mathbf{t}=\mathbf{N}-1,..,2,1$ ):

$$\hat{x}_{\mathbf{t}-1}(x_{\mathbf{t}})=\arg\min_{x_{\mathbf{t}-1}}\left[\gamma_{\mathbf{t}}(x_{\mathbf{t}-1},x_{\mathbf{t}})+\tilde{J}_{\mathbf{t}-1}(x_{\mathbf{t}-1})\right] \qquad (5)$$

## 3. MULTI-QUADRATIC DYNAMIC PROGRAMMING PROCEDURE FOR IMAGE PROCESSING

In this paper, we propose a new non-convex type of pair-wise potential functions, allows more flexibility to set a priori preferences, using different coefficients of penalty for various ranges of differences between the values of adjacent image elements:

$$\gamma_t(x_t,x_{t-1}) = \min\left[\gamma_t^1(x_t,x_{t-1}),\ldots,\gamma_t^L(x_t,x_{t-1})\right] \quad (6)$$

The developed image analysis procedure can significantly extend the class to solve applied problems, to take into account the presence of heterogeneities and discontinuities in the original data, while retaining the high computational efficiency of procedures of dynamic programming and Kalman filter-interpolator (Kalman and Bucy, 1961).
It can be proven, that if the pair-wise Gibbs potentials are selected as a minimum of a finite set of quadratic functions, and node functions are in quadratic form, the procedure breaks down at each step into several parallel procedures, according to the number of quadratic functions forming the intermediate optimization problems of one variable. The Bellman function at each step of the dynamic programming will has a minimum of a finite set of quadratic functions:

$$\tilde{J}_t(x_t) = \min\left[\tilde{J}_t^1(x_t),\tilde{J}_t^2(x_t),\ldots,\tilde{J}_t^L(x_t)\right] \quad (7)$$

where

$$\tilde{J}_t^{\,i}(x_t) = \tilde{q}_t^i(x_t-\tilde{x}_t^i)^2+\tilde{d}_t^i, \tilde{q}_t^i \ge 0, i=1,..,L\,. \quad (8)$$

Let $\tilde{F}_t(x_t) = \min\limits_{x_{t-1}}\left[\gamma_t(x_{t-1},x_t)+\tilde{J}_{t-1}(x_{t-1})\right]$

then

$$\tilde{F}_t(x_t) = \min\begin{bmatrix}\min\limits_{x_{t-1}}\left\{\gamma_t^1(x_{t-1},x_t)+\psi_{t-1}(x_{t-1})+\tilde{F}_{t-1}(x_{t-1})\right\},\\ \cdots \\ \min\limits_{x_{t-1}}\left\{\gamma_t^L(x_{t-1},x_t)+\psi_{t-1}(x_{t-1})+\tilde{F}_{t-1}(x_{t-1})\right\}\end{bmatrix} \quad (9)$$

$$\tilde{F}_t(x_t) = \min\left(\tilde{F}_t^1(x_t),\tilde{F}_t^2(x_t),\ldots,\tilde{F}_t^K(x_t)\right) \quad (10)$$

$$\tilde{F}_t^{\,j}(x_t) = \tilde{q}_t^j(x_t-\tilde{x}_t^j)^2+\tilde{d}_t^j, \tilde{q}_t^j \ge 0, j=1,..,K \quad (11)$$

The backward recurrent relation $\hat{x}_{t-1}(x_t)$ has the following form:

$$\hat{x}_{t-1}(x_t) = \arg\min_{x_{t-1}}\left[F(x_{t-1})\right] \quad (12)$$

$$F(x_{t-1}) = \min\left[\gamma_t^i(x_{t-1},x_t)+\psi_{t-1}(x_{t-1})+\tilde{F}_{t-1}^{\,j}(x_{t-1})\right] \quad (13)$$

It is easy to see that if the values $\tilde{x}_1,\tilde{x}_2,..,\tilde{x}_N$ are the minimum points of criteria (1), then $\tilde{x}_t \in [a,b]$ for every $\mathbf{t}=1,..,\mathbf{N}$, where $a=\min(Y)$ and $a=\min(Y)$. It can be noted that not all of the quadratic function will participate in forming of the final function, because their values are not minimum for any point. Such functions can be dropped using enough simple procedure that takes into account the position of the minimum point and the points of intersection of quadratic functions with each other.
At the beginning, sort by ascending values $d_t^j$ of array $\tilde{F}_t^j$. At each step, looking for the minimum constant and discard all others constant. Discard all functions than have minimum greater than or equal to this constant.
After that, we find necessary and sufficient condition of intersection of quadratic function by following equations:

$$\tilde{q}_t^{\,i}(x_t-\tilde{x}_t^{\,i})^2+\tilde{d}_t^{\,i} = \tilde{q}_t^{\,j}(x_t-\tilde{x}_t^{\,j})^2+\tilde{d}_t^{\,j} \quad (14)$$

where $i,j=1..K; i\ne j$
The coordinates of the intersection points on the real axis defined by the following relations:

$$x\_c1_t^{(ij)} = \frac{1}{\tilde{q}_t^{(i)}-\tilde{q}_t^{(j)}}(\tilde{q}_t^{(i)}\tilde{x}_t^{(i)}-\tilde{q}_t^{(j)}\tilde{x}_t^{(j)}+\sqrt{\Lambda}) \quad (15)$$

$$x\_c2_t^{(ij)} = \frac{1}{\tilde{q}_t^{(i)}-\tilde{q}_t^{(j)}}(\tilde{q}_t^{(i)}\tilde{x}_t^{(i)}-\tilde{q}_t^{(j)}\tilde{x}_t^{(j)}-\sqrt{\Lambda}) \quad (16)$$

The points of intersection $x\_c1_t^{(ij)}$ and $x\_c1_t^{(ij)}$ have real coordinates, if expression $\Lambda$ under the square root is greater than or equal to zero:

$$\Lambda = \tilde{q}_t^{(i)}\tilde{q}_t^{(j)}(\tilde{x}_t^{(j)}-\tilde{x}_t^{(i)})^2+(\tilde{d}_t^{(j)}-\tilde{d}_t^{(i)})(\tilde{q}_t^{(i)}-\tilde{q}_t^{(j)}) \ge 0 \quad (17)$$

After that, select functions that have points of intersection with satisfying $a \le x\_c1_t^{(ij)}, x\_c2_t^{(ij)} \le b$. Discard all the functions for which there is no intersection. Among the tested functions, select the function with the smallest minimum and leave it. Check its intersection with other functions. Discard all functions for which there is no intersection. Repeat as long as there will be functions for which have no decision on acceptance or discarding. Then the reduction of the amount Bellman functions at each step is determined by the formula:

$$\tilde{F}_t(x_t) = \min\left(\tilde{F}_t^1(x_t),\tilde{F}_t^2(x_t),\ldots,\tilde{F}_t^H(x_t)\right), H<K \quad (18)$$

The number of Bellman functions can be reduced according to the expression (17). It was proven, that in the case of signal processing the number of quadratic functions that are required for representation of a Bellman function, generally, does not increase by more than one at each step. Nevertheless, using this approach for processing data based on the two-dimensional the tree approximation of lattice graph neighborhood (Mottl et.al, 1998), the number of quadratic functions on the Bellman functions may be too large and leads to a lack of effective implementation of the procedure.
The basic idea of the proposed procedure is to find the groups of closest, in the appropriate sense, quadratic functions using k-means clustering algorithm, and to replace each of these groups by one quadratic function having the smallest minimun value. We consider quadratic functions in (17) as a points in three-dimensional space with coordinates $\left(\tilde{q}_t^i,\tilde{x}_t^i,\tilde{d}_t^i\right), i=1,\ldots,H$.
Using k-means clustering algorithm, the distance between quadratic functions can be calculated as the sum of the squares of the difference on the definitive range [a, b]:

$$D_{ij} = \int_a^b \left[ \tilde{q}_t^i \cdot (x_t - \tilde{x}_t^i)^2 + \tilde{d}_t^i - \tilde{q}_t^j \cdot (x_t - \tilde{x}_t^j)^2 - \tilde{d}_t^j \right] dx_t \quad (19)$$

$$D_{ij} = c_1 s_2^2 + c_2 s_3^2 + c_3 \cdot (2 \cdot s_1^2 + s_2 s_3) + c_4 s_1 s_3 + c_5 s_1 s_2 \quad (20)$$

where $c_1 = \frac{(b^5 - a^5)}{5}$; $c_2 = (b - a)$; $c_3 = \frac{2(b^3 - a^3)}{3}$; $c_4 = 2(a^2 - b^2)$; $c_5 = (a^4 - b^4)$; $s_1 = \tilde{q}_t^i \tilde{x}_t^i - \tilde{q}_t^j \tilde{x}_t^j$; $s_2 = \tilde{q}_t^i - \tilde{q}_t^j$; $s_3 = \tilde{q}_t^i (\tilde{x}_t^i)^2 - \tilde{q}_t^j (\tilde{x}_t^j)^2 + \tilde{d}_t^i - \tilde{d}_t^j$; $i \neq j; i, j = 1, \ldots, H$

Assume that the number $k$ is a predetermined number of groups for k-means clustering algorithm. To preserve the quality of image processing, we do not use directly the final cluster centers which derived by the k-means algorithm. For each of the derived final groups we choose a point that have the lowest third coordinate.

Using the algorithm for reduction in the number of quadratic functions allows you to get the effective implementation of image processing procedure on the basis multi quadratic dynamic programming procedure. Experimental studies show that the vast majority of the original data sets of two or three square functions that are quite fully reflect each function in the Bellman criteria (7).

## 4. EXPERIMENTAL RESULTS

In this paper, all the experiments are run on MATLAB 7.14. The test images are 8-bit grayscale brain MR images from Simulated Brain Database (SBD)[1]. In our experiments we compare proposed approaches with the other filters for MR images as Wiener filtering (WF), non – linear total variation (TV), anisotropic diffusion filter (ADF), fast bilateral filter (FBF), the Bayesian least squares with Gaussians Scale Mixture (BLS-GSM). Results are quantified by the Mean Structure Similarity Index (MSSIM), peak-to-signal ratio (PSNR), Signal to Noise Ratio (SNR), and Peak to Signal Noise Ratio (PSNR). MSSIM is defined by the mean of values SSIM:

$$SSIM(P,Q) = \frac{(2\mu_P \mu_Q + c_1)(2\sigma_{P,Q} + c_2)}{(\mu_P^2 + \mu_Q^2 + c1)(\sigma_P^2 + \sigma_Q^2 + c_2)} \quad (21)$$

PSNR (dB) is defined as:

$$PSNR = 10\log_{10}\left(\frac{255^2}{MSE}\right) \quad (22)$$

SNR (dB) is defined as:

$$SNR = 10\log_{10}\left(\sum_{i=1}^{M}\sum_{j=1}^{N} P(i,j)^2 \frac{1}{MSE \times M \times N}\right) \quad (23)$$

where $MSE = \frac{1}{M \times N}\sum_{i=1}^{M}\sum_{j=1}^{N}(P(i,j) - Q(i,j))^2$; $\mu_P, \mu_Q$ - means of images; $\sigma_P, \sigma_Q$ - standard deviations (the square root of variance) of images; $\sigma_{P,Q}$ - covariance of images $P(i,j)$ and $Q(i,j)$; $c_1$=6.5025, $c_2$=58.5225; $P(i,j)$ and $Q(i,j)$ denote pixel values of the original image and the reconstructed or noisy image accordingly.

In the experiments, the node functions selected quadratic form $\psi_{t_1 t_2}(x_{t_1 t_2}) = (x_{t_1 t_2} - y_{t_1 t_2})^2$, and edge functions are at least a set of quadratic functions instead of a single quadratic function. Node functions $\psi_{t_1 t_2}(x_{t_1 t_2}) = (x_{t_1 t_2} - y_{t_1 t_2})^2$ are the same for both horizontal $\gamma_h(x_{t_1,t_2-1}, x_{t_1,t_2})$ and vertical $\gamma_v(x_{t_1-1,t_2}, x_{t_1,t_2})$ adjacency. For edge functions, we chose values $L = 2, L = 3$. With $L = 2$, if the second function is a constant function, we have forms of function Blake and Zisserman:

$$\gamma_h = u_h \min[(x_{t_1,t_2-1} - x_{t_1,t_2})^2, \Delta^2] \quad (24)$$

$$\gamma_v = u_v[(x_{t_1-1,t_2} - x_{t_1,t_2})^2, \Delta^2] \quad (25)$$

and with $L = 3$ we have following forms of edge functions:

$$\gamma_h = u_h \min[(x_{t_1,t_2-1} - x_{t_1,t_2})^2, \lambda(x_{t_1,t_2-1} - x_{t_1,t_2})^2 + d, \Delta^2] \quad (26)$$

$$\gamma_v = u_v[(x_{t_1-1,t_2} - x_{t_1,t_2})^2, \lambda(x_{t_1-1,t_2} - x_{t_1,t_2})^2 + d, \Delta^2] \quad (27)$$

On figure 1 show example results of reducing noise MRI with Racian noise level $\sigma = 10$. On figure 2 show example results of reducing noise MRI with Gaussian noise level $\sigma = 15$.

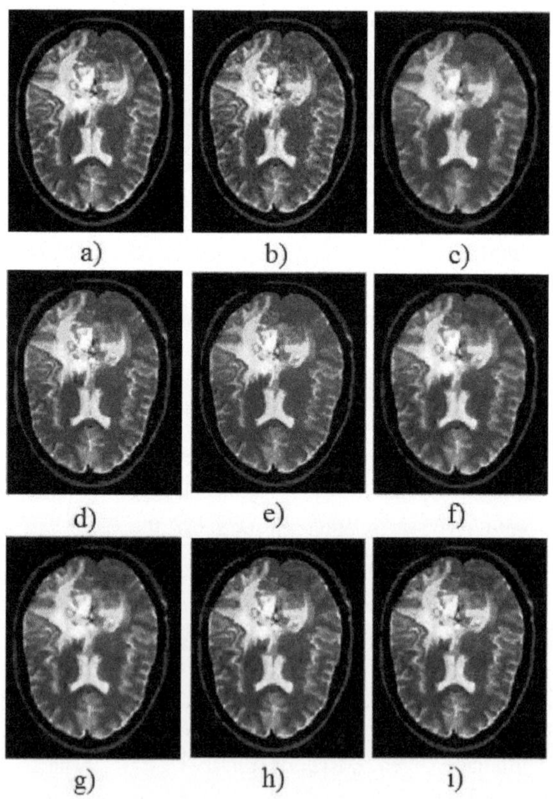


Figure 1. Original image (a); Racian noise image (b) with noise level $\sigma = 10$; denoising results with TV (c), BLS-GSM (d), FBF (e), ADF (f), WF (g), L=2 (h), L=3 (i)

[1] Available at http://brainweb.bic.mni.mcgill.ca/brainweb/

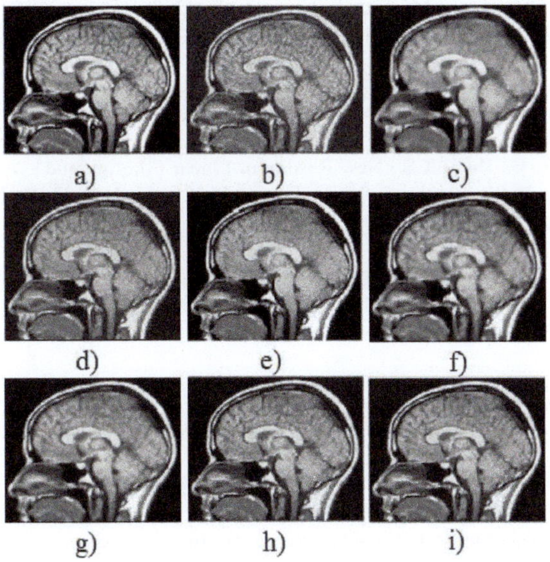


Figure 2. Original image (a); Gaussian noise image (b) with noise level $\sigma = 15$ ; denoising results with TV (c), BLS-GSM (d), FBF (e), ADF (f), WF (g) , L=2 (h), L=3 (i)

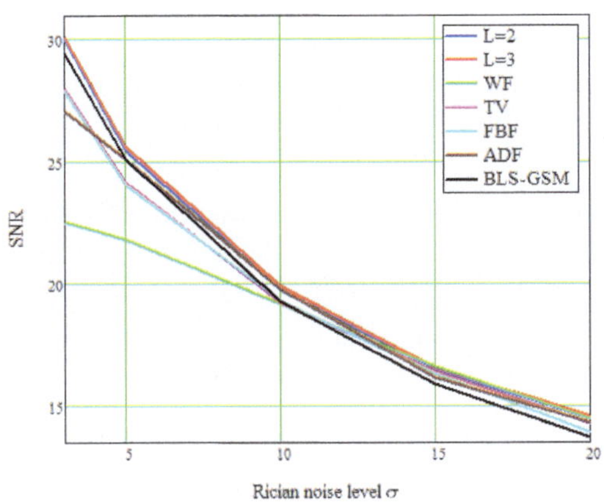


Figure 3. Comparison of measurements *SNR* of methods with different Rician noise levels

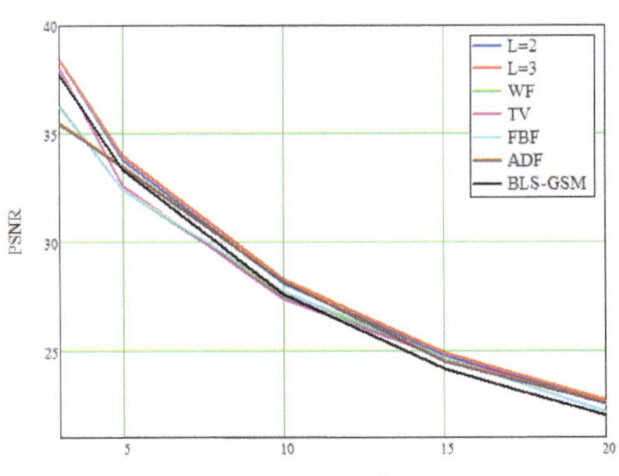


Figure 4. Comparison of measurements *PSNR* of methods with different Rician noise levels

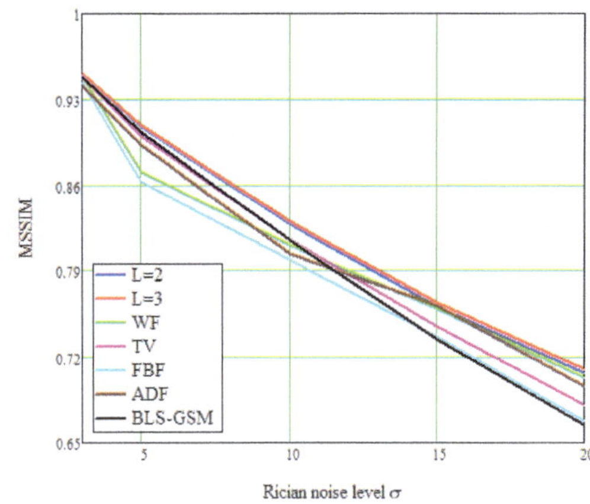


Figure 5. Comparison of measurements *MSSIM* of method with different Rician noise levels

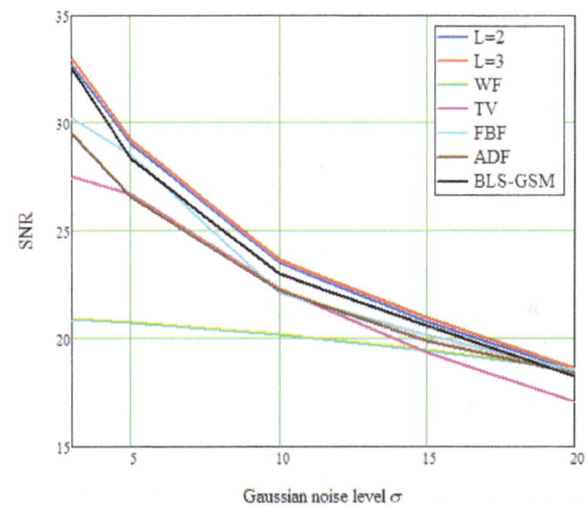


Figure 6. Comparison of measurements *SNR* of method with different Gaussian noise levels

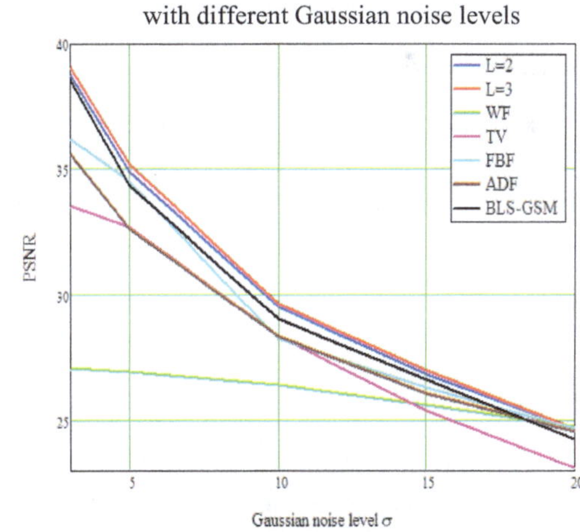


Figure 7. Comparison of measurements *PSNR* of methods with diffrent Gaussian noise levels

Results of measurements in figure 3-8 are has been averaged over ten tests. Results show, that our algorithm has the best accuracy to speed ratio, especially in the case of high-resolution images. Details of quantified measurements are showed on figure 3-8 with different Gaussian and Racian noise levels

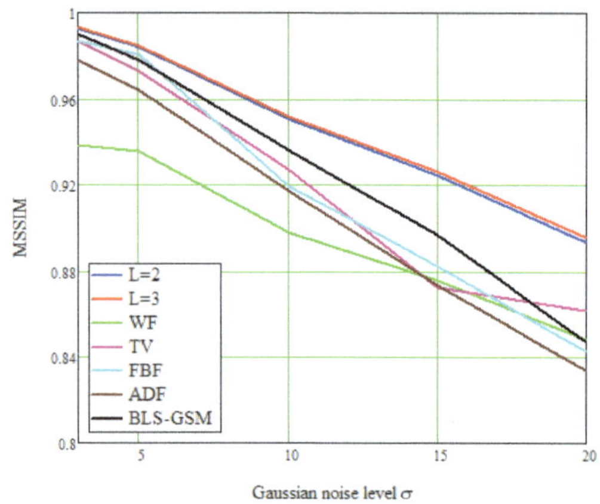


Figure 8. Comparison of measurements *MSSIM* of method with diffrent gaussian noise levels

## 5. CONCLUSION

Edge preserving MR denoising has become an urgent step in medical imaging to remove noise and to preserve local image features for an accurate and effective diagnosis. In this paper, we proposed a new approach to achieve these aims. The experimental results show that muti-quadratic dynamic programming procedure with the application of the algorithm k-means for reduction of the number of executed quadratic functions, allows get the high computational efficiency of dynamic programming for image processing, especially in the case of high-resolution images.

## ACKNOWLEDGEMENTS

This work was supported by Russian Foundation for Basic Research (RFBR), project № 13-07-00529-a.

## REFERENCES

Blake A., Zisserman A., 1987. Visual Reconstruction. MIT Press, Cambridge, Massachusetts, 1987, 232 pages.

Brown, R. and Hwang, P., 1996. Introduction to Random Signals and Applied Kalman Filtering (3 ed.). New York: John Wiley & Sons, 248 pages.

Chaudhury, K., Sage, D. and Unser, M., 2011. , *In: IEEE Transactions on Image Processing*, Vol. 20(12), pp. 3376 – 3382.

Dekker A.J., Sijbers J., 2014. Data distributions in magnetic resonance images: A review, *Physica Medica* (30),Elsevier, pp. 725-741

Drapaca, C., 2009. A nonlinear total variation-based denoising method with two regularization parameters. In: *IEEE Transactions on Biomedical* Engineering, Vol. 56(3), pp. 582 – 586.

Gerig, G. et al., 1992. Nonlinear anisotropic filtering of MRI data. In: *IEEE Transaction on Medical Imaging*, Vol. 11 (2), pp. 221–232.

Geman S., German D. (1984) Stochastic Relaxation, Gibbs Distributions, and the Bayesian Restoration of Images. In: *Pattern Analysis and Machine Intelligence*, IEEE Transactions, 1984. PAMI-6(6): p. 721 - 741.

Hammersley J. M., Clifford P.E., Markov random fields on finite graphs and lattices, 1971.

Kalman R. E., Bucy R.S., New Results in Linear Filtering and Prediction Theory. Journal of Basic Engineering, 1961. Vol. 83: p. 95-108

Kopylov, A. et al., 2010. Signal Processing Algorithm Based on Parametric Dynamic Programming. *Lecture Notes in Computer Science*, Vol. 6134, pp. 280–286.

MacQueen J., Some methods for classification and analysis of multivariate observations // Proceedings of 5-th berkeley symposium on mathematical statistics and probability, berkeley, university of california press, 1967.1(281-297).

Mottl V., et al., 1998. Optimization techniques on pixel neighborhood graphs for image processing. In: *Graph-Based Representations in Pattern Recognition.* Computing, Supplement 12. Springer–Verlag/Wien, pp. 135-145.

Nikolova M., Michael K., and Tam C.P., 2010. Fast Nonconvex Nonsmooth Minimization Methods for Image Restoration and Reconstruction. In: *IEEE Transactions on Image Processing*, Vol. 19, no. 12, pp. 3073-3088.

Perona, P. and Malik, J., 1990. Scale space and edge detection using anisotropic diffusion. In: *IEEE Transactions on Pattern Analysis and Machine Intelligence*, Vol. 12. pp. 629–639.

Portilla, J., et al., 2003. Image denoising using scale mixtures of gaussians in the wavelet domain. In: *IEEE Transactions On Image Processing*, Vol. 12, pp. 1338– 1351.

Rudin, L., Osher, S., and Fatemi, E., 1992. Nonlinear total variation based noise removal algorithms. In: *Journal Physica D*, Vol. 60, pp. 259-268.

Tomasi, C. and Manduchi, R, 1998. Bilateral filtering for gray and color images. In: *proceedings of Sixth International Conference on Computer Vision*, pp. 839–846.

Xiaojuan G., Li G., 2009. A new method for parameter estimation of edge-preserving regularization in image restoration. In: *Journal of Computational and Applied Mathematics*, 225, Elsevier, pp. 478-486

# 6

# STATISTICAL ANOMALY DETECTION FOR MONITORING OF HUMAN DYNAMICS

K. Kamiya[a], T. Fuse[b]

[a] Dept. of Civil Engineering, The University of Tokyo, 7-3-1 Hongo Bunkyo Tokyo 113-8656, Japan
kamiya@trip.t.u-tokyo.ac.jp
[b] Dept. of Civil Engineering, The University of Tokyo, 7-3-1 Hongo Bunkyo Tokyo 113-8656, Japan
fuse@civil.t.u-tokyo.ac.jp

**Commission V, WG V/4**

**KEY WORDS:** Anomaly detection, Human dynamics monitoring, Dirichlet process, Hidden markov model, time-series data

**ABSTRACT:**

Understanding of human dynamics has drawn attention to various areas. Due to the wide spread of positioning technologies that use GPS or public Wi-Fi, location information can be obtained with high spatial-temporal resolution as well as at low cost. By collecting set of individual location information in real time, monitoring of human dynamics is recently considered possible and is expected to lead to dynamic traffic control in the future. Although this monitoring focuses on detecting anomalous states of human dynamics, anomaly detection methods are developed ad hoc and not fully systematized. This research aims to define an anomaly detection problem of the human dynamics monitoring with gridded population data and develop an anomaly detection method based on the definition. According to the result of a review we have comprehensively conducted, we discussed the characteristics of the anomaly detection of human dynamics monitoring and categorized our problem to a semi-supervised anomaly detection problem that detects contextual anomalies behind time-series data. We developed an anomaly detection method based on a sticky HDP-HMM, which is able to estimate the number of hidden states according to input data. Results of the experiment with synthetic data showed that our proposed method has good fundamental performance with respect to the detection rate. Through the experiment with real gridded population data, an anomaly was detected when and where an actual social event had occurred.

## 1 INTRODUCTION

Understanding of human dynamics such as people's mobility or distribution in a city has drawn attention to various areas; for example, urban planning and marking. Recently, due to the wide spread of positioning technologies that use GPS or public Wi-Fi, location information can be obtained with high spatial-temporal resolution as well as at low cost. Therefore, by collecting set of individual location information in real time, monitoring of human dynamics is considered possible and is expected to lead to dynamic traffic control in the future. This monitoring focuses on detecting anomalous states of human dynamics as well as normal states. Because there are uncountable possible factors that may trigger the anomalous states, it would not possible to monitor all of the factors. In contrast, if anomalous states are detected by analysis on observation data of human dynamics monitoring such as mesh population data or set of individual GPS trajectories, this will be very useful. The characteristics that human dynamics may drastically change depending on regions and times, and the massive observation data through 24-hour-monitoring would however make manual anomaly detection difficult. On the other hand, statistical anomaly detection method can learn normal states from regularly acquired data and automatically detect anomalies as a different state from normal states.

Several researches on statistical anomaly detection of human dynamics have been conducted. Candia (Candia et al., 2008) has used extensive cell phone records resolved in both time and space, focused on the occurrence of anomalous events and discussed how these spatial-temporal anomalies can be described using standard percolation theory tools. Horanont (Horanont, 2010) in terms of hot spots extraction and visualization, has interpolated aggregate cell phone Call Detail Records by using an implemented platform. As an example of researches focusing on GPS trajectory data, Pan (Pan et al., 2013) has identified anomalies according to drivers' routing behavior on an urban traffic network. Horiguchi (Horiguchi et al., 2013) has implemented a real time monitoring system that provides the visual comprehension of regional traffic situations in terms of the fluidity and the singularity indices. Moreover, there are other researches that try to detect an anomalous trajectory caused by external factors such as accidents or vicious drivers by collecting the vehicle trajectories over a long period of time.

Although these researches on statistical anomaly detection of human dynamics have been conducted so far, these methods are developed ad hoc due to the diversity of data and anomalies. Anomaly detection methods are not fully systematized yet. Additionally it is considered difficult even for a well-suited method for certain application domain to apply to another domain because of the discrepancy in the definition of anomalies between the domains.

For these reasons, this research aims to firstly define an anomaly detection problem with regard to the human dynamics monitoring. Because of the anonymity and the high accessibility, this research targets gridded population data that is aggregated individual GPS data into each grid to estimate population of each grid for each instance of time. Secondly, based on the definition, we develop a suitable anomaly detection method for the human dynamics monitoring, and finally evaluate the fundamental performance of the proposed method. Therefore, development of anomaly detection method and clarification of the fundamental performance of the proposed method will make it possible to detect anomalous states behind the human dynamics. In the future, the human dynamics monitoring is expected to lead an effective traffic control and a better understanding of social mechanisms.

The paper is organized as follows: In section 2, we explain how the anomaly detection of human dynamics monitoring can be interpreted with respect to the factors that characterize anomaly detection problem. Additionally, we give an explanation that the state-space model, especially the Hidden Markov Model based on Hierarchical Dirichlet Process, is considered possible to apply. In section 3, after introducing the properties of nonparametric Bayesian model such as Dirichlet process, detail of the proposed anomaly detection method is presented. We demonstrate our proposed method with synthetic and real data and discuss the fundamental performance in section 4. Finally, section 5 concludes with a summary and some directions for future works.

## 2 PROBLEM SETUP FOR ANOMALY DETECTION OF HUMAN DYNAMIC MONITORING

### 2.1 Interpretation of anomaly detection factors

Statistical anomaly detection aims to identify data that are not consistent with a pattern that most other data follow (Chandola et al., 2009). It is also referred to as novelty detection (Pimentel et al., 2014) or outlier detection (Hodge and Austin, 2004) in different application domains although the main principle is common. The importance of anomaly detection attributes to value of detected anomalies in data; anomalies are often important trends behind huge data or critical and significant data. Combined with the progress of data acquisition techniques and computer performance, this great importance makes anomaly detection gain much research attention in variety of application domains; intrusion detection, image processing, structural health monitoring, et cetera.

Firstly, we have comprehensively reviewed the previous studies of anomaly detection in variety of areas and a result of the review shows that there are four factors that characterize anomaly detection problem; the nature of the input data, the output of anomaly detection, availability or unavailability of the data labels, and the type of anomalies (Chandola et al., 2009). This section provides a clear explanation of these factors as well as the interpretation of human dynamics monitoring targeting gridded population data.

**2.1.1 Nature of input data:** The first factor that characterizes anomaly detection is the nature of the input data. This can be roughly divided into binary, categorical and continuous. Additionally, each input data might consist of univariate or multivariate. Input data can also be categorized in terms of relationship between each input data; for example, temporal data, spatial data, graph data, et cetera. In the case of this study with gridded population data, input data can be considered as temporal data of continuous values because each gridded population can be regarded as input value. Moreover, since it is commonly considered that urban population distribution has a greater influence with closer region, this temporal data may also have some spatial correlation. Therefore, development to handle temporal anomaly detection is necessary and spatial extension may be preferred.

**2.1.2 Output of anomaly detection:** The second important factor for any anomaly detection method is the manner in which the anomalies are reported. There are typically two ways: One is to calculate anomaly score for each instance, and the another is to assign "normal" label or "anomalous" label for each instance. If the anomaly score of every grid is computed, it can be utilized as an indicator of the traffic control priority. Also, if normal or anomalous label that represents what kind of state the grid is, it can be beneficial. In this manner, the human dynamics monitoring is considered to accept both anomaly score and labels.

**2.1.3 Data labels:** The normal or anomalous labels associated with a input data instance that is used to learn anomaly detection models is the third factor to characterize anomaly detection. Since it is difficult to obtain all of possible states, especially anomalous states, anomaly detection techniques can be divided into three modes based on the extent to which the labels are available; supervised anomaly detection with the need of both normal and anomalous labels in training dataset, semi-supervised anomaly detection with only normal labels, and unsupervised anomaly detection without any prior information with regard to input dataset. In the case of human dynamics monitoring, normal labels can be obtained in a relatively simple way by using observational data of the day when and the place where big social events such as serious accidents and natural disasters are not reported. On the other hand, the availability of anomalous labels is a big issue for not only human dynamics monitoring but any anomaly detection problem. For these reasons, semi-supervised or unsupervised anomaly detection is thought able to be employed.

**2.1.4 Type of anomalies:** Anomalies can be classified into following three categories; point anomaly, contextual anomaly and collective anomaly. An individual data is regarded as a point anomaly if the data is considered anomalous with respect to the rest of the data, a data instance is regarded as a contextual anomaly if anomalous in the specific context (but not separately), and a set of data instance is regarded as a collective anomaly if anomalous with respect to the entire data set. As examples of anomalies that are possibly detected in gridded population data, there is an extreme increase or decrease caused by traffic congestion and a change of the population pattern affected by a change of the traffic demands. Figure 1 illustrates these possible anomalies in a gridded population temporal data. Aimed at the detection of these temporal anomalies as well as spatial anomalies due to the spatial correlations of gridded population data, contextual anomaly detection techniques are considered suitable to this study.

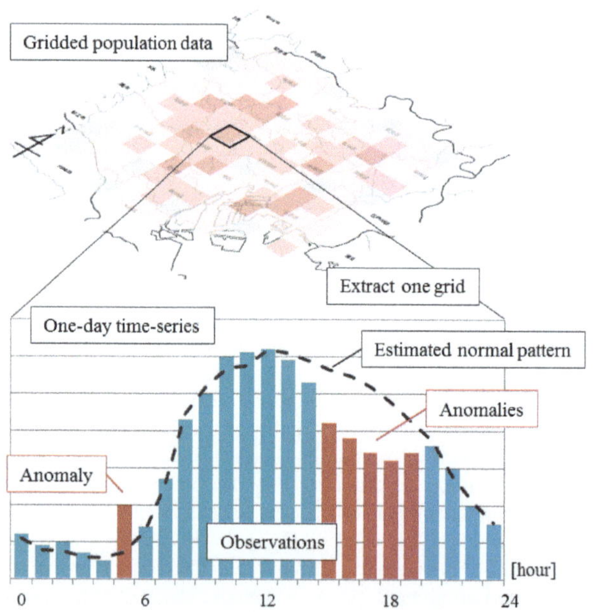


Figure 1: Anomalies in gridded population data

Based on these analyses showed in Table 2, we define the anomaly detection of human dynamics monitoring as a problem that detects contextual anomalies in time-series data of gridded population data by learning normal states with training dataset.

Table 2: Characteristics of the anomaly detection of human dynamics monitoring

| Factors | Interpretation of human dynamics monitoring |
|---|---|
| Input data | Temporal data of continuous value of an individual gridded population<br>Temporal data with spatial correlation |
| Output | Both anomaly scores and labels |
| Data labels | Semi or unsupervised anomaly detection<br>Normal labels obtained from daily data |
| Type of anomaly | Contextual anomaly<br>An extreme increase or decrease caused by traffic congestion<br>A change of the population pattern by a change of the traffic demands |

### 2.2 State-space model for anomaly detection

According to the problem setup defined above, anomaly detection techniques based on state-space models are considered possible to be applied. State-space models are models that consist of hidden (or latent) state variables and observed variables, and assume that hidden variables emit observed variables while evolving through time. This representation makes it possible to model complex temporal data, and this is the reason why state-space models are often used for anomaly detection in time-series data (Pimentel et al., 2014). Additionally, the high capability of addition of new variables such as neighbor population can make spatial extension of models possible.

There is said to be two most common techniques that use state-space models for anomaly detection. One assumes that anomalies are detected by calculating likelihood or emission probability of each observation, while the another assumes that anomalies are assigned to state variables which indicate anomalous states.

When considering the application for the anomaly detection of human dynamics monitoring, the construction of data generation models can be seriously problematic with regard to the former technique. Since there are few studies with respect to prediction or simulation of gridded population data, it is considered difficult to develop system models and observation models of which accuracy is enough for anomaly detection.

On the other hand, the later technique often suffers from the decision of the number of the hidden states because it generally needs to be fixed beforehand. Unfortunately, the accurate number of the hidden states that gridded population data possibly come into is unknown. This problem, however, can be conquered when the number is estimated from the collection of observation data.

This research regards that Hidden Markov Models based on Hierarchical Dirichlet Process, which are ones of nonparametric Bayesian models, can estimate the number of states of temporal data, and can be applied to the anomaly detection of human dynamics monitoring. Especially, we use a sticky Hierarchical Dirichlet Process - Hidden Markov Model (sHDP-HMM) which allows more robust learning of smoothly varying dynamics by limiting rapid state transitions. In the next section, an anomaly detection method based on a sHDP-HMM is proposed.

## 3 THE PROPOSED ANOMALY DETECTION METHOD

### 3.1 Nonparametric Bayesian models

In this section, following a brief explanation about the Dirichlet Process that is important aspect for understanding of nonparametric Bayesian models, a sticky HDP-HMM is referred which is applied Dirichlet Process to Hidden Markov Model.

**3.1.1 Dirichlet process:** The dirichlet process (DP) is called a distribution over a distribution or a countably infinite probability measure, uniquely defined by a base measure $H$ on a parameter space $\Theta$ and a concentration parameter $\alpha$. We denote it by $\mathrm{DP}(\alpha, H)$. DP is also regarded as a stochastic process in which a probability distribution called as a base measure is approximated by a random discrete distribution. A random draw $G_0 \sim \mathrm{DP}(H, \alpha)$ can be expressed as

$$G_0 = \sum_{k=1}^{\infty} \pi_k \delta(\theta_k), \quad \theta_k \sim H \tag{1}$$

where the $\delta(\theta_k)$ indicates a Dirac delta, and the weights $\pi_k$ are sampled via a stick-breaking construction, denoted below:

$$\begin{aligned} \nu_k &= \mathrm{Beta}(1, \alpha) \\ \pi_k &= \nu_k \prod_{l=1}^{k-1} (1 - \nu_l) \end{aligned} \tag{2}$$

Here, let them be denoted by $\pi = (\pi_1, \pi_2, ..., \pi_\infty) \sim \mathrm{GEM}(\alpha)$. The stick-breaking process can be interpreted as dividing a stick of 1 in length, which is a total sum of probability, by weights $\pi_k$ drawn from a Beta distribution. These infinitely divided sticks make an infinite multinomial distribution $\pi$. Additionally, $\delta(\theta_k)$ is called an atom and assigned to parameter drawn from the parameter space $\Theta$. This representation gives us an insight into how the parameter $\alpha$ controls the model complexity of $G_0$ in terms of the expected numbers of components.

The DP is commonly used as a prior on the parameters of a mixture model with a random number of components. Such a model is called a dirichlet process mixture model and its graphical model is depicted in Figure 3(a). Here, let $N$ be the number of the observations $y_i$'s and $z_i$ be an indicator for the assignment of observations to the parameters. An indicator $z_i$ is drawn from the countably infinite multinomial distribution $\pi$ constructed by the stick-breaking process $\mathrm{GEM}(\alpha)$, a parameter $\theta_k$ is assigned by the indicator $z_i$, and the observation $y_i$ is emitted from the likelihood distribution $F$ with the parameter $\theta_k$.

**3.1.2 Hierarchical dirichlet process:** The hierarchical dirichlet process (HDP) defines first a global base measure $G_0$, drawn by a DP prior $\mathrm{DP}(\gamma, H)$ as an average distributions of the group specific distributions $G_j$, which is sampled from a $\mathrm{DP}(\alpha, G_0)$. Since $G_0$ is discrete due to the characteristics of the DP, the group specific distributions $G_j$ will share the same atoms $\delta(\theta_k)$. This ensures that the mixture models in the different groups share mixture components.

With respect to the representation of the stick-braeking process, the HDP can be interpreted that the global parameter $\beta$ is sampled from $\mathrm{GEM}(\gamma)$, then the group specific distribution $\pi_j$ is sampled from the $\mathrm{DP}(\alpha, \beta)$, and finally the group specific indicator $z_{ji}$ and the $i$th observation are generated in the same way as the DP, as depicted in Figure 3(b).

**3.1.3 Hierarchical dirichlet process - hidden markov model:** The model that hierarchical dirichlet process applies to a hidden

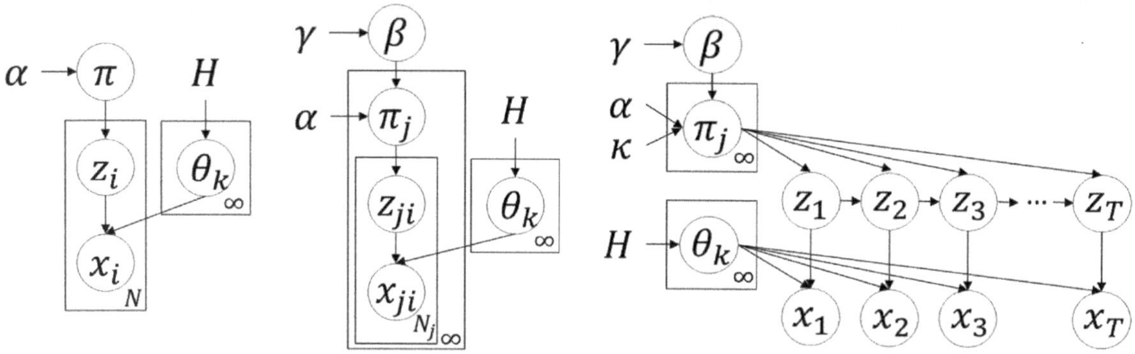


Figure 3: Dirichlet process mixture model(a), Hierarchical Dirichlet Process(b) and sticky HDP-HMM(c)

markov model is called hierarchical dirichlet process - hidden markov model (HDP-HMM) (Teh et al., 2006). In HDP-HMMs, the inditator $z_t$ acts as the hidden state at the time $t$, $\pi_j$ means the state specific transition distribution for state $j$, and $\beta$ is called as the global transition distribution. Due to the fact that each group specific ditribution $G_j$ shares the common components, each state specific distribution $\pi_j$ share the common countably infinite states. This also means the state transition matrix $\pi$ can be regarded to have raw and column of countably infinite.

The ability of HDP-HMMs to have countably infinite states theoretically make it possible to estimate the number of states according to input data, and has leads to some application such as clustering and anomaly detection, especially where the number of clusters or states is unknown (Lello et al., 2012).

An extension of the HDP-HMM, nemed a sticky HDP-HMM or sHDP-HMM, has been deveploed by Fox (Fox et al., 2009) to solve the issue which the standard HDP-HMM learns a model with unrealistically fast dynamics that causes the reduction of the model precision. The sHDP-HMM introduces the hyper-parameter $\kappa > 0$ that controls the expected probability of the self-transition to the stage where the state specific ditribution $\pi_j$ is sampled. The graphical representation of the sHDP-HMM is depicted in Figure 3(c). The resulting generative model is given by:

$$\begin{aligned} \beta|\gamma &\sim \text{GEM}(\gamma) \\ \theta_k|H &\sim H \\ \pi_j|\alpha,\beta,\kappa &\sim \text{DP}(\alpha+\kappa, \tfrac{\alpha\beta+\kappa\delta_j}{\alpha+\kappa}) \\ z_t|z_{t-1},\{\pi_j\}_{j=1}^{\infty} &\sim \pi_{z_{t-1}} \\ x_t|z_t,\{\theta_j\}_{j=1}^{\infty} &\sim F(x_t|\theta_{z_t}) \end{aligned} \quad (3)$$

In terms of the human dynamics monitoring, since it is considered that hidden states behind human dynamics gradually transit through time, setting the accurate expected probability of tself-transition can lead to the model with high predictive performance. Therefore, we consider that the adequate number of states behind human dynamics data can probably be estimated by using a sHDP-HMM for the an anomaly detection model.

### 3.2 THE PROPOSED ANOMALY DETECTION METHOD

The proposed method consists of the following three steps: It firstly learns hidden state that represent normal states sequence and the hyper-parameters from training data (step 1). Secondly, the sHDP-HMM with the learned hyper-parameters infers hidden states of test data (step 2). By comparing the two estimated states for each instant of time, an anomaly will finally be detected if they are unequal (step 3).

In the step 1, a "normal" training temporal data $\{x_t\}_{t=1}^{T}$ is set as the observation variables of a sHDP-HMM, where $T$ indicates the length of $d$ dimensional temporal data. We use $d$ dimensional Gaussian distribution $\text{N}(\mu_k, \Sigma_k)$ for the likelihood distribution of the $k$th hidden state, uniquely defined by parameter $\theta_k = (\mu_k, \Sigma_k)$. Also, a Gaussian distribution $\text{N}(\mu_0, \Sigma_0)$ and a Inverse-Wishart distribution $\text{IW}(\nu, \Delta)$ are used for the prior distribution of the mean $\mu_k$ and the variance-covariance matrix $\Sigma_k$ each. Since hyper-parameters are unknown, we give vague prior distributions and try infer them by sampling algorithm. For the sHDP-HMM explained above, the Blocked Gibbs Sampler is employed to infer all of the hyper-parameters and the parameters including the state sequence $\{z_t\}_{t=1}^{T}$.

The step 2 applies the inferred sHDP-HMM to training temporal data $\{x'_t\}_{t=1}^{T}$ and estimates its hidden state sequence $\{z'_t\}_{t=1}^{T}$ which may contain anomalies. With respect to the parameter setting, the step 2 uses the expected values which are calculated from posterior distributions of the hyper-parameters sampled the in the step 1. Moreover, the initialization of the parameters such as $\beta, \pi, \theta$ is given not by new sampled values, but by using the last ones in step 1.

Finally in the step 3, the normal state sequence $\{z_t\}_{t=1}^{T}$ and the test state sequence $\{z'_t\}_{t=1}^{T}$ are compared. To handle the problem that the indicators assigned to each state can be exchanged due to re-estimation of $\beta, \pi, \theta$, the proposed method employs the Hungarian Algorithm to solve assignment problem of which cost function is given by Hamming distance between the normal and test state sequences. After the state assignment, an anomaly at time $t$ is detected if $z_t \neq z'_t$.

Based on the characteristics of anomaly detection problem defined in section 2, this proposed method is categorized to a semi-supervised anomaly detection technique that uses normal time-series data and detects contextual anomalies with output of anomaly labels. This proposed method especially aims to detect contextual anomalies that are anomalous at specific time, but normal otherwise.

## 4 EXPERIMENTS AND RESULTS

In this section, we apply our proposed anomaly detection method to both synthetic and real data, and discuss the fundamental performance. In implementing the inference algorithm, we use a MATLAB toolbox supporting several inference algorithms based on Gibbs sampling for the sticky HDP-HMM, which has been made available by Fox (Fox et al., 2009).

### 4.1 Experiment with synthetic data

For an experiment with synthetic data, we generate 1 dimensional training and test time-series data with length of $T = 400$, depicted in Figure 5(a). We manually set twelve states represented by the numbers in Figure 5 and obtain i.i.d. samples from Gaussian distribution defined for each state. This synthetic temporal data imitates an average one-day pattern of a real gridded population data discussed later, by adjusting means and variances of Gaussian distributions. The test data, which is also sampled from the same distributions, is manually added anomalous values on 10% of the test data as follows; (i)$5\sigma$ anomaly at $t = 46$, (ii)$3\sigma$ anomalies at $t = 98 \sim 108$, (iii)$3\sigma$ anomalies at $t = 196 \sim 210$, (iv)$3\sigma$ anomalies at $t = 311 \sim 328$ and (v)$10\sigma$ anomaly at $t = 381$.

We used 1 dimensional Gaussian emission distribution, and placed priors $N(\mu_0, \Sigma_0)$ and $IW(\nu, \Delta)$ on the space of mean and variance parameters. The $\mu_0$ and $\Sigma_0$ are given by empirical mean and variance, degree of freedom $\nu$ is set as 4 and $\Delta = I$ where $I$ indicates the identity matrix. To realize a steady inference, we normalize the input data to make its mean be 0 and variance be 1.

The upper side of Figure 5(b) depicts the learned state sequence $\{z_t\}_{t=1}^{T}$ where green solid and dotted lines represent the estimated mean and standard variation of each state. The estimated state sequence is also visualized. Although we have initially set twelve states, the final 10,000th sampling gave six states because a couple of states were unified into one common state. Yet, significant differences between the real Gaussian parameters and the estimated parameters were not found.

After the inference of the test data in the step 2, hidden states and their Gaussian parameters were estimated as illustrated in the lower side of Figure 5(b). The number of the hidden states, which was six at the time of training, was estimated seven. Besides, our anomaly detection method regarded 64 points as anomaly pointed by red circles in the figure. Table 4 shows a confusion matrix in which a result of this anomaly detection experiment is listed. The fact that results scored 80.0% of the detection rate shows that our proposed method has good fundamental performance of anomaly detection for pseudo gridded population data with respect to the detection rate. On the other hand, the precision was 50.0% because a number of anomalies including false positives were often detected near borders of states.

Table 4: Confusion matrix of the result with the synthetic data

| | | Result of anomaly detection | |
|---|---|---|---|
| | | Anomaly 64 | Normal 336 |
| True | Anomaly 40 | True Positive 32 | False Negative 8 |
| | Normal 360 | False Positive 32 | True Negative 328 |

### 4.2 Experiment with real data

As a real gridded population data for this experiment, we use 500m-1hour resolved population of 30 October 2012 and 6 November 2012 in Tokyo, provided by ZENRIN DataCom Co., Ltd. Because a serious accident on the Tobu Isesaki Line has been recognized at about 6 am on 6 November to end up an hour-and-a-half service interruption between Asakusa Sta. and Kita-Senju Sta., we use the data of Octber 30 for training and the another for test. Especially, we target on a grid where Asakusa Sta. is located (Figure 6) because it was probably considerably impacted

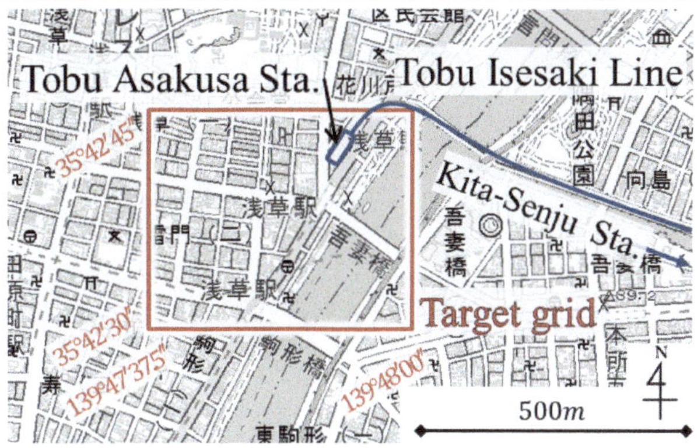


Figure 6: Location of the target grid

by the accident. We employ our proposed anomaly method with the normalized temporal data for observation variables and the same sHDP-HMM setting as before.

The upper side of Figure 7 illustrates a result of the inference of training data and shows that the number of normal states was estimated three, each of which stands for morning, afternoon and evening. After the test-data inference, new two states have been assigned for data in morning and evening, and then, two anomalies were detected as pointed by red circles in the bottom of Figure 7. This result shows that an anomaly was detected at the time when train services had been stopped. On the other hand, while the proposed method detected the another anomaly in evening, there is not recognized serious events such as traffic accidents or social events around the target area. An extensive analysis is required on the reason of this detection, whether it is result of latent anomaly or a false alarm.

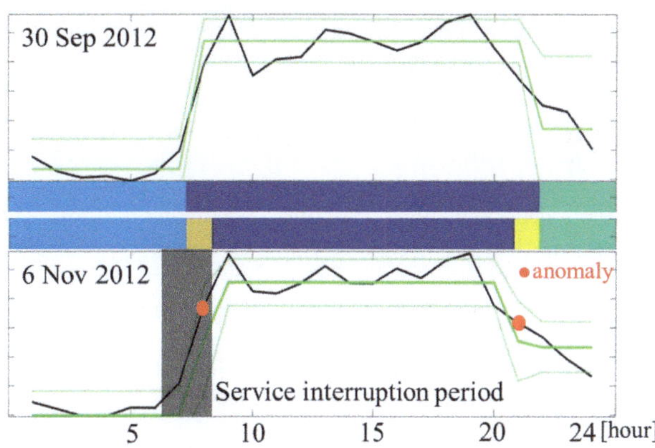


Figure 7: learning result of training data(upper), anomaly detection result(lower) and compared states(middle)

## 5 CONCLUSIONS

In this study, we firstly defined an anomaly detection problem of the human dynamics monitoring with respect to gridded population data. Based on the result of review we have comprehensively done, we discussed how the anomaly detection of human dynamics monitoring can be interpreted in terms of the four factors; nature of data, output of anomaly detection, data labels and type of anomalies. Additionally, we gave an explanation that state-space model, especially a HDP-HMM is considered possible to apply due to its property that a HDP-HMM can estimate the number of hidden states according to input data. Besides, we developed an anomaly detection method based on a sticky HDP-HMM, which

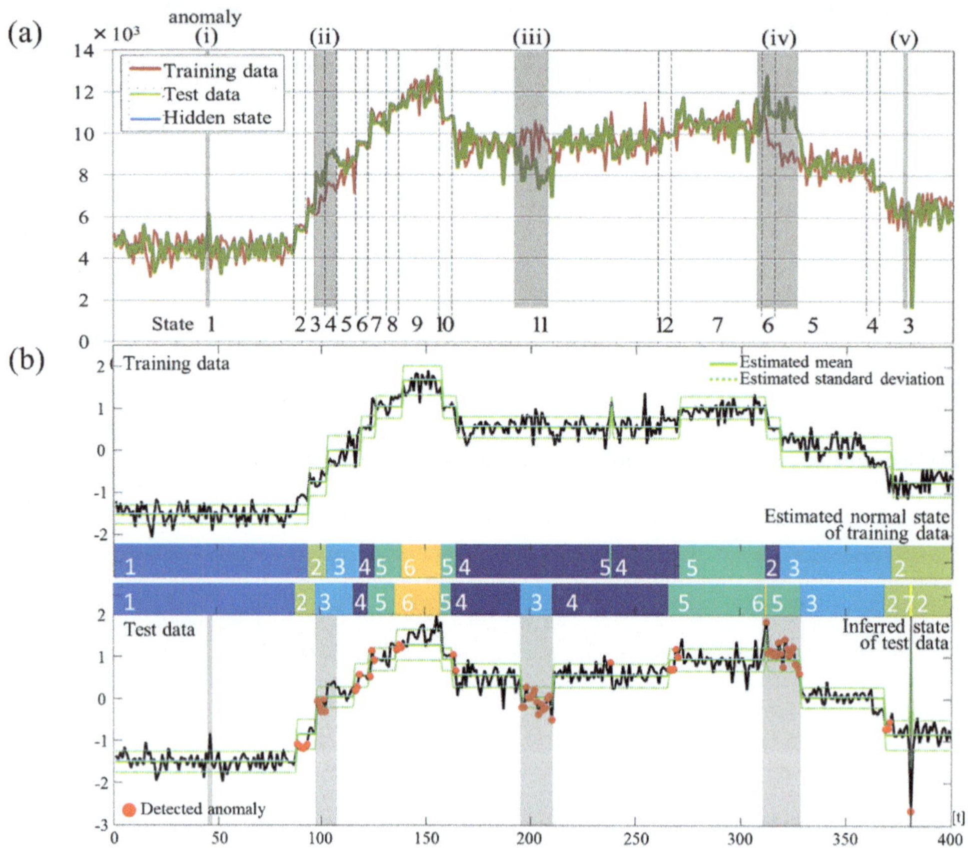


Figure 5: Synthetic training / test time-series data(a), inference and anomaly detection results(d)

is categorized to a semi-supervised anomaly detection technique that uses normal time-series data and detects contextual anomalies with output of anomaly labels. Results of the experiment with synthetic data showed that our proposed method has good fundamental performance with respect to the detection rate. Through the experiment with real gridded population data, an anomaly was detected at the location where a station was located and the time when train services had been stopped.

Future works include the improvement of the precision of the proposed method, the application of the proposed method to a bigger dataset of different gridded population data, and the deepening the interpretation of the estimated states and the detected anomalies on the traffic networks. The final goal of this research is to construct a dynamic traffic control model by integrating the spatial extension of the proposed method, the on-line anomaly detection method and the control theory.

## REFERENCES

Candia, J., González, M. C., Wang, P., Schoenharl, T., Madey, G. and Barabási, a. L., 2008. Uncovering individual and collective human dynamics from mobile phone records. Journal of Physics A: Mathematical and Theoretical 41(22), pp. 1–11.

Chandola, V., Banerjee, A. and Kumar, V., 2009. Anomaly detection: a survey. ACM Computing Surveys 41(3), pp. 15:1–15:58.

Fox, E. B., Sudderth, E. B., Jordan, M. I. and Willsky, A. S., 2009. Nonparametric bayesian learning of switching dynamical systems. Advances in Neural Information Processing Systems 21, pp. 457–464.

Hodge, V. J. and Austin, J., 2004. A survey of outlier detection methodologies. Artificial Intelligence Review 22(2), pp. 85–126.

Horanont, T., 2010. A study on urban mobility and dynamic population estimation by using aggregate mobile phone sources. CSIS Discussion Paper.

Horiguchi, R., Iijima, M., Kobayashi, M. and Hanabusa, H., 2013. Traffic anomaly detection for surface street networks with the mesh-wised traffic indices on macroscopic fundamental diagram. OPTIMUM 2013 International Symposium on Recent Advances in Transport Modelling pp. 1–7.

Lello, E. D., Laet, T. D. and Bruyninckx, H., 2012. Hierarchical dirichlet process hidden markov models for abnormality detection in robotic assembly. Workshop on Bayesian Nonparametric Models (BNPM) For Reliable Planning And Decision-Making Under Uncertainty pp. 1–8.

Pan, B., Zheng, Y., Wilkie, D. and Shahabi, C., 2013. Crowd sensing of traffic anomalies based on human mobility and social media. Proceedings of the 21st ACM SIGSPATIAL International Conference on Advances in Geographic Information Systems pp. 344–353.

Pimentel, M. A., Clifton, D. A., Clifton, L. and Tarassenko, L., 2014. A review of novelty detection. Signal Processing 99, pp. 215–249.

Teh, Y. W., Jordan, M. I., Beal, M. J. and Blei, D. M., 2006. Hierarchical dirichlet processes. Journal of the american statistical association 101(476), pp. 1–30.

7

# ESTIMATION OF CHLOROPHYLL CONCENTRATION IN MAIZE USING SPECTRAL REFLECTANCE

L. J. Martinez M [a] *, A. Ramos [a]

[a] Universidad Nacional de Colombia, Facultad de Ciencias Agrarias, Bogotá, Colombia (ljmartinezm, aramosg)@unal.edu.co

**KEYWORDS:** crops, spectral reflectance, chlorophyll content

**ABSTRACT:**

The objective of this research was to identify the relationship between chlorophyll contents and spectral measurements in a maize crop. Spectral measurements were taken and the chlorophyll content was determined in leaf samples in a field experiment with different N rates. Analysis of variance showed differences in the chlorophyll content and spectral indices for the treatments and strong correlations between some reflectance indices and chlorophyll content. Due to a relationship between the N and chlorophyll contents, the results proved to be important for making decisions related to the nitrogen supply for crops, which is crucial in agriculture

## 1. INTRODUCTION

Maize is an important crop in Colombia, the production is about 1,400,000 tons per year and imports almost reach 2,000,000 tons per year (Venegas, 2010). In Colombia, the application of fertilizers to crops is done, in many cases, without technical criteria; sometimes soil chemical analysis is used and, in only a few cases, a plant tissue analysis is carried out as a basis for nutritional diagnosis. Laboratory analyses require a lot of time to get the results and are expensive; traditional sampling schemes are not suitable for representing the spatial variability of the crop. Currently, it is important to have reliable and fast information on the condition of crops to make input applications in a timely manner and at appropriate amounts. Measurements of spectral responses are an important alternative for identifying plant status to make decisions on input applications and for monitoring the crop status.

Spectral measurements can be extracted from remote sensing images, photographs taken with cameras mounted on unmanned aerial vehicles and proximal sensors, and, therefore, are alternatives with a lot of potential for generating the information required for more efficient crop management.

Chlorophyll is a pigment that has a clear impact on the spectral responses of plants, mainly in the visible spectrum portion. N is a key element in chlorophyll, therefore is usually a high correlation between them (Schlemmer, Francis, Shanahan, & Schepers, 2005). Knowledge on the relationship between spectral reflectance and chlorophyll content is an important alternative for making decisions related to the nitrogen supply for crops, which is crucial in agriculture.

Nitrogen is one of the most important nutrients in agriculture due to the role it plays in growth and crop production and due to the amount that plants require. Nitrogen is a component of chlorophyll and, therefore, is essential for photosynthesis. It is also the basic element of plant and animal proteins, including the genetic material DNA and RNA, and is important in periods of rapid plant growth. In agriculture, nitrogen applications represent an important percentage of production costs; therefore, decisions about the application rate, time and source are key to agricultural competitiveness and sustainability. Currently, N and/or chlorophyll estimations are important to a range of applications, such as precision agriculture and the global carbon cycle.

Several researchers have evaluated the role of spectral measurements in crop diagnosis and monitoring. Recent studies have demonstrated the feasibility of estimate chlorophyll content from hyperspectral vegetation indices composed by the reflectance of specific bands (Schlemmer et al., 2005)(Peñuelas, Garbulsky, & Filella, 2011).The position of the inflexion point in the red edge region (680 to 780 nm) of the spectral reflectance signature, termed the red edge position (REP), is affected by biochemical and biophysical parameters and has been used as a means to estimate foliar chlorophyll and nitrogen content (Cho & Skidmore, 2006)(Curran, Dungan, Macler, & Plummer, 1991).

Remote sensing images have been evaluated in several studies as a source of data for crop mapping and monitoring. In the case of Colombia, elevated cloudiness is an important characteristic that prevents the use of these images in an efficient manner. Therefore, it is necessary to identify and develop new ways to take low-rise, low cost photographs that can be processed quickly. UAV cameras are important alternatives that can be tailored to the needs of the user, which requires knowledge of the spectral responses of different species and crop conditions in order to know the optimal spectral zone for determining the nutritional status of the plants.

To make efficient and reliable use of data from satellite and proximal sensors, it is necessary to know the spectral response of each crop at different growth stages as this depends on the biochemical and physiological characteristics of the plant. Then, indices can be developed that will later be used for identification and characterization of the crops. These indices can serve as indicators of stress, senescence, and disease in higher plants. Research is needed to assess these relationships and to predict leaf nitrogen and chlorophyll contents for a wide variety of plant species.

The objective of this study was to evaluate the relationship between the chlorophyll contents of maize crop with spectral measurements as a basis to improve the nutritional diagnosis of this crop in Colombia. The research included nitrogen measurements, crop yield evaluation, satellite images and local photographs processing, however, in this paper only the results of chlorophyll and spectral measurements are presented.

## 2. METHODS AND MATERIALS

### 2.1 Study areas

The study area for the maize (*Zea mays*) experiment was located in the Meta Department, on the eastern plains of Colombia, at the CORPOICA Taluma research Station, 4°22'24.39"N and 72°13'5.30"W. Soils are fine and medium-textured, nutrient-poor, and acid Typic Hapludox, with a high content of aluminum, which is toxic for several crops (Instituto Geográfico Agustin Codazzi, 1998) .The climate is an Am type according to Köppen's classification, with a short dry season in January and February. The average annual rainfall varies between 2500 and 2800mm throughout the area. The mean annual temperature is about 26°C. Cattle grazing is the traditional, principle land use, but now agriculture with rubber plantations, maize, rice and soybean is being promoted.

### 2.2 Data collection and analysis

An experiment, was carried out with randomized blocks with four repetitions. There were 5 N treatments, including 0 as the control and 50, 100, 150, and 200 kg of N ha $^{-1}$. Spectral measurements were taken at 36 days, 56 days and 108 days after sowing, with a leaf clip using a Fieldspec 4 spectroradiometer (ASD Inc.) and one measurement from the canopy. This radiometer measures in the range of 350-2500nm, with a sampling interval of 1.4 nm at the spectral range of 350 - 1050 nm, while it is 2 nm at the spectral range of 1050 - 2500 nm. The chlorophyll (a and b) was estimated in a laboratory, measuring the absorbance at 647 nm and 665nm for pooled extracts using Lichtenthaler equations (Lichtenthaler, 1987)

Several spectral vegetation indices were calculated (table 1) to identify the better indices that are related to the chlorophyll content. Analysis of variance (ANOVA procedure) was used to determine the effect of the treatments on the spectral measurements. Simple and multiple correlation and regression analyses were employed to identify the relationships between the studied variables. The relationships that were explored mainly concerned the chlorophyll content in the leaves in relation to spectral reflectance and vegetation indices.

**Table 1: Spectral vegetation indices used in the analysis**

| Index | Name | Formula | Source |
|---|---|---|---|
| REP-LE | Red edge position: linear extrapolation method | Linear extrapolation of straight lines on the far-red (680 to 700 nm) and NIR (725 to 760 nm) flanks of the first derivative reflectance spectrum | Cho and Skidmore (2006) |
| REP-LI | Red edge position: linear interpolation method | 700+40(Rre−R700)/(R740−R700)<br>Rre: (R670+R780)/2 | Guyot et al. (1988) |
| Ch-RE | Chlorophyll index red edge | (R750 - R700)/R700 | (Gitelson, Merzlyak, & Lichtenthaler, 1996) |
| REPIG | Red edge position: Inverted Gaussian fitting | RMAX-(RMAX-RMIN)*EXP(-((780(nm)-660(nm) )^2)/((2*30^2))) | (Pu, Gong, Biging, & Larrieu, 2003) |
| PRI | Photochemical reflectance index | (R531 - R570)/(R531 + R570) | (Peñuelas et al., 2011) |
| NDVI | normalized difference vegetation index | Measure with GreenSeeker (Trimble) | |
| REP-P | Polynomial fitting | $R(\lambda) = a_0 + \sum_{i=1}^{5} a_i \lambda^i$ | (Pu, Gong, Biging, & Larrieu, 2003) |

## 3. RESULTS

The figure 1 shows the leaf spectral signature at three growing stages for the treatments without fertilization. Although the general trend of the spectral responses were similar, differences were found mainly in the 850 nm to 2350 nm, where water content has more effect in reflectance and in the visible range, where the chlorophyll had more influence. Higher values of reflectance were found at the 108 days after sowing when the crop is near to senescence. In the visible range the maximum reflectance was found at 560nm in the three growing stages (figure 2). According to Liang (2004), in the visible part of the spectrum, absorption is due to pigments such as chlorophyll a+b and carotenoids, while the near-IR and middle-IR absorption by water is more important in the variation of reflectance.

Figures 3 and 4 show the spectral responses for the different rates of N in the maize at 108 and 56 days after sowing in part of the visible spectrum. The higher reflectance values were found in the control without N and dropped to the lower values with the 200 kh $^{-1}$ N rate.

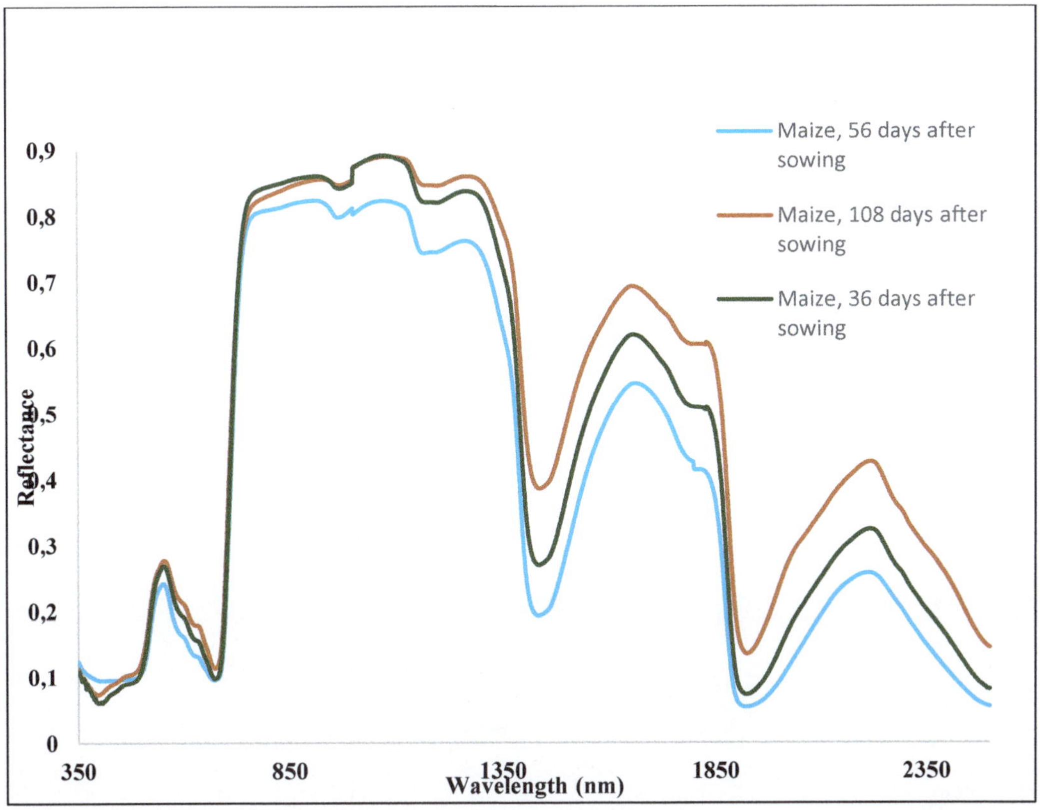


Figure 1: spectral reflectance for leafs in maize at different growing stag

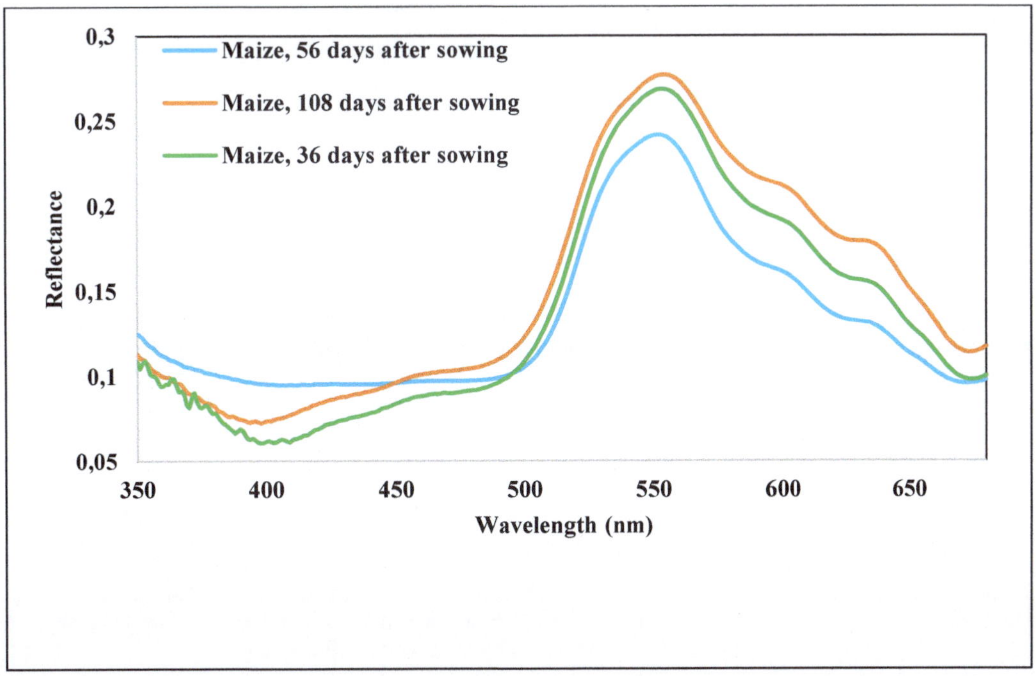


Figure 2: Spectral reflectance in the visible range for leafs in maize at different growing stages

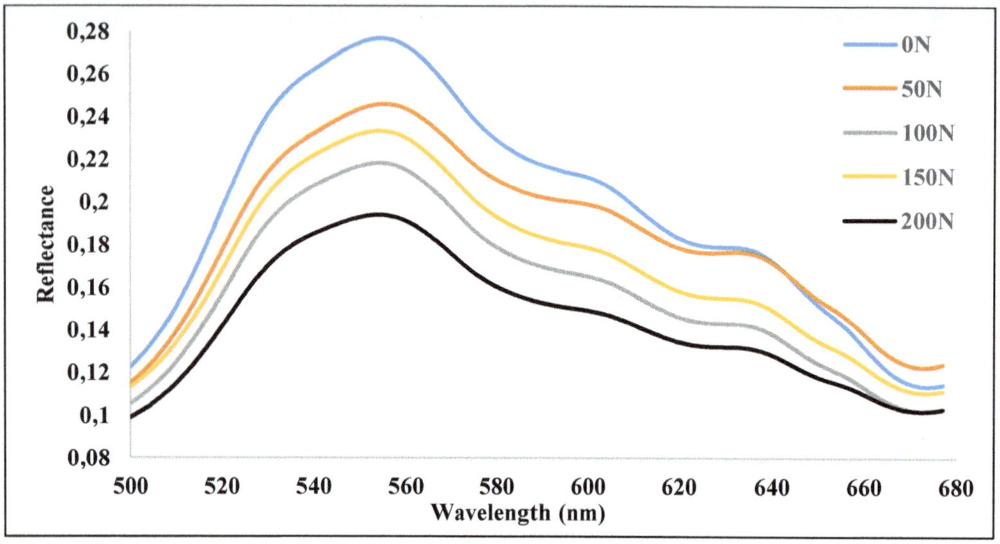


Figure 3: Spectral reflectance for maize, at 108 days after sowing, with different N rates.

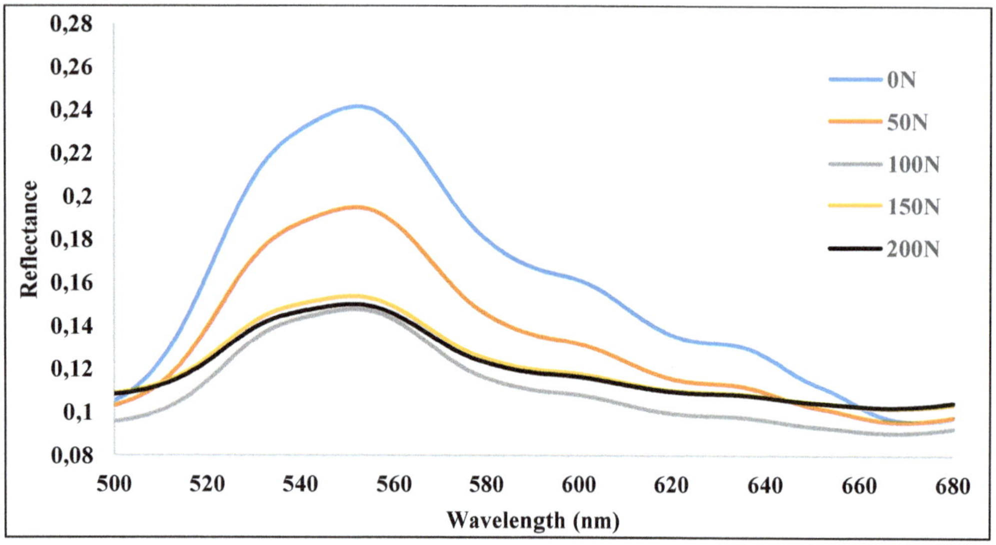


Figure 4: Spectral reflectance for maize, at 56 days after sowing, with different N rates.

Likewise, differences were found between the periods. The higher reflectance was found at 108 days after planting, which may have been related to the incidence of some brown pigments that increase when a plant is in senescence.

Tables 2, 3 and 4 show the Pearson correlation coefficients between the chlorophyll and some spectral vegetation indices for the maize at 36 and 56 days after sowing. The higher correlations in the first stage presented with the red edge position, as determined by the linear interpolation method (REP-LI) (Guyot, G., & Baret, 1988) for chlorophyll b, a and a+b. . The low correlation for NDVI in tables 2 and 3 is because this indice was not calculated but was measured with the Green Seeker sensor (Trimble) and is a canopy measurement. When compared NDVI with canopy reflectance the correlation with chlorophyll increases

The differences in spectral responses due to N rates are important for nutritional diagnosis of the crop as they are the basis for identifying nitrogen deficiencies in plants and making timely decisions for inputs application

In this way can be designed cameras to take images in the portion of the visible and infrared spectrum that most closely correlated to the chlorophyll content and generate information about the status of the crop. Similarly, when cloud conditions permit, remote sensing images can be used to have an approach to the general state of the crops

Table 2: correlation coefficients between chlorophyll and leaf spectral indices for maize, 36 days after sowing (*n=39)*

| | NDVI | Ch-RE | PRI | REP - L | REP -P | REPIG | REP-LE |
|---|---|---|---|---|---|---|---|
| Chlorophyll a | 0,40* | 0,57** | 0,49** | 0,97** | 0,93** | 0,93** | 0,54** |
| Chlorophyll b | 0,40* | 0,57** | 0,49** | 0,97** | 0,93** | 0,93** | 0,54** |
| Chlorophyll a+b | 0,44** | 0,60** | 0,52** | 0,97** | 0,92** | 0,92** | 0,55** |

* $p < 0.01$, ** $p < 0.001$

Table 3: correlation coefficients between chlorophyll and leaf spectral indices for maize, 56 days after sowing *(n=120)*

| | NDVI | Ch -RE | PRI | REP - L | REP- P | REPIG | REP-LE |
|---|---|---|---|---|---|---|---|
| Chlorophyll a | 0,12 | 0,45** | 0,32** | 0,91** | 0,86** | 0,86** | 0,93** |
| Chlorophyll b | 0,25** | 0,56** | 0,47** | 0,82** | 0,77** | 0,77** | 0,85** |
| Chlorophyll a+b | 0,22* | 0,55** | 0,45** | 0,87** | 0,82** | 0,82** | 0,89** |

* $p < 0.01$, ** $p < 0.001$

Table 4: correlation coefficients between chlorophyll and canopy spectral indices for maize, 56 days after sowing (n=120)

| | NDVI | Ch -RE | PRI | REP - L | Rre | REP-LE |
|---|---|---|---|---|---|---|
| Chlorophyll a | 0,81** | 0,70** | 0,47* | 0,85** | 0,36 | 0,80** |
| Chlorophyll b | 0,71** | 0,57** | 0,42 | 0,76** | 0,33 | 0,74** |
| Chlorophyll a+b | 0,75** | 0,61** | 0,43 | 0,80** | 0,33 | 0,77** |

* $p < 0.01$, ** $p < 0.001$

For 56 days (table 3), the better correlations were found with the REP calculated by the linear extrapolation method (Cho and Skidmore, 2006), also other indices such as REP-L, REP-P and REPIG showed good correlation indices. For the canopy measures, the better correlations with the chlorophyll were found with REP-LE, NDVI and REP-L.

Figures 5 and 6 represent the relationship between chlorophyll content and REP-LI on 56 days plants, and indicates a good estimation of the chlorophyll content. Other indices showed god results, although results are not presented here. Regression analysis between chlorophyll content and some indices indicates the potential of these indices to estimate the chlorophyll content in the states of crop growth when plants are still young and indirectly estimate the nitrogen content. This information is useful for correcting the deficiencies of this nutrient by applying fertilizer.

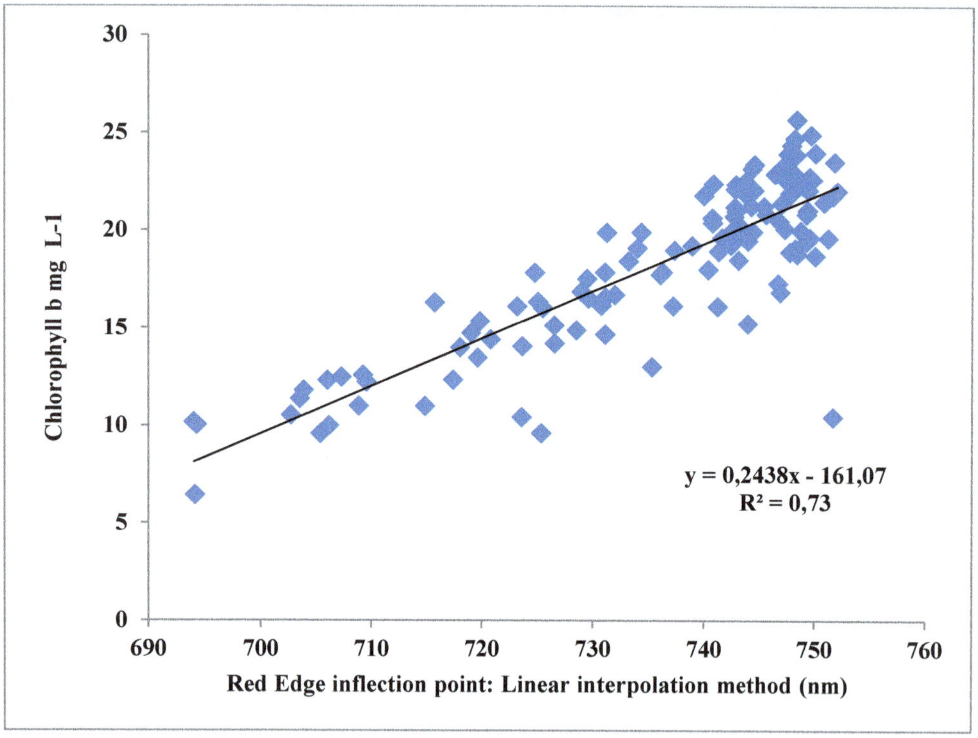


Figure 5: Regression analysis for chlorophyll b content and REP-LI at 56 days after sowing

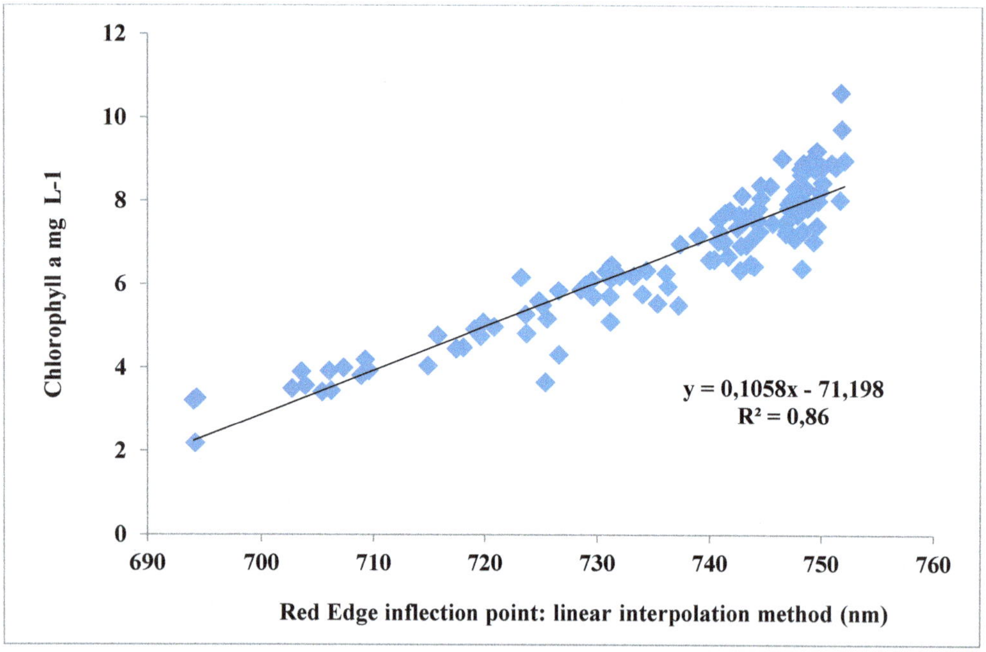


Figure 6: Regression analysis for chlorophyll a content and REP-LI at 56 days after sowing

## 4. CONCLUSIONS

The nitrogen rates applied had a significant effect on chlorophyll content of maize and on spectral responses. Various vegetation indices had a highly significant correlation with chlorophyll content, which makes them potentially suitable for estimating chlorophyll and indirectly identify N deficiencies in crops to help decisions on fertilizer application.

## 5. ACKNOWLEDGEMENTS

The authors thank the Research Division (DIB) of the Universidad Nacional de Colombia and Corporación Colombiana de Investigación Agropecuaria – CORPOICA for providing financial support for this research.

## 6-REFERENCES

Cho, M. A., & Skidmore, A. K. (2006). A new technique for extracting the red edge position from hyperspectral data: The linear extrapolation method. *Remote Sensing of Environment*, *101*(2), 181–193. doi:10.1016/j.rse.2005.12.011

Curran, P. J., Dungan, J. L., Macler, B. A., & Plummer, S. E. (1991). The effect of a red leaf pigment on the relationship between red edge and chlorophyll concentration. *Remote Sensing of Environment*, *35*(1), 69–76. doi:10.1016/0034-4257(91)90066-F

Gitelson, A. A., Merzlyak, M. N., & Lichtenthaler, H. K. (1996). Detection of Red Edge Position and Chlorophyll Content by Reflectance Measurements Near 700 nm. *Journal of Plant Physiology*.

Guyot, G., & Baret, F. (1988). Utilisation de la haute résolution spectrale pour suivre l'état des couverts végétaux. In *Proceedings of the 4th International colloquium on spectral signatures of objects in remote sensing* (p. 279–286).

Instituto Geográfico Agustin Codazzi. (1998). *Estudio general de suelos del departamento del Meta*. Memorias y mapas

Liang, S. (2004). *Quantitative Remote Sensing of Land Surfaces*. (Wiley & Sons, Ed.)*Wiley Series on Remote Sensing*. Wiley & Sons. doi:10.1002/047172372X

Lichtenthaler, H. K. (1987). [34] Chlorophylls and carotenoids: Pigments of photosynthetic biomembranes. *Methods in Enzymology*, *148*(C), 350–382.

Peñuelas, J., Garbulsky, M. F., & Filella, I. (2011). Photochemical reflectance index (PRI) and remote sensing of plant CO2 uptake. *New Phytologist*.

Pu, R., Gong, P., Biging, G. S., & Larrieu, M. R. (2003). Extraction of red edge optical parameters from hyperion data for estimation of forest leaf area index. *IEEE Transactions on Geoscience and Remote Sensing*.

Schlemmer, M. R., Francis, D. D., Shanahan, J. F., & Schepers, J. S. (2005). Remotely measuring chlorophyll content in corn leaves with differing nitrogen levels and relative water content. *Agronomy Journal*, *97*(1), 106–112.

Venegas, H. (2010). El Cultivo del maíz,historia e importancia. *El Cerealista*, *93*, 19.

# FACE RECOGNITION USING LOCAL QUANTIZED PATTERNS AND GABOR FILTERS

V. Khryashchev [a], A. Priorov [a], O. Stepanova [a], A. Nikitin [b]

[a] P.G. Demidov Yaroslavl State University, Yaroslavl, Russia – (vhr, andcat, olga1stepanova)@yandex.ru
[b] PicLab LLC, Yaroslavl, Russia – nikitin@piclab.ru

**Commission VI, WG VI/4**

**KEY WORDS:** Face Recognition, Eye Center Localization, Local Quantized Patterns, Gabor Filters, Histogram Comparing

**ABSTRACT:**

The problem of face recognition in a natural or artificial environment has received a great deal of researchers' attention over the last few years. A lot of methods for accurate face recognition have been proposed. Nevertheless, these methods often fail to accurately recognize the person in difficult scenarios, e.g. low resolution, low contrast, pose variations, etc. We therefore propose an approach for accurate and robust face recognition by using local quantized patterns and Gabor filters. The estimation of the eye centers is used as a preprocessing stage. The evaluation of our algorithm on different samples from a standardized FERET database shows that our method is invariant to the general variations of lighting, expression, occlusion and aging. The proposed approach allows about 20% correct recognition accuracy increase compared with the known face recognition algorithms from the OpenCV library. The additional use of Gabor filters can significantly improve the robustness to changes in lighting conditions.

## 1. INTRODUCTION

The problem of face recognition in a natural or artificial environment has received a great deal of researchers' attention over the last few years. Despite of many efforts devoted to this problem, we have to admit that it is still far from being resolved (Vizilter et. al., 2010).

Being one of biometric-based technologies face recognition are less reliable than other alternative biometric systems of identification. However, face recognition systems are widely used in different applications such as parental control system, interface interaction of humans and robots, sorting of photos and video (Szeliski, 2010).

Different methods are used for person identification using facial image. For example, multilayer neural network shows good results when the number of people is fixed while the task of searching of a particular person in a crowded place requires more complex methods including usage of different feature sets, classifiers, etc.

Despite the variety of different methods of face recognition, a typical algorithm consists of three main components (Vizilter et. al., 2010; Zhao et. al., 2003): (i) projection of face data to a lower dimensional representation; (ii) feature extraction; (iii) classification stage: neural network, metrics, cluster analysis, etc. In addition, a priori information (facial features characteristics) is used for algorithm constructing and experimental information is used for developing and correcting of algorithm.

Different approaches are used for face recognition in computer vision systems: using key facial points and measuring facial characteristics, exploiting geometrical information (Brunelli et. al., 1992), principal component analysis (Turk et. al., 1991), linear discriminant analysis (Belhumeur et. al., 1997), Gabor wavelets based methods (Wiskott et. al., 1997), algorithms using discrete cosine transform (Messer et. al., 2006).

In the recent years a number of approaches based on local binary patterns (LBP) has gained increasing attention from researches in the field of image processing (Ojala et. al., 1994; Ahonen et. al., 2004; Huang et. al., 2011; Petruk et. al., 2011).

The basic idea of the local binary patterns is to avoid image representation as a high-dimensional vector which contains a lot of redundant information, and describe the local structure of an image. Extracted features will have lower dimension.

The original LBP was introduced by T. Ojala in 1994. This theoretically very simple yet efficient multiresolution operator is defined for each pixel by thresholding the $3\times3$ neighborhood pixel value with the center pixel value. In this way, it can give us a binary sequence defined by local structure of an image.

LBP is computationally efficient because it works with integer arithmetic (it can achieve real-time performance in some tasks), and it is invariant to changes in the brightness of the image caused by shooting in different lighting conditions.

## 2. FACE RECOGNITION ALGORITHM

In this paper novel face recognition algorithm based on modified LBP is presented. The following steps are offered to apply this algorithm to the problem of face recognition (Fig. 1):

### 2.1 Normalization of face image using eyes coordinates

The detected face is rotated, scaled and cropped according to the known coordinates of eyes (determined using eye localization algorithm). The chosen resolution of the face image is 90x150 pixels, coordinates of the left eye center are (75, 55), coordinates of the right eye center are (15, 55).

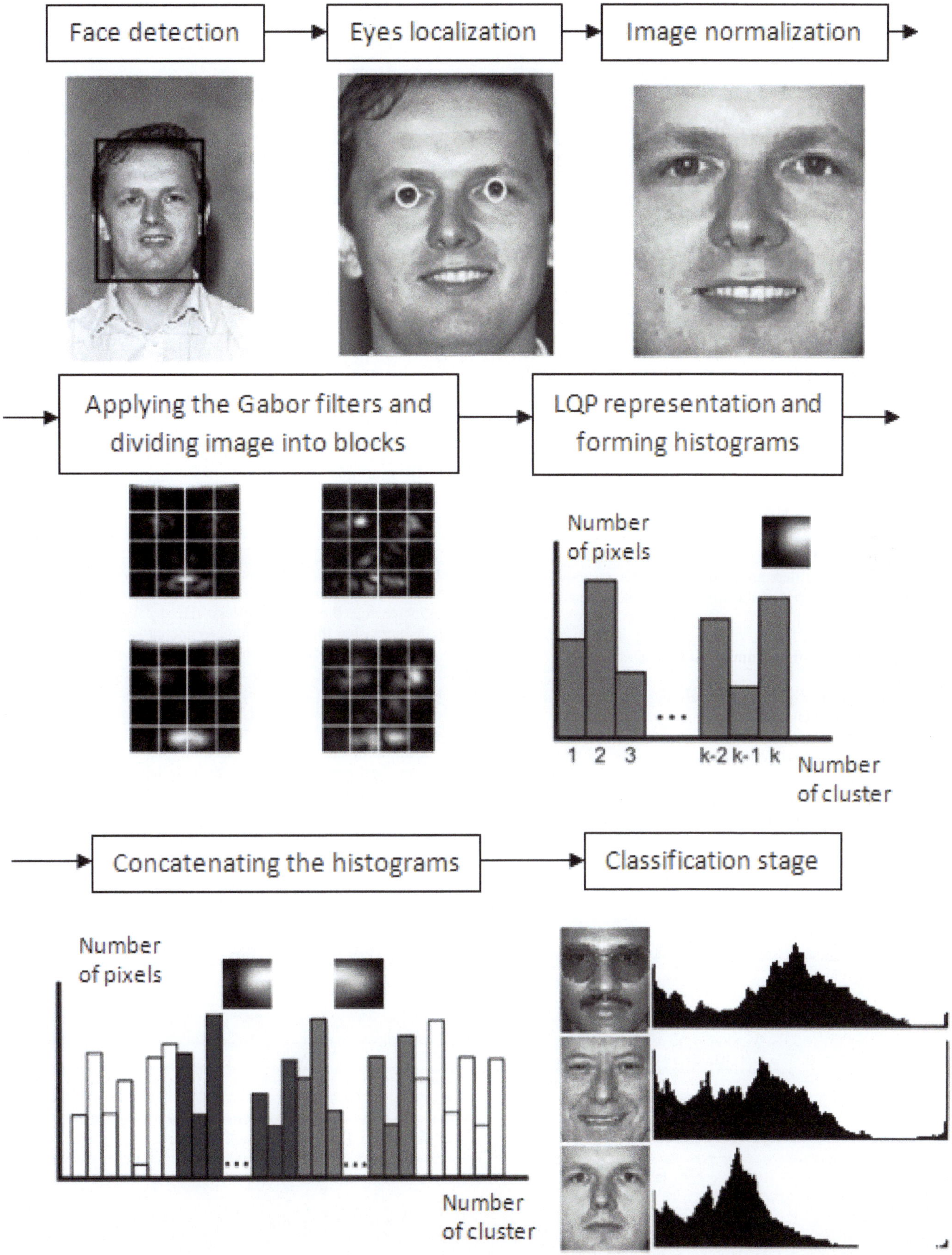


Figure 1. Block diagram of face recognition algorithm

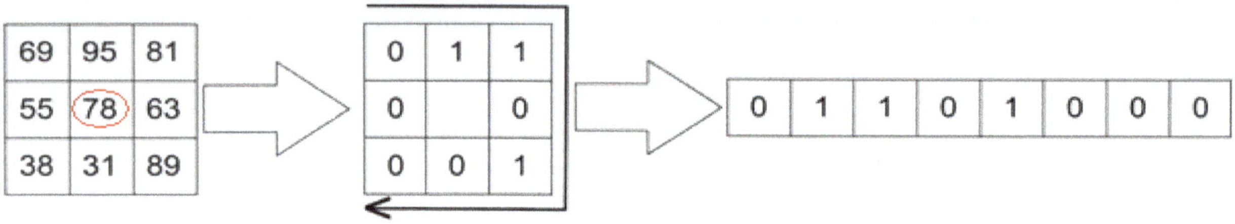


Figure 2. Example of counting of LBP

## 2.2. Gabor filters applying

40 Gabor filters which are determined by (1) are applied to normalized image:

$$\psi_{u,v} = \frac{\|k_{u,v}\|^2}{\sigma^2} e^{\left(-\frac{\|k_{u,v}\|^2\|z\|^2}{2\sigma^2}\right)} \left[e^{i\vec{k}_{u,v}z} - e^{-\frac{\sigma^2}{2}}\right], \quad (1)$$

where $k_{u,v} = k_v e^{i\phi_u}$, $k_v = \frac{k_{max}}{f^v}$ = the frequency

$\phi_u = \frac{u\pi}{8}$, $\phi_u \in (0, \pi]$ = the orientation

$z = (x, y)$.

It should be noted that variable $u$ in (1) determines the scale of Gabor filters; $v$ controls the Gabor filters orientation. The following parameters are chosen for Gabor filters: $\sigma = 2\pi, k_{max} = \frac{\pi}{2}, f = \sqrt{2},$ five scales $v \in \{0,1,2,3,4\}$ and eight orientations $v \in \{0,1,2,3,4,5,6,7\}$. These Gabor kernels form a bank of 40 different filters and exhibit desirable characteristics of spatial frequency, special locality, and orientation selectivity (Shan et. al., 2004).

## 2.3. Dividing image into a number of blocks

Each image is divided into $p$ non-overlapping rectangle blocks each of size $L \times L$ pixels. In this paper $p = 9 \times 15 = 135$.

## 2.4. Constructing a dictionary for each block

Describing the local structure of an image by LBP operator we avoid image representation as a high-dimensional vector which contains a lot of redundant information. Extracted features have lower dimension.

LBP operator is defined for each pixel by thresholding the $L \times L$ neighborhood pixel value with the center pixel value. An output value of the LBP operator can be obtained as follows (Ojala et. al., 1994):

$$LBP = \sum_{p=0}^{P-1} 2^p s(i_p - i_c), \quad (2)$$

where $i_c$ = the average intensity of the center pixel

$i_p$ = the intensity of its neighborhood pixel $p$

$s$ = sigma function:

$$s(x) = \begin{cases} 1, x \geq 0 \\ 0, x < 0 \end{cases}.$$

Basic LBP thresholds eight neighborhood pixel values with the center pixel value (see Fig. 2 for an example). In this way, it can give us a binary sequence defined by local structure of an image (Ahonen et. al., 2004).

Despite their encoding efficiency and simplicity, local binary pattern features remain very local and hence somewhat myopic, where increasing the neighborhood size by including more pixels or increasing the circle diameter increases the histogram size (number of codes) exponentially, e.g. increasing the number of surrounding pixels from 8 to 16 in LBP increases the number of codes from 256 to 65,536. Furthermore, these local patterns use hard-wired codings and fixed layouts, which are most probably not well adapted to the underlying dataset and/or the application domain. These shortcomings limit the encoding capacity of local patterns and prevent them from harnessing all the information available locally.

In this paper perspective approach described in (Hussain, Triggs, 2012; Hussain, Napoléon, Jurie, 2012) is offered to solve many of the above mentioned problems of local

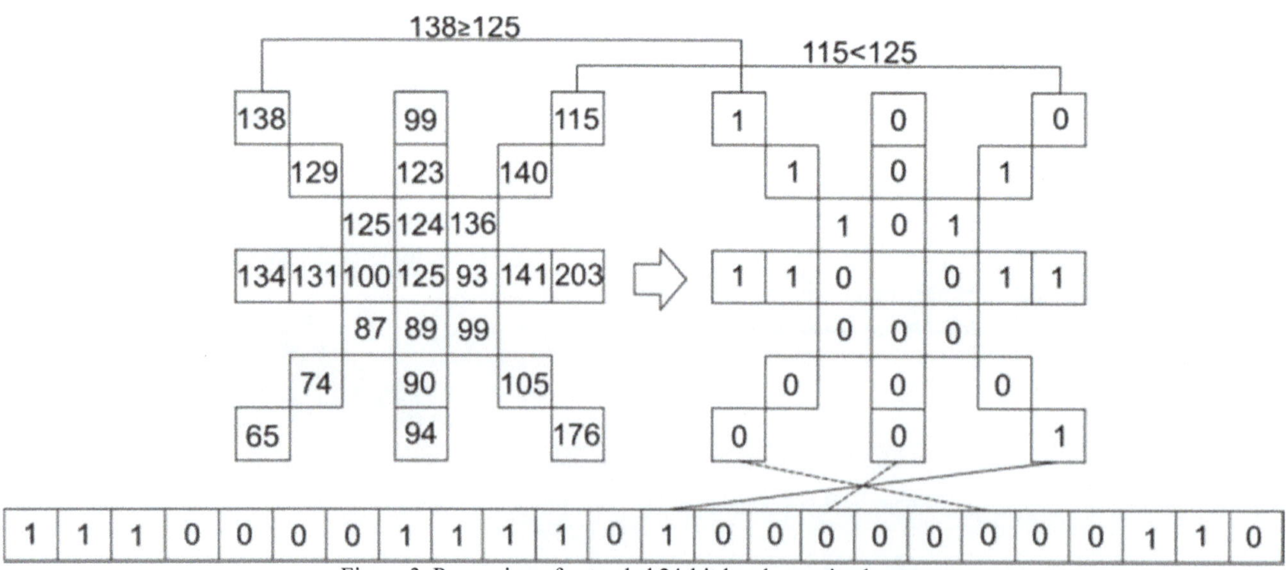


Figure 3. Processing of extended 24-bit local quantized pattern

patterns. To maintain the speed and simplicity of local pattern features and to make the process of vector quantization fast, local quantized patterns (LQP) are used. The basic idea of LQP is to extend processing region and to divide distinct codes into groups by clustering (Mandel, 1988). Groups of codes form new feature set of lower dimension.

Many different neighborhood geometries are possible for LQP and one of its main advantages is its flexibility in this respect. In this paper 24 pixel local pattern neighborhood is used. After standard thresholding this neighborhood with the center pixel value we get 24-bit binary code for each pixel of image (see Fig. 3).

These 24-bit codes form huge array of binary sequences. For example, 655,360 codes are formed from training database that consists of 10 images with resolution 256x256 pixels. To divide these codes into groups we should count the number of occurrences of each code over suitable image regions and take into account only codes that occur more than $N$ times. The resulting codes can be easily stored in a table. LQP builds these tables by mapping all the codes to the nearest cluster centers using K-Means, and thus guarantees fast vector quantization with no extra overhead at runtime. As a result codes that are divided into $k$ clusters form a dictionary. Example of dictionary for $k = 100$ is presented in table 1.

| 24-bit binary code | Number of cluster |
|---|---|
| 0000 0000 0000 0000 0000 0000 | 53 |
| 0000 0000 0000 0000 0000 0001 | 53 |
| .... .... .... .... .... .... | ... |
| 0110 0011 0100 1000 0110 0001 | 1 |
| 0110 0011 0100 1000 0110 0010 | 1 |
| 0110 0011 0100 1000 0110 0011 | 27 |
| .... .... .... .... .... .... | ... |
| 1010 0001 0110 1011 0111 0011 | 78 |
| 1010 0001 0110 1011 0111 0100 | 100 |
| 1010 0001 0110 1011 0111 0101 | 100 |
| .... .... .... .... .... .... | ... |
| 1111 1111 1111 1111 1111 1111 | 91 |

Table 1. Example of dictionary for $k = 100$

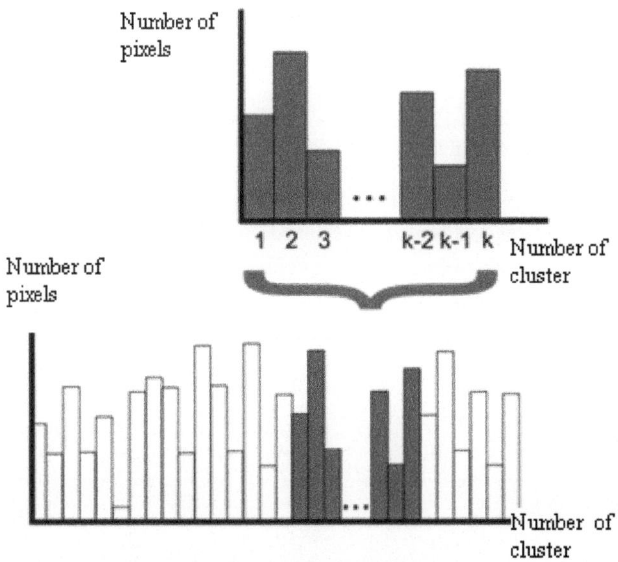


Figure.4. Concatenating the histograms

After the first Gabor filter is applied the histogram with $k$ bins and with height proportional to the number of pixels that fall into each cluster is formed for each block of every image from training dataset. The same procedure is repeated after applying the rest Gabor filters.

### 2.5. Concatenating the histograms

Histograms for all blocks of every Gabor image are concatenated (see Fig. 4). The resulting histogram may have quite big amount of bins. For example, histogram has 135x200x40 = 1 080 000 bins for $p = 9x15 = 135$ blocks and $k = 200$ clusters.

### 2.6. Classification stage

For recognition, histogram comparing is used to measure the similarity of different facial images and exploited for final classification. From the statistical hypothesis point of view, it cannot be said that any of the metrics would be the best one with a high (>0.95) probability.

However, histogram intersection and $\chi^2$ measures are clearly better than log-likelihood when the average number of labels per histogram bin is low but log-likelihood performs better when this number increases. The log-likelihood measure has been preferred for texture images (Wiskott et. al., 1997) but because of its poor performance on small windows in our experiments it is not appealing for face recognition. The $\chi^2$ measure performs slightly better than histogram intersection so we chose to use it despite the simplicity of the histogram intersection.

## 3. EVALUATION

To illustrate the performance of the proposed LQP-based algorithm, we conduct experiments on FERET database. It consists of 14,000 images with resolution 384x256 pixels. The FERET images are all taken indoors, with good resolution, image quality, and limited variations in lighting, size and position of faces in images. Pose of the faces in these images is typically very close to frontal.

The FERET database is subdivided into categories: The gallery images are 1196 Fa images captured during different photo sessions, with one image per person. The Fb probes consist of 1195 images with alternative facial expression; The Fc probes contain 194 images taken with a different camera and lighting; The duplicate I probes involve 722 images with corresponding gallery image taken from on a different day and in a different place, and thus they might contain "unite variation" due to illumination, facial expression, accessories, and aging, etc. The duplicate II probes, a subset of duplicate I, is considered the most challenging probe, since they contain 234 images from subjects whose gallery match was taken more than 18 months beforehand.

The experimental results for algorithm based on LQP and LQP-G (which uses Gabor filters) comparing with other popular algorithms from Open Source Computer Vision Library are shown in Table 2.

| Algorithm | Testing set | | | | |
|---|---|---|---|---|---|
| | fb | fc | dup-I | dup-II | Average |
| LQP (without using Gabor filters) | 98,6% | 45,4% | 61,6% | 50,4% | 64,0% |
| LQP-G (using Gabor filters) | 98,6% | 96,4% | 73,4% | 71,4% | 85,0% |
| LDA (OpenCV) | 76,1% | 9,8% | 36,3% | 19,2% | 35,4% |
| PCA (OpenCV) | 75,8% | 6,7% | 34,1% | 15,4% | 33,0% |
| LBP (OpenCV) | 76,7% | 21,1% | 38,1% | 20,9% | 39,2% |

Table 2. The experimental results on FERET database.

The results lead to the following conclusions:

- The proposed algorithm LQP-G allows about 20% correct recognition accuracy increase compared with the known face recognition algorithms from the OpenCV library. Medium rate of recognition accuracy is 85% (98,6% on FB dataset) when number of classes in training is 1196.
- The additional use of Gabor filters can significantly improve the robustness to changes in lighting.
- Using Gabor filters allows processing the most challenging dataset dup-II. Standard algorithms show poor recognition accuracy on this dataset.

## 4. CONCLUSION

This paper proposes novel LQP-based algorithm. The effectiveness of the proposed method comes from several aspects including the multi-resolution and multi-orientation Gabor decomposition, the local quantized patterns, and the local spatial histogram modeling. Experimental evaluations of the proposed approach on the FERET face database have evidently illustrated the effectiveness and robustness of algorithm to the general variations of lighting, expression, occlusion and aging.

## ACKNOWLEDGEMENTS

This work was supported by project #1060 in the framework of the base part of the state of R&D jobs of P.G. Demidov Yaroslavl State University and RFBR grants (№ 15-07-08674 and № 15-08-99639).

## REFERENCES

Ahonen T., Hadid A., Pietikäinen M., 2004. Face recognition with local binary patterns. In: *European Conference on Computer Vision*, pp. 469-481.

Belhumeur, P.N., Hespanha, J., Kriegman, D., 1997. Eigenfaces vs. fisherfaces: recognition using class specific linear projection In: *IEEE Transactions on Pattern Analysis and Machine Intelligence 19*. pp. 711–720. (ECCV).

Brunelli, R., Poggio, T., 1992. Face recognition through geometrical features. In: *European Conference on Computer Vision (ECCV)*. pp. 792–800.

Huang, D., Shan, C., Ardabilian, M., Wang, Y., Chen, L., 2011. Local binary patterns and its application to facial image analysis: a survey. In: *IEEE transactions on systems, man, and cybernetics* – part C: applications and reviews. V. 41, no. 6. pp. 765–781.

Hussain, S., Triggs, B., 2012. Visual recognition using local quantized patterns. In: *European Conference on Computer Vision (ECCV)*. pp. 716–729.

Hussain, S., Napoléon, T., Jurie, F., 2012. Face recognition using local quantized patterns. In: *British Machine Vision Conference*, pp. 1–11.

MacQueen, J.B., 1967. Some methods for classification and analysis of multivariate observations. In: *5th Berkeley Symposium on mathematical statistics and probability*. University of California Press, 1967. pp. 281–297.

Mandel, I.D., 1988. *Cluster analysis*. M.: Finance and Statistics,. p. 176.

Messer, K., Kittler, J., Short, J., 2006. Performance characterization of face recognition algorithms and their sensitivity to severe illumination changes. In: *International Conference on Biometrics (ICB)*. pp. 1–11.

Nikitin, A.E., Khryashchev, V.V., Priorov, A.L., Matveev, D.V., 2014. Development and analysis of face recognition algorithm based on local quantized patterns. In: *Nonlinear world*. № 8. pp. 35-42.

Petruk, V.I., Samorodov, A.V., Spiridonov, I.N., 2011. Application of Local binary patterns to solving the problem of face recognition. In: *Bulletin of the Moscow Bauman State Technical University*. Instrument. 2011. Special. vol. Biometric technology. pp. 58-63.

Shan, S.G, Gao, W., Chang, Y.Z., Cao, B., Yang, P., 2004. Review the strength of Gabor features for face recognition from the angle of its robustness to mis-alignment. In: *International Conference on Pattern Recognition (ICPR)*. pp. 338–341.

Szeliski, R., 2010. *Computer Vision: Algorithms and Applications*. Springer.

Turk, M., Pentland, A., 1991. Eigenfaces for recognition. *Journal of Cognitive Neuroscience*, no. 3. pp. 71–86.

Vizilter, Y.V., Zheltov, S.Y., Bondarenko, A.V., Ososkov, M.V., Morzhin, A.V., 2010. *Image processing and analysis tasks in machine vision: Lectures and practical classes*. M .: Fizmatkniga.

Wiskott, L., Fellous, J., Krüger, N., Malsburg, C., 1997. Face recognition by elastic bunch graph matching. In: *IEEE Transactions on Pattern Analysis and Machine Intelligence* 19. pp. 775–779.

Zhao, W., Chellappa, R., Phillips, P., Rosenfeld, A., 2003. Face recognition: A literature survey. In: *ACM Computing Surveys (CSUR)*, V. 35. № 4. pp. 399-458.

# THE TASSELED CAP TRANSFORMATION FOR RAPIDEYE DATA AND THE ESTIMATION OF VITAL AND SENESCENT CROP PARAMETERS

M. Schönert [a], E. Zillmann [a], H. Weichelt [a], J. U. H. Eitel [b], T. S. Magney [b], H. Lilienthal [c], B. Siegmann [d], T. Jarmer [d]

[a] BlackBridge, Dept. of Application Research, 10719 Berlin, Germany - (maurice.schoenert, erik.zillmann, horst.weichelt)@blackbridge.com

[b] Geospatial Laboratory for Environmental Dynamics, University of Idaho, Moscow, ID 83844-1135, USA (jeitel, tmagney)@uidaho.edu

[c] Julius-Kühn-Institut (JKI), Federal Research Centre for Cultivated Plants, 38116 Braunschweig, Germany – (holger.lilienthal@jki.bund.de)

[d] Institute for Geoinformatics and Remote Sensing, University of Osnabrueck, 49076 Osnabrueck, Germany – (tjarmer@igf.uni-osnabrueck.de)

**KEY WORDS: Tasseled Cap, RapidEye, crop biophysical parameters, Greenness, Yellowness, NDVI, PSRI, LAI, chlorophyll, nitrogen**

**ABSTRACT:**

The retrieval of crop biophysical parameters using spectral indices obtained from high temporal and spatial resolution satellite data, is a valuable tool to monitor crop growth and status. Tasseled Cap Features (TCFs) for RapidEye data were derived from spectral variances typically present in agricultural scenes. The TCF Greenness (GRE) was aligned to the spectral variance of vital vegetation, and therefore, it represents the typical reflectance characteristics of green vegetation, with relatively higher reflectance at the near-infrared (NIR) range. The TCF Yellowness (YEL) was aligned to correspond to the reflectance characteristics of senescent crops, with relatively higher reflectance in the visible portion of the spectrum due to chlorophyll breakdown, and lower reflectance in the NIR range due to cell structure decomposition compared to vital green vegetation. The goal of this work was to assess the potential of RapidEye's TCFs for the prediction of green leaf area index (LAI), plant chlorophyll (Chl), and nitrogen (N) concentration, as well as the identification of senescence patterns. The linear relationships between the biophysical parameters and the TCFs were compared to the performance of the widely used indices NDVI and PSRI. Preliminary results indicate that GRE is strongly related to LAI in vital crops and suggests a higher predictive power than NDVI. YEL demonstrated a strong linear relation and a higher potential to estimate Chl and N concentration in senescent soft white winter wheat (*Triticum aestivum L.*) in comparison to PSRI. PSRI showed a stronger correlation to Chl in senescent soft white spring wheat (*Triticum aestivum L.*), compared to YEL. Results indicate that YEL may be used to characterize the variability in senescence status within fields. This information, in conjunction with soil fertility and yield maps, can potentially be used to designate precision management zones.

## 1. INTRODUCTION

The high temporal (< 5 days) and spatial resolution (5 m) of RapidEye satellite imagery allows for the assessment of in-field variability of crop biophysical characteristics, and therefore is a valuable tool for site specific management to increase the productivity and quality of crops.

Leaf area index (LAI), the ratio of one-sided green leaf area to ground area, is a key variable for estimating foliage cover, biomass, and yield in agricultural systems (Gitelson et al. 2003; Haboudane et al. 2004). LAI is a valuable parameter to assess the temporal and spatial variability of crop status and thereby can support farmer's management decisions during the growing season (Wu et al. 2007; Kross et al. 2015).

Information on the chlorophyll (Chl) and/or nitrogen (N) concentration of wheat during senescence can be useful to predict plant water status and dry-down rates. Mapping variations of senescence (crop drying patterns) within a field can potentially provide additional information to designate precision management zones, e.g. to highlight areas that are more or less sensitive to abiotic conditions, which could ultimately drive spatial variations in grain quality and quantity. Recent studies pointed out that site specific harvesting bears potential to increase wheat quality and net return rates in quality incentive markets (Morari et al. 2013; Diacono et al. 2012; Meyer-Aurich et al. 2008; Wright et al. 2004; Freeman et al. 2003). In accordance with previous studies, Diacono et al. (2012) reported that climatic conditions, especially available soil water, play a major role in affecting the spatial patterns of yield as well as other quality parameters of wheat crops in rain fed agricultural systems (Buttafuoco et al. 2010; Ibrahim & Huggins 2011; Basso et al. 2007). Practically, the longer the duration and slower the rate of senescence, the greater is the probability for higher yields (Herbek & Lee 2009). Further, remotely estimating canopy Chl and/or N at these late stages of plant growth could provide information on salvageable N, N use efficiency, N status, and rate of senescence, and might be useful to indicate spatial patterns related to yield or wheat quality conditions.

Senescence is the final stage of annual plant development and it is characterized by well-orchestrated degradation and remobilization processes. Leaf N and other soluble constituents are mobilized from the degrading photosynthetic apparatus (Chl breakdown) to the seeds after anthesis (Hörtensteiner & Kräutler 2010; Bieker & Zentgraf 2013; Li-Hong et al. 2007). N retained from wheat leaves, in particular from the flag leaf and the upper canopy, accounts for up to 90% of the total grain N content (Kichey et al. 2007; Wang et al. 2005). Senescence is triggered by endogenous signals, e.g. plant growth regulators and hormone levels, but also by exogenous factors, such as drought, waterlogging, high soil salinity, mineral nutrition deficiency, high or low solar radiation, and extreme temperatures (Thomas & Stoddart 1980; Smart 1994; Edwards et al. 2012).

The optical properties of vital crop canopies in the spectrum of 400 to 1200 nm are mainly governed by photosynthetic active pigments (e.g. Chl), and leaf and canopy structure (Gates et al. 1965; Knipling 1970). Green vegetation typically reflects less radiation in the visual spectrum, due to the absorbance characteristics of pigments, while in the NIR portion of the spectrum, relatively high amounts of radiation are reflected due to light scattering induced by cell structure, lack of absorbing features, and light transmission feedback of lower leaf layers (Mestre 1935; Knipling 1970; Penuelas & Filella 1998).

With the onset and progress of senescence, Chl is degraded much faster than carotenoids, which causes reflectance to increase in the red spectrum, while in the blue spectrum only a small increase of reflectance is observed (Knipling 1970; Penuelas et al. 1995; Wilson et al. 2014; Sanger 1971; Verdebout et al. 1994; Gitelson et al. 2002; Zur et al. 2000). With the advancement of senescence, degradation processes form metabolites (polyphenols, denatured proteins, "brown pigments"), which visually evoke a brownish color and which were reported to absorb light in the visual and red-edge/NIR region (400-800 nm) (Verdebout et al. 1994; Penuelas & Filella 1998). The decrease of NIR reflectance during senescence is induced by the breakdown of cell structures, which decreases light scattering properties, e.g. mesophyll and intercellular space, and furthermore reduces the feedback of transmitting radiation in the subsequent canopy layers (Knipling 1970).

The Normalized Difference Vegetation Index (NDVI) is based on the above mentioned distinctive reflectance characteristics of plants in the red and near-infrared (NIR) region of the spectrum, thereby showing strong correlations to the amount of photosynthetically active vegetation and biomass (Tucker 1979). NDVI has been shown to be a good predictor of LAI (Viña et al. 2011; Gitelson et al. 2003; Colombo et al. 2003). However, it is known that NDVI saturates at moderate to high LAI values (ca. >2-3), exhibiting asymptotic relationships between these two variables (Viña et al. 2011, Hatfield et al. 2008; Serrano et al. 2000).

Merzlyak et al. (1999) developed the Plant Senescence Reflectance Index (PSRI), which was found to be responsive to changes in the carotenoid-chlorophyll ratio of senescing leaves, and therefore was proposed to be used to estimate the onset, the stage, and the relative rates and kinetics of senescence or ripening processes. Hatfield et al. (2010) found PSRI very suitable to detect differences in plant senescence across fields. Further, by comparing the change rate of the canopy from green to senescent, PSRI can be used to determine the rate of senescence in different crops.

An alternative approach to the use of indices for deriving biophysical properties of surface features from satellite imagery is the Tasseled Cap Transformation (Kauth & Thomas 1976). The Tasseled Cap Transformation assigns fixed weights to the original reflectance values of the individual spectral bands to generate transformed features, which can directly be associated with biophysical properties of surface features, e.g. the brightness or moisture variation of surface objects. TCFs were developed for several sensors, such as Landsat (Kauth & Thomas 1976; Crist & Cicone 1984; Huang et al. 2002; Baig et al. 2014), SPOT (Ivits et al. 2008), IKONOS (Horne 2003) and RapidEye (Schönert et al. 2014).

The TCFs for RapidEye data were derived from spectral variances typically present in agricultural scenes (Schönert et al. 2014). The first TCF, Brightness, was aligned to correspond with the variance of soil brightness. The second TCF, Greenness (GRE), corresponds to the direction of the spectral variation of vital vegetation, and therefore, represents the typical reflectance characteristics of green vegetation, with relatively higher reflectance at the NIR range. Yellowness (YEL), the third TCF, shows a contrast between the Blue + Green bands, and the Red + Red Edge + NIR bands. YEL corresponds to the reflectance characteristics of senescent crops, which generally show higher reflectance at the visual spectrum due to chlorophyll breakdown, and lower reflectance at the NIR range due to cell structure decomposition.

The aim of this work was first, to assess the suitability of RapidEye's TCFs GRE and YEL for estimating the biophysical parameters LAI, Chl, and N of agricultural crops; and second, to evaluate the performance of TCFs in comparison with the commonly used indices PSRI and NDVI.

## 2. MATERIALS AND METHODS

### 2.1 Study Area

The analysis was carried out in two study areas. The first study area, located in the federal state of Saxony-Anhalt, Germany (11°54′E, 51°47′N), is intensively used for agriculture due to the predominance of Chernozem soils. Within the study area, one winter wheat (*Triticum aestivum L*) field with a size of 90 ha was selected. The second study area was in the Palouse region of Northern Idaho (46°30′N, 116°49′W) and Eastern Washington (46°46′N, 117°05′W; 46°47′N, 117°26′W) USA. These study sites are managed as no-till and have silt load soils developed in loess deposits with some ash. All three USA farms are completely rain fed and have precipitation ranging from 330-610 mm annually (Soil Survey Staff 2014).

Table 1. Ground reference data of leaf area index (LAI), chlorophyll (SPAD) and nitrogen (N) concentration of winter (WW) and summer wheat (SW) and corresponding RapidEye imagery. Site-ID is coded according to country, field and date, e.g. G1.1 means: Germany, field 1, 1st date; U=USA

| Site-ID | Country | Crop | Sampling Date | No. of Samples | Min. / max. of | | | Tile-ID | Image Date |
|---|---|---|---|---|---|---|---|---|---|
| | | | | | LAI | SPAD | N | | |
| G1.1 | Germany | WW | 8th May '11 | 24 | 0.50/3.32 | 32.0/47.6 | | 3363006 | 7th May '11 |
| G1.2 | Germany | WW | 22nd June '11 | 18 | 0.53/4.67 | 31.8/52.4 | | 3363006 | 27th June '11 |
| U1.1 | USA | WW | 18th July '12 | 12 | 2.05/3.34 | 7.1/55.6 | 0.86/3.09 | 1160613 | 13th July '12 |
| U1.2 | USA | WW | 31st July '12 | 12 | 1.74/3.65 | 0.5/20.8 | | 1160613 | 3rd Aug. '12 |
| U2.1 | USA | SW | 18th July '12 | 12 | 2.13/4.34 | 23.2/52.6 | 1.96/3.79 | 1160515 | 15th July '12 |
| U2.2 | USA | SW | 24th July '12 | 12 | 1.80/4.36 | 4.9/42.3 | | 1160515 | 4th Aug. '12 |
| U3.1 | USA | WW | 13th July '12 | 16 | 1.90/4.98 | 35.7/54.6 | 2.65/4.15 | 1160614 | 13th July '12 |
| U3.2 | USA | WW | 25th July '12 | 16 | 1.09/4.06 | 6.8/38.4 | | 1160614 | 31st July '12 |

### 2.2 Satellite Imagery

The RapidEye satellite system is a constellation of five identical earth observation satellites with the capability to provide large area, multi-spectral images with frequent revisits in high resolution (6.5 m at nadir). In addition to the Blue (B) (440-510 nm), Green (G) (520-590 nm), Red (R) (630-685 nm) and near-infrared (NIR) (760-850 nm) bands, the sensor has a Red-edge (RE) (690-730 nm) band, especially suitable for vegetation analysis (s. Eitel et al. 2007). The RapidEye level 3A standard product covers an area of 25x25 km, is radiometrically calibrated to radiance values (Anderson et al. 2013), and is ortho-rectified and resampled to 5 m spatial resolution. All the images used in this study (Tab. 1) were calibrated to top of atmosphere reflectance. Due to atmospheric influences the Blue band of the scene corresponding to site-ID G1.1 was noisy and therefore treated with a noise removal based on the Maximum Noise Fraction approach.

### 2.3 Field Measurements and Data Extraction

In Germany, field data collection included leaf area index (Licor LAI-2200©, Delta-T Sunscan©) and leaf chlorophyll meter readings at the upper canopy (Minolta SPAD-502©). SPAD measurements represent unit-less relative values, which have been proven to be positively correlated to chlorophyll content of wheat (Reeves et al. 1993) and other crops (Zhu et al. 2012). Ground reference data were gathered at winter wheat stem elongation (8th May) and early ripening stages (22nd June). Sample locations were defined aiming at covering the entire crop variability within the field as described in Siegmann et al. (2013) (Fig. 1). Due to extreme dry weather conditions in early summer 2011, some areas of the field were affected by premature ripening (Fig. 1).

Figure 1. Test site with the sampling locations measured on 8th of May (dark points) and 27th of June 2011 (light points). The background image from the 27th of June shows the early ripening areas in brownish tones.

In the USA, LAI and SPAD measurements were collected similarly and sample dates nearest to image acquisition were chosen. The average of twelve SPAD values from fully-expanded upper leaves and four LAI measurements within a 1 $m^2$ location at each ground point were used in analysis. Leaf nitrogen concentration (%N) was also averaged from twelve top of canopy leaf samples processed by dry combustion. Due to the highly heterogeneous nature of the hilly Palouse, wide differences in SPAD, %N and LAI were observed during these time periods (Tab. 1).

All sampling points were buffered with a 10-meter-radius circle, and the image reflectance values extracted from these areas were used to calculate mean GRE, YEL, PSRI, and NDVI values for each point.

### 2.4 Spectral Indices

GRE, YEL (Schönert et al. 2014), NDVI (Rouse et al. 1973), and PSRI (Merzlyak et al. 1999) were calculated by

$$GRE = -0.2216 \times B - 0.2319 \times G - 0.4622 \times R - 0.2154 \times RE + 0.7981 \times NIR \quad (1)$$

$$YEL = -0.7564 \times B - 0.3916 \times G + 0.5049 \times R + 0.14 \times RE + 0.0064 \times NIR \quad (2)$$

$$NDVI = \frac{NIR-R}{NIR+R} \quad (3)$$

$$PSRI = \frac{R-G}{NIR} \quad (4)$$

where B, G, R, RE and NIR represent the values of Blue, Green, Red, Red-edge, and NIR bands respectively.

NDVI and GRE were compared to one another regarding their relationship to measured LAI, Chl, and N in vital wheat (site G1 and U3.1). The comparison of the PSRI and YEL focused on assessing the performance of estimating Chl and N in senescent wheat (G1.2, U1, U2 and U3.2).

## 3. RESULTS AND DISCUSSION

All correlations (r) related to LAI were statistically significant ($p = 0.01$), and ranged from 0.79 to 0.91 (Tab. 2). In Germany (sites G1.1 and G1.2), LAI was strongly correlated with NDVI and GRE, while GRE showed higher coefficients of determination ($R^2$) at both dates (Tab. 2). In the USA (site U3.1), NDVI and GRE were only moderately correlated to LAI, with GRE performing slightly better than NDVI (Tab. 2). As an example, Figure 2 shows the linear relationship of GRE (A) and NDVI (B) to LAI at test site G1.1. The higher correlation of NDVI and GRE with LAI observed at the German test sites (G1.1, G1.2) compared to USA (U3.1), can most likely be attributed to the lack of the saturation effect in the Germany sites. In Germany, LAI ranged from 1.4 to 2.3; whereas in the USA, it ranged from 1.9 to 5. Based on the observations made, GRE seems to provide similar, if not better estimations of LAI compared to NDVI in vital crops.

Table 2. Coefficient of determination of leaf area index (LAI), chlorophyll (SPAD) and nitrogen (N) concentration, Greenness (GRE) vs. NDVI.

| Site-ID | Crop | Tested Index | Coefficient of Determination ($R^2$) for | | |
|---|---|---|---|---|---|
| | | | LAI | SPAD | N |
| G1.1 | WW | GRE | 0.82 | 0.25* | |
| | | NDVI | 0.78 | 0.25* | |
| G1.2 | WW | GRE | 0.79 | 0.23* | |
| | | NDVI | 0.76 | 0.31* | |
| U3.1 | WW | GRE | 0.63 | 0.70 | 0.52 |
| | | NDVI | 0.62 | 0.69 | 0.56 |

* not significantly correlated at level of confidence p=0.01

Further, in Germany (G1), no significant correlations of GRE and NDVI to Chl concentration were observed at both dates (Tab. 2). In the USA (U3.1), statistically significant correlations were observed for GRE and NDVI to Chl and N concentration (U3.1). Resulting coefficients of determination of $R^2 = 0.70$ and $R^2 = 0,69$ for estimating Chl concentration indicated a slight advantage of

GRE over NDVI, while NDVI performed better at the estimation of N at site U3.1 (Tab. 2). The poor performance of GRE and NDVI as estimators of Chl measurements in Germany might be explained by the small range of SPAD values.

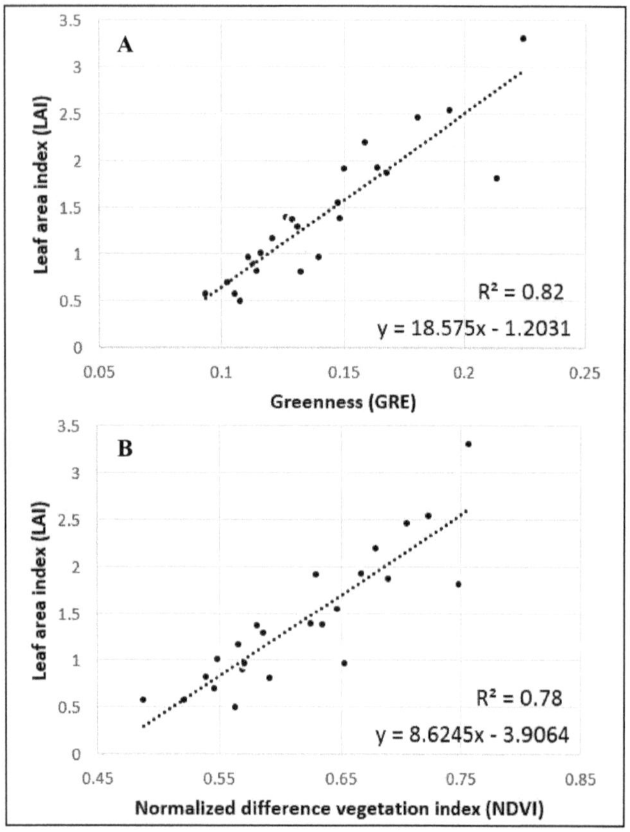


Figure 2. Linear regression of GRE (A) and NDVI (B) to LAI measurements at sample site G1.1.

In most cases, YEL and PSRI showed a strong linear relationship to Chl concentration in senescent wheat (Tab. 3). YEL yielded higher $R^2$ values than PSRI for all fields with winter wheat (U1, U3 and G1), whereas PSRI showed higher predictive potential at test site U2, where summer wheat was sown (Tab. 3). Generally, correlations were significant, except for PSRI at G1.2 and YEL at U2.1. At site U1.1 (winter wheat location), YEL explained approximately 16% more of the total variance compared to the PSRI (Fig. 3), whereas, at the latest senescence stage (U1.2) the estimation capacity of both indices were comparable (Tab. 3). At the summer wheat farm (U1), both indices performed poorly at the first date (U1.1), although PSRI yielded a significant correlation, while YEL did not ($p = 0.01$) (Tab. 3). It has to be mentioned that at site U2.2, images were acquired 11 days after the ground sampling. Changes in crop senescence during this time could have been drastic enough to make the satellite data information inconsistent with the ground measurements.

In some areas of test site G1.2, crops were in a premature ripening stage, while others were in a vital condition (Fig. 1). After excluding nine vital crop samples, YEL showed a significant correlation of $r = -0.84$ in contrast to the PSRI, and was linearly related to Chl concentration with an $R^2 = 0.71$. Similarly, two green samples were excluded from the correlation analysis at U3.2. However, only moderate $R^2$ values were observed (Tab. 3).

Table 3. Coefficient of determination of leaf area index (LAI), chlorophyll (SPAD) and nitrogen (N) concentration, Yellowness (YEL) vs. PSRI.

| Site-ID | Crop | Tested Index | Coefficient of Determination ($R^2$) for | | |
|---|---|---|---|---|---|
| | | | LAI | SPAD | N |
| G1.2** | WW | YEL | 0.59* | 0.71 | |
| | | PSRI | 0.68 | 0.61* | |
| U1.1 | WW | YEL | 0.24* | 0.77 | 0.77 |
| | | PSRI | 0.16* | 0.61 | 0.57 |
| U1.2 | WW | YEL | 0.04* | 0.84 | |
| | | PSRI | 0.01* | 0.83 | |
| U2.1 | SW | YEL | 0.19* | 0.45* | 0.74 |
| | | PSRI | 0.21* | 0.53 | 0.37* |
| U2.2 | SW | YEL | 0.05* | 0.78 | |
| | | PSRI | 0.10* | 0.89 | |
| U3.2** | WW | YEL | 0.06* | 0.60 | |
| | | PSRI | 0.12* | 0.56 | |

*not significantly correlated at level of confidence p=0.01

** Some vital vegetation ground reference measurements excluded

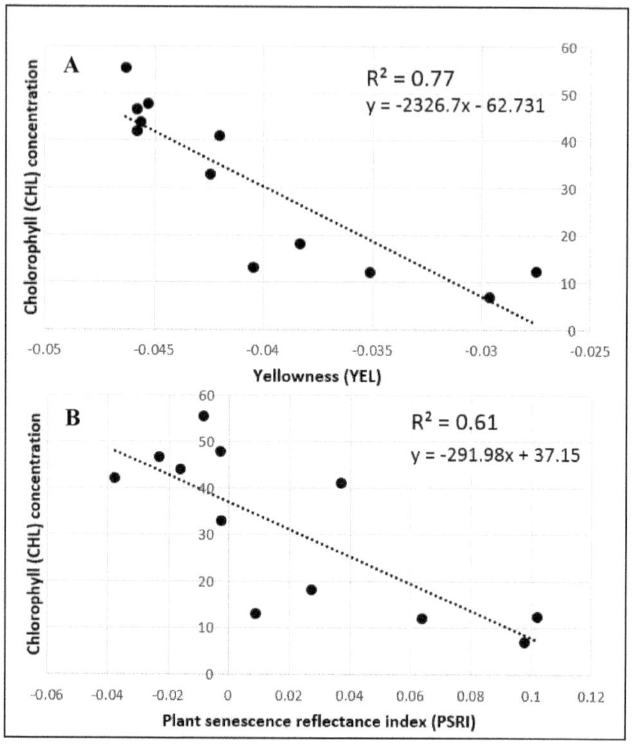


Figure 3. Linear regression of YEL (A) and PSRI (B) to Chl measurements at sample site U1.1.

For the estimation of N concentration in senescent crops, YEL yielded significant correlations at site U1.1 and U2.1 with $r = -0.88$ and $r = -0.86$, respectively, while the PSRI was only significantly correlated to N concentration ($r = -0.76$) at site U1.1. Results revealed that YEL, compared to the PSRI, provided higher $R^2$ values for the estimation of N concentration (Tab. 3, Fig. 4).

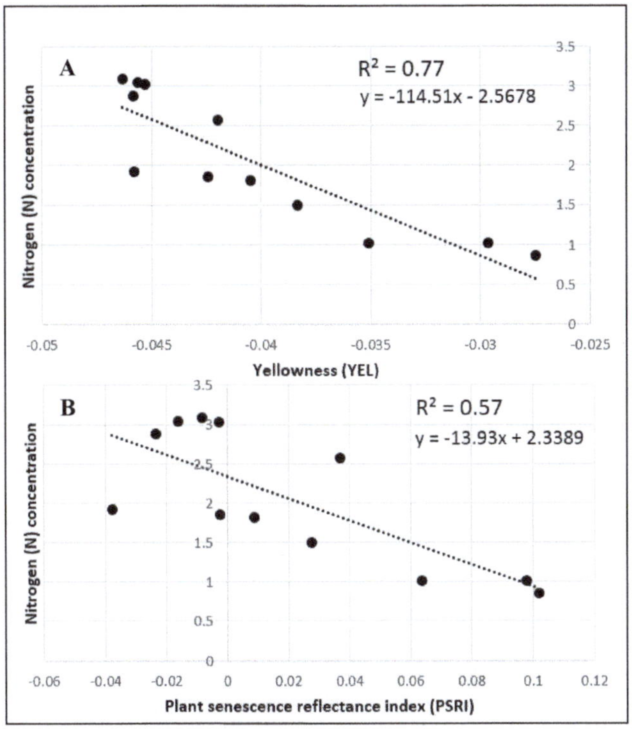


Figure 4. Linear regression of YEL (A) and PSRI (B) to N measurements at sample site U1.1.

Studies investigating the influence of varying N concentration on green canopy spectral reflectance properties have shown that variations of N concentration positively correlate with Chl concentration, which in turn is one of the key drivers of the reflectance properties of plants in the visible part of the spectrum (Evans 1983; Moorby & Besford 1983; Thomas & Oerther 1972; Knipling 1970). Thus, the N status of vital, green crops has been estimated using the relationship between vegetation indices and LAI, which corresponds to canopy Chl content (LAI × Chl) (Filella et al. 1995; Johnson 2001; Penuelas et al. 1994; Serrano et al. 2000; Zhao et al. 2005).

Even though the senescent crops investigated in this study showed different optical properties than vital crops, the field measurements indicated that the N-Chl dependency mentioned above was still present, explaining the correlations found. The significant r of YEL to N observed at site U2.1, despite the non-significant r to Chl concentration ($p = 0.01$), might indicate that other optical properties, which might have been masked by the strong absorbing features of Chl in a vital status, contribute to the YEL-N relationship (Tab. 3). The greater potential of YEL for the estimation of N might be attributed to the inclusion of the Red-edge band, which is more sensitive to a wider range of Chl than the Red and Green bands used in the PSRI. Red reflectance is a poor predictor of low to high levels of Chl or N, since it saturates at low Chl levels (Gitelson & Merzlyak 1998; Gitelson et al. 1996; Eitel et al. 2011). Also, several studies suggested that the wavelength in the red-edge region is superior to the green in regards to its responsiveness to stress induced changes in Chl (e.g. Carter & Knapp 2001; Eitel et al. 2011).

No significant correlations of the indices YEL and PSRI to LAI were observed at the level of confidence $p = 0.01$, with the exception of site G1.2, where the PSRI showed a significant correlation of $r = -0.82$ and a high $R^2$ of 0.68 (Tab. 3). These results could be explained by the changing reflectance characteristics of senescent wheat. In many studies, the retrieval of LAI from remotely sensed data was based on indices or related spectral derivations, which make use of the red and NIR bands, e.g. NDVI or Enhanced Vegetation Index (EVI), because strongest correlations of LAI to these spectral regions were observed (Holben et al. 1980; Kimes et al. 1981; Houborg & Boegh 2008, Viña et al. 2011). With the progress of senescence, crops increase their reflectance in the red electromagnetic spectrum due to Chl degradation and decrease reflectance in the NIR region due to cell structure decomposition, which in turn might negatively influence the sensitivity of the spectral indices to LAI. Additionally, even though LAI decreases a bit, it does not change significantly when crops become senescent and therefore, no correlations of YEL and PSRI to LAI were observed.

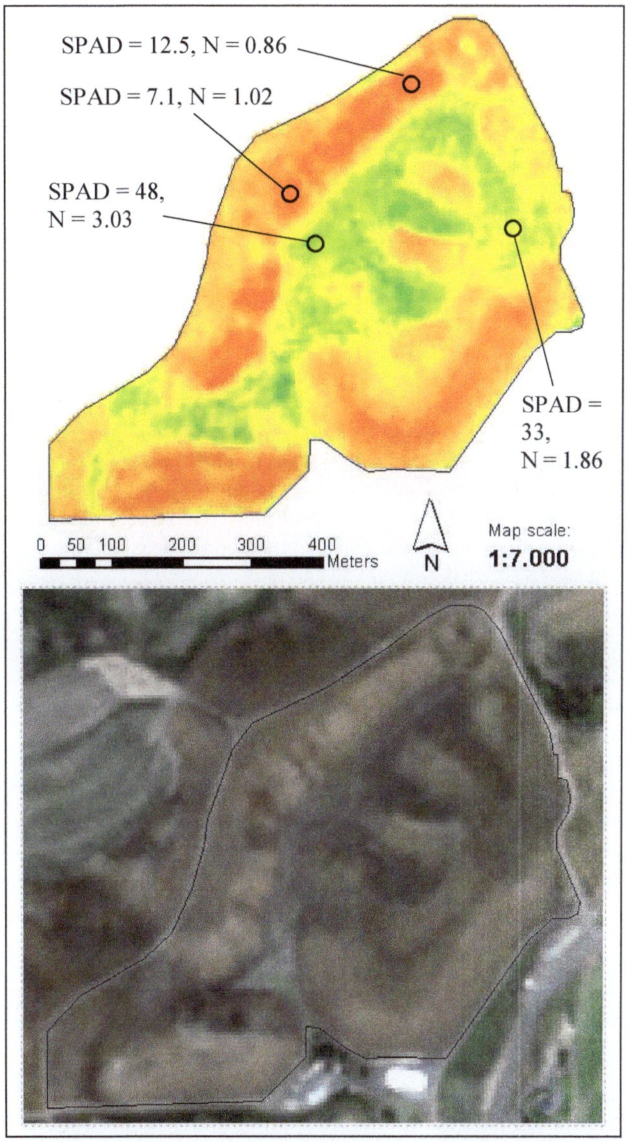


Figure 5. Map of potential within field senescence patterns (above) and corresponding RGB RapidEye image (below) at site U1.1. A selection of ground reference samples are denoted by black circles. Red and yellow colors indicate an advanced state of senescence (lower chlorophyll (SPAD) and nitrogen (N) concentration), while green colors indicate a less advanced senescence status.

Hatfield et al. (2010) showed that PSRI was sensitive to changes in LAI at the onset and during the senescence phase, and therefore, could be used to detect premature loss of LAI. The significant correlation between LAI and PSRI observed at the site

G1.2 (Tab. 3), which was affected by premature ripening, confirms the findings of Hatfield et al. (2010). However, at all other senescence stages present at the sites U1, U2, and U3, no significant relationship between LAI and PSRI was observed.

Interestingly, PSRI and YEL showed an inverse relationship to Chl and LAI at site G1.2 (Tab. 3), suggesting that PSRI responds closely to structural properties, while YEL, to pigment-based biophysical properties. Also, regression analysis at site U1.1 revealed that measurement points were closer to the line of best fit for YEL, compared to PSRI (Fig. 3), suggesting that YEL is more sensitive to capture changes in pigment concentration.

The TCFs are basically a weighted sum of all bands, which means that potentially all information contained in the discrete bands, characterizing a particular feature, e.g. vital or senescent vegetation, are considered. Vegetation indices on the other hand tend to use only few bands. This might explain the relative advantage of the TCFs compared to simple band ratios, such as NDVI or PSRI.

As mentioned in the introduction, determining Chl and/or N post anthesis could be useful to estimate senescence variations of crops within a field (crop drying patterns), which might be indicative of yield and crop quality conditions. Spatial variations of senescence within fields can potentially be mapped and used as additional information for the delineation of harvesting zones. Figure 5 illustrates an example of a map of YEL values at site U1.1, potentially depicting senescence patterns within this field. For comparison, the corresponding RGB image is displayed below the map. Black circles show the values of some ground reference measurements of Chl (SPAD) and nitrogen (N) concentration. In the map, areas coded with red and yellow colors are relatively low in Chl and N, and therefore, most likely to be in a more advanced senescence state compared to areas marked by green colors (relatively high Chl and N concentration).

## 4. CONCLUSION

The results of this preliminary study demonstrated that the TCFs of the RapidEye sensor offer great potential for the retrieval of crop biophysical parameters, such as LAI, Chl and N. GRE has been proven to be a good estimator of LAI in vital crops, especially for LAI values below 3, and it yielded a higher coefficient of determination than the NDVI for test sites in Germany and the USA. YEL and PSRI showed strong linear relationships to the Chl concentration in senescent wheat. However, the prediction power of YEL was greater for winter wheat, while for summer wheat, PSRI provided better results. For the prediction of N in senescent crops, YEL outperformed PSRI, showing statistically significant, strong correlations, while PSRI provided only moderate or non-significant relationships.

Results demonstrated the capability of YEL to predict both Chl and N concentration, probably attributable to the additional information provided by the Red-edge band. This makes RapidEye a useful tool for understanding and mapping the variability of Chl or N concentrations, and senescence patterns within individual fields. In conjunction with soil fertility and yield maps, senescence patterns might aid in the assignment of precision harvesting zones related to yield and wheat quality parameters.

Further studies on the relationship between YEL and Chl and N concentration are needed to corroborate these results, and make them extensive to other crop types or cultivars. Furthermore, research should be focused on in-field senescence pattern variations, microclimatic conditions, and their impact on the amount of yield and yield quality parameters.

## 5. ACKNOWLEDGEMENT

The authors wish to express their appreciation to the Julius-Kühn-Institut (Federal Research Centre for Cultivated Plants in Germany) and the Institute for Geoinformatics and Remote Sensing of the University of Osnabrueck for kindly supplying the field data used in this study, based on the "Hyland" project supported under code 50 EE 1014. Further, we kindly thank the Geospatial Laboratory for Environmental Dynamics, University of Idaho for the provision of ground reference data. The ground data collection in Idaho was supported by USDA-NIFA Award Nos. 2011-67003-3034 and 2011-68002-30191.

## REFERENCES

Anderson, C.; Thiele, M.; Brunn, A. 2013. Calibration and validation of the RapidEye constellation. In: *Anais XVI Simpósio Brasileiro de Sensoriamento Remoto* - SBSR, Foz do Iguaçu, PR, Brasil, 13 a 18 de abril de 2013.

Baigab, M. H. A.; Zhanga, L.; Shuaiab, T.; Tonga, Q. 2014. Derivation of a tasselled cap transformation based on Landsat 8 at-satellite reflectance. In: Remote Sensing Letters 5(5): 423-431.

Basso, B., Bertocco, M., Sartori, L., Martin, E.C., 2007. Analyzing the effects of climate variability on spatial pattern of yield in a maize-wheat-soybean rotation. In: Eur. J. Agron. 26, 82–91.

Bieker, Stefan; Zentgraf, Ulrike 2013. Plant Senescence and Nitrogen Mobilization and Signaling. In: Senescence and Senescence-Related Disorders, 53-83.

Buttafuoco, G., Castrignanò, A., Colecchia, A.S., Ricca, N., 2010. Delineation of management zones using soil properties and a multivariate geostatistical approach. In: Ital. J. Agron. 4, 323–332.

Carter, G. A. & Knapp, A. K. 2001. Leaf optical properties in higher plants: linking spectral characteristics to stress and chlorophyll concentration. American Journal of Botany 88, 677-684.

Colombo, R.; Bellingeri, D.; Fasolini, D.; Marino, C. M. 2003. Retrieval of leaf area index in different vegetation types using high resolution satellite data. In: Remote Sensing of Environment 86(1): 120-131.

Crist, Eric P. & Cicone, Richard C. 1984. A Physically-Based Transformation of Thematic Mapper Data – The TM Tasseled Cap. In: Transactions on Geoscience and Remote Sensing GE-22(3): 256-263.

Diaconoa, M.; Castrignanòb, A.; Troccolic, A.; De Benedetto, D.; Basso, B.; Rubino, P. 2012. Spatial and temporal variability of wheat grain yield and quality in a Mediterranean environment: A multivariate geostatistical approach. In: Field Crops Research 131: 49-62.

Edwards, Kieron D.; Humphry, Matt; Sanchez-Tamburrino, Juan P. 2012. Advances in Plant Senescence. In: Senescence, 117-136.

Eitel, J.U.H.; Long, D.S.; Gessler, P.E.; Smith, a. M.S. 2007. Using in-situ measurements to evaluate the new RapidEye ™ satellite series for prediction of wheat nitrogen status. In: Int. J. Remote Sens. 28, 4183–4190.

Eitel, J.U.H.; Vierling, L.A.; Litvak, M.E.; Long, D.S.; Schulthess, U.; Ager, A.A.; Krofcheck, D.J.; Stoscheck, L. 2011.

Broadband, red-edge information from satellites improves early stress detection in a New Mexico conifer woodland. In: Remote Sens. Environ. 115, 3640–3646.

Evans, J. R. 1983. Nitrogen and Photosynthesis in the Flag Leaf of Wheat (Triticum aestivum L.). In: Plant Physiology 72(2): 297–302.

Filella, I.; Serrano, L.; Serra, J.; Peñuelas, J. 1995. Evaluating Wheat Nitrogen Status with Canopy Reflectance Indices and Discriminant Analysis. In: Crop Science 35(5): 1400-1405.

Freeman, K. W.; Raun, W. R.; Johnson, G. V.; Mullen, W. R. ; Stone, M. L.; Solie, J. B. 2003. Late-Season Prediction of Wheat Grain Yield and Grain Protein. In: COMMUNICATIONS IN SOIL SCIENCE AND PLANT ANALYSIS 34(13&14): 1837-1852.

Gates, David M.; Keegan, Harry, J.; Schleter, John C.; Weidner Victor R. 1965: Spectral Properties of Plants. In: Applied Optics 4(1): 11-20.

Gitelson, A. A.; Merzlyak, M. N.; Lichtenthaler, H. K. 1996. Detection of red edge position and chlorophyll content by reflectance measurements near 700 nm. In: Journal of Plant Physiology, 148, 501-508.

Gitelson, A. A.; Merzlyak, M. N.; Grits, Y. 1998. Remote sensing of chlorophyll concentration in higher plant leaves. In: Advances in Space Research, 22, 689–692.

Gitelson, A. A.; Zur, Y.; Chivkunova, O. B.; Merzlyak, M. N. 2002. Assessing Carotenoid Content in Plant Leaves with Reflectance Spectroscopy. In: Photochemistry and Photobiology 75(3): 272-281.

Gitelson, Anatoly A.; Viña, Andrés; Arkebauer, Timothy J.; Rundquist, Donald C.; Keydan, Galina; Leavitt, Bryan 2003. Remote estimation of leaf area index and green leaf biomass in maize canopies. In: Geophysical Research Letters 30(5): 52-1 - 52-4.

Haboudane, Driss; Miller, John R.; Pattey, Elizabeth; Zarco-tejada, Pablo J.; Strachan, Ian B. 2004. Hyperspectral vegetation indices and novel algorithms for predicting green LAI of crop canopies: Modeling and validation in the context of precision agriculture. In: Remote Sensing of Environment 90(3): 337-352.

Hatfield, J. L.; Gitelson, A. A.; Schepers, J. S.; Walthall, C. L. 2008. Application of Spectral Remote Sensing for Agronomic Decisions. In: Agronomy Journal 100(3): S-117-S-131.

Hatfield, J.L.; Prueger, J.H. 2010. Value of Using Different Vegetative Indices to Quantify Agricultural Crop Characteristics at Different Growth Stages under Varying Management Practices. In: Remote Sensing 2(2): 562-578.

Hörtensteiner, S.; Kräutler B. 2010. Chlorophyll breakdown in higher plants. In: Biochimica et Biophysica Acta (BBA) - Bioenergetics 1807(8): 977-988.

Holben, B.N., Tucker, C.J., Fan, C.J., 1980. Spectral assessment of soybean leaf area and leaf biomass. Photogrammetric Engineering and Remote Sensing 46 (5), 651–656.

Horne, James H. (2003): A Tasseled Cap Transformation for IKONOS Images. In: ASPRS 2003 Annual Conference Proceedings, Anchorage, Alaska, USA.

Houborg, R. & Boegh, E. 2008. Mapping leaf chlorophyll and leaf area index using inverse and forward canopy reflectance modeling and SPOT reflectance data. In: Remote Sensing of Environment 112(1): 186-202.

Huang, Chengquan; Wylie, Bruce; Yang, Limin; Homer, Collin; Zylstra, Gregory 2002. Derivation of a Tasseled Cap Transformation based on Landsat 7 at-satellite reflectance. In: International Journal of Remote Sensing 23(8): 1741-1748.

Ibrahim, Hesham M.; Huggins, David R. 2011. Spatio-temporal patterns of soil water storage under dryland agriculture at the watershed scale. In: Journal of Hydrology 404(3–4): 186-197.

Ivits, Eva; Lamb, Alistair; Langar, Filip; Hemphill, Scott, Koch, Barbara 2008. Orthogonal Transformation of Segmented SPOT5 Images: Seasonal and Geographical Dependence of the Tasseled Cap Parameters. In: Photogrammetric Engineering and Remote Sensing 74(11): 1351-1364.

Johnson, L. 2001. Nitrogen Influence on Fresh-leaf NIR Spectra. In: Remote Sensing of Environment 78(3): 312-318.

Kauth, R. J. & Thomas, G. S. 1976. The Tasseled Cap – a graphic description of the spectral-temporal development of agricultural crops as seen by LANDSAT. In: Proceedings of the Symposium on Machine Processing of Remotely Sensed Data. Pages 4B41-4B51.

Kichey, T.; Hirel, B.; Heumez, E.; Dubois, F.; Le Gouis, J. 2007. In winter wheat (Triticum aestivum L.), post-anthesis nitrogen uptake and remobilisation to the grain correlates with agronomic traits and nitrogen physiological markers. In: Field Crop Research 102(1): 22-32.

Kimes, D.S.; Markham, B.L.; Tucker, C.J.; McMurtrey, J.E. III. 1981. Temporal relationships between spectral response and agronomic variables of a corn canopy. In: Remote Sensing of Environment 11(5): 401-411.

Knipling, Edward B. 1970. Physical and Physiological Basis for the Reflectance of Visible and Near-Infrared Radiation from Vegetation. In: Remote Sensing of Environment 1(3): 155-159.

Kross, A.; McNairn, H.; Lapen, D.; Sunohara, M.; Champagne, C. 2015. Assessment of RapidEye vegetation indices for estimation of leaf area index and biomass in corn and soybean crops. In: International Journal of Applied Earth Observation and Geoinformation 34: 235-248.

Lukina, E. V.; Freeman, K. W.; Wynn, K. J.; Thomason, W. E.; Mullen, R. W.; Stone, M. L.; Solie, J. B.; Klatt, A. R.; John, G. V.; Elliott, R. L.; Raun, W. R. 2001. Nitrogen Fertilization Optimization Algorithm Based on In-Season Estimates of Yield and Plant Nitrogen Uptake. In: Journal of Plant Nutrition 24(6): 885-898.

Merzlyak, J. "Non-destructive Optical Detection of Pigment Changes During Leaf Senescence and Fruit Ripening." In: Physiologia Plantarum 106(1999): 135-141.

Mestre, Harold 1935. The absorption of radiation by leaves and algae. In: Photochemical Reactions. Volume 3: 191-209.

Meyer-Aurich, A.; Gandorfer, M.; Weersink, A.; Wagner, P. 2008. Economic analysis of site-specific wheat management with respect to grain quality and separation of the different quality fractions. In: International Congress, August 26-29, 2008, Ghent, Belgium from European Association of Agricultural Economists.

Moorby J. & Besford R. T. 1983. Mineral nutrition and growth. In: Encyclopedia of Plant Physiology Vol. 15 B: 481-515.

Morari, F.; Loddo, S.; Berzaghi, P.; Ferlito, J.C.; Berti, A.; Sartori, L.; Visioli, G.; Marmiroli, N.; Piragnolo, D.; Mosca, G. 2013. Understanding the effects of site-specific fertilization on yield and protein content in durum wheat. In: Precision agriculture '13. 321-327.

Penuelas, J.; Gamon, J.A.; Freeden, A.; Merino, J.; Field, C. 1994. Reflectance indices associated with physiological changes in N- and water-limited sunflower leaves. In: Remote Sens. Environ. 46: 100–118.

Penuelas, J.; Baret, F.; Filella, I. 1995. Semi-empirical indices to assess carotenoids/chlophyll a ratio from leaf spectral reflectance. In: Photosynthetica 31(2): 221-230.

Penuelas, J. & Filella, I. 1998. Visible and near-infrared reflectance techniques for diagnosing plant physiological status. In: Trends in Plant Science 3(4): 151-156.

Reeves, D.W.; Mask, P.L.; Wood, C.W.; Delano, D.P. 1993. Determination of Wheat Nitrogen Status with Hand-held Chlorophyll Meter: Influence of Management Practices. In: Journal Plant Nutrition, 16: 781–796.

Rouse, J.W., Haas, R.H., Schell, J.A., Deering, D.W., 1973. Monitoring vegetation systems in the Great Plains with ERTS, Third ERTS Symposium, NASA SP-351 I, 309-317.

Sanger, Jon E. 1971. Quantitative Investigations of Leaf Pigments From Their Inception in Buds Through Autumn Coloration to Decomposition in Falling Leaves. In: Ecology 52(6): 1075-1089.

Schönert, M.; Weichelt, H.; Zillmann, E.; Jürgens, C. 2014. Derivation of Tasseled Cap Coefficients for RapidEye data. In: Michel, Ulrich; Schulz, Karsten (Ed.): Earth Resources and Environmental Remote Sensing/GIS Applications V. Proc. SPIE 9245.

Serrano, L.; Filella, I.; Peñuelas, J. 2000. Remote Sensing of Biomass and Yield of Winter Wheat under Different Nitrogen Supplies. In: Crop Science 40(3): 723-731.

Siegmann, B., Jarmer, T., Lilienthal, H., Richter, N., Selige, T., Höfle, B., 2013. Comparison of narrow band vegetation indices and empirical models from hyperspectral remote sensing data for the assessment of wheat nitrogen concentration. In: *Proc. 8th EARSeL SIG IS workshop*, Nantes, France.

Smart, C. M. (1994) Gene expression during leaf senescence. In: New Phytologist 126(3): 419-448.

Soil Survey Staff, Natural Resources Conservation Service, United States Department of Agriculture. Web Soil Survey. Available online at http://websoilsurvey.nrcs.usda.gov/. Accessed [03/20/2014].

Strong, W.M. 1981. Nitrogen requirements of irrigated wheat in the Darling Downs. In: Australian Journal of Experimental Agriculture and Animal Husbandry 21(111): 424-431.

Thomas, H. & Stoddart, J. L. 1980. Leaf senescence. In: Annual Review of Plant Physiology 31: 83–111.

Thomas J. R & Oerther, G. F. 1972. Estimating nitrogen content of sweet pepper leaves by reflectance measurements. In: Agron. J. 64: 11-13.

Tucker, J. C. 1979. Red and photographic infrared linear combinations for monitoring vegetation. In: Remote Sensing of Environment 8(2): 127-150.

Verdebout, Jean; Jacquemoud, Stephane; Schmuck, Guido 1994. Optical Properties of Leaves: Modelling and Experimental Studies. In: Imaging Spectrometry - a Tool for Environmental Observations: 169-191.

Viña, A.; Gitelson, A. A.; Nguy-Robertson, A. L.; Peng, Y. 2011. Comparison of different vegetation indices for the remote assessment of green leaf area index of crops. In: Remote Sensing of Environment 115(12): 3468-3478.

Wang, Z.; Wang, J.; Zhao, C.; Zhao, M.; Huang, W.; and Wang, C. 2005. Vertical Distribution of Nitrogen in Different Layers of Leaf and Stem and their Relationship with Quality of Winter Wheat. In: Journal of Plant Nutrition 28(1): 73-91.

Wilson, J.H.; Zhang, C.; Kovacs, J.M. 2014. Separating Crop Species in Northeastern Ontario Using Hyperspectral Data. In: Remote Sensing 6(2): 925-945.

Wright, D. L.; Rasmussen, V. P.; Ramsey, R. D.; Baker, D. J.; Ellsworth J. W. 2004. Canopy Reflectance Estimation of Wheat Nitrogen Content for Grain Protein Management. In: GIScience and Remote Sensing 41(4): 287-300.

Wu, J.; Wang, D.; Bauer, M. E. 2007. Assessing broadband vegetation indices and QuickBird data in estimating leaf area index of corn and potato canopies. In: Field Crops Research 102(1): 33-42.

Zhao, D.; Reddy, K.R.; Kakani, V.G.; Read, J.J.; Koti, S. 2005. Selection of optimum reflectance ratios for estimating leaf nitrogen and chlorophyll concentrations of field-grown cotton. Agron. J. 97(1): 89-98.

Zhu, J.; Tremblay, N.; Liang, Y.; 2012. Comparing SPAD and atLEAF values for chlorophyll assessment in crop species In: Canadian Journal of Soil Science 92: 645-648.

Zur, Y.; Gitelson A. A.; Chivkunova, O. B.; Merzlyak M. N. 2000. The spectral contribution of carotenoids to light absorption and reflectance in green leaves. In: Proceedings of the 2nd International Conference Geospatial Information in Agriculture and Forestry: 1-7.

# RECOGNITION OF SINGLE AND OVERLAY OF OBJECTS ON A CONVEYOR BELT

S.V. Savicheva [a]

[a] Murom Institute (branch) or the Federal State Budget Institution of Higher Professional Education "Vladimir State University Alexander G. and Nicholai G. Stoletovs", Russia – savicheva.svetlana@gmail.com

**KEY WORDS:** Recognition, Flat object, Binary image, Circuit, Curvature, Signature, Sign.

**ABSTRACT:**

Proposed a method for detection of flat objects when they overlap condition. The method is based on two separate recognition algorithms flat objects. The first algorithm is based on a binary image of the signature of the object plane. The second algorithm is based on the values of the discrete points in the curvature contour of a binary image. The results of experimental studies of algorithms and a method of recognition of individual superimposed flat objects.

## 1. MANUSCRIPT

When sorting and control parts, assembling products, which are, for example, on an assembly line, the problem of recognition of individual and superimposed objects [1, 2].

If the problem of recognition of individual objects already well solved [3], the question of recognition of individual objects as they imposed a state yet to be resolved.

Development of algorithms for recognizing objects overlap several research centers in Japan and the United States [4]. However, not algorithms and methods suitable for practical use. In Russia, research in this area is not yet underway.

In this regard, the development and practical application of new algorithms for fast and accurate recognition of individual and superimposed objects is an urgent research task that would create an industrial system of automatic recognition.

In this paper, we consider two separate recognition algorithm and a method for recognition of superimposed (two) flat objects.

Under the flat object refers to real objects, products, details of height h is much smaller than its length.

## 2. RECOGNITION ALGORITHMS INDIVIDUAL FLAT OBJECTS

The first algorithm is based on the use of signatures of the binary object image plane.

At the stage of learning implements the following steps:

1. Generation of 2000 options for the location of each flat object in the field of view of the system.

Basic information (parameters) for the generation of options for the location of flat objects are the coordinates of the center of gravity $x_ц$ and $y_ц$ and angle of the objects. The parameter values are specified using the random number. The values of the angle change in the range from 0 to 360 degrees (one step is equal degrees), and the coordinates $x_ц$ and $y_ц$ within the $x_1 < x_ц < x_2 ; y_1 < y_ц < y_2$, where $(x_1;y_1)$, $(x_2;y_2)$ - the coordinates of the diagonal of the image, which should not go for the redistribution of the field, the size 1023x1023. For research, the size of the working area was chosen to 1024x1024 pixels. The size of the workspace is 1023x1023 points. 20000 variants is generated separate flat objects 10 (Figure 1). They are placed in the database.

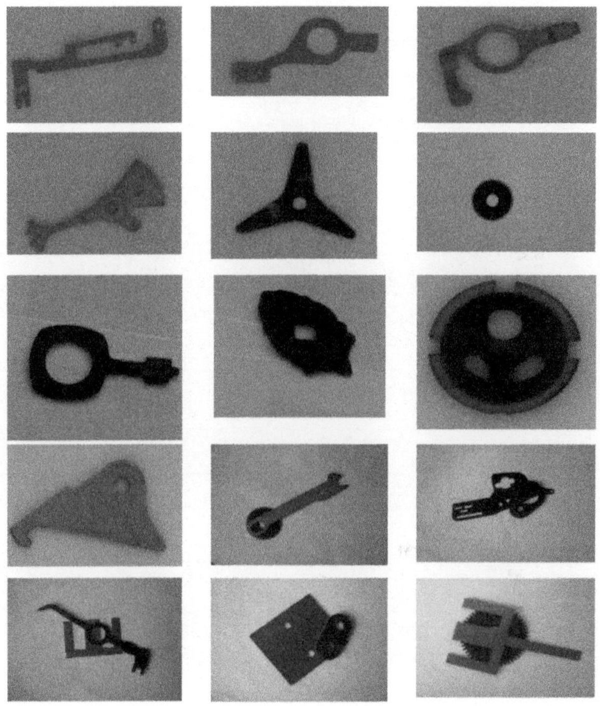

Figure 1. Examples of individual and superimposed flat objects

2. Determination of the frequency of occurrence of each variant location in the field of vision recognition systems. The results are also placed in the database

3. Pretreatment

The proposed method of image pre-processing of flat objects comprises the following steps [4]: 1) reduction of the noise (mostly background) on the image of the object to the desired; 2) binarization of images; 3) marking objects in the image; 4) removing the objects from the memory of non-integer; 5) edge detection of binary images of objects; 6) determination of the center of gravity of binary images.

Results of operations 5 and 6, the provisional image processing flat objects stored in the database.

4. Formation of R-functions for each binary image object

The algorithm is used as an indication of the function $R(r_i)$. It represents a set of values $\{r_i\}$ distances from the center of the object to the points of its outer contour. This function will be called R-function.

The process of forming an R-function binary image on its contour is shown in Figure 2.

Formation of R-function object is to arrange an array with values $r_i$ ($i$=1, 2, 3, …, $m$) in the order they are defined in the bypass circuit.

It should be noted that, when the R-specs, its first element r1 is the distance from the center point to the first circuit, the first found during the scanning of the image field from left to right and top to bottom. R-functions of all 20,000 images of objects 10 are entered into the database.

Figure 2 shows examples of R-functions of certain objects (1, 2, 3, 4) in Figure 1.

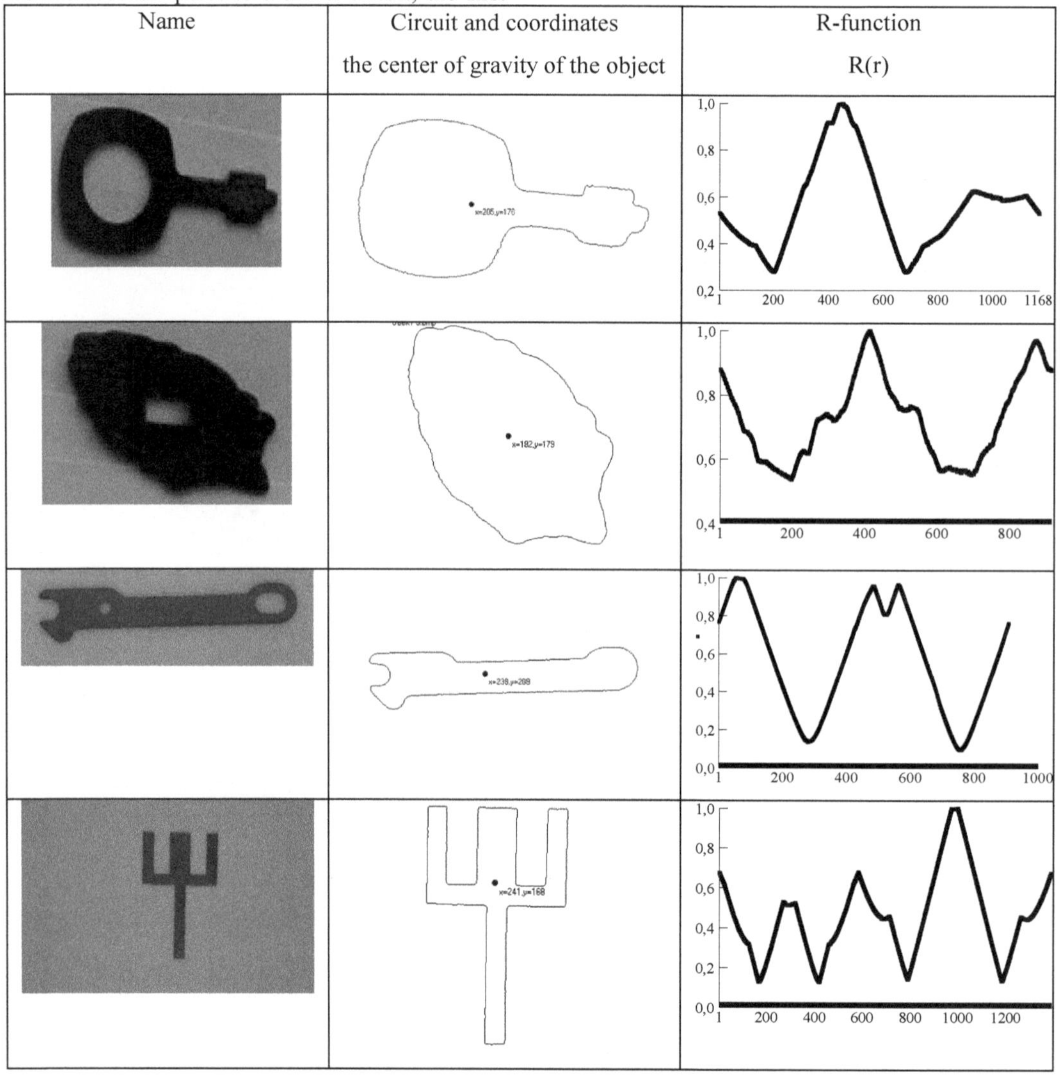


Figure 2. Charts R-functions of some real flat objects

2. Recognition of planar objects and determining the required number of standards

In recognition of its object R-function compared with standard rs-functionality of objects based on the following expression

$$Z_{kl} = \sqrt{\frac{1}{n+1}\sum_{i=1}^{n}(r_{ik} - r_{il})^2} \qquad (1)$$

where

$k$ - number of the reference object;

$l$ - number of recognizable object;

$r_{ik} = \{r_i(k)$, if i $i \le m_k$; 0 - otherwise$\}$;

$r_{il} = \{r_i(l)$, if $i \le m_l$; 0 - otherwise$\}$;

$m_k$, $m_l$ - the number of R-function number $k$ and the number of recognizable object $l$, respectively;

$Z_{kl}$ - the standard deviation of the values of R-functions identify objects with a number of reference $l$ rs-function with $k$.

$$n = \max_{k,l}\{m_k, m_l\}.$$

The values of the missing elements, compared R- and Re-functions assumed to be zero. Unlike $m_k$ and $m_l$ due to the fact that different objects have different number of contour elements.

The decision to identify a recognizable object is made based on the following expression:

$$\min_{k,l} Z^{o}_{kl} = \min_{k}\{\min_{l} Z_{kl}\} \qquad (2)$$

where

$\min_{k} Z_{kl}\ \{*\}$ - minimum value $Z'kl$ among all $Z'kl$ for k = 1, 2, ..., N and the constant value $l$;

Studies have been conducted with all versions 20000 10 separate images of flat objects with the selected amount of standards.

Studies have confirmed the results obtained during the training phase. Figure. 5 and Figure. 6 are graphs of the probability of correct recognition and time on all 10 original objects on the number of plane standards.

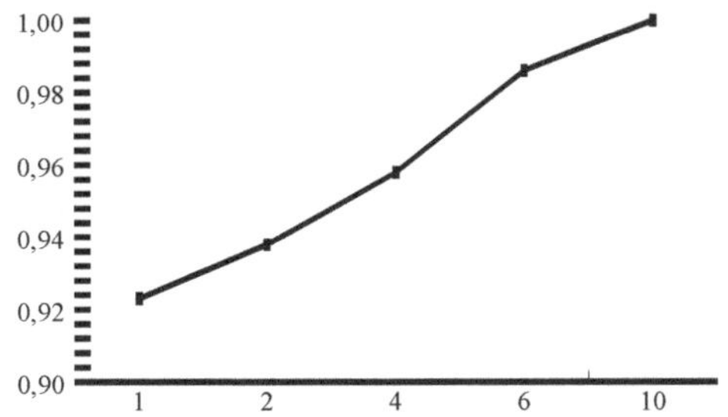


Figure 5. A plot of the probability of correct recognition of a separate flat object on the number of standards

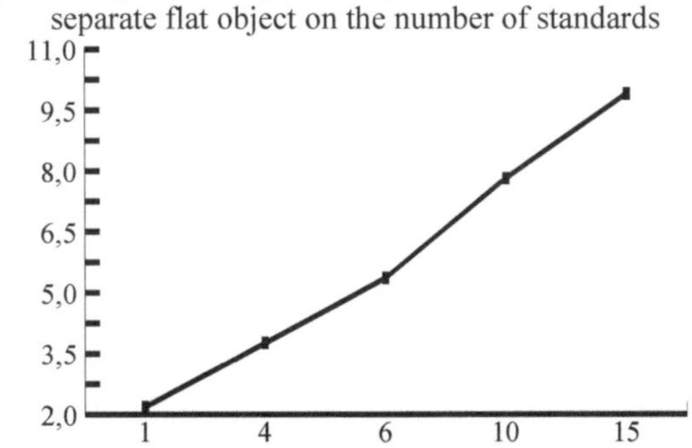


Figure. 6. Schedule time change detection separate flat object on the number of standards

Analysis of the results shows that the detection accuracy of certain embodiments of real planar object using standards of 10 to 100% and further increase the number of reference only leads to increase in the recognition time for each object.

The second algorithm is based on the values of the discrete points in the curvature contour of a binary image of a flat object.

All the stages and steps of the algorithm are almost similar to the steps and the first steps of the algorithm.

The difference function is formed from the curvature values at discrete points of the binary contour image of the object plane. Therefore, the process of determining the values of a curvature of the contour and curvature of the shaping function will be called A-function.

For one-point for seamless contours of binary images of flat objects used algorithm, which forms a discrete contour line of 4 and d-connected points

This means a single-point circuit consists of points, the curvature of which can take only three values - 0°, 90° and 135°. Therefore, the determination of the value of curvature at such a circuit can be done in a logical way with masks, which greatly simplifies and accelerates the process of formation of signs as opposed to analytical algorithms for computing curvature points to the discrete circuit.

In order to determine the curvature at points one-element discrete line 16, it is sufficient to scan the mask size of 3x3 elements.

In Figure 7 shows examples of normalized A-functions of certain objects (1-4) of Figure 1.

| Name | Circuit and coordinates the center of gravity of the object | A-function $A(\alpha)$ |
|---|---|---|
|  | 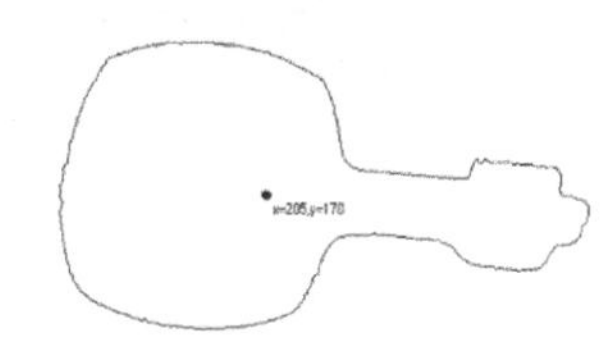 |  |
| 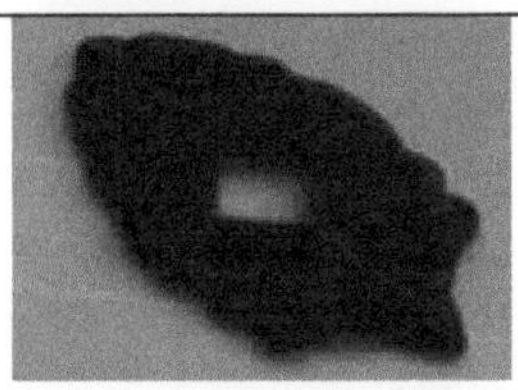 |  | 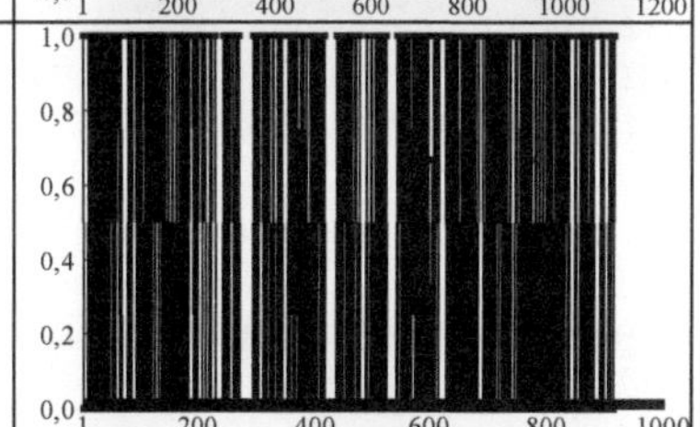 |
| 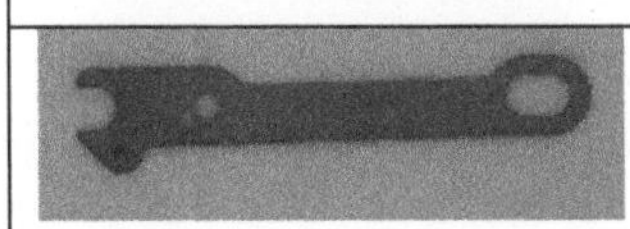 | 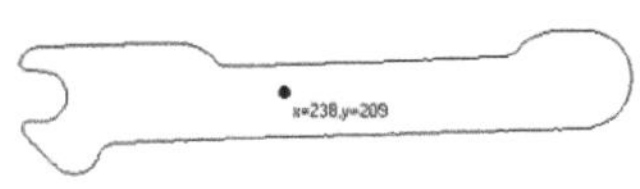 |  |
|  | 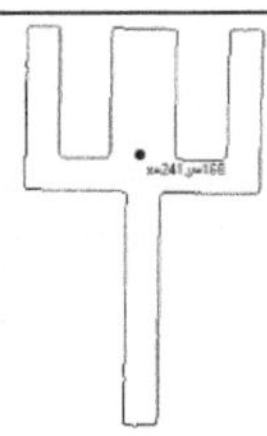 | 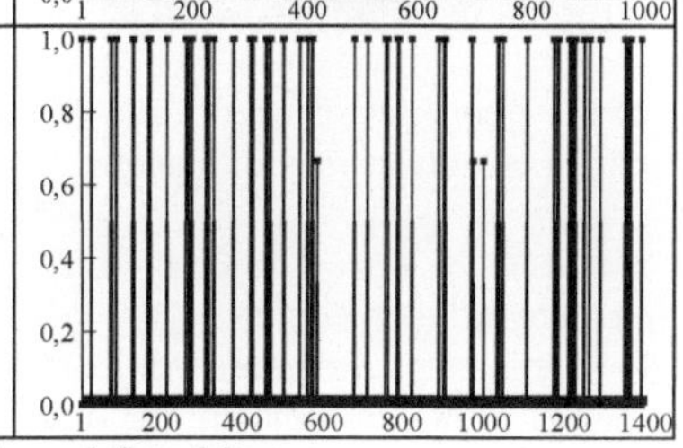 |

Figure. 7. Graphs and functions of some specific real flat objects

Figure. 8 and Figure 9 are plots of the probability of correct recognition and time on all 10 original flat objects on the amount of standards.

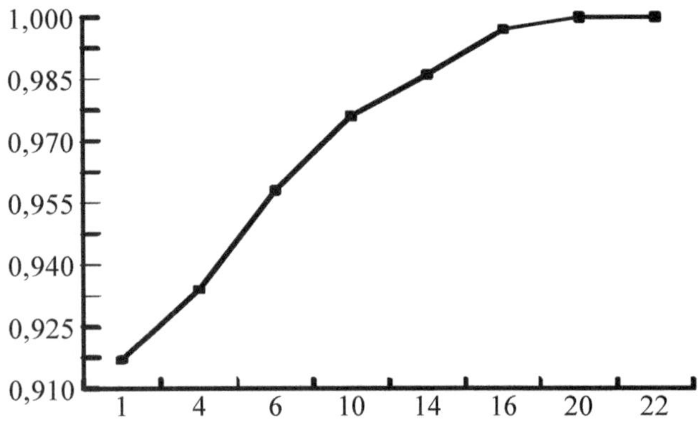


Figure. 8. A plot of the probability of correct recognition of individual flat objects on the number of standards

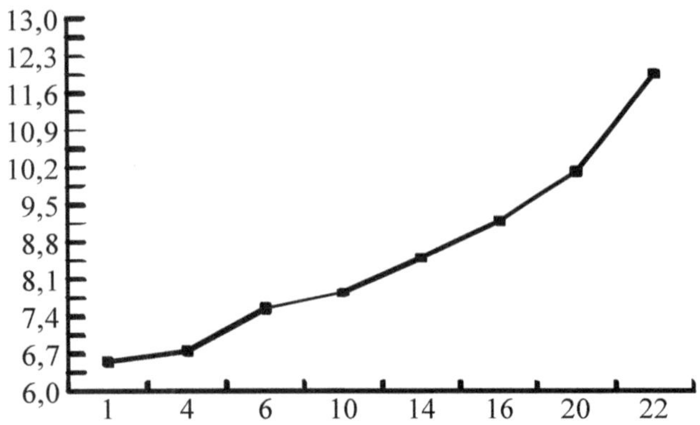


Figure. 9. Schedule time change detection of individual flat objects on the number of standards

As seen from the graph in Figure 8 maximum error rate of 10 etalons is 4%, and at 20 and more reference error becomes zero.

## 3. RECOGNITION OF INDIVIDAUL FLAT OBJECTS IN THEIR CONDITION IMPOSED

This method is like any recognition method consists of two main phases - training and exam.

During the training phase, the following steps:

1. Generate imposed objects (let's call a complex object) of two flat objects. 10 flat objects are formed 55 pairs imposed on 2 sites. The creation of complex objects is to generate versions of the image of each individual flat object and their placement on the same field of view of the recognition system.

The imposition of any two types of objects in the field of view is determined by the terms of the marking:

- If the field of view of one connected component, the objects are superimposed or concern (zero overlay);
- If the sight of two (or more) of the connected components, the objects are not imposed.

2. Determination of the frequency of occurrence of a complex object with a specific percentage of the superposition of two objects.
3. Pre-treatment images of complex objects - the same pre-treatment in the first recognition algorithm.
4. Formation of R-functions of a complex object. function - similar to the formation of R-functions in the first algorithm recognition as a binary image of a complex object in the field is a separate object.

Figure 10, Figure 11 shows plots of the probability of correct recognition, and time for all 55 types of complex objects on the amount of standards.

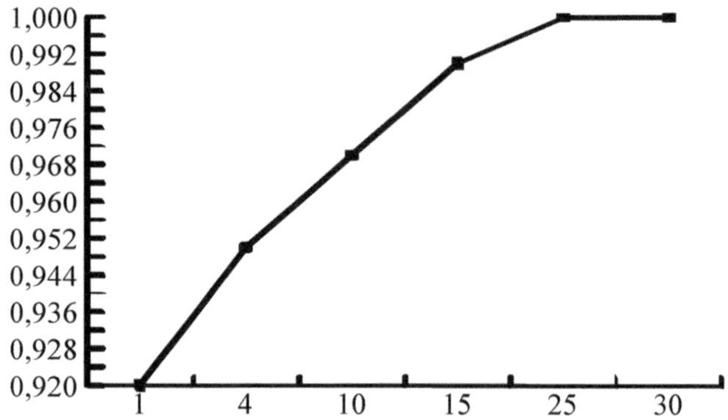


Figure. 10. A plot of the probability of correct recognition of complex object on the number of standards

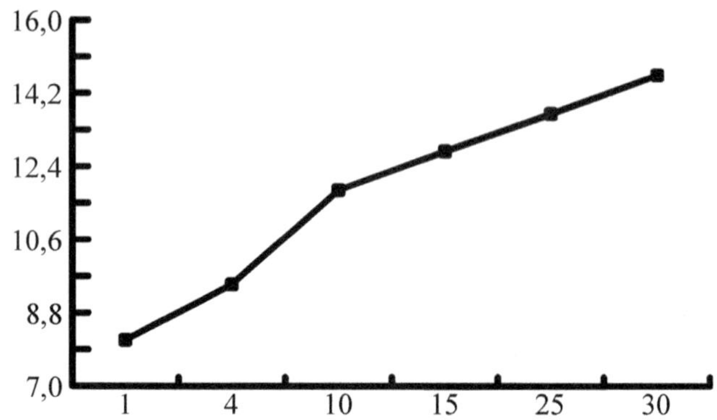


Figure. 11. Schedule change the time of a complex object recognition on the number of standards

Analysis of the results shows that at 25 measurement standards ensures 100% detection of all 110,000 options 55 pairs of complex objects. This means we can assume that the probability of error detection of any complex object using the first recognition algorithm is less than $1 \cdot 10^{-5}$.

Complex object recognition and identification of its facilities lets you know of any individual flat objects is a complex object. Now you need to determine what part of a complex object is one and the other separate flat object. This problem is solved by using the second pattern recognition algorithm separate flat objects in the following order:

1. Forming reference A-function initial 10 separate flat objects.
2. Generate A-feature detects a complex object.
3. A comparison of the reference function of a single flat object and A-function complex object.

A similar comparison of AE function of the second subject is an individual with all the A-functions of complex objects.

Figure. 12 shows a graph of the recognition accuracy, separate flat objects in each of the superimposed object 55 depending on the number of standards for each complex object. Figure 13 is a graph of change of probability of correct recognition of certain objects in the plane of complex objects when the number of standards from 1 to 14.

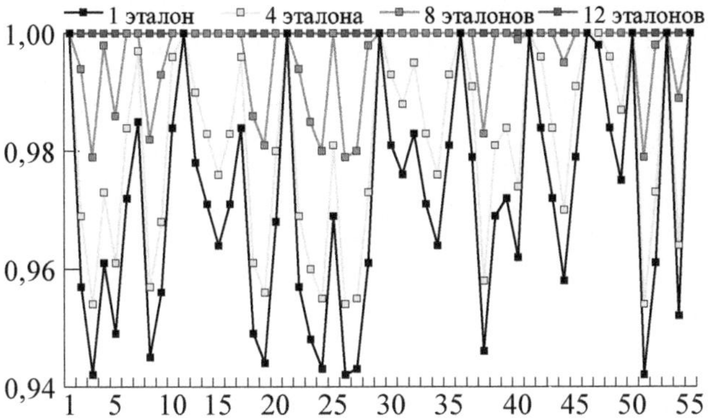


Figure. 12. Schedule changes in the probability of correct recognition of individual flat objects in each of the 55 complex objects with different number of reference

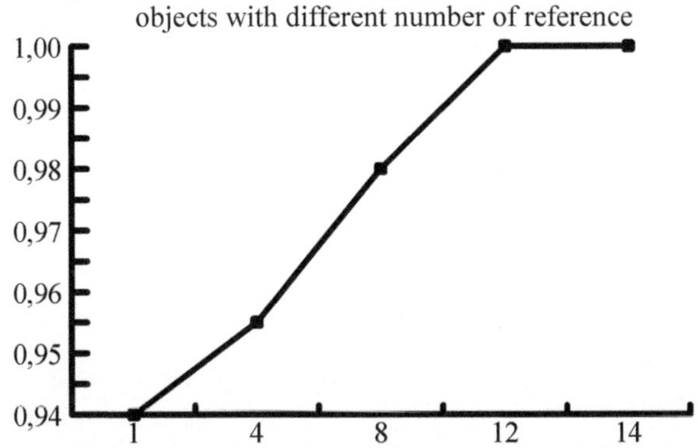


Figure. 13. A plot of the probability of correct recognition of individual flat objects in a complex object on the number of standards

Time graph of the recognition of the number of standards is shown in Figure 14.

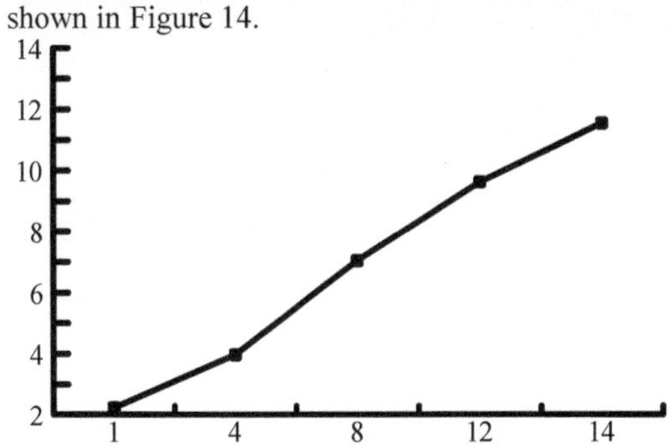


Figure. 14. A plot of the time of recognition of individual flat objects in a complex on the number of references to an object

Analysis of the graphs in Figure. 12-14 shows that the maximum error rate of recognition of individual objects in the complex plane based on the A-function in a plurality of reference is: at 4 standard - 5.5%; time -3.968 ms; etalons at 8 - 2%; time - 7.038 ms; at 12 or more etalons - 0%; time - 9.615 ms.

## CONCLUSION

1. R-algorithm provides for a small number of reference for each object and a small-time high accuracy of recognition of individual flat objects With it successfully solved the problem of determining the class of a complex object formed by the superposition of two separate flat objects.
2. A recognition algorithm allows individual flat objects, but with a large number of references to an object, and longer than the R-algorithm. Detection of individual objects in the complex plane is possible by using only A-algorithm.
3. The method of recognition of individual flat objects in a complex object, based on the R- and A-functions provides high recognition accuracy with a small number of reference as a complex object, and a separate flat objects.

## REFERENCES

Sadykov, S.S., Savicheva, S.V. 2012. Pretreatment images of flat objects in vision systems. *Instrument*, 2, pp. 19-24.

Sadykov, S.S., Savicheva, S.V. 2012Evaluation methods recognition capabilitiestion of flat objects. *Algorithms, methods, and data processing systems.* 4 (22). pp. 110-119.

Sadykov, S.S., Savicheva, S.V. 2011. Identification of the actual plane of objects based on a single feature of their points of external circuits. *Information Technology.* 8. pp. 13-16.

Sadykov, S.S., Savicheva, S.V. 2012. Study imposing flat objects in sight STZ. *Instrument.* 2. pp. 14-19

# 11

# HUMAN ACTION POSELETS ESTIMATION VIA COLOR G-SURF IN STILL IMAGES

M. Favorskaya *, D. Novikov, Y. Savitskaya

Institute of Informatics and Telecommunications, Siberian State Aerospace University, 31 Krasnoyarsky Rabochy av., Krasnoyarsk, 660014 Russian Federation - (favorskaya, novikov_dms)@sibsau.ru, juliklapa@yandex.ru

**Commission WG V/5, WG III/3**

**KEY WORDS:** Human Action, Poselets, Gauge-SURF, Random Forest, Still Image

**ABSTRACT:**

Human activity is a persistent subject of interest in the last decade. On the one hand, video sequences provide a huge volume of motion information in order to recognize the human active actions. On the other hand, the spatial information about static human poses is valuable for human action recognition. Poselets were introduced as latent variables representing a configuration for mutual locations of body parts and allowing different views of description. In current research, some modifications of Speeded-Up Robust Features (SURF) invariant to affine geometrical transforms and illumination changes were tested. First, a grid of rectangles is imposed on object of interest in a still image. Second, sparse descriptor based on Gauge-SURF (G-SURF) invariant to color/lighting changes is constructed for each rectangle separately. A common Spatial POselet Descriptor (SPOD) aggregates the SPODs of rectangles with following random forest classification in order to receive fast classification results. The proposed approach was tested on samples from PASCAL Visual Object Classes (VOC) Dataset and Challenge 2010 providing accuracy 61-68% for all possible 3D poses locations and 82-86% for front poses locations regarding to nine action categories.

## 1. INTRODUCTION

Recognition of human active actions is employed in many tasks of computer vision, e.g., video surveillance, human detection, activity analysis, scene analysis, image annotation, image and video retrieval, augmented reality, human-computer interaction, among others. Conventional methods of human action recognition use motion information in videos (Laptev and Lindeberg T, 2003, Zhen et al., 2013), which can be applied for object capturing and then for trajectory motion analysis as the temporal component in human action classification. During action classification, the spatial component plays a significant role that makes reasonable to develop methods for human action recognition in still images.

Methods used to estimate the human poses can be classified in four categories: model-based, example-based, pictorial structure-based, and poselet-based approaches. The model-based methods predefine a parametric body model and find the pose matching based on labelled extracted features. A graphical model of human-object interactions was developed by Gupta et al. (Gupta et al., 2009) including reach motions, manipulation motions, and object reactions. Such models are often built on silhouette-based representation of body parts or edge information. The main disadvantage deals with difficulties in design of parametric body model and pose sub-models.

The example-based methods do not use a global modelling structure and store a set of images with corresponding pose descriptions. In this framework, two problems appear as relevant descriptors and fast search. Three shape descriptors – Fourier descriptors, shape contexts, and Hu moments were compared by Poppe and Poel (Poppe and Poel, 2006) for representation of human silhouettes. They experimented with deformed silhouettes robustness to body sizes, viewpoint, and noise. Shakhnarovich et al (Shakhnarovich et al., 2003) proposed interesting hashing-based search technique for pose estimation relevant to pose examples in a large database. A fast pruning method based on shape contexts in order to speed up the search for similar body poses was presented by Wang et al. (Wang et al., 2006). High computational cost of searches in the high-dimensional spaces inside large datasets is the major drawback of example-based methods.

The pictorial structure-based methods represent poses as the cues using prior information of a human body structure (Felzenszwalb and Huttenlocher, 2005). In this approach, the histogram-based methods prevail. It may be circular histograms of spatial and orientation binning (Ikizler et al., 2008) or the most popular Histogram of Oriented Gradients (HOG) (Dalal, and Triggs, 2005) with multiple modifications. The last research was the pioneer investigation in pose descriptor construction based on non-negative matrix factorization. The action classes were represented by the HOGs of pose primitives with following simple histogram comparison for action recognition. This technique works well in typical cases but it fails in occlusions or significant changes of camera viewpoints. To overcome these problems, Delaitre et al. (Delaitre et al., 2010) proposed a Bag-of-Features (BoF) approach for human action recognition in still images in combination with Support Vector Machine (SVM) classification. They combined the statistical and part-based representations integrating a person-centric description in cluttered background.

The poselet-based methods provide rich information about locations of body parts. They are built on 3D pose images. The poselets were introduced by Bourdev and Malik (Bourdev and Malik, 2009) for person detection in natural framework. The detailed literature review of methods from the fourth category is

* Corresponding author

presented in following section. Better pose descriptor means a wide invariance to various geometric transforms, lighting, viewpoints, and image warping in general case. This difficult task is in the field of current and following investigations. Our contribution deals with the proposed color Gauge-SURF with selection of special imposed grids in human body image for human action recognition. The original Gauge-SURF was extended by invariance to color and lighting changes.

In the following, Section 2 gives a brief review of high-level and low-level evaluations of poselets. The G-SURF background is presented in Section 3. The proposed methodology of poselets estimation is explained in Section 4. The experimental results are presented and analyzed in Section 5. The paper is concluded in Section 6.

## 2. RELATED WORK

The poselet as a subject of interest is 2D still non-segmented image of configured body parts in 3D space (head, shoulders, arms, torso, legs) capturing a part of neighbouring background. Thus, the poselet of single pose is a collection of 2D still images received from various shooting viewpoints, in different scales and lighting conditions. Sometimes, the presence of articulation makes the pose estimation harder. Although the pose estimation is used for human action recognition, one may consider the pose estimation as a separate task of computer vision. Figure 1 depicts some examples of human actions.

Figure 1. Human pose examples

Since only the spatial information is available in still images, one can represent information as high-level cues and low-level cues/features. Description of human body, body parts, action-related objects, human object interaction, and scene context are included in high-level cues. Typical low-level features are a Dense sampling of Scale Invariant Feature Transform (DSIFT), HOG, Shape Context (SC), GIST, and some other features. Short surveys of high-level and low-level evaluation methods are situated in Sections 2.1 and 2.2, respectively, while existing 3D-based poselet methods are discussed in Section 2.3.

### 2.1 High-level Evaluation

A human body image is an important cue in human action recognition, which can be detected automatically or manually labelled. Usually a bounding box is used to indicate a location of a person. Some approaches extract features in areas within or surrounding the human bounding boxes. Delaitre et al. (Delaitre et al., 2010) defined a person setting in each image in one and a half time more that the sizes of human bounding box. Then these regions are resized up to 300 pixels (in larger size) and analyzed using low-level features.

Some methods extract contour information of human body and body parts from still images. Wang et al. (Wang et al., 2006) exploited the overall coarse shape of human body as a collection of edge points obtained via Canny edge detector. Then the received features were classified and categorized into different actions. Also semantic features can be used to describe the actions in images with the human body (Yao et al., 2011). The attributes were related to verbs in a human language and remained visual words from annotation system.

Yang et al. (Yang et al., 2010) developed a coarse example-based poselets representation, when each body part may have more than 20 poselets concerning to different body poses. They constructed a set of four corresponding body parts $L = \{l_0, l_1, ..., l_{k-1}\}$ denoting the upper-body, legs, left-arm, and right-arm. Often a graph is a good model to represent the relations between different body parts. Raja et al. (Raja et al., 2011) constructed a graphical model containing six nodes: the action label and five body parts correspond to head *H*, right-hand *RH*, left-hand *LH*, right-foot *RF*, and left-foot *LF*. The links between nodes encode action-dependent constraints on the relative positions of body parts.

In many scenarios, a person relates with other objects, e.g. phone, ball, animal, etc. These objects serve as a source of reliable information about category of human action. Some approaches analyse individual objects separately while other methods consider them as a scene context. As a result, some methods were developed as a Human Object Interaction (HOI), for example, weakly supervised method proposed by Prest et al. (Prest et al., 2012).

### 2.2 Low-level Evaluation

The high-level evaluation is usually based on various low-level scores. The DSIFT features are extracted from many image patches with following clustering to obtain a limited number of "keywords", which are grouped in a codebook. Many methods use the DSIFT features due to their possibility for direct classification of human actions. One can mention the researches of Delaitre et al. (Delaitre et al., 2010), Yao et al. (Yao et al., 2011), etc.

The HOG descriptor is very popular for pedestrian detection (Dalal and Triggs, 2005). The HOG descriptor counts the occurrences of discrete gradient orientations within a local image patch similar to the edge orientation histogram, the SIFT, and the SC. The SC is useful to detect and segment the human contour; however, this technique is crucial for high-level cue representation of human body silhouettes.

The spatial envelop or GIST was proposed by Oliva and Torralba (Oliva and Torralba, 2001). A set of spatial properties in a scene can be computed by the GIST method, which provides the abstract category representations of a scene based on integrated background information. This approach has been used by Gupta et al. (Gupta et al., 2009), Prest et al. (Prest et al., 2012), among others. One can mention the development of other approaches based on SURF (Bay et al., 2008), Circular

Histogram of Oriented Rectangles (CHORs) (Ikizler et al., 2008), Adaboost classifiers (Gupta et al., 2009).

### 2.3 Towards to 3D Representation

All variety of poselet-based methods may be classified according to various criteria. A viewpoint dependence/ independence in 2D/2D spaces, respectively, can be considered the main criterion. Bourdev and Malik (Bourdev and Malik, 2009) were the first, who formulated the task of human pose estimation and recognition as 3D object representation. They constructed the body part detectors trained from annotated data of joint locations of people and based on patches similarities. As a result, the poselet activation vector consisting of poselets inside the bounding box was introduced. The SVM classifier was used to recognize these patches. The distribution of these joints and personal bounding boxes can be obtained to each poselet.

Russakovsky et al. (Russakovsky et al., 2012) developed an Object-Centric spatial Pooling (OCP) approach for detection an object of interest. The local OCP information is used to pool the foreground and background features. Khan et al. (Khan et al., 2013) applied a comprehensive evaluation of color descriptors with combination of shape features.

Some methods deal with detection of view-independent objects using 3D object models. Glasner et al. (Glasner et al., 2011) used a viewpoint estimation method for rigid 3D objects from 2D images by voting method for efficient accumulation of evidence. This method was tested on rigid car data. Fidler et al. (Fidler et al., 2012) developed a method for localizing objects in 3D space by enclosing them within tightly oriented 3D bounding boxes. This model represents an object class as a deformable 3D cuboid by anchors of body parts in 3D box. Hejrati and Ramanan (Hejrati and Ramanan, 2012) developed a two-stage model, when, first, a large number of effective views and shapes are modelled using a small number of local view-based templates and, second, these estimates are refined by an explicit 3D model of the shape and viewpoint.

Another issue is a computational speed required for view-independent object detection because the most methods generate the classifiers at multiple locations and scales. Sometimes, the shared features are extracted to reduce a number of classifiers and the runtime complexity correspondingly. Razavi et al. (Razavi et al., 2010) used an extension of Hough-based object detection and built a shared codebook by jointly considering several viewpoints. Tosato et al. (Tosato et al., 2010) modelled a human image as a hierarchy of fixed overlapping parts. Each part was trained using a boosted classifier learned using Logicboost algorithm. Velaldi et al. (Velaldi et al., 2009) introduced a three-stage SVM classifier combining linear, quasi-linear, and non-linear kernels for object detection.

## 3. G-SURF BACKGROUND

The original SURF defines the determinant of the approximate Hessian matrix in the points, where the determinant has maximum values (Bay et al., 2008). The Hessian matrix $\mathbf{H}(\mathbf{p}; \sigma)$ is defined by Equation 1:

$$\mathbf{H}(\mathbf{p};\sigma)=\begin{bmatrix} L_{xx}(\mathbf{p},\sigma) & L_{xy}(\mathbf{p},\sigma) \\ L_{xy}(\mathbf{p},\sigma) & L_{yy}(\mathbf{p},\sigma) \end{bmatrix} \quad (1)$$

where $\mathbf{p} = (x, y)^{\mathrm{T}}$ = a point in an image $I$

$\sigma$ = a scale factor

$L_{xx}(\mathbf{p}, \sigma)$ = a convolution of an image $I(\mathbf{p})$ in a point $\mathbf{p}$ with a Gaussian second order derivative along direction OX

A convolution $L_{xx}(\mathbf{p}, \sigma)$ is determined by Equation 2:

$$L_{xx}(\mathbf{p},\sigma)=I(\mathbf{p})*\frac{\partial^2}{\partial x^2}g(\sigma) \quad (2)$$

and similarly for $L_{xy}(\mathbf{p}, \sigma)$ and $L_{yy}(\mathbf{p}, \sigma)$ along diagonal and OY directions, respectively.

Alcantarilla et al. (Alcantarilla et al., 2013) developed a novel family of multi-scale local feature descriptors called as Gauge-SURF (G-SURF). In this case, every pixel in the image is described by 2D local structure. The multi-scale gauge derivatives are invariant to rotations and shifts. Additionally, they describe the non-linear diffusion processes. The use of G-SURF makes blurring locally adaptive to the region so that noise becomes blurred, whereas details or edges remain unaffected. Such local structures are described by Equation 3:

$$\begin{aligned} \vec{w} &= \left(\frac{\partial L(\mathbf{p},\sigma)}{\partial x},\frac{\partial L(\mathbf{p},\sigma)}{\partial y}\right)= \\ &\frac{1}{\sqrt{L_x^2(\mathbf{p},\sigma)+L_y^2(\mathbf{p},\sigma)}}\cdot\left(L_x(\mathbf{p},\sigma),L_y(\mathbf{p},\sigma)\right); \\ \vec{v} &= \left(\frac{\partial L(\mathbf{p},\sigma)}{\partial y},-\frac{\partial L(\mathbf{p},\sigma)}{\partial x}\right)= \\ &\frac{1}{\sqrt{L_x^2(\mathbf{p},\sigma)+L_y^2(\mathbf{p},\sigma)}}\cdot\left(L_y(\mathbf{p},\sigma),-L_x(\mathbf{p},\sigma)\right) \end{aligned} \quad (3)$$

where $\vec{w}$ = a gradient vector

$\vec{v}$ = a perpendicular direction vector

$\sigma$ = a kernel's standard deviation or scale parameter

$L(\mathbf{p}, \sigma)$ = a convolution of image $I(\mathbf{p})$ with 2D Gaussian kernel $g(\mathbf{p}, \sigma)$

Using gauge coordinates, one can obtain a set of derivatives invariant to any order and scale. The second-order gauge derivatives $L_{vv}(\mathbf{p}, \sigma)$ and $L_{ww}(\mathbf{p}, \sigma)$ are in special interest. They can be obtained as a product of gradients in $\vec{w}$ and $\vec{v}$ directions and the $2 \times 2$ second order derivatives or Hessian matrix provided by Equation 4:

$$\begin{aligned} L_{ww}(\mathbf{p},\sigma) &= \frac{1}{L_x^2(\mathbf{p},\sigma)+L_y^2(\mathbf{p},\sigma)}\begin{pmatrix} L_x(\mathbf{p},\sigma) & L_y(\mathbf{p},\sigma)\end{pmatrix} \\ &\begin{pmatrix} L_{xx}(\mathbf{p},\sigma) & L_{xy}(\mathbf{p},\sigma) \\ L_{yx}(\mathbf{p},\sigma) & L_{yy}(\mathbf{p},\sigma)\end{pmatrix}\begin{pmatrix} L_x(\mathbf{p},\sigma) \\ L_y(\mathbf{p},\sigma)\end{pmatrix}; \\ L_{vv}(\mathbf{p},\sigma) &= \frac{1}{L_x^2(\mathbf{p},\sigma)+L_y^2(\mathbf{p},\sigma)}\begin{pmatrix} L_y(\mathbf{p},\sigma) & -L_x(\mathbf{p},\sigma)\end{pmatrix} \\ &\begin{pmatrix} L_{xx}(\mathbf{p},\sigma) & L_{xy}(\mathbf{p},\sigma) \\ L_{yx}(\mathbf{p},\sigma) & L_{yy}(\mathbf{p},\sigma)\end{pmatrix}\begin{pmatrix} L_y(\mathbf{p},\sigma) \\ -L_x(\mathbf{p},\sigma)\end{pmatrix} \end{aligned} \quad (4)$$

The G-SURF descriptor is based on the original SURF descriptor. Mention some modifications such as Modified Up-right SURF (MU-SURF) descriptor based on Haar wavelets responses and two Gaussian weighting steps, the Center Surround Extremas (CenSurE) for approximation the bi-level Laplacian of Gaussian using boxes and octagons, the Speeded Up Surround Extrema (SUSurE), which is a fast modification of the MU-SURF, and the CenSurE descriptors for mobile devices. These modifications well describe edges, angles, boundaries of unknown objects in an image. However, they poorly concern color information, which is useful in analysis of still images.

## 4. ESTIMATION OF POSELETS

The poselets estimation is concerned to identification task, when a number of classes is restricted by a finite set of human poses describing in feature space by some descriptors. The complexity deals with another issue – the great variants of 3D images of a single pose mapping in 2D still images. Nowadays, this problem has not been solved, and many authors develop heuristic algorithms, more or less successful.

The structure of bounding boxes, in common case, grid, for poselet detection is discussed in Section 4.1. The proposed color G-SURF family is represented in Section 4.2. Section 4.3 provides a classification procedure based on random forest in order to identify a testing poselet.

### 4.1 Imposed Grid for Poselet Detection

For poselet recognition, it is required to capture a salient part of one's pose from a given viewpoint and impose a set of corresponding rectangle boxes at given orientation, position, and scale. Many authors use the predetermined aspect ratios sometimes with normalization of distance between hips and shoulders. Bourdev and Malik (Bourdev and Malik, 2009) used the poselets with following of aspect ratios 96 × 64, 64 × 64, 64 × 96, and 128 × 64 pixels. Their algorithm was trained using 300 poselets of each type of pose. Then a model predicting a bounding box for each poselet was fitted. Additionally, the overlapping bounding boxes were considered with overlapping area more 20%. These scores were added to the basic set. The 1200 dimensional vector to estimate the human pose was constructed, besides the poses of head and torso were considered separately. This approach was developed by Ko et al. (Ko et al., 2015), when five aspect ratios such as 96 × 64, 64 × 64, 64 × 96, 64 × 128, and 128 × 64 pixels were used in order to consider variations of human poses. These authors modified the selecting algorithm of the action poselets using the modified Hausdorff distance with Epanechnikiv kernel. The descriptor is based on Oriented Center-Symmetric Local Binary Patterns (OCS-LBPs) due to their low computational complexity. Such approach proposes the overlapping of aspect ratios with preliminary rough body parts segmentation. Background is involved in this grid. Often segmentation is implemented manually.

It is mindless to use the predefined aspect ratios as well as the random aspect ratios. It will be reasonable to find Regions Of Interest (ROIs) or pose primitives accurately. If a video sequence is available, then this issue is the solved problem, when the ROIs are detected from the temporal domain in previous frames (Favorskaya, 2012; Favorskaya et al., 2015). Also the hybrid methods based on optical flow and learning of salient regions are possible (e.g., Eweiwi et al. 2015).

If a single still image is available, then one of pixel-based segmentation methods can extract large areas with identical texture under assumption that such large areas are the body parts. These areas representing the bounding boxes are the basis of the SPOD constructions, which later are aggregated in a common SPOD for classification. This task is similar to automatic image annotation. Our recommendations deal with the use of J-SEG algorithm or similar ones. If this assumption is not fulfilled, then additional human segmentation methods are required in order to receive the cropped image of human body.

Let us suppose that cropped still image of human body is obtained, and six aspect ratios are used including head, torso, left arm, right arm, left leg, and right leg. Each of the arm and leg boxes is composed from two sub-boxes including hands and feet. A number of imposed aspect ratios can be reduced due to human action and human position. Moreover, a number of imposed aspect ratios serve as a weak classifier for human action categorization. For example, the active action "phoning" may include an image of head and any arm while the active action "running" may involve images of torso and both legs.

### 4.2 The Proposed Color G-SURF Family

Under various conditions of color/lighting changes in human pose images, it is important to develop a descriptor invariant to these changes. Our contribution deals with the development of original G-SURF, which is invariant to geometrical (affine) distortions. Family of the proposed color G-SURF descriptors includes the following components.

The rg G-SURF (rgG-SURF) includes the chromaticity components $r$ and $g$ invariant to scale and light changes. The rg histogram is based on the normalized RGB color model, where the components $r$ and $g$ describe the color information by Equation 5 ($b$ is redundant as $r + g + b = 1$):

$$\begin{pmatrix} r \\ g \\ b \end{pmatrix} = \begin{pmatrix} \dfrac{R}{R+G+B} \\ \dfrac{G}{R+G+B} \\ \dfrac{B}{R+G+B} \end{pmatrix} \tag{5}$$

where $R, G, B$ = Red, Green, Blue components in RGB color space, respectively
$r, g, b$ = normalized components

Because of normalization, the components $r$ and $g$ are scale-invariant and invariant to lighting changes, shadows and shading (Gevers et al., 2006). This descriptor has 45 dimensions.

The Opponent G-SURF (OppG-SURF) analyzes three channels in the opponent color space using G-SURF descriptor (Equation 6):

$$\begin{pmatrix} O_1 \\ O_2 \\ O_3 \end{pmatrix} = \begin{pmatrix} \dfrac{R-G}{\sqrt{2}} \\ \dfrac{R+G-2B}{\sqrt{6}} \\ \dfrac{R+G+B}{\sqrt{3}} \end{pmatrix} \tag{6}$$

where $O_1$, $O_2$ = two channels in the opponent color space providing color invariance
$O_3$ = the channel in the opponent color space providing intensity invariance

The Hue G-SURF (HueG-SURF) is constructed by concatenation of a hue histogram in Hue Saturation Value (HSV) color space with a histogram of G-SURF descriptors. Such descriptor is scale-invariant and shift-invariant with respect to light intensity due to a Hue histogram. In a Hue histogram, the hue is weighted by the saturation of a pixel and reflected the instabilities in hue. This histogram has 36 dimensions.

The transformed RGB color G-SURF (RGBG-SURF) descriptor is computed for each normalized RGB channel by Equation 7:

$$\begin{pmatrix} R' \\ G' \\ B' \end{pmatrix} = \begin{pmatrix} \frac{R-\mu_R}{\sigma_R} \\ \frac{G-\mu_G}{\sigma_G} \\ \frac{B-\mu_B}{\sigma_B} \end{pmatrix} \quad (7)$$

where $\mu_R$, $\mu_G$, $\mu_B$ = means of the distribution in RGB channels computed in a chosen region of image, respectively
$\sigma_R$, $\sigma_G$, $\sigma_B$ = standard deviations of the distribution in RGB channels computed in a chosen region of image, respectively

The RGBG-SURF is invariant to scale and shift with respect to light intensity (Van de Sande et al., 2009), while a classic RGB histogram is not invariant to changes in lighting conditions The histogram has 45 dimensions.

Also one can mention additional color descriptors with some invariance to color/lighting conditions such as the color moment histogram, the hue-saturation descriptor, the color names, the discriminative color descriptor, among others. However, the experiments show that the rg G-SURF, the OppG-SURF, the HueG-SURF, and the RGBG-SURF provide better results in poselets estimation.

### 4.3 Classification Procedure

Various classifiers have been applied for object classification such as SVM, Boosting-algorithms, and random forest. The SVM classifier is the well proved technique for general classification. However, SVM is not suitable, when the features have high dimensionality and a set of analysed images is huge. The Boosting-algorithms such as AdaBoost or GentlBoost are the popular machine learning methods but their performance depends critically on the choice of weak classifiers.

A random forest is one of the most popular tree-based classification approaches, which is effective in a large variety of high-dimensional tasks such as object detection and object tracking (Ko et al., 2013). The random forest is an ensemble classifier of several randomized decision trees, which have a capacity to analyze big data at high training and runtime speeds. Each tree is grown using some type of randomization. The structure of each tree is binary, and all trees are created in a top-down manner (Breiman, 2001). During the training stage, the random forest uses a random subset from training data initially. In common case, at node $n$ the training data $D_n$ are split interactively into left and right subsets using a threshold and the split function according to Equation 8:

$$D_l = \{ i \in D_n | F(v_i) < th \}; \quad D_r = D_n \setminus D_l \quad (8)$$

where $D_l$, $D_r$ = left and right subsets, respectively
$th$ = a threshold
$F(v_i)$ = a split function
$v_i$ = $i$th feature vector

The threshold $T$ is selected randomly in the range $T \in (\min F(v_i), \max F(v_i))$. The use of an ensemble of trees trained with small random subsets increases a speed of training and reduces amount of overfitting. Because a random forest dismisses the spatial information of local regions (patches) within a detection window, it produces some false positives, especially when a background has a similar appearance.

In this study, the common SPOD in a view of the concatenated histograms (OX and OY) of color G-SURF descriptors in each bounding box is used as a feature for random forest classification. In training stage, each tree $T$ is constructed based on a set of examples $Ex_i$ of a poselet $Ex_i = (e_i, c_i)$, where $e_i$ is a poselet example and $c_i$ is a class label of a poselet example. The positive poselet examples are marked by a class label $c_i = 1$, and the negative poselet examples gain a class label $c_i = 0$. Samples of the other body parts including a background are concerned to the negative poselet examples. During tree construction, each leaf node $L$ stores the class information $C_L$ of the examples. When only positive poselet examples are reached at node $N$, then $C_L = 1$. Value of $C_L$ is proportional to a number of positive poselet examples. A split function is assigned to each non-leaf node in such manner that the uncertainties in class labels ought to be reduced towards leaves. In testing stage, the current element descriptor of a poselet is checked using the created binary trees by sequential comparison with value of split function in each node.

## 5. EXPERIMENTS AND DISCUSSIONS

The PASCAL Visual Object Classes (VOC) Dataset and Challenge 2010[1] was used in training and testing stages. It includes 10,103 and 9,637 images for training and testing, and also 23,374 and 22,992 objects for training and testing, respectively. For experiments, a part of dataset called Action Classification Taster Challenge 2010 was applied. In this part, each person image with 2D pose was annotated by "layouts" of the person with bounding boxes: head, hands, and feets. In total, 322 images containing 439 people were annotated as the training/validation set. The test set involves 441 images with a single person. Figure 2 depicts example images for nine action categories (phoning, walking, running, taking photo, playing instrument, riding bike, riding horse, reading, and using computer) with detected G-SURFs in bounding boxes. Sometimes, a non-significant number of SURFs and G-SURFs can be detected in bounding box due to very small sizes of box, e.g., hand and foot boxes. In these cases, a Kuwahara filter with radius $R = 3$ was applied in order to receive increased number of feature points. As one can see in Figure 3, where the scaled fragments of images are represented, this additional procedure helps to decide this problem partly.

[1]http://pascallin.ecs.soton.ac.uk/challenges/VOC/

Figure 2. The processed person layouts with detected G-SURF

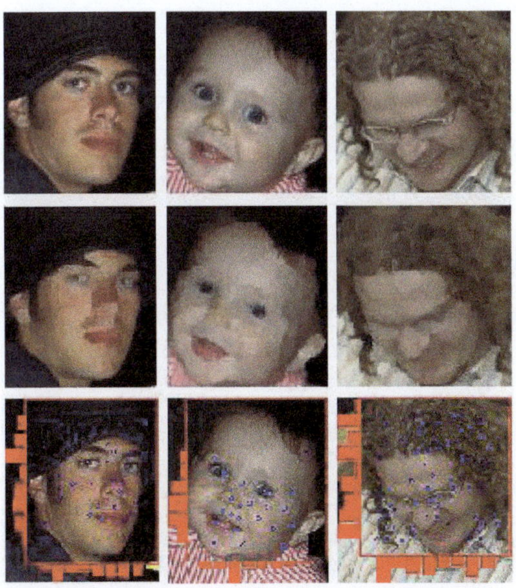

Figure 3. Scaled fragments with increased number of G-SURFs

Then histogram descriptors had been built in bounding boxes concerning to various body parts. They were classified by random forest technique. During training stage, the separate random forest for heads, hands, and feet were constructed. An example of random forest training (one from a set of trees) is located in Figure 4.

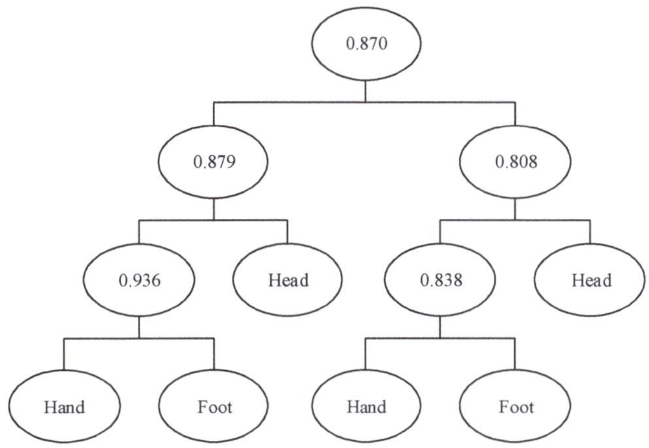


Figure 4. Random forest training

Except SURF and G-SURF, all four types of color G-SURF were calculated in corresponding color spaces. The experimental results show that the rgG-SURF and the HueG-SURG demonstrated the worst results of recognition and false ratios because they are not robust to light intensity scaling and the G-SURF provided a partial robustness. Table 1 includes the precision evaluation for nine action categories.

| Category | SURF | G-SURF | OppG-SURF | RGBG-SURF |
|---|---|---|---|---|
| Phoning | 73.7 | 75.1 | 76.8 | 76.2 |
| Walking | 62.2 | 63.5 | 64.6 | 64.1 |
| Running | 59.6 | 60.4 | 61.3 | 61.0 |
| Taking photo | 77.8 | 78.1 | 78.9 | 78.4 |
| Playing instrument | 61.1 | 62.3 | 63.1 | 62.9 |
| Riding bike | 58.3 | 59.7 | 60.2 | 59.7 |
| Riding horse | 76.6 | 78.2 | 78.4 | 78.2 |
| Reading | 68.2 | 69.7 | 70.3 | 69.3 |
| Using computer | 83.4 | 85.1 | 86.0 | 85.4 |

Table 1. Precision results for nine action categories (%)

The analysis of received results shows the common tendency: more sizes of bounding boxes provide large number of feature points that leads to better recognition results. Thus, phoning, playing photo, riding horse, and using computer are characterized with better precision results.

Tables 2 and 3 contain false rejection rates and false acceptance rates. Also one can see that errors have high values. This may be explained by restriction of current statement of problem. In future, additional procedures, e.g., a graph of body parts, skin detection, skeleton representation, Kinect data analysis, and analysis of surrounding objects, among others, will permit the promising results against current results. The current research shows that the analysis of bounding boxes limits estimators significantly.

| Category | SURF | G-SURF | OppG-SURF | RGBG-SURF |
|---|---|---|---|---|
| Phoning | 38.9 | 37.5 | 36.1 | 35.9 |
| Walking | 40.1 | 39.2 | 38.6 | 38.3 |
| Running | 40.2 | 38.1 | 37.2 | 37.1 |
| Taking photo | 39.0 | 38.3 | 36.8 | 36.4 |
| Playing instrument | 49.1 | 47.8 | 45.4 | 45.1 |
| Riding bike | 50.3 | 49.6 | 47.9 | 47.7 |
| Riding horse | 51.4 | 50.2 | 48.7 | 48.4 |
| Reading | 34.8 | 33.7 | 32.5 | 32.3 |
| Using computer | 37.2 | 36.4 | 35.2 | 35.0 |

Table 2. False rejection rates for nine action categories (%)

| Category | SURF | G-SURF | OppG-SURF | RGBG-SURF |
|---|---|---|---|---|
| Phoning | 28.2 | 27.3 | 25.6 | 25.2 |
| Walking | 25.1 | 24.4 | 23.1 | 22.8 |
| Running | 28.2 | 27.1 | 25.7 | 25.3 |
| Taking photo | 21.8 | 21.2 | 20.2 | 20.1 |
| Playing instrument | 49.1 | 47.9 | 46.0 | 45.6 |
| Riding bike | 38.9 | 37.4 | 36.3 | 35.8 |
| Riding horse | 34.4 | 33.2 | 32.5 | 32.1 |
| Reading | 29.8 | 28.5 | 27.3 | 26.9 |
| Using computer | 27.2 | 26.1 | 25.2 | 24.8 |

Table 3. False acceptance rates for nine action categories (%)

False rejection rates are higher in comparison to false acceptance rates. It can be explained by usual difficulties in object recognition using still images relative to video-based object recognition in the spatio-temporal domain.

## 6. CONCLUSION

In this study, the poselets estimation is proposed for recognition of human active actions. Our approach deals with accurate segmentation of body parts in a still image by a possibility of the temporal data extraction from a video sequence or pixel-based segmentation method in order to receive the images of body parts. The experiments with G-SURF lead to the descriptor invariant to color/lighting conditions. The SPOD is based on color G-SURF family. Better results were received using OppG-SURF and RGBG-SURF. For classification of common SPOD, a random forest classification was used as a fast and effective procedure during work with big data representing various poselets of various human active actions. PASCAL VOC Dataset and Challenge 2010 provided the test material for the training and the testing stages. The precision for nine action categories such as phoning, walking, running, taking photo, playing instrument, riding bike, riding horse, reading, and using computer achieves on the average of 61-68% for all 3D poses locations and 82-86% for front poses locations regarding to action categories.

## REFERENCES

Alcantarilla, P.F., Bergasa, L.M., Davison, A.J., 2013. Gauge-SURF descriptors. *Image and Vision Computing*, 31(1), pp. 103-116.

Bay, H., Tuytelaars, T., Van Gool, L., 2008. Surf: Speeded up robust features. *J. Computer Vision and Image Understanding*, 110(3), pp. 346-359.

Bourdev, L., Malik, J., 2009. Poselets: body part detectors trained using 3D human pose annotations. In: *The IEEE Conference on Computer Vision*, Kyoto, Japan, pp. 1365-1372.

Breiman, L., 2001. Random forests. *Mach. Learn.* 45(1), pp. 5-32.

Dalal, N., Triggs, B., 2005. Histograms of oriented gradients for human detection. In: *The IEEE Conference on Computer Vision and Pattern Recognition*, San Diego, CA, USA, pp. 886-893.

Delaitre, V., Laptev, I., Sivic, J., 2010. Recognizing human action in still images: a study of bag-of-features and partial-based representations. In: *The British Machine Vision Conference*, Aberystwyth, UK, pp. 1-11.

Eweiwi, A., Cheema, M.S., Bauckhage, C., 2015. Action recognition in still images by learning spatial interest regions from videos. *Pattern Recognition Letters*, 51, pp. 8-15.

Favorskaya, M., 2012. Motion estimation for objects analysis and detection in videos. In: Kountchev, R., Nakamatsu, K. (Eds) *Advances in Reasoning-Based Image Processing Intelligent Systems*, ISRL vol. 29, Springer-Verlag Berlin Heidelberg, pp. 211-253.

Favorskaya, M., Pyankov, D., Popov, A., 2015. Accurate motion estimation based on moment invariants and high order statistics for frames interpolation in stereo vision. In: Tweedale, J.W., Jain, L.C., Watada, J., Howlett, R.J. (Eds) *Knowledge-Based Information Systems in Practice*, SIST vol. 30, Springer International Publishing Switzerland, pp. 329-351.

Felzenszwalb, P.F., Huttenlocher, D.P., 2005. Pictorial structures for object recognition. *Int. J. Comput. Vis.*, 61(1), pp. 55-79.

Fidler, S., Dickinson, S., Urtasun, R., 2012. 3D object detection and viewpoint estimation with a deformable 3D cuboid model. In: *The Conference on Neural Information Processing Systems*, Lake Tahoe, Nevada, USA, pp .620-628.

Gevers, T., van de Weijer, J., Stokman, H., 2006. Color feature detection: an overview. In Lukac, R., Plataniotis, K.N. (Eds) *Color image processing: methods and applications*. University of Toronto, Ontario, Canada: CRC press, pp. 203-226.

Glasner, D., Galun, M., Alpert, S., Basri, R., Shakhnarovich, G., 2011. Viewpoint-aware object detection and pose estimation., In: *The IEEE Conference on Computer Vision*, Barcelona, Spain, pp.1275-1282.

Gupta, A., Kembhavi, A., Davis, L.S., 2009. Observing human-object interactions: using spatial and functional compatibility for recognition. *IEEE Trans. on Pattern Anal. Mach. Intell.*, 31(10), pp. 1775-1789.

Hejrati, M., Ramanan, D., 2012. Analyzing 3D objects in cluttered images, In: *The Annual Conference on Neural Information Processing Systems*, Lake Tahoe, Nevada, USA, pp.602-610.

Ikizler, N., Cinbis, R.G., Pehlivan, S., Duygulu P., 2008. Recognizing actions from still images. In: *The International Conference of Pattern Recognition*, Tampa, Florida, USA, pp. 1-4.

Khan, F.S., Anwer, R.M., van de Weijer, J., Bagdanov, A.D., Lopez A.M., Felsberg M. 2013. Coloring Action Recognition in Still Images. *Int. J. of Computer Vision*, 105(3), pp. 205-221.

Ko, B.C., Kwak, J.Y., Nam, J.Y., 2013. Human tracking in thermal images using adaptive particle filters with online random forest learning. *Opt. Eng.* 52(11), pp. 1-14.

Ko, B.C., Hong, J.H., Nam, J.Y., 2015. View-independent object detection using shared local features. *J. of Visual Languages and Computing*, 28, pp. 56-70.

Laptev, I., Lindeberg, T., 2003. Space-time interest points. In: *The IEEE International Conference on Computer Vision*, Nice, France, pp.432-439.

Oliva A., Torralba A., 2001. Modeling the shape of the scene: a holistic representation of the spatial envelope. *Int. J. Comput. Vis.*, 42(3), pp. 145-175.

Poppe, R., Poel, M., 2006. Comparison of silhouette shape descriptors for example-based human pose recovery. In: *The International Conference on Automatic Face and Gesture Recognition*, Southampton, UK, pp.541-546.

Prest, A., Schmid, C., Ferrari, V., 2012. Weakly supervised learning of interactions between humans and objects. *IEEE Trans. on Pattern Anal. Mach. Intell.*, 34(3), pp. 601-614.

Raja, K., Laptev, I., Pérez, P., Oisel, L., 2011. Joint pose estimation and action recognition in image graphs. In: *The IEEE International Conference on Image Processing*, Brussels, Belgium, pp.25-28.

Razavi, N., Gall, J., Gool, L.V. ,2010. Back projection revisited: scalable multi-view object detection and similarity metrics for detections. In: *The European Conference on Computer Vision*, Heraklion, Crete, Greece, pp.620-633.

Russakovsky O., Lin Y., Yu K., Fei-Fei L., 2012. Object-centric spatial pooling for image classification, In: *The European Conference on Computer Vision*, Florence, Italy, pp.1-15.

Shakhnarovich, G., Viola, P., Darrell, R., 2003. Fast pose estimation with parameter-sensitive hashing. In: *The IEEE International Conference on Computer Vision*, Nice, France, pp.750-757.

Tosato, D., Farenzena, M., Cristani, M., Murino, V., 2010. Part-based human detection on Riemannian manifolds. In: *The IEEE International Conference on Image Processing*, Hong Kong, China, pp. 3469-3472.

Van de Sande, K.E.A., Gevers, T., Snoek, C.G.M., 2009. Evaluating color descriptors for object and scene recognition. *IEEE Trans. on Pattern Anal. Mach. Intell.*, 32(9), pp. 1582-1596.

Velaldi, A., Gulshan, V., Varma, M., Zisserman, A., 2009. Multiple kernels for object detection. In: *The IEEE Conference on Computer Vision*, Kyoto, Japan, pp.606-613.

Wang, Y., Jiang, H., Drew, M.S., Li, Z.N., Mori, G., 2006. Unsupervised discovery of action classes. In: *The IEEE Conference on Computer Vision and Pattern Recognition*, New York, USA, pp. 1654-1661.

Yang, W., Wang, Y., Mori, G., 2010. Recognizing human actions from still images with latent poses. In: *The IEEE Conference on Computer Vision and Pattern Recognition*, San Francisco, USA, pp. 2030-2037.

Yao, B., Jiang, X., Khosla, A., Lin, A., Guibas, L., Fei-Fei, L., 2011. Human action recognition by learning bases of action attributes and parts. In: *The IEEE International Conference on Computer Vision*, Barcelona, Spain, pp.1331-1338.

Zhen, X., Shao, L., Tao, D., Li, X., 2013. Embedding motion and structure features for action recognition. *IEEE Trans. on Circuits Syst. Video Technol.*, 23(7), pp. 1182-1190.

# 12

# THE INFLUENCE OF THE TIME EQUATION ON REMOTE SENSING DATA INTERPRETATION

B. Fichtelmann*, E. Borg, E. Schwarz

German Aerospace Center (DLR), Earth Observation Center, German Remote Sensing Data Center, National Ground Segment, 17235 Neustrelitz, Germany – Bernd.Fichtelmann@dlr.de

**KEY WORDS:** Equator crossing time, Equation of time, Land surface temperature, Trend analysis, Validation

**ABSTRACT:**

The interpretation of optical Earth observation data (remote sensing data from satellites) requires knowledge of the exact geographic position of each pixel as well as the exact local acquisition time. But these parameters are not available in each case. If a satellite has a sun-synchronous orbit, equator crossing time (ECT) can be used to determine the local crossing time (LCT) and its corresponding solar zenith distance. Relation between local equator crossing time (LECT) and LCT is given by orbit geometry. The calculation is based on ECT of satellite. The method of actual ECT determination for different satellites on basis of the two-line-elements (TLE), available for their full lifetime period and with help of orbit prediction package is well known. For land surface temperature (LST) studies mean solar conditions are commonly used in the relation between ECT given in Coordinated Universal Time (UTC) and LECT given in hours, thus neglecting the difference between mean and real Sun time (MST, RST). Its difference is described by the equation of time (ET). Of particular importance is the variation of LECT during the year within about ±15 minutes. This is in each case the variation of LECT of a satellite, including satellites with stable orbit as LANDSAT (L8 around 10:05 a.m.) or ENVISAT (around 10:00 a.m.). In case of NOAA satellites the variation of LECT is overlaid by a long-term orbital drift. Ignatov et al. (2004) developed a method to describe the drift-based variation of LECT that can be viewed as a formal mathematical approximation of a periodic function with one or two Fourier terms. But, nevertheless, ET is not included in actual studies of LST.

Our paper aims to demonstrate the possible influence of equation of time on simple examples of data interpretation, e.g. NDVI.

## 1. INTRODUCTION

The exact derivation of geophysical and biophysical parameters based on optical Earth observation data (remote sensing data from satellite) requires the calibration and transformation of digital number (DN) to top-of-atmosphere reflectance ρ, as described for LANDSAT 7/ETM+ - data in the LANDSAT handbook (LANDSAT Project Science Office, 1998). This calibration assumes the exact determination of solar illumination, depending on the solar zenith angle. The variation of the solar zenith angle $X$ at each geographical point can be calculated in dependence of its latitude and longitude, as function of day number of the year, of declination of Earth axis and local time. Furthermore the solar radiation is modified by variable distance between Sun and Earth. The zenith angle has its daily minimum at local noon. During a year the zenith distance has at noon a minimum or maximum at summer or winter solstice. If not available, the coordinates of each pixel of an Earth observation scene can be calculated with orbit geometry and the given corner coordinates of the scene, an ideal sphere assumed (e.g. Borg et al., 2015). For instance the sun elevation angle $(90^o\text{-}X)$ of the centre coordinate is given for a Landsat TM or ETM+ product. For detailed studies based on reflectance it is usual to determine the Local Time (LT) for each pixel of the image. Such an algorithm is based on a relation between the Local Equator Crossing Time (LECT) of sun-synchronous orbit and Local Crossing Time (LCT) of each orbit coordinate, meaning local time when the satellite is in the zenith of this point (crossing) (Ignatov et al., 2004, Wu and McAvaney, 2006). LT of the other pixels of the same image line can be calculated with help of longitudinal difference. One problem arises from the fact that the specification of LECT for different sun-synchronous satellites in different papers and documents suggests a nearly constant value apart from drift effects of orbit, connected with variation of ECT in hours (LECT for mean Sun time). For example, in the Aster project (Aster Science Project, 2001) an equatorial crossing at a local time of 10:30 a.m. is given considering that the local time depends on both, the latitude and the off-nadir-angle of an observation point. Only in a table of orbit characteristics a variation of ±15 min is included. The same local solar time can be found in the ERS-1 documents (ESA, 1992). For LANDSAT 7 an Equator Crossing Time of 10:00 to 10:15 a.m. is given (LANDSAT Project Science Office, 1998). For IRS-P6/RESOURCESAT-1 a Local Time of Equator Crossing of 10:30 a.m. can be found (ISRO, 2015) without the hint at a mean value. This is resulting in a second problem: A difference exists between real and mean Sun time, also called as real and mean Sun, described by the Equation of Time (ET). It is a relatively small variation of about ±15 minutes within a year. LECT, LT and LCT are connected for real solar conditions with this variation that can have influence on further results of optical satellite sensor data.

Nevertheless, it was found that in the used relation between LECT in LT and ECT in GMT (Greenwich Mean Time) in connection with studies of land surface temperature e.g. on basis of NOAA data, the term with Equation of Time is not included (Ignatov et al., 2004, Wu and McAvaney, 2006). That means that a relation for mean Sun time is used and therefore, the used relation between LECT and LCT as function of latitude is also based on mean Sun time. Moreover, Maers et al. (2002) have shown that the drift of LECT (the orbit of NOAA satellites is not so stabilized as of e.g. ERS or ENVISAT satellite) is not superimposed by the characteristic annual variation caused by ET. Therefore, it is a presentation of mean ECT. This is of advantage in determination of long-term drift of equator crossing. But with respect to variation of LECT (mean LECT is used) Maers et al. (2002) present a diurnal adjustment of temperature showing the characteristic variation of ET, which adds the same periodic signal with a constant offset to all (NOAA) satellites. According Mo (2009) "any variation of the local equator crossing time (LECT) of a satellite will have impact on the time series of data." Also other authors (e.g. Julien and Sobrino, 2012, Sobrino and Julien, 2013) use Ignatov et al., (2004) as reference, without hint at use of LECT for real solar conditions.

The objective of this paper is to indicate the need for the correct application of Equation of Time for calculation of local time. It is shown that the use of imprecise times leads to imprecise results of satellite data analysis, e.g. deviations in Normalized Difference Vegetation Index (NDVI). This is especially important for the analysis of time series.

## 2. MATERIAL AND METHODS

It is well known that the crossing of the same latitude at the same local time is one of the most important characteristics of sun-synchronous orbiting satellites. In comparison with other satellite orbits an advantage consists in the much smaller variation of the solar zenith angle between mid-latitudes in south and north of a single path. It was demonstrated e.g. for NOAA-11 satellite that the time offset between ±60 degrees varies not more than ±60 minutes (Johnson et al., 1994). A further advantage consists in very similar conditions of illumination in the same geographic latitude of neighbouring passes within a relatively short time interval, perhaps within the 16 days of LANDSAT repetition rate. Corresponding image composites are characterised by relatively small deviations of illumination from image to image only.

If the actual time of each image pixel is not given for further calculations, e.g. Sun's zenith angle, the use of the mean LECT for corresponding satellite is helpful, as given for some satellites above.

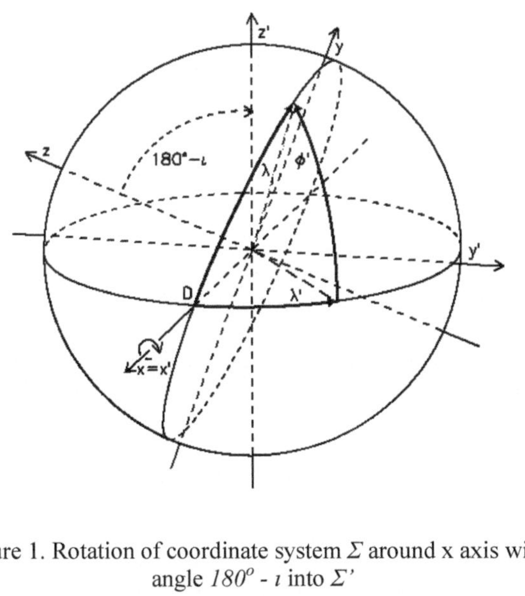


Figure 1. Rotation of coordinate system $\Sigma$ around x axis with angle $180^o$ - $\iota$ into $\Sigma'$

An orbit can be described by a Cartesian system $\Sigma$ with one axis from the centre point through $x=r$, $y=0$, $z=0$ (r – semi major axis of orbit). This means in spherical coordinates that for crossing of x-axis by the orbit latitude and longitude are zero. The orbit is located in the oblique x-y plane (circle) of Figure 1 and its orientation to Sun is nearly constant during a day. When the Cartesian system $\Sigma$ rotates only around the x-axis with $180^o$ - $\iota$ ($\iota$ - inclination of orbit) according Figure 1 the matrix elements for transformation (Mathematik, 1970) of equation (1) are reduced to the values shown in Table 2. In this case the orbit is crossing the $x=x'$ axis at the same "equator crossing" points in $\Sigma$ and $\Sigma'$: $(r;0;0)$ and $(-r;0;0)$, respectively.

$$\begin{bmatrix} x' \\ y' \\ z' \end{bmatrix} = \begin{bmatrix} a_{11} & a_{12} & a_{13} \\ a_{21} & a_{22} & a_{23} \\ a_{31} & a_{32} & a_{33} \end{bmatrix} \begin{bmatrix} x \\ y \\ z \end{bmatrix} \quad (1)$$

| $a_{ik}$ | k=1 | k=2 | k=3 |
|---|---|---|---|
| i=1 | 1 | 0 | 0 |
| i=2 | 0 | -cos ι | sin ι |
| i=3 | 0 | -sin ι | -cos ι |

Table 2. Matrix elements of transformation equation of a Cartesian system $\Sigma$ rotating with angle $180^o$ - $\iota$ around the x-axis into $\Sigma'$

The Cartesian coordinates can be substituted by using spherical coordinates according equation (2):

$$x = r\, cos\varphi\, cos\lambda \\ y = r\, cos\varphi\, sin\lambda \\ z = r\, sin\varphi \quad (2)$$

where x, y, z = Cartesian coordinates in $\Sigma$
r, $\varphi$, $\lambda$ = spherical coordinates in $\Sigma$

In the same way the coordinates $x'$, $y'$, $z'$ of $\Sigma'$ can be expressed in spherical (geographical) coordinates angles $\varphi'$ (latitude) and $\lambda'$ (longitude) and the relation $r = r'$. Introducing these values including the coefficient $a_{ik}$ of Table 2 into equation (1) the following 3 relations can be derived:

$$cos\varphi'\, cos\lambda' = cos\varphi\, cos\lambda \\ cos\varphi'\, sin\lambda' = -\, cos\iota\, cos\varphi\, sin\lambda + sin\iota\, sin\varphi \\ sin\varphi' = -\, sin\iota\, cos\varphi\, sin\lambda - cos\iota\, sin\varphi \quad (3)$$

where r', $\varphi'$, $\lambda'$ = spherical (geographical) coordinates in $\Sigma'$
$\iota$ = angle between $\Sigma$ and $\Sigma'$ after rotation around x-axis (inclination of satellite orbit)

When using the ratio between second and third relation of equation (3) with the assumption that the latitude of the orbit in $\Sigma$ is $\varphi=0^o$ the following relation can be found

$$cos\varphi'\, sin\lambda' / sin\varphi' = cot\iota \quad (4)$$

where $\varphi'$, $\lambda'$ = spherical (geographical) coordinates of satellite orbit in $\Sigma'$

in which the longitude $\lambda'$ can be separated as given in equation (4).

$$\lambda' = arcsin\,(tan\varphi'\, cot\iota) \quad (5)$$

Equation (5) can be solved for each longitude $\lambda' = \lambda'_{LC}$ of local crossing including the special case $\lambda' = \lambda'_{LEC} = 0$ of equator crossing of descending node (point D) of the orbit in Figure 1. $\lambda'_{LEC}$ and $\lambda'_{LC}$ has to be used to define a difference $\Delta\lambda'$.

$$\Delta\lambda' = \lambda'_{LEC} - \lambda'_{LC} \quad (6)$$

where $\lambda'_{LEC}$ = geographical longitude of local equator crossing
$\lambda'_{LC}$ = geographical longitude of local crossing

The difference of the longitude in equation (6) is independent from Earth rotation, from geographic longitude of equator crossing and can be expressed as a time offset from the point of equator crossing to the orbit point of local crossing with equation (7).

$$t_{off} = t_{LEC} - t_{LC} = \Delta\lambda' / 15 = \arcsin(\tan\varphi' \cot\iota) / 15 \quad (7)$$

where $t_{LEC}$ = LECT
$t_{LC}$ = LCT

After rearrangement equation (7) results in

$$t_{LC} = t_{LEC} - \arcsin(\tan\varphi' \cot\iota) / 15 \quad (8)$$

Maximum and minimum of geographical latitude φ' are determined by inclination of orbit. Its variation can be described by varying of $\lambda$ in the third part of equation (3) with $\varphi=0$. Equation (8) describes the relation for descending node (+ in case of ascending node) of the orbit as can be found already in Ignatov et al. (2004), Wu and McAvaney (2006). Equation (8) assumes that the daily actual value for LECT is used. In connection with equation (8) a relation of Local Time (LT) was used by Ignatov et al. (2004) and Wu and McAvaney (2006) at each geographic point for mean solar conditions given by e.g. Duffett-Smith (1988):

$$t_L = t_{GM} + \lambda_G / 15 \quad (9)$$

where $t_L$ = LT at geographic point with $\lambda_G$ in hours
$t_{GM}$ = GMT at geo-graphic point with $\lambda_G$ in hours
$\lambda_G$ = geographical longitude (λ' in equations before)

Because LCT is a synonym of LT in relation to orbit geometry, the combination of equation (8) and (9) gives the central geographic sampling positions of the LCT sampling method (Wu and McAvaney, 2006). Furthermore, if GMT and geographic longitude $\lambda$ of the equator crossing are known or calculated (Ignatov et al., 2004), the result of equation (9) is a constant (mean) LECT, only influenced in case of NOAA satellites by drift of orbit connected with drift of LECT. Longitude of equator crossing and the corresponding crossing time in UTC (for our purpose UTC = GMT) are determined by Ignatov et al. (2004) for different satellites (NOAA, ERS, EOS) and for each day on basis of the two-line-elements (TLE), available for their full lifetime period and with help of Brouwer-Lyddane orbit prediction package (Kidwell, 1998, Goodrum et al., 2003). On basis of these results and use of equation (9) an equation for LECT will be determined describing the drift of LECT, with changing parameters from satellite to satellite.

But as already mentioned, a difference exists between true and mean Sun. For example, local noon will change within -14 and +16 minutes within a year when using real Sun conditions. Neglecting the change of orbit parameters, LECT for real Sun will change for each satellite in the same order. This relation described by Equation of Time (ET) can also be found in Duffett-Smith (1988), and is caused by tilt of Earth axis and elliptical orbit of Earth around Sun. More details about this complex term are available e.g. in Duffet-Smith (1988) and Smart (1949). For this purpose equation (10) for real Sun conditions has to be used instead of mean Sun conditions used in equation (9).

$$t_L = t_{LC} = t_{GM} + t_E + \lambda_G / 15 \quad (10)$$

where $t_E$ = ET

The ET variation computed for year 2000 is shown in Figure 3.

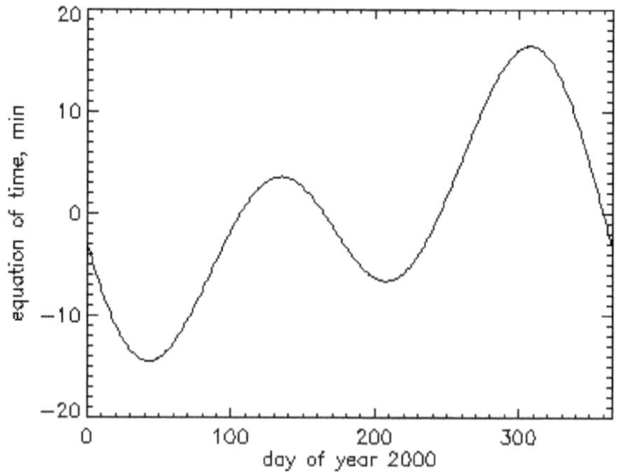


Figure 3. Variation of the difference between true and mean solar time calculated for year 2000

Because the calculation of ET is very complex, equation (11) represents an approximation for comparable purposes given in Baldocchi (2012) and Meeus (2000):

$$ET = [-104.7\sin(f) + 596.2\sin(2f) + 4.3\sin(3f) - 12.7\sin(4f) - 429.3\cos(f) - 2.0\cos(2f) + 19.3\cos(3f)]/3600 \quad \textbf{(11)}$$

where $f = \pi/180(279.5 + 0.9856d)$
$d$ = day number of year
$ET$ and $f$ here in units of radians

Nominal ECTs (ECT at launch) including orbital drift rate of different NOAA satellites are given by Waliser and Zhou (1997) and Lucas et al. (2001). Mears et al., (2002) and Mears and Wentz (2009) have determined a global diurnal correction adjustment of temperature in their trend analysis, obtaining a similar annual behaviour as for the Equation of Time, but with different amplitudes. A variation of global adjustment of about 0.06 K within a year is shown in the corresponding results. It was shown that it is necessary to consider the long term drift of LECT, caused by slow changes in orbital parameters of each platform. Compared with these trends in changes of more than 3 hours during their life time e.g. NOAA-11 from 1989 to 1995 the influence of ET is small. Bush et al. (1999) show, that the long term trend of the local solar time of the ascending node of the NOAA-9 orbit is overlapped by the characteristic cycles of ET.

In many fields of geophysics the diurnal behaviour of time in Earth system is described with relation (10), for example when modelling the behaviour of minor constituents in neutral atmosphere (Fichtelmann and Sonnemann, 1989) for a geographic position depending on solar zenith angle. If LECT ($t_{LEC}$) of a date is known (index 0), it is possible to describe its annual variation. Solving equation (8) and (10) for two different times (one with index 0), combining the corresponding two equations, and with view to orbit characteristic that $t_{GM} + \lambda_G /15$ for sun-synchronous orbit of a satellite, the difference between the two equations results after rearrangement in eq. (12).

$$t_{LEC} = t_{LEC\_0} + (t_E - t_{E\_0}) \qquad (12)$$

where $t_{LEC\_0}$, $t_{E\_0}$ = LECT, ET for real Sun and defined day 0
$t_{LEC}$, $t_E$ = LECT, ET for real Sun and a day of year

Possible disturbances as orbital drift have to be included into equation (12) additionally. The inclusion of $t_{LEC}$ of equation (12) into equation (8) results in equation (13).

$$t_{LC} = t_{LEC\_0} + (t_E - t_{E\_0}) - \arcsin(\tan(\varphi_N) \cot\iota) / 15 \qquad (13)$$

where $t_{LC}$ = LCT, ET for real Sun
$\varphi_N$ = geographic latitude in nadir view of satellite
$\iota$ = inclination of orbit

As shown before for some examples LECT for sun-synchronous satellites is given only for a mean value. This equation can be used to determine the actual ECT for real Sun conditions when using for $t_{LEC\_0}$ the corresponding mean value $t_{LEC\_MEAN}$ together with equation of time $t_{E\_0}$ of the same day.
Equation (13) only applies to the nadir point of the satellite orbit. Therefore, the local time for all pixels of an image line have to be corrected by using their respective length difference $\Delta\lambda = \lambda - \lambda_N$ to the nadir point of this line by an additional term, converting length difference into time difference according equation (14).

$$t_{LC} = t_{LEC\_0} + (t_E - t_{E\_0}) - \arcsin(\tan(\varphi_N) \cot(\iota)) / 15 + \Delta\lambda / 15 \qquad (14)$$

where $\varphi_N$ = geographic latitude in nadir view of satellite
$\Delta\lambda$ = difference of longitude from pixel in nadir view to a pixel in the same line

When using the ratio between first and second relation of equation (3) with the assumption that the longitude of the orbit in $\Sigma$ is $\lambda = 0^o$ the following equation (15) can be found.

$$\lambda' = \arctan(\tan\varphi \sin\iota) \qquad (15)$$

where $\varphi$ = geographic latitude in $\Sigma$ of Figure 1
$\lambda' = \Delta\lambda$

For determination of mean LECT of LANDSAT 7 satellite the available TLEs and the STK orbit determination software (AGI, 2006) were used.
For demonstration of possible influence of ET to satellite products a LANDSAT 7 scene of 2nd June 2000 was used, as well as an atmospheric correction procedure according Holzer-Popp et al. (2004) in which the correct ET could be modified.

## 3. RESULTS

The result of calculation of $t_{off}$ in equation (7) for a complete LANDSAT 7 orbit with an inclination $\iota$=*98.2degrees* is given with respect to the equator crossing of the descending node in Figure 4. The variation of $t_{off}$ is not only similar to results described already by Johnson et al. (1994) for NOAA satellites. It is a characteristic feature of a sun-synchronous orbit. Knowing the actual (mean, real) ECT of descending node, the value can be added to the offset of this figure. Equation (8) describes the dependency of (mean, real) LCT as function of actual latitude and the two orbit parameters $t_{LEC}$ and inclination $\iota$.

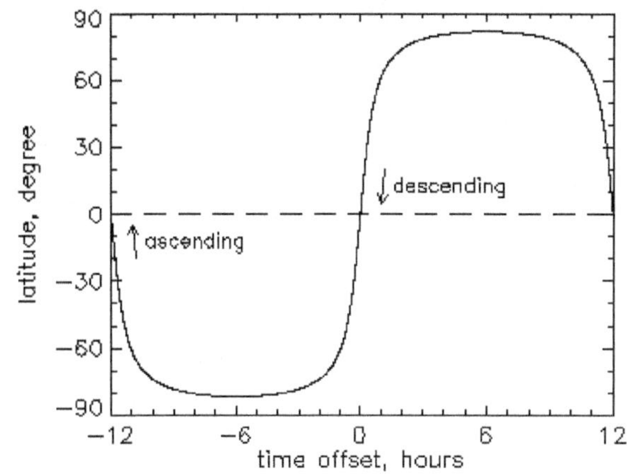


Figure 4. Time offset from LECT of descending node of LANDSAT 7 with an inclination of *98.2 degrees*

The STK orbit determination software (AGI, 2006) was used to calculate mean LECT of LANDSAT 7 satellite in UTC on basis of the TLEs. The regular use of this software at acquisition facility in Neustrelitz has demonstrated over many years that a good agreement exists between calculated and real orbit parameters. Therefore we presupposed that the computation accuracy is well for calculating ECT. Ignatov et al. (2004) have pointed out, that their reconstruction of different past ECTs (TIROS-N, NOAA-6 through -17) was within an accuracy of ±2 min, using the same orbit determination software as mentioned before. The annual variation of equator crossing (mean LECT) of LANDSAT 7 platform was simulated for this paper by using NORAD orbit elements provided by CelesTrak ,within the System Tool Kit Software developed by Analytical Graphics, Inc. (AGI) for each 21st of all twelve months over year 2000. Equation (10) was used to transform ECT given in UTC (it corresponds to mean LECT) into LECT for real Sun.

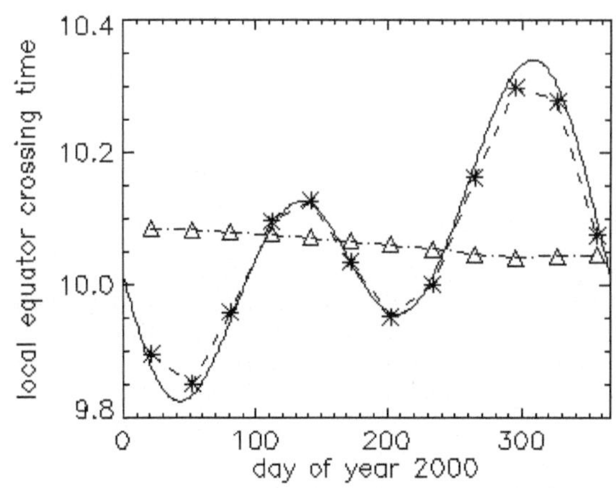


Figure 5. Variation of LECT determined for each 21st day of a month, calculated by means of STK software of acquisition facility for mean (triangles) and real (asterisks, connected by dashed line) Sun of descending node of Landsat 7 for year 2000. Variation of ET according equation (12) with ET of 21st June as $t_{LEC\_0}$ (solid line).

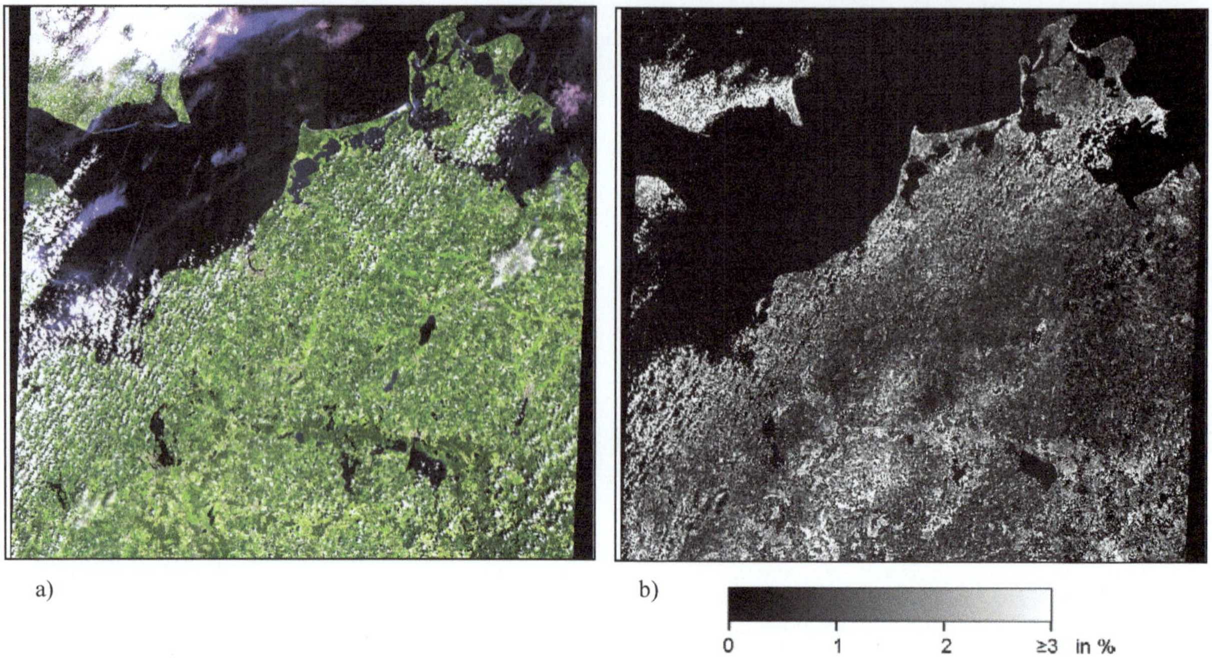


Figure 6a and b. The region of Mecklenburg-Western Pomerania in an atmospheric corrected Landsat-7 dataset (rgb – ETM+ bands 7,5,3) on 2nd June 2000 with LECT for real Sun (a) in comparison with the relative deviation of NDVI when using a time offset for ET of 30 min .

The calculated values of real LECT for descending node of LANDSAT 7 are shown by asterisks in Figure 5. Already with help of these 12 points it is obvious, that the variation of LECT for real Sun is very similar to that of ET in Figure 3. The nearest value to the 10:06 a.m. mean of these 12 values is given for the 21st of June. If using the $t_{LEC\_2106}$ and $t_{E\_2106}$ for this date as basis (index 2106 instead of 0), it is possible to describe the annual variation of $t_{LEC}$ with equation (12). The result is given as solid line in Figure 5. There is a good correspondence between these two curves but with increasing difference in the second part of the year. The third curve, the mean LECT (triangles) determined without consideration of ET shows a small decrease of LECT with a minimum in October and an increase in the last two months of the year. The maximum of deviation in year 2000 is around 2.5 minutes.

As can be seen in Figure 5, ET crosses four times (on days 107, 164, 245 and 360) the mean value of year 2000. It should be mentioned here, that a small change of crossing dates (also ET of a selected day) is possible, comparable with change of e.g. vernal equinox from year to year in Gregorian calendar (Duffet-Smith, 1988). In case, that LECT for real Sun ($t_{LEC\_0}$) is available or can be determined for a day of year the variation of LECT for real Sun can be described by equation (13) without additional use of orbit determination software if the drift of orbit during life time of a satellite or time interval is with 2.5 min as found in our example with LANDSAT 7 in an acceptable order of magnitude for data users.

An automatic atmospheric correction (Holzer-Popp et al., 2004) was used to demonstrate in a single experiment the order of magnitude of influences of LECT variations caused by ET to Normalized Difference Vegetation Index (NDVI). For the purpose, two computations were carried out with the same LANDSAT-7 ETM+ dataset for the region of Mecklenburg-Western Pomerania (Germany) on 2nd June 2000. The first one was made under real LECT conditions including an ET of 1.93 minutes calculated for this date. The second one was carried out under simplified assumption. At first, to avoid changes of resulting parameters, e.g. NDVI, caused by changes of vegetation within a year, the data set above was used as reference. Secondly a simulated time offset of 30 minutes based on real variability of ET between -14 min and + 16 min within a year was used. As mentioned before, if using one data set in minimum and another one in maximum of ET it is difficult to compare the results, e.g. on basis of NDVI, due to seasonal changes.

Figure 6 gives an estimation of the order of magnitude of the deviations of NDVI when using a mean LECT. Figure 6a shows the result in the RGB combination of ETM+ bands 7,5,3 after the automatic atmospheric correction using a pre-classification and real atmospheric data (ozone, water vapour and aerosol) of the same date. The relative difference of the NDVI ($|NDVI_{offset}$-$NDVI|/NDVI$) calculated with a hypothetical and without time offset, is shown in Figure 6b. Within the large and relatively dark region on the right side of Figure 6b the deviation (with some small exceptions) is consistently in the range of 0.5 to 1.5 %. In the lower centre, the deviation is in the order of 2.0 to 3.0 % and the cloud-free region in the upper left part (Denmark) shows deviations of up to 8 %.

## 4. DISCUSSION

The nearly constant ECT or corresponding LCT of a selected latitude is one of the important characteristics of sun-synchronous orbits. The term "nearly constant" includes small disturbances of the satellite orbit around the Earth and disturbances caused by the characteristics of Earth orbit around the Sun. The latter effect is connected with a seasonal variation of LECT of about ±15 minutes described by the Equation of Time. It was shown with the simple assumption of a 30 minutes time offset that the use of a mean LECT can have not only a strong influence to the results of automatic atmospheric correction but also on results based on these corrected optical data. It is assumed that atmospheric correction commonly includes the correct use of ET. Nevertheless, the inconsistence in determination of LCT was found with the use of the above

formulas but without consideration of Equation of Time when no temporal information of a dataset was available.
It was already shown in studies about troposphere temperature that the long time drift of LECT can have an important influence on trend analyses of climate. Bush et al.(1999) are showing in their long term trend analysis of the local solar time of the ascending node of the NOAA-9 orbit the characteristic cycles of the equation of time. Mears et al. (2002) present in their studies about correction of MSU (Microwave Sounding Units) middle tropospheric temperature for diurnal drifts a global diurnal correction to 9 NOAA satellites between 1978 and 1999 with the characteristic behaviour of the ET. Therefore, concerning LST analysis, it is necessary to check whether the neglect of ET is resulting in a part of diurnal correction parameters. This part is in the order of magnitude up to 0.08 Kelvin as can be found in the results of Mears et al. (2002).
The use of the equations for determination of LCT without consideration of ET was found in connection with simulation of radiances with the method of local-crossing-time sampling. For example, it is necessary to examine whether it is possible that the combination of the measured radiances with a small difference in LCT can result in small deviations of simulated brightness temperatures.
The results of a simplified change of correct use of ET show for example, that the influence on a value added product on basis of bottom of atmosphere reflectance as vegetation index NDVI is relatively small, but in some cases up to 8 %. The background is that the deviations for two used bands are going regularly into the same direction. On one side this paper shall convey an impression about the dimension of ET caused deviations in NDVI, which is connected with the atmospheric corrected values of one band. It should be noted here that already the calculation of top of atmosphere reflectance is required by more accurate calculation of Sun elevation. Studies of Jiang et al. (2008) show, that there is a dependency of NDVI magnitude on satellite ECT in the unadjusted global NDVI data set. Furthermore it was shown by Jiang et al. (2008) that one of the significant problems in the unadjusted NDVI time series on basis of NOAA satellites is the obvious dependency on satellite ECTs from different polar orbiters. But in some of the presented time series of unadjusted and adjusted NDVI the typical variation of ET within a year is obvious.
On the other side no reason was found why in the field of climatology the Equation of Time is not included. A lot of work has been done to determine the drift of LECT, e.g. Ignatov et al. (2004) have fitted the LECT drift by a mathematical approximation of a periodic function with one or two Fourier terms. It should be of advantage to use the complete formula of e.g. Duffet-Smith (1988) describing additionally the deviation between real and mean Sun.
A simple formula can be used to determine the actual LECT for real Sun within a year on basis of its value on a well defined date or on basis of a mean value given as orbit parameter and without inclusion of additional orbit determination software in the processing of data. In case of satellites with a drift during life time as NOAA satellites it is of advantage to eliminate ET from the mathematical description of such long time drift effects of equator crossing.
The determination of LECT of satellites with stabilized orbits as e.g. LANDSAT 7 shows for year 2000 a variation of about 2.5 minutes. Because the accuracy of time calculations is essential in Earth observation it would be of advantage when the actual LECT for real Sun will be included in the metadata of such a remote sensing product.

**ACKNOWLEDGEMENTS**

The authors wish to thank C. Wloczyk for her suggestions and advices.

**REFERENCES**

AGI, 2006, Analysis software for land, sea, air & space. 220 Valley Creek Blvd., Exton, PA 19341 USA, https://www.agi.com/ (23 Oct. 2012).

Aster Science Project, 2001. User's guide, ERSDAC. http://www.science.aster.ersdac.jspacesystems.or.jp/en/documnt s/users_guide/part1/pdf/Part1_4E.pdf (23 Mar. 2015).

Baldocchi, Dennis. "Lecture 7, Solar Radiation, Part 3, Earth-Sun Geometry" September 10, 2012., http://nature.berkeley.edu/biometlab/espm129/notes/Lecture%207%20Solar%20Radiation%20Part%203%20Earth%20Sun%20Geometry%20notes.pdf (20 Mar. 2015).

Borg, E., Fichtelmann, B., Asche, H., 2012. Design and implementation of data usability processor into an automated processing chain for optical remote sensing data. 5th International Symposium On Leveraging Applications of Formal Methods, Verification and Validation 15-18 October 2012 Amirandes, Heraklion, Crete, (oral presentation).

Bush, K.A., Smith, G.L., and Young, D.F., 1999. The NOAA-9 earth radiation budget experiment wide field-of-view data set. 10th Conference on Atmospheric Radiation, June 1999, NASA 1999 Technical Docs.

Duffet-Smith, P., 1988. Practical astronomy with your calculator. 3rd edn (Cambridge, N.Y., Port Melbourne, Madrid, Cape Town: Cambridge University Press).

ESA, 1992. ERS-1 Satellite Concept. Orbit Information https://earth.esa.int/web/guest/missions/esa-operational-eo-missions/ers/satellite (23 Mar. 2015).

Fichtelmann, B. and Sonnemann, G., 1989. On the variation of ozone in the upper mesosphere and lower thermosphere: A comparison between theory and observation, Z. Meteorol., 39, 297–308,

Goodrum, G, Kidwell, K., and Winston, W. (editors), 2003. NOAA–KLM User's Guide. US Department of Commerce, NOAA/NESDIS. Available from National Climatic Data Center, 151 Patton Ave, Rm.120, Asheville, NC 28801-5001, USA.

Holzer-Popp, T., Bittner, M., Borg, E., Dech, S., Erbertseder, T., Fichtelmann, B., and Schroedter, M., 2004. Method for correcting atmospheric influences in multispectral optical teledetection data. European Patent Office, EP 1 091 188 B1,.

Ignatov, A., Lazlo, I., Harrod, E.D., Kidwell, K.B., and Goodrum, G.P., 2004. Equator crossing times for NOAA, ERS and EOS sun-synchronous satellites, International Journal of Remote Sensing, 25, 5255-5266.

ISRO, 2015. IRS-P6 / RESOURCESAT-1, http://www.isro.gov.in/Spacecraft/irs-p6-resourcesat-1 (23 Mar 2015).

Jiang, L., Tarpley, J.D., Mitchell, K.E. Zhou, S., Kogan, F.N., and Guo, W., 2008. Adjusting for long-term anomalous trends in NOAA's global vegetation index data sets, IEEE Transactions on Geoscience and Remote Sensing, 46, pp. 409-422.

Johnson, D.B., Flament, P., and Bernstein, R.L., 1994. High-resolution satellite imagery for mesoscale meteorological studies. Bulletin of the American Meteorological Society, 75, 5-34.

Julien, Y. & Sobrino, J. A. (2012). Correcting Long Term Data Record V3 estimated LST from orbital drift effects, Remote Sensing of Environment, 123 (2012) 207-219.

Julien, Y., Sobrino, J. A. & Jiménez-Muñoz, J.-C. (2011). Land use classification from multitemporal Landsat imagery using the Yearly Land Cover Dynamics (YLCD) method, International Journal of Applied Earth Observations and Geoinformation, 13 (2011), 711-720.

Kidwell, K. (editor), 1998. NOAA Polar Orbiter Data User's Guide (TIROS–N, NOAA-6, -7, -8, -9, -10, -11, -12, -13, and -14. US Department of Commerce, NOAA/NESDIS, 394 pp. Available from National Climatic Data Center, 151 Patton Ave, Rm.120.

LANDSAT Project Science Office, 1998. NASA's Goddard Space Flight Center, Greenbelt, Maryland http://landsathandbook.gsfc.nasa.gov/data_prod/prog_sect11_3.html/ (20 Mar. 2015).

Lucas, L.E., Waliser, D.E., Xie, P., Janowiak, J.E., and Liebmann, B., 2001. Estimating the satellite equatorial crossing time biases in the daily, global outgoing longwave radiation dataset. Journal of Climate, 14, 2583-2605.

Mathematik, 1970. *Kleine Enzyklopädie Mathematik.* Ed.: Gellert, A., Küstner, H., Hellwich, M., Kaestner, H., 5th Ed. Verlag Enzyklopädie Leipzig.

Mears, A.M., Schabel, M.C., Wentz, F.J., Santer, B.D., and Govindasamy, B., 2002. Correcting the MSU middle tropospheric temperature for diurnal drifts. Proc. Int. Geophysics and Remote Sensing Symp., Volume III, Toronto, ON, Canada, IEEE, pp. 1839-1841.

Mears, A.M., Wentz, F.J., 2009. Construction of the remote sensing systems V3.2 atmospheric temperature records from the MSU and AMSU microwave sounders. Journal of Atmospheric and Oceanic Technology, 2009, Volume 26, pp. 1040-1056.

Meeus, Jean. 2000. *Mathematical Astronomy Morsels*, 2nd Ed. Willmann-Bell Inc.: Richmond, pp. 337-346

Mo, T., 2009. A study of the NOAA-15 AMSU-A brightness temperatures from 1998 through 2007. J. Geophys. Res. 114 D11110 doi:10.1029/2008JD011267 pp. 10.

Smart, W.M., 1949. Text-Book on Spherical Astronomy. Cambridge University Press, first edition 1931.

Sobrino, J.A., Julien, Y., 2013. Time series corrections and analyses in thermal remote sensing. In *Thermal Infrared Remote Sensing: Sensors, Methods, Applications*, Remote Sensing and Digital Image Processing, C. Kuenzer and S. Dech (Eds.), DOI 10.1007/978-94-007-6639-6_14, Springer Science+Business Media Dordrecht, 267-285.

Waliser, D.E., and Zhou, W., 1997, Removing satellite equator crossing time biases from the OLR and HRC datasets. Journal of Climate, 10, 2125-2146.

Wu, Z.-J., and McAvaney, B., 2006. Sampling methods for climate model calculated brightness temperatures. BMRC Research Report No. 118, Bureau of Meteorology Australia, pp. 1-43.

# 13

# THE GREENHOUSE GAS PROJECT OF ESA'S CLIMATE CHANGE INITIATIVE (GHG-CCI): OVERVIEW, ACHIEVEMENTS AND FUTURE PLANS

M. Buchwitz [a], M. Reuter [a], O. Schneising [a], H. Boesch [b], I. Aben [c], M. Alexe [d], R. Armante [e], P. Bergamaschi [d], H. Bovensmann [a], D. Brunner [f], B. Buchmann [f], J. P. Burrows [a], A. Butz [g], F. Chevallier [h], A. Chédin [e], C. D. Crevoisier [e], S. Gonzi [i], M. De Mazière [j], E. De Wachter [j], R. Detmers [c], B. Dils [j], C. Frankenberg [k], P. Hahne [g], O. P. Hasekamp [c], W. Hewson [b], J. Heymann [a], S. Houweling [c], M. Hilker [a], T. Kaminski [l], G. Kuhlmann [f], A. Laeng [g], T. T. v. Leeuwen [c], G. Lichtenberg [m], J. Marshall [n], S. Noël [a], J. Notholt [a], P. Palmer [i], R. Parker [b], M. Scholze [o], G. P. Stiller [g], T. Warneke [a], C. Zehner [p]

[a] Institute of Environmental Physics (IUP), University of Bremen, Bremen, Germany - (Michael.Buchwitz, mreuter, schneising, bovensmann, burrows, heymann, mhilker, noel, jnotholt, warneke)@iup.physik.uni-bremen.de

[b] University of Leicester, Leicester, United Kingdom - (hb100, wh76, rjp23)@leicester.ac.uk

[c] SRON Netherlands Institute for Space Research, Utrecht, Netherlands - (E.A.A.Aben, R.G.Detmers, O.P.Hasekamp, S.Houweling, T.T.van.Leeuwen)@sron.nl

[d] European Commission Joint Research Centre (EC-JRC), Institute for Environment and Sustainability (IES), Air and Climate Unit, Ispra, Italy - (Peter.Bergamaschi, Mihai.Alexe)@jrc.ec.europa.eu

[e] Laboratoire de Météorologie Dynamique (LMD), Palaiseau, France - (Raymond.Armante, Chedin, Cyril.Crevoisier)@lmd.polytechnique.fr

[f] Swiss Federal Laboratories for Materials Science and Technology (Empa), Dübendorf, Switzerland - (Dominik.Brunner, Brigitte.Buchmann, Gerrit.Kuhlmann)@empa.ch

[g] Karlsruhe Institute of Technology (KIT), Karlsruhe, Germany - (Andre.Butz, Philipp.Hahne, Alexandra.Laeng, Gabriele.Stiller)@kit.edu

[h] Laboratoire des Sciences du Climat et de l'Environnement (LSCE), Gif-sur-Yvette, France - Frederic.Chevallier@lsce.ipsl.fr

[i] University of Edinburgh, Edinburgh, United Kingdom (UK) - (sgonzi, Paul.Palmer)@ed.ac.uk

[j] Belgian Institute for Space Aeronomy (BIRA), Brussels, Belgium - (Martine.DeMaziere, Evelyn.DeWachter, Bart.Dils)@aeronomie.be

[k] Jet Propulsion Laboratory (JPL), Pasadena, California, United States of America - Christian.Frankenberg@jpl.nasa.gov

[l] The Inversion Lab, Hamburg, Germany - Thomas.Kaminski@Inversion-Lab.com

[m] Deutsches Zentrum für Luft- und Raumfahrt (DLR), Oberpfaffenhofen, Germany - Guenter.Lichtenberg@dlr.de

[n] Max-Planck-Institute for Biogeochemistry (MPI-BGC), Jena, Germany - marshall@bgc-jena.mpg.de

[o] Lund University, Lund, Sweden - Marko.Scholze@nateko.lu.se

[p] European Space Agency (ESA), ESRIN, Frascati, Italy - Claus.Zehner@esa.int

**KEY WORDS:** Climate change, carbon cycle, greenhouse gases, emissions, satellites, remote sensing, SCIAMACHY, GOSAT

## ABSTRACT

The GHG-CCI project (http://www.esa-ghg-cci.org/) is one of several projects of the European Space Agency's (ESA) Climate Change Initiative (CCI). The goal of the CCI is to generate and deliver data sets of various satellite-derived Essential Climate Variables (ECVs) in line with GCOS (Global Climate Observing System) requirements. The "ECV Greenhouse Gases" (ECV GHG) is the global distribution of important climate relevant gases – namely atmospheric $CO_2$ and $CH_4$ - with a quality sufficient to obtain information on regional $CO_2$ and $CH_4$ sources and sinks. The main goal of GHG-CCI is to generate long-term highly accurate and precise time series of global near-surface-sensitive satellite observations of $CO_2$ and $CH_4$, i.e., $XCO_2$ and $XCH_4$, starting with the launch of ESA's ENVISAT satellite. These products are currently retrieved from SCIAMACHY/ENVISAT (2002-2012) and TANSO-FTS/GOSAT (2009-today) nadir mode observations in the near-infrared/shortwave-infrared spectral region. In addition, other sensors (e.g., IASI and MIPAS) and viewing modes (e.g., SCIAMACHY solar occultation) are also considered and in the future also data from other satellites. The GHG-CCI data products and related documentation are freely available via the GHG-CCI website and yearly updates are foreseen. Here we present an overview about the latest data set (Climate Research Data Package No. 2 (CRDP#2)) and summarize key findings from using satellite $CO_2$ and $CH_4$ retrievals to improve our understanding of the natural and anthropogenic sources and sinks of these important atmospheric greenhouse gases. We also shortly mention ongoing activities related to validation and initial user assessment of CRDP#2 and future plans.

## 1. INTRODUCTION

Carbon dioxide ($CO_2$) is the most important anthropogenic greenhouse gas responsible for global warming (*IPCC, 2013*). Despite its importance, our knowledge of the $CO_2$ sources and sinks is inadaquate and does not meet the needs for attribution, mitigation and the accurate prediction of future change (e.g., *Ciais et al., 2010*; *Canadell et al., 2010; IPCC, 2013; CEOS, 2014; Ciais et al., 2014*), and despite efforts to reduce $CO_2$ emissions, atmospheric $CO_2$ continues to increase with approximately 2 ppm/year (Fig. 1; *Le Quéré et al., 2014*).

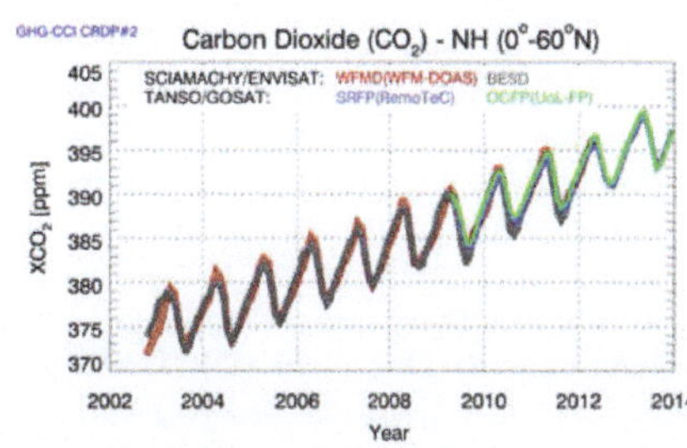


Fig. 1: GHG-CCI CRDP#2 $XCO_2$ Northern Hemisphere 2002-2013 (see Tab. 1 for details).

Figure 1 shows Northern Hemispheric $XCO_2$, i.e., the column-averaged $CO_2$ dry air mole fraction (in ppm), as retrieved from SCIAMACHY/ENVISAT and TANSO-FTS/GOSAT using four different GHG-CCI retrieval algorithms (see Sect. 2). Clearly visible is the $CO_2$ seasonal cycle - primarily caused by uptake and release of $CO_2$ by the terrestrial biosphere - and the atmospheric $CO_2$ increase with time, which is primarily caused by burning of fossil fuels (fraction not taken up by the terrestrial biosphere or the oceans). Also visible is the good agreement of the different GHG-CCI CRDP#2 $XCO_2$ data products. Perfect agreement is not expected due to different spatio-temporal sampling and different altitude sensitivities (averaging kernels).

Appropriate knowledge about the $CO_2$ sources and sinks is needed for reliable prediction of the future climate of our planet (*IPCC, 2013*). This is also true for methane ($CH_4$; e.g., *IPCC, 2013; Kirschke et al., 2013*). The goal of the GHG-CCI project (*Buchwitz et al., 2013a*), which is one of several projects of ESA's Climate Change Initiative (CCI, *Hollmann et al., 2013*), is to generate global satellite-derived $CO_2$ and $CH_4$ data sets as needed to improve our understanding of the regional sources and sinks of these important atmospheric gases.

Global near-surface-sensitive satellite observations of $CO_2$ and $CH_4$ combined with inverse modeling yields information on the regional sources and sinks of these gases. The goal of the GHG-CCI project is to generate the Essential Climate Variable (ECV) Greenhouse Gases (GHG) as required by GCOS. The GCOS definition of this ECV is (*GCOS, 2011*): "Product Number A.8.1: Retrievals of greenhouse gases, such as $CO_2$ and $CH_4$, of sufficient quality to estimate regional sources and sinks".

Currently multi-year measurements from two satellite instruments can be used to retrieve information on $CO_2$ and $CH_4$ with sufficient near-surface-sensitivity: SCIAMACHY on ENVISAT (2002 - April 2012) (*Burrows et al., 1995; Bovensmann et al., 1999*) and TANSO-FTS on-board GOSAT (launched in 2009) (*Kuze et al., 2009*). Both instruments perform (or have performed) nadir observations in the near-infrared/short-wave-infrared (NIR/SWIR) spectral region covering the relevant absorption bands of $CO_2$, $CH_4$ and $O_2$ (needed to obtain the "dry-air column" used to compute GHG column-averaged dry-air mole fractions, i.e., $XCO_2$ (in ppm) and $XCH_4$ (in ppb)). These two instruments are therefore currently the two main sensors used within GHG-CCI. The corresponding retrieval algorithms are referred to as "ECV Core Algorithms" (ECAs) within GHG-CCI.

In addition, a number of other sensors are also used within GHG-CCI (e.g., MIPAS/ENVISAT and IASI/MetOp-A) as they provide additional constraints for atmospheric layers above the planetary boundary layer. The corresponding retrieval algorithms are referred to as "Additional Constraints Algorithms" (ACAs) within GHG-CCI.

Even moderate to strong $CO_2$ and $CH_4$ sources and sinks only result in quite small changes of the column-averaged mole fractions relative to their background concentration. High relative accuracy of the satellite retrievals is required because even very small (regional) biases can lead to significant errors of the inferred surface fluxes. One of the first activities within GHG-CCI was to establish the user requirements, e.g., in terms of required accuracy and precision of the different data products. The result of this activity was the initial version of the GHG-CCI User Requirements Document (URD) (*Buchwitz et al., 2011*), which has recently been updated (*Chevallier et al., 2014b*). Note that the GHG-CCI URD requirements are more detailed and often also more demanding compared to the GCOS requirements (*GCOS, 2011*).

The GHG-CCI data products and related documentation are freely available via the GHG-CCI website and yearly updates generated with improved retrieval algorithms and covering (where possible) longer time series are foreseen.

Here we present an overview about the latest data set - Climate Research Data Package No. 2 (CRDP#2) (Sect. 2) - and summarize key findings from using satellite $CO_2$ and $CH_4$ retrievals to improve our understanding of the natural and anthropogenic sources and sinks of these important greenhouse gases (Sect. 3). We also shortly mention ongoing activities related to the validation and initial user assessment of CRDP#2 and future plans (Sect. 4).

## 2. CLIMATE RESEARCH DATA PACKAGE 2 (CRDP#2)

In this section, we present an overview about the GHG-CCI CRDP#2. CRDP#2 consists of several satellite-derived $CO_2$ and $CH_4$ data products and related documentation (freely available from http://www.esa-ghg-cci.org -> CRDP (Data)).

Currently (mid March 2015) a preliminary unvalidated version of CRDP#2 is already publicly available. Validation and initial user assessments as conducted by the GHG-CCI validation team and Climate Research Group (CRG) are ongoing activities. The final validated data set is planned to be ready end of March 2015 (see Sect. 4 for details).

Via the GHG-CCI website also the previous data set CRDP#1 and related documentation is available. Note that for CRDP#2 an improved data format has been defined focusing on harmonization of the ECA products (*Buchwitz et al., 2014*). An overview about the various satellite-derived data products stored in the CRDP#2 data base is shown in Tab. 1 (for ECA products) and Tab. 2 (for ACA products).

Table 1 lists the GHG-CCI ECV core data products $XCO_2$ and $XCH_4$ as retrieved from SCIAMACHY/ENVISAT and TANSO-FTS/GOSAT. Note that more details for each product are available on the GHG-CCI website including spatio-temporal coverage, detailed documentation (e.g., Algorithm Theoretical Basis Documents (ATBDs)), point of contact information, information on data access, figures, etc.

As can be seen from Tab. 1 typically the same product (e.g., $XCO_2$ from SCIAMACHY) has been generated using different retrieval algorithms. We encourage users of our data products to make use of the fact that several different methods are available to generate a given product. This gives users the possibility to find out if important conclusions drawn by using one product are robust with respect to the method used to generate that product. This however may require significant effort and is therefore not always possible. For users who only want to use one product but do not know which one to choose, we aimed at defining one recommended "baseline product" generated with a baseline algorithm (see Tab. 1). The other products are called "alternative products". Note that the quality of an alternative product may be (at least on average) equivalent to the corresponding baseline product. Typically different methods have different strengths and weaknesses and therefore which product to use for a given application is expected to depend on the application. For our products we found (typically quite) small but potentially still significant differences between the baseline and the alternative products but have not yet always been able to clearly identify which of the products is better (e.g., due to the limited number of ground-based validation sites). For this reason we have not yet defined a baseline product for all products (see Tab. 1).

As can also be seen from Tab. 1, the $XCH_4$ algorithms / products are typically classified as "Full Physics" (FP) or "Proxy" (PR). The PR algorithms are using simultaneously retrieved $CO_2$ columns and model $CO_2$ columns to convert the retrieved methane columns (in molecules/area) to $XCH_4$ (in ppb), whereas the FP algorithms do not rely only modelled $CO_2$. The advantage of the PR algorithms is that scattering related errors (due to aerosols and clouds) cancel to a large extent when computing the $CH_4$ to $CO_2$ column ratio. As a consequence, the PR algorithms are typically simpler and faster and typically deliver a larger number of quality filtered (i.e., "good") observations. See, e.g., *Schepers et al., 2012*, for a discussion of $XCH_4$ FP and PR methods.

Note that we have also generated a merged $XCO_2$ product via the EMMA algorithm (*Reuter et al., 2013*) by combining the individual SCIAMACHY and GOSAT $XCO_2$ products. Currently however the EMMA CRDP#2 product covers only a limited time period (see Tab. 1). However, also a recently updated product (EMMA v2.0) is available via the GHG-CCI website covering 4 years. Within GHG-CCI the EMMA $XCO_2$ product is also used as a comparison tool for the individual products.

In line with the GHG-CCI user requirements (*Chevallier et al., 2014b*) the GHG-CCI ECA data products listed in Tab. 1 are (non-gridded) Level 2 products, i.e., they contain $XCO_2$ and $XCH_4$ values for each single observation along with information on time and location, uncertainty, quality flag, etc. (see *Buchwitz et al., 2014*, for details).

For illustration, seasonal averages of CRDP#2 products are shown in Fig. 2 for $XCO_2$ and Fig. 3 for $XCH_4$.

## 3. OVERVIEW SCIENTIFIC ACHIEVEMENTS

In this section a short overview is presented on scientific publications related to $CO_2$ and $CH_4$ satellite retrievals, which have been published during approximately the first four years of the GHG-CCI project (until mid March 2015). Focus is on publications related to GHG-CCI retrieval algorithms and corresponding data products and their use to address important scientific questions related to the natural and anthropogenic sources and sinks of $CO_2$ and $CH_4$. In this context also some other (non-GHG-CCI) publications are mentioned (in the References GHG-CCI-related publications are marked with (*); currently the number of peer-reviewed publications with GHG-CCI/CCI funding explicitly acknowledged is 38).

| GHG-CCI CRDP#2: ECV Core Algorithm (ECA) Products | | | | |
|---|---|---|---|---|
| **Algorithm / Product ID (version)** | **Product** | **Sensor Satellite** | **Algorithm Institute** | **Comment (Reference)** |
| CO2_SCI_BESD (v02.00.08) | $XCO_2$ | SCIAMACHY ENVISAT | BESD IUP | *SCIAMACHY $XCO_2$ baseline product (Reuter et al., 2011)* |
| CO2_SCI_WFMD (v3.8) | $XCO_2$ | SCIAMACHY ENVISAT | WFM-DOAS IUP | *SCIAMACHY $XCO_2$ alternative product (Schneising et al., 2011)* |
| CO2_GOS_OCFP (v5.1) | $XCO_2$ | TANSO GOSAT | UoL-FP UoL | *GOSAT $XCO_2$ product (baseline not yet decided) (Cogan et al., 2012)* |
| CO2_GOS_SRFP (v2.3.6) | $XCO_2$ | TANSO GOSAT | RemoTeC SRON/KIT | *GOSAT $XCO_2$ product (baseline not yet decided) (Butz et al., 2011)* |
| CO2_EMMA (v1.7) | $XCO_2$ | Merged SCIA and GOSAT | EMMA IUP (lead) | *Short time period only (6.2009-7.2010) (Reuter et al., 2013) (*)* |
| | | | | |
| CH4_SCI_WFMD (v3.7) | $XCH_4$ | SCIAMACHY ENVISAT | WFM-DOAS IUP | *SCIAMACHY $XCH_4$ proxy product (baseline not yet decided) (Schneising et al., 2011)* |
| CH4_SCI_IMAP (v7.0) | $XCH_4$ | SCIAMACHY ENVISAT | IMAP SRON/JPL | *SCIAMACHY $XCH_4$ proxy product (baseline not yet decided) (Frankenberg et al., 2011)* |
| CH4_GOS_OCPR (v5.1) | $XCH_4$ | TANSO GOSAT | UoL-PR UoL | *GOSAT $XCH_4$ proxy baseline product (Parker et al., 2011)* |
| CH4_GOS_SRPR (v2.3.6) | $XCH_4$ | TANSO GOSAT | RemoTeC SRON/KIT | *GOSAT $XCH_4$ proxy alternative product (Butz et al., 2010)* |
| CH4_GOS_SRFP (v2.3.6) | $XCH_4$ | TANSO GOSAT | RemoTeC SRON/KIT | *GOSAT $XCH_4$ full physics baseline product (Butz et al., 2011)* |
| Details (temporal coverage, etc.): http://www.esa-ghg-cci.org -> CRDP (Data) | | | | |

Tab. 1: Overview GHG-CCI core ("ECA") data products. (*) The latest version, EMMAv2.0, covers 4 years and is also available on the GHG-CCI website.

| GHG-CCI CRDP#2: Additional Constraints Algorithm (ACA) Products | | | | |
|---|---|---|---|---|
| **Algorithm / Product ID** | **Product** | **Sensor** | **Algorithm / Institute** | **Reference** |
| CO2_AIR_NLIS (*) | Mid/upper tropospheric column | AIRS | NLIS / LMD | *Crevoisier et al., 2004* |
| CO2_IAS_NLIS | Mid/upper tropospheric column | IASI | NLIS / LMD | *Crevoisier et al., 2009* |
| CO2_ACE_CLRS | Upper trop. / stratospheric profile | ACE-FTS | CLRS / LMD | *Foucher et al., 2009* |
| CH4_IAS_NLIS | Upper trop. / stratospheric profile | IASI | NLIS / LMD | *Crevoisier et al., 2013* |
| CH4_MIP_IMK (*) | Upper trop. / stratospheric profile | MIPAS | MIPAS / KIT-IMK | *Laeng et al., 2014* |
| CH4_SCI_ONPD | Stratospheric profile | SCIAMACHY | ONPD / IUP | *Noël et al., 2011* |
| CO2_SCI_ONPD | Stratospheric profile | SCIAMACHY | ONPD / IUP | *Noël et al., 2011* |
| Details (temporal coverage, etc.): http://www.esa-ghg-cci.org -> CRDP (Data) | | | | |

Tab. 2: Overview GHG-CCI ACA products providing information on $CO_2$ and $CH_4$ in atmospheric layers above the planetary boundary layer. (*) CRDP#1 product (no update for CRDP#2).

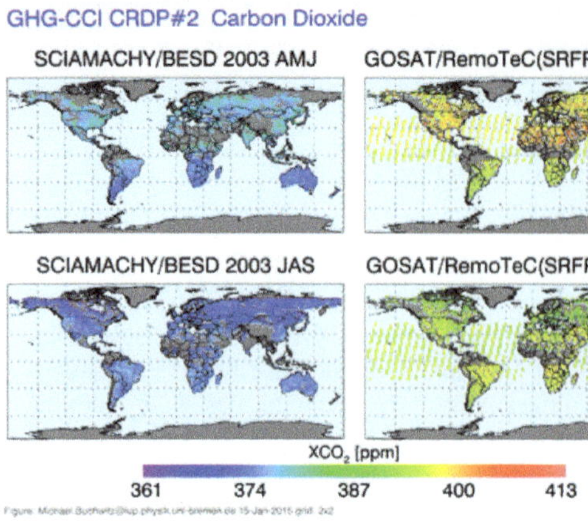


Fig.2: GHG-CCI CRDP#2 $XCO_2$ products.

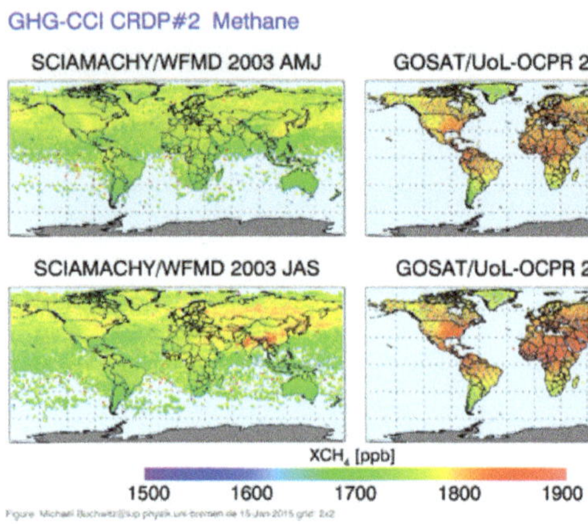


Fig.3: GHG-CCI CRDP#2 $XCH_4$ products.

The list of all GHG-CCI publications is available via the GHG-CCI website (http://www.esa-ghg-cci.org -> Publications), where also links to the publications are given. Please visit this website for the most up-to-date list of all GHG-CCI publications. Several publications are addressing improvements of the retrieval algorithms, e.g.,

- *Reuter et al., 2011*, presents first results from the application of the advanced BESD algorithm (*Reuter et al., 2010*) to SCIAMACHY $XCO_2$ retrievals. BESD has been developed to improve the accuracy and precision compared to the simpler but much faster WFMD algorithm and as shown in, e.g., *Dils et al., 2014*, this goal has been achieved.
- The WFMD $XCO_2$ retrieval algorithm has also been significantly improved during GHG-CCI as shown in *Heymann et al., 2012a, 2012b,* and *Schneising et al., 2011, 2012,* and used to address important $CO_2$ science issues as described below (e.g., *Schneising et al., 2013, 2014a*). This is also true for the WFMD $XCH_4$ retrieval algorithm (*Schneising et al., 2011, 2012, 2014b*).
- GHG-CCI GOSAT $XCO_2$ and $XCH_4$ algorithm improvements are also reported in a number of publications: *Butz et al., 2011*, *Cogan et al., 2012*, *Guerlet et al., 2013a, 2013b*, *Parker et al., 2011*, and *Schepers et al., 2012*.
- Recently, *Heymann et al., 2015*, has used the BESD algorithm to retrieve $XCO_2$ from GOSAT. This new product is being generated within the framework of the European MACC project (https://www.gmes-atmosphere.eu/), the predecessor of the upcoming operational European Copernicus Atmosphere Monitoring Service (CAMS), and not further discussed here.
- Retrieval algorithm related aspects for ACA products are also presented in a number of publications (e.g., *Noël et al., 2011, Laeng et al., 2014*).

Some publications are addressing related aspects, e.g.,

- *Dils et al., 2014*, presents a detailed validation of the initial ("Round Robin exercise", see *Buchwitz et al., 2013a*) GHG-CCI data products by comparisons with ground-based Total Carbon Column Observing Network (TCCON) (*Wunch et al., 2011*) $XCO_2$ and $XCH_4$ retrievals.
- Specific aspects related to the validation of the satellite $XCH_4$ products are presented in *Sussmann et al., 2011, 2013.*
- *Reuter et al., 2013*, developed the ensemble algorithm EMMA which uses the individual satellite Level 2 $XCO_2$ data products as input data to generate a new Level 2 data product where essentially outliers have been identified and removed to generate a potentially more robust "median product" exploiting the availability of an ensemble of individual products exists generated with algorithms each having different strengths and weaknesses. The EMMA method is used within GHG-CCI as a comparison tool for the individual global products. The advantage here is that products can be compared with the median product without relying on model simulations and without being limited to (sparse) ground-based validation sites. Note that for EMMA not only the European GHG-CCI data products are being used but also non-European GOSAT products generated in Japan at NIES (*Yoshida et al., 2013, Oshchepkov et al., 2011, 2013*) and the NASA ACOS product (*O'Dell et al., 2012*).
- *Reuter et al., 2012a,* used the $XCO_2$ retrieval algorithm BESD to study to what extent information on $CO_2$ isotopologues can be retrieved from GOSAT data and in *Reuter et al., 2012b*, a simple model ("SECM") is described which can be (and is) used to obtain atmospheric $CO_2$ background concentrations to be used as *a priori* information for satellite $XCO_2$ retrievals.

The main goal of the GHG-CCI project is to generate satellite-derived data products required to improve our knowledge on the sources and sinks of $CO_2$ and $CH_4$. Having delivered improved data products, these are then combined with knowledge of wind fields, with sophisticated atmospheric models or data assimilation techniques to determine, assess and constrain surface fluxes of $CO_2$ and $CH_4$. Relevant publications are described briefly below starting with publications addressing natural $CO_2$ fluxes:

- Using global GOSAT $XCO_2$ retrievals *Basu et al., 2013,* presented first $CO_2$ surface flux inverse modeling results for various regions. Their analysis suggests a reduced global land sink and a shift of the carbon uptake from the tropics to the extra-tropics. Their results also imply that Europe is a stronger carbon sink than expected.
- *Chevallier et al., 2014a,* used an ensemble of inversion methods and GOSAT $XCO_2$ retrievals to also derive regional $CO_2$ surface fluxes. They also found a significantly larger European carbon sink. They conclude that the derived sink is unrealistically large and they argue that this may be due to modelling issues related to long-range transport modelling and biases of the satellite retrievals. In particular they argue that errors of the satellite data outside of Europe may adversely influence the European results.
- *Reuter et al., 2014a,* investigated this European carbon sink issue in detail using an ensemble of SCIAMACHY and

GOSAT $XCO_2$ data products and a new inversion method which is not, or significantly less, sensitive to the potential issues discussed in *Chevallier et al., 2014a*. For example, *Reuter et al., 2014a,* only used satellite $XCO_2$ retrievals over Europe to rule out that non-European satellite data adversely influence the results related to the European carbon sink and they also only used short-term (days) transport modelling to minimize long-range transport errors. *Reuter et al., 2014a,* also performed several sensitivity tests to investigate and ensure the robustness of their results and to establish a reliable error budget. Based on an extensive analysis they conclude: "We show that the satellite-derived European terrestrial carbon sink is indeed much larger (1.02 +/- 0.30 GtC/year in 2010) than previously expected". The value they derived is larger compared to earlier inversion estimates using in-situ observations of 0.47 +/- 0.50 ("LSCE-39-insitu inversion") or 0.42 +/- 0.25 ("UoE-insitu") GtC/year for 2010 (*Chevallier et al., 2014a*), or 0.40 +/- 0.42 GtC/year for 2001-2004 (*Peylin et al, 2013*), which is reported in the recent IPCC report (*IPCC, 2013*) (see also: http://www.esa.int/Our_Activities/Observing_the_Earth/Is_Europe_an_underestimated_sink_for_carbon_dioxide). The disagreement with bottom-up estimates is even larger and significant: *Schulze et al., 2009,* report 0.235 +/- 0.05 GtC/year between 2000 and 2005.

- The findings of *Reuter et al., 2014a,* stimulated additional research (*Feng et al., 2015*).
- Focussing on Canadian and Siberian boreal forests, *Schneising et al., 2011,* computed longitudinal $XCO_2$ gradients from SCIAMACHY $XCO_2$ retrievals during the vegetation growing season over Canadian and Siberian boreal forests and compared the gradients with outputs from NOAA's $CO_2$ inversion system CarbonTracker (*Peters et al., 2007*). They found good agreement for the total boreal region and for inter-annual variations. For the individual regions, however, they found systematic differences suggesting a stronger Canadian boreal forest growing season $CO_2$ uptake and a weaker Siberian forest uptake compared to CarbonTracker.
- Focussing on hemispheric data and on carbon-climate feedbacks, *Schneising et al., 2013b,* used SCIAMACHY $XCO_2$ to study aspects related to the terrestrial carbon sink by looking at co-variations of $XCO_2$ growth rates and seasonal cycle amplitudes with near-surface temperature. They found $XCO_2$ growth rate changes of 1.25+/-0.32 ppm/year/K (approximately 2.7+/-0.7 GtC/year/K; indicating less carbon uptake in warmer years, i.e., a positive carbon-climate feedback) for the Northern Hemisphere in good agreement with CarbonTracker.
- *Reuter et al., 2013,* computed $CO_2$ seasonal cycle amplitudes using various satellite $XCO_2$ data products (using GHG-CCI products but also GOSAT $XCO_2$ products generated in Japan at NIES (*Yoshida et al., 2013, Oshchepkov et al., 2011, 2013*) and the NASA ACOS product (*O'Dell et al., 2012*)) and compared the amplitudes with TCCON and CarbonTracker. They found that the satellite products typically agree well with TCCON but they found significantly lower amplitudes for CarbonTracker suggesting that CarbonTracker underestimates the $CO_2$ seasonal cycle amplitude by approx. 1.5+/-0.5 ppm (see also *Buchwitz et al., 2013a,* for a discussion of these findings).
- *Guerlet et al., 2013b,* analyzed GOSAT $XCO_2$ retrievals focusing on the Northern Hemisphere. They identified a reduced carbon uptake in the summer of 2010 and found that this is most likely due to the heat wave in Eurasia driving biospheric fluxes and fire emissions. Using a joint inversion of GOSAT and surface data, they estimated an integrated biospheric and fire emission anomaly in April–September of 0.89±0.20 PgC over Eurasia. They found that inversions of surface measurements alone fail to replicate the observed $XCO_2$ inter-annual variability (IAV) and underestimate emission IAV over Eurasia. They highlighted the value of GOSAT $XCO_2$ in constraining the response of land-atmosphere exchange of $CO_2$ to climate events.
- *Basu et al., 2014,* studied seasonal variations of $CO_2$ fluxes during 2009-2011 over Tropical Asia using GOSAT, CONTRAIL and IASI data. They found an enhanced source for 2010 and concluded that this is likely due to the biosphere response to above-average temperatures in 2010 and unlikely due to biomass burning emissions.
- *Parazoo et al., 2013,* used GOSAT $XCO_2$ and solar induced chlorophyll fluorescence (SIF) retrievals to better understand the carbon balance of southern Amazonia.
- *Ross et al., 2013,* used GOSAT data to obtain information on wildfire $CH_4$:$CO_2$ emission ratios.
- For flux inversions not only the retrieved greenhouse gas values are relevant but also their error statistics, in particular the reported uncertainties. *Chevallier and O'Dell, 2013,* analyzed this aspect in the context of $CO_2$ flux inversions using GOSAT $XCO_2$ retrievals.

Despite the fact that none of the existing satellite missions has been optimized to obtain information on anthropogenic $CO_2$ emissions this important aspect has been addressed in several recent publications using existing satellite $XCO_2$ products:

- *Schneising et al., 2013a,* present an assessment of the satellite data over major anthropogenic $CO_2$ source regions. They used a multi-year SCIAMACHY $XCO_2$ data set and compared the regional $XCO_2$ enhancements and trends with the emission inventory EDGAR v4.2 (*Olivier et al., 2012*). They found no significant trend for the Rhine-Ruhr area in central Europe and the US East Coast but a significantly increasing trend for the Yangtze River Delta in China of about 13+/-8%/year, in agreement with EDGAR (10+/-1%/year).
- *Reuter et al., 2014,* studied co-located SCIAMACHY $XCO_2$ and $NO_2$ retrievals over major anthropogenic source regions. For East Asia they found increasing emissions of $NO_x$ (+5.8%/year) and $CO_2$ (+9.8%/year), i.e., decreasing emissions of $NO_x$ relative to $CO_2$ indicating that the recently installed and renewed technology in East Asia, such as power plants and transportation, is cleaner in terms of $NO_x$ emissions than the old infrastructure, and roughly matches relative emission levels in North America and Europe (see also: http://www.esa.int/Our_Activities/Observing_the_Earth/Space_for_our_climate/Good_and_bad_news_for_our_atmosphere).

Methane:

SCIAMACHY data have already been extensively used to improve our knowledge on regional methane emissions prior to the start of the GHG-CCI project (e.g., *Bergamaschi et al., 2009*). A more recent research focus has been to investigate the unexpected renewed atmospheric methane increase since 2007 using ground-based and satellite data (e.g., *Rigby et al., 2008, Dlugokencky et al., 2009, Bergamaschi et al., 2009, 2013, Schneising et al., 2011, Frankenberg et al., 2011, Sussmann et al., 2012, Crevoisier et al., 2013*). Based on an analysis of SCIAMACHY year 2003-2009 retrievals an increase of 7-9 ppb/year (0.4-0.5%/year) has been found with the largest increases in the tropics and northern mid latitudes (*Schneising et al., 2011*) but a particular region responsible for the increase has not been identified (*Schneising et al., 2011; Frankenberg et al., 2011*). *Bergamaschi et al., 2013,* used SCIAMACHY retrievals and NOAA surface data for 2003-2010 and inverse modelling to address this aspect. They concluded that the main reason for the increase are increasing anthropogenic emissions with wetland and biomass burning emissions being responsible for most of the inter-annual variations.

- Methane emission estimates have also been obtained from GOSAT as shown in a number of recent publications, e.g., *Fraser et al., 2013, 2014, Monteil et al., 2013, Cressot et al., 2014, Alexe et al., 2015*. In these studies often $CH_4$ retrievals from several satellites have been used (as well as NOAA data), e.g., *Monteil et al., 2013*, and *Alexe et al., 2015*, used SCIAMACHY and GOSAT retrievals and *Cressot el al., 2014,* used GOSAT, SCIAMACHY and IASI.
- Several publications focused on (relatively localized) methane sources in the United States: For example, *Schneising et al., 2014*, analyzed SCIAMACHY data over major US "fracking" regions and quantified methane emissions and leakage rates. For two of the fastest growing production regions in the US, the Bakken and Eagle Ford formations, they estimated that emissions increased by 990±650 $ktCH_4$/year and 530±330 $ktCH_4$/year between the periods 2006–2008 and 2009–2011. Relative to the respective increases in oil and gas production, these emission estimates correspond to leakages of 10.1%±7.3% and 9.1%±6.2% in terms of energy content, calling immediate climate benefit into question and indicating that current inventories likely underestimate the fugitive emissions from Bakken and Eagle Ford. Others also used SCIAMACHY data over the US to identify and quantify localized anthropogenic methane emission sources (*Kort et al., 2014, Wecht et al., 2014*).
- The SCIAMACHY $XCH_4$ retrievals have also been used to improve chemistry-climate models (*Shindell et al., 2014, Hayman et al., 2014*).

## 4. ONGOING ACTIVITIES AND FUTURE PLANS

Currently (mid March 2015) a preliminary unvalidated version of CRDP#2 is available via the GHG-CCI website (http://www.esa-ghg-cci.org -> CRDP (Data)) as validation and initial user assessments as conducted by the GHG-CCI validation team and Climate Research Group (CRG) are ongoing activities. The validation results will be reported in a document called "Product Validation and Intercomparison Report, version 3.x" (PVIRv3.x) and the user assessments will be reported in the "Climate Assessment Report, version 2.x" (CARv2.x). PVIRv3.x will be ready end of March 2015 and CARv2.x in April 2015 and both documents will be made publicly available along with the final validated CRDP#2 data products via the GHG-CCI website. These documents are updates of the corresponding CRDP#1 documents PVIRv2.0 (*Notholt et al., 2013*) and CARv1.1 (*Chevallier et al., 2013*).

Based on the outcome of the quality assessments the retrieval algorithms will be further improved and the satellite data will be reprocessed (if necessary) and, where possible, the time series will be extended. Yearly updates are foreseen and it is planned to release CRDP#3 in April 2016. GHG-CCI retrieval experts are also members of the OCO-2 Science Team and involved in the development of retrieval algorithms for Sentinel-5-Precursor and the data products of these sensors will also be considered by GHG-CCI. For OCO-2 it is initially planned to perform detailed comparisons to determine the consistency of the $XCO_2$ data products and to perform initial retrievals.

GHG-CCI team members are also involved in the specification of future GHG satellites, in particular CarbonSat (*Bovensmann et al., 2010, Buchwitz et al., 2013b*). CarbonSat, if selected for ESA's Earth Explorer 8 satellite, will continue the time series of greenhouse gas observations from space presented in this manuscript but will also address many important new aspects which cannot (or only with severe limitations) be addressed with other existing or planned satellites in particular the detection of localized $CO_2$ and $CH_4$ sources and the quantification of their emissions. Like SCIAMACHY, GOSAT and OCO-2, sun induced chlorophyll fluorescence, SIF, will be a secondary data product from CarbonSat (*Buchwitz et al., 2013b*) suitable to obtain Gross Primary Production (GPP; e.g., *Parazoo et al., 2013*, and references given therein) and for investigating the impact of stress on vegetation and the $CO_2$ uptake at the few $km^2$ spatial resolution scale of CarbonSat. The main goal of CarbonSat is to advance our knowledge on the natural and man-made sources and sinks of the two most important anthropogenic greenhouse gases carbon dioxide and methane from the global via the sub-continental to the local scale. CarbonSat will be the first satellite mission to image small scale emission hot spots of $CO_2$ (e.g., cities, volcanoes, industrial areas) and $CH_4$ (e.g., fossil fuel production, landfills, seeps) and to quantify their emissions and discriminate them from surrounding biospheric fluxes. In this context see also *Ciais et al., 2014,* and *CEOS, 2014,* for an overview about current capabilities and limitations and future needs for establishing a global carbon observing system.

## ACKNOWLEDGEMENTS

We thank JAXA and NIES for support and for providing us with GOSAT Level 1 data. We thank NIES and the NASA/ACOS team for supporting us and for providing us with GOSAT Level 2 data products. We thank ESA and DLR for providing us with SCIAMACHY Level 1 data products and ESA for funding this project.

## REFERENCES

*Note: (*) indicates publications with acknowledged GHG-CCI funding*

Alexe, M., Bergamaschi, P., Segers, A., et al., Inverse modeling of $CH_4$ emissions for 2010–2011 using different satellite retrieval products from GOSAT and SCIAMACHY, *Atmos. Chem. Phys.*, 15, 113–133, www.atmos-chem-phys.net/15/113/2015/, doi:10.5194/acp-15-113-2015, 2015. (*)

Basu, S., Guerlet, S., Butz, A., et al., Global $CO_2$ fluxes estimated from GOSAT retrievals of total column $CO_2$, *Atmos. Chem. Phys.*, 13, 8695-8717, 2013. (*)

Basu, S., Krol, M., Butz, A., et al., The seasonal variation of the $CO_2$ flux over Tropical Asia estimated from GOSAT, CONTRAIL and IASI, *Geophys. Res. Lett.*, doi: 10.1002/2013GL059105, 2014. (*)

Bergamaschi, P., Frankenberg, C., Meirink, J. F., et al., Inverse modeling of global and regional $CH_4$ emissions using SCIAMACHY satellite retrievals, *J. Geophys. Res.*, 114, D22301, doi:10.1029/2009JD012287, 2009.

Bergamaschi, P., Houweling, H., Segers, A., et al., Atmospheric $CH_4$ in the first decade of the 21st century: Inverse modeling analysis using SCIAMACHY satellite retrievals and NOAA surface measurements, *J. Geophys. Res.*, 118, 7350-7369, doi:10.1002/jgrd.50480, 2013.

Bovensmann, H., Burrows, J. P., Buchwitz, M., et al., SCIAMACHY - Mission objectives and measurement modes, *J. Atmos. Sci.*, 56 (2), 127-150, 1999.

Bovensmann, H., Buchwitz, M., Burrows, J. P., et al., A remote sensing technique for global monitoring of power plant $CO_2$ emissions from space and related applications, *Atmos. Meas. Tech.*, 3, 781–811, doi:10.5194/amt-3-781-2010, 2010.

Buchwitz, M., Chevallier, F., Bergamaschi, P., et al., User Requirements Document for the GHG-CCI project of ESA's Climate Change Initiative, version 1 (URDv1), 3. February 2011, http://www.esa-ghg-cci.org/?q=webfm_send/21, 2011. (*)

Buchwitz, M., Reuter, M., Schneising, O., et al., The Greenhouse Gas Climate Change Initiative (GHG-CCI): comparison and quality assessment of near-surface-sensitive satellite-derived $CO_2$ and $CH_4$ global data sets, *Remote Sensing of Environment*, doi:10.1016/j.rse.2013.04.024, pp. 19, 2013a. (*)

Buchwitz, M., Reuter, M., Bovensmann, H., et al., Carbon Monitoring Satellite (CarbonSat): assessment of atmospheric $CO_2$ and $CH_4$ retrieval errors by error parameterization, *Atmos. Meas. Tech.*, 6, 3477-3500, 2013b.

Buchwitz, M., Detmers, R., Boesch, H., et al., Product Specification Document for the GHG-CCI project of ESA's Climate Change Initiative, version 3 (PSDv3), 6. June 2014, http://www.esa-ghg-cci.org/index.php?q=webfm_send/160, 2014. (*)

Burrows, J. P., Hölzle, E., Goede, A. P. H., Visser, H., and Fricke, W., SCIAMACHY—Scanning Imaging Absorption Spectrometer for Atmospheric Chartography, *Acta Astronaut.*, 35(7), 445–451, doi:10.1016/0094-5765(94)00278-t, 1995.

Butz, A., Hasekamp, O. P., Frankenberg, C., Vidot, J., and Aben, I., $CH_4$ retrievals from space-based solar backscatter measurements: Performance evaluation against simulated aerosol and cirrus loaded scenes, *J. Geophys. Res.*, Volume 115, Issue D24, 27, DOI: 10.1029/2010JD014514, 2010.

Butz, A., Guerlet, S., Hasekamp, O., et al., Toward accurate $CO_2$ and $CH_4$ observations from GOSAT, *Geophys. Res. Lett.*, doi:10.1029/2011GL047888, 2011. (*)

Canadell, J. G., Ciais, P., Dhakal, S., et al., Interactions of the carbon cycle, human activity, and the climate system: a research portfolio, *Curr. Opin. Environ. Sustainabil.*, 2, 301–311, 2010.

CEOS, CEOS Strategy for Carbon Observations from Space. The Committee on Earth Observation Satellites (CEOS) Response to the Group on Earth Observations (GEO) Carbon Strategy., April 2014 (issued date: September 30, 2014), http://ceos.org/document_management/Publications/WGClimate_CEOS-Strategy-for-Carbon-Observations-from-Space_Apr2014.pdf, 2014.

Chevallier, F., Bergamaschi, P., Kaminiski, T., Scholze, M., Climate Assessment Report (CAR) for the GHG-CCI project of ESA's Climate Change Initiative, version 1.1 (CARv1.1), 18. Nov. 2013, http://www.esa-ghg-cci.org/index.php?q=webfm_send/153, 2013. (*)

Chevallier, F., and O'Dell, C. W., Error statistics of Bayesian $CO_2$ flux inversion schemes as seen from GOSAT, *Geophys. Res. Lett.*, doi: 10.1002/grl.50228, 2013. (*)

Chevallier, F., Palmer, P.I., Feng, L., Boesch, H., O'Dell, C.W., Bousquet, P., Towards robust and consistent regional $CO_2$ flux estimates from in situ and space-borne measurements of atmospheric $CO_2$, *Geophys. Res. Lett.*, 41, 1065-1070, DOI: 10.1002/2013GL058772, 2014a. (*)

Chevallier, F., Buchwitz, M., Bergamaschi, et al., User Requirements Document for the GHG-CCI project of ESA's Climate Change Initiative, version 2 (URDv2), 28. August 2014, http://www.esa-ghg-cci.org/?q=webfm_send/173, 2014b. (*)

Ciais, P., Dolman, A. J., Dargaville, R., Barrie, L., Butler, J., Canadell, J., & Moriyama, T., GEO Carbon Strategy. Rome: GEO Secretariat, Geneva / FAO, 2010.

Ciais, P., Dolman, A. J., Bombelli, A., et al., Current systematic carbon cycle observations and needs for implementing a policy-relevant carbon observing system, *Biogeosciences*, 11, 3547-3602, www.biogeosciences.net/11/3547/2014/, doi:10.5194/bg-11-3547-2014, 2014.

Cogan, A. J., Boesch, H., Parker, R. J., et al., Atmospheric carbon dioxide retrieved from the Greenhouse gases Observing SATellite (GOSAT): Comparison with ground-based TCCON observations and GEOS-Chem model calculations, *J. Geophys. Res.*, 117, D21301, doi:10.1029/2012JD018087, 2012.

Cressot, C., F. Chevallier, P. Bousquet, et al., On the consistency between global and regional methane emissions inferred from SCIAMACHY, TANSO-FTS, IASI and surface measurements, *Atmos. Chem. Phys.*, 14, 577-592, 2014.

Crevoisier C., Heilliette S., Chédin A., et al., Midtropospheric $CO_2$ concentration retrieval from AIRS observations in the tropics, *Geophys. Res. Lett.*, 31, L17106, 2004.

Crevoisier, C., Chédin, A., Matsueda, H., et al., First year of upper tropospheric integrated content of $CO_2$ from IASI hyperspectral infrared observations, *Atmos. Chem. Phys.*, 9, 4797-4810, 2009.

Crevoisier, C., Nobileau, D., Armante, R., et al., The 2007–2011 evolution of tropical methane in the mid-troposphere as seen from space by MetOp-A/IASI, *Atmos. Chem. Phys.*, 13, 4279-4289, 2013. (*)

Dils, B., Buchwitz, M., Reuter, M., et al., The Greenhouse Gas Climate Change Initiative (GHG-CCI): comparative validation of GHG-CCI SCIAMACHY/ENVISAT and TANSO-FTS/GOSAT $CO_2$ and $CH_4$ retrieval algorithm products with measurements from the TCCON, *Atmos. Meas. Tech.*, 7, 1723-1744, 2014. (*)

Dlugokencky, E. J., Bruhwiler, L., White, J. W. C., et al., Observational constraints on recent increases in the atmospheric $CH_4$ burden, *Geophys. Res. Lett.*, 36, L18803, doi:10.1029/2009GL039780, 2009.

Feng, L., Palmer, P. I., R. Parker, R. J., et al., Elevated uptake of $CO_2$ over Europe inferred from GOSAT $XCO_2$ retrievals: a real phenomenon or an artefact of the analysis?, *Atmos. Chem. Phys. Discuss.*, 15, 1989-2011, 2015. (*)

Foucher, P. Y., Chédin, A., Dufour, G., et al., Technical Note: Feasibility of $CO_2$ profile retrieval from limb viewing solar occultation made by the ACE-FTS instrument, *Atmos. Chem. Phys.*, 9, 2873–2890, 2009.

Frankenberg, C., Aben, I., Bergamaschi, P., et al., Global column-averaged methane mixing ratios from 2003 to 2009 as derived from SCIAMACHY: Trends and variability, *J. Geophys. Res.*, doi:10.1029/2010JD014849, 2011.

Fraser, A., Palmer, P. I., Feng, L., et al., Estimating regional methane surface fluxes: the relative importance of surface and GOSAT mole fraction measurements, Atmos. Chem. Phys., 13, 5697-5713, doi:10.5194/acp-13-5697-2013, 2013. (*)

Fraser, A., Palmer, P. I., Feng, L., et al., Estimating regional fluxes of $CO_2$ and $CH_4$ using space-borne observations of $XCH_4$:$XCO_2$, *Atmos. Chem. Phys.*, 14, 12883-12895, www.atmos-chem-phys.net/14/12883/2014/, doi:10.5194/acp-14-12883-2014, 2014. (*)

GCOS (Global Climate Observing System): SYSTEMATIC OBSERVATION REQUIREMENTS FOR SATELLITE-BASED DATA PRODUCTS FOR CLIMATE - 2011 Update - Supplemental details to the satellite-based component of the "Implementation Plan for the Global Observing System for Climate in Support of the UNFCCC (2010 Update)", GCOS-154, 2011.

Guerlet, S., Basu, S., Butz, A., et al., Reduced carbon uptake during the 2010 Northern Hemisphere summer from GOSAT, *Geophys. Res. Lett.*, doi: 10.1002/grl.50402, 2013b. (*)

Guerlet, S., Butz, A., Schepers. D., et al., Impact of aerosol and thin cirrus on retrieving and validating $XCO_2$ from GOSAT shortwave infrared measurements, *J. Geophys. Res.*, doi: 10.1002/jgrd.50332, 2013a. (*)

Hayman, G. D., O'Connor, F. M., Dalvi, M., et al., Comparison of the HadGEM2 climate-chemistry model against in-situ and SCIAMACHY atmospheric methane data, *Atmos. Chem. Phys.*, 14, 13257-13280, doi:10.5194/acp-14-13257-2014, 2014. (*)

Heymann, J., Schneising, O., Reuter, M., et al., SCIAMACHY WFM-DOAS $XCO_2$: comparison with CarbonTracker $XCO_2$ focusing on aerosols and thin clouds, *Atmos. Meas. Tech.*, 5, 1935-1952, 2012a. (*)

Heymann, J., Bovensmann, H., Buchwitz, M., et al., SCIAMACHY WFM-DOAS $XCO_2$: reduction of scattering related errors, *Atmos. Meas. Tech.*, 5, 2375-2390, 2012b. (*)

Heymann, J., Reuter, M., Hilker, M., et al., Consistent satellite $XCO_2$ retrievals from SCIAMACHY and GOSAT using the BESD algorithm, *Atmos. Meas. Tech. Discuss.*, 8, 1787-1832, 2015. (*)

Hollmann, R., Merchant, C. J., Saunders, R., et al., The ESA Climate Change Initiative: satellite data records for essential climate variables, *Bulletin of the American Meteorological Society (BAMS)*, 0.1175/BAMS-D-11-00254.1, 2013. (*)

IPCC, Climate Change 2013: The Physical Science Basis, Working Group I Contribution to the Fifth Assessment Report of the Intergovernmental Report on Climate Change, http://www.ipcc.ch/report/ar5/wg1/, 2013.

Kirschke, S., Bousquet, P., Ciais, P., et al., Three decades of global methane sources and sinks, *Nat. Geosci.*, 6, 813–823, doi:10.1038/ngeo1955, 2013.

Kort, E. A., Frankenberg, C., Costigan, K. R., et al., Four corners: The largest US methane anomaly viewed from space, *Geophys. Res. Lett.*, 41, doi:10.1002/2014GL061503, 2014.

Kuze, A., Suto, H., Nakajima, M., and Hamazaki, T., Thermal and near infrared sensor for carbon observation Fourier-transform spectrometer on the Greenhouse Gases Observing Satellite for greenhouse gases monitoring, *Appl. Opt.*, 48, 6716–6733, 2009.

Laeng, A., Plieninger, J., von Clarmann, T., et al., Validation of MIPAS IMK/IAA methane version V5R_CH4_222 profiles, *in preparation for AMT*, 2014. (*)

Le Quéré, G., Peters, G. P., Andres, R. J., et al., Global carbon budget 2013, *Earth Syst. Sci. Data*, 6, 235–263, doi:10.5194/essd-6-235-2014, www.earth-syst-sci-data.net/6/235/2014/, 2014.

Monteil, G., Houweling, S., Butz, A., et al., Comparison of $CH_4$ inversions based on 15 months of GOSAT and SCIAMACHY observations, *J. Geophy. Res.*, doi: 10.1002/2013JD019760, Vol 118, Issue 20, 11807-11823, 2013. (*)

Noël, S., Bramstedt, K., Rozanov, A., et al., Stratospheric methane profiles from SCIAMACHY solar occultation measurements derived with onion peeling DOAS, *Atmos. Meas. Tech.*, 4, 2567-2577, 2011. (*)

Notholt, J., Dils, B., Buchwitz, M., et al., Product Validation and Intercomparison Report (PVIR) for the GHG-CCI project of ESA's Climate Change Initiative, version 2 (PVIRv2), 4. Nov. 2013, http://www.esa-ghg-cci.org/index.php?q=webfm_send/152, 2013. (*)

O'Dell, C. W., Connor, B., Boesch, H., et al., The ACOS $CO_2$ retrieval algorithm – Part 1: Description and validation against synthetic observations, *Atmos. Meas. Tech.*, 5, 99–121, 2012.

Olivier, J. G. J., Janssens-Maenhout, G., and Peters, J. A. H. W., Trends in global $CO_2$ emissions, 2012 Report, PBL Netherlands Environmental Assessment Agency, The Hague, Joint Research Centre, Ispra, ISBN 978-92-79-25381-2, 2012.

Oshchepkov, S., Bril, A., Maksyutov, S., and Yokota, T., Detection of optical path in spectroscopic space-based observations of greenhouse gases: Application to GOSAT data processing, *J. Geophys. Res.*, 116, D14304, doi:10.1029/2010JD015352, 2011.

Oshchepkov, S., Bril, A., Yokota, T., et al., Effects of atmospheric light scattering on spectroscopic observations of greenhouse gases from space. Part 2: Algorithm intercomparison in the GOSAT data processing for $CO_2$ retrievals over TCCON sites, *J. Geophys. Res.*, 118, 1493–1512, doi:10.1002/jgrd.50146, 2013. (*)

Parazoo, N. C., Bowman, K., Frankenberg, C., et al., Interpreting seasonal changes in the carbon balance of southern Amazonia using measurements of $XCO_2$ and chlorophyll fluorescence from GOSAT, *Geophys. Res. Lett.*, 40, 2829–2833, doi:10.1002/grl.50452, 2013. (*)

Parker, R., Boesch, H., Cogan, A., et al., Methane Observations from the Greenhouse gases Observing SATellite: Comparison to ground-based TCCON data and Model Calculations, *Geophys. Res. Lett.*, doi:10.1029/2011GL047871, 2011.

Peters, W., Jacobson, A. R., Sweeney, C., et al.: An atmospheric perspective on North American carbon dioxide exchange: CarbonTracker, *Proceedings of the National Academy of Sciences (PNAS)* of the United States of America, 27 Nov. 2007, 104(48), 18925-18930, 2007.

Peylin, P., Law, R. M., Gurney, et al., Global atmospheric carbon budget: results from an ensemble of atmospheric $CO_2$ inversions, Biogeosciences, 10, 6699–6720, doi:10.5194/bg-10-6699-2013, URL http://www.biogeosciences.net/10/6699/2013/, 2013.

Reuter, M., Buchwitz, M., Schneising, O., et al., A method for improved SCIAMACHY $CO_2$ retrieval in the presence of optically thin clouds, *Atmos. Meas. Tech.*, 3, 209-232, 2010.

Reuter, M., Bovensmann, H., Buchwitz, M., et al., Retrieval of atmospheric $CO_2$ with enhanced accuracy and precision from SCIAMACHY: Validation with FTS measurements and comparison with model results, *J. Geophys. Res.*, 116, D04301, doi:10.1029/2010JD015047, 2011. (*)

Reuter, M., Bovensmann, H., Buchwitz, M., et al., On the potential of the 2041-2047 nm spectral region for remote sensing of atmospheric $CO_2$ isotopologues, *Journal of Quantitative Spectroscopy and Radiative Transfer*, 113(16), 2009-2017, doi:10.1016/j.jqsrt.2012.07.013, 2012a. (*)

Reuter, M., Buchwitz, M., Schneising, O., et al., A simple empirical model estimating atmospheric $CO_2$ background concentrations, *Atmos. Meas. Tech.*, 5, 1349-1357, 2012b. (*)

Reuter, M., Boesch, H., Bovensmann, H., et al., A joint effort to deliver satellite retrieved atmospheric $CO_2$ concentrations for surface flux inversions: the ensemble median algorithm EMMA, *Atmos. Chem. Phys.*, 13, 1771-1780, 2013. (*)

Reuter, M., Buchwitz, M., Hilker, M., et al., Satellite-inferred European carbon sink larger than expected, *Atmos. Chem. Phys.*, 14, 13739-13753, www.atmos-chem-phys.net/14/13739/2014/, doi:10.5194/acp-14-13739-2014, 2014a. (see also http://www.esa.int/Our_Activities/Observing_the_Earth/Is_Europe_an_underestimated_sink_for_carbon_dioxide) (*)

Reuter, M., Buchwitz, M., Hilboll, A., et al., Decreasing emissions of NOx relative to $CO_2$ in East Asia inferred from satellite observations, *Nature Geoscience*, 28 Sept. 2014, doi:10.1038/ngeo2257, pp.4, 2014b. (see also http://www.esa.int/Our_Activities/Observing_the_Earth/Space_for_our_climate/Good_and_bad_news_for_our_atmosphere) (*)

Rigby, M., Prinn, R. G., Fraser, P. J., et al., Renewed growth of atmospheric methane, *Geophys. Res. Lett.*, 35, L22805, doi:10.1029/2008GL036037, 2008.

Ross, A. N., Wooster, M. J., Boesch, H., Parker, R., First satellite measurements of carbon dioxide and methane emission ratios in wildfire plumes, *Geophys. Res. Lett.*, 40, 1-5, doi:10.1002/grl.50733, 2013. (*)

Schepers, D., Guerlet, S., Butz, A., et al., Methane retrievals from Greenhouse Gases Observing Satellite (GOSAT) shortwave infrared measurements: Performance comparison of proxy and physics retrieval algorithms, *J. Geophys. Res.*, 117, D10307, doi:10.1029/2012JD017549, 2012.

Schneising, O., Buchwitz, M., Reuter, M., et al., Long-term analysis of carbon dioxide and methane column-averaged mole fractions retrieved from SCIAMACHY, *Atmos. Chem. Phys.*, 11, 2881-2892, 2011. (*)

Schneising, O., Bergamaschi, P., Bovensmann, H., et al., Atmospheric greenhouse gases retrieved from SCIAMACHY: comparison to ground-based FTS measurements and model results, *Atmos. Chem. Phys.*, 12, 1527-1540, 2012. (*)

Schneising, O., Heymann, J., Buchwitz, M., Reuter, M., Bovensmann, H., and Burrows, J. P., Anthropogenic carbon dioxide source areas observed from space: assessment of regional enhancements and trends, *Atmos. Chem. Phys.*, 13, 2445-2454, 2013. (*)

Schneising, O., Reuter, M., Buchwitz, M., Heymann, J., Bovensmann, H., and Burrows, J. P., Terrestrial carbon sink observed from space: variation of growth rates and seasonal cycle amplitudes in response to interannual surface temperature variability, *Atmos. Chem. Phys.*, 14, 133-141, 2014a. (*)

Schneising, O., Burrows, J. P.,, Dickerson, R. R., Buchwitz, M., Reuter, M., Bovensmann, H., Remote sensing of fugitive methane emissions from oil and gas production in North American tight geologic formations, *Earth's Future*, 2, DOI: 10.1002/2014EF000265, pp. 11, 2014b. (*)

Schulze, E. D., Luyssaert, S., Ciais, P, et al., Importance of methane and nitrous oxide emissions for Europe's terrestrial greenhouse gas balance, *Nat. Geosci.*, 2, 842–850, doi:10.1038/ngeo686, 2009.

Shindell, D. T., Pechony, O., Voulgarakis, A., et al., Interactive ozone and methane chemistry in GISS-E2 historical and future climate simulations, *Atmos. Chem. Phys.*, 13, 2653–2689, doi:10.5194/acp-13-2653-2013, 2013.

Sussmann, R., Forster, F., Rettinger, M, and Jones, N., Strategy for high-accuracy-and-precision retrieval of atmospheric methane from the mid-infrared FTIR network, *Atmos. Meas. Tech.*, 4, 1943-1964, 2011. (*)

Sussmann, R., Forster, F., Rettinger, M., and Bousquet, P., Renewed methane increase for five years (2007-2011) observed by solar FTIR spectrometry, *Atmos. Chem. Phys.*, 12, 4885-4891, 2012. (*)

Sussmann, R., Ostler, A., Forster, F., et al., First intercalibration of column-averaged methane from the Total Carbon Column Observing Network and the Network for the Detection of Atmospheric Composition Change, *Atmos. Meas. Tech.*, 6, 397-418, 2013. (*)

Wecht, K. J, Jacob, D. J., Sulprizio, M. P., et al., Spatially resolving methane emissions in California: constraints from the CalNex aircraft campaign and from present (GOSAT, TES) and future (TROPOMI, geostationary) satellite observations, *Atmos. Chem. Phys.*, 14, 8173-8184, www.atmos-chem-phys.net/14/8173/2014/, doi:10.5194/acp-14-8173-2014, 2014. (*)

Wunch, D., Toon, G. C., Blavier, J.-F., et al., The Total Carbon Column Observing Network, *Phil. Trans. R. Soc. A*, 369, 2087–2112, doi:10.1098/rsta.2010.0240, 2011.

Yoshida, Y., Kikuchi, N., Morino, I., et al., Improvement of the retrieval algorithm for GOSAT SWIR $XCO_2$ and $XCH_4$ and their validation using TCCON data, *Atmos. Meas. Tech.*, 6, 1533–1547, doi:10.5194/amt-6-1533-2013, 2013.

# 14

# NDVI DYNAMICS OF THE TAIGA ZONE IN CONNECTION WITH MODERN CLIMATE CHANGES

A. Bobkov [a], E. Panidi [b], N. Torlopova [c], V. Tsepelev [d, *]

[a] Saint-Petersburg State University, Institute of Earth Sciences, Department of Physical Geography and Landscape Design, 33-35 10-th line V.O., 199178 Saint-Petersburg, Russia
[b] Saint-Petersburg State University, Institute of Earth Sciences, Department of Cartography and Geoinformatics, 33-35 10-th line V.O., 199178 Saint-Petersburg, Russia
[c] Russian Academy of Sciences, Ural Branch, Komi Scientific Center, Institute of Biology, 28 Kommunisticheskaya st., 167982 Syktyvkar, Russia
[d] Russian State Hydrometeorological University, Meteorological Department, 98 Malookhtinsky prospect, 195196 Saint-Petersburg, Russia – v0010200@mail.ru

**KEY WORDS:** Climate Change, Vegetation Cover, Dynamics of Vegetation, Taiga Zone

**ABSTRACT:**

This research is dedicated to the investigation of the relations between the XXI century climate changes and Normalized Difference Vegetation Index (NDVI) variability of the taiga zone. For this purposes was used the observations of vegetation variability on the test area located nearby Syktyvkar city (Komi Republic, Russia), 16-day averages of NDVI data derived from TERRA/MODIS space imagery (spatial resolution is about 250 meters), and the air temperature and precipitation observations from Syktyvkar meteorological station.
The research results confirmed the statistically significant positive correlation between NDVI and air temperature for all vegetation types of the test area, for both spring and autumn seasons. The weakest correlation was found for coniferous forest, namely, pine forest on poor soils, and the strongest correlation was found for meadows and bogs.
Additionally the map of NDVI trends of the test area shows that the sectors of greatest positive trend located on the territories with non-forest cover, and as a result, the positive trend of air temperature is indicated most brightly on vegetation of non-forest lands. Thereby these lands can serve as climate changes indicator in the investigated region.
The study was partially supported by Russian Foundation for Basic Research (RFBR), research project No. 14-05-00858 a.

## 1. INTRODUCTION

Climate change in northern Eurasia appears in the seasonal variation of the precipitation, air temperature, sea ice cover and vegetation distribution (Kattsov, 2004; Wookey at al., 2009). However, the hydrometeorological network of the Russian northern territories has decreased its density over the past 30 years. Therefore, the quality of climate change monitoring within the continental areas is not high enough.
Nevertheless, there is a long series of satellite observations of vegetation at these regions, which we can use to study the climate change dynamics. The remote sensing data, in particular vegetation index data (NDVI), allow us to study vegetation cover change of arctic and subarctic regions (Goetz at al., 2007; Jia at al., 2009; Stow at al., 2004). A gradual change in the vegetation type, especially at the areas occupied by annuals, should follow the changing of temperature and humidity conditions. Some studies showed that changes at the areas of shrub vegetation types in Siberian and Canadian Arctic tundra has a positive correlation with summer air temperature (Jia at al., 2006; Jia at al., 2009; Walker at al., 2006; Wookey at al., 2009). That is why we can try to use the vegetation dynamics as an indicator of the climate change impact on terrestrial ecosystems (Goetz at al., 2007; Krankina at al., 2010).
The main objective of our study was the determination of the techniques, which may indicate climatic changes in the mainland areas of northern regions of Russia. For this purpose, we studied the relationships between vegetation index for different types of vegetation, derived from remote sensing materials, and terrestrial observations of air temperature and precipitation.

## 2. METHODOLOGY

To investigate the relationships between the vegetation characteristics and climate change, we attempted to identify the types of vegetation that are most affected by the temperature and humidity conditions. This study was conducted on the ground test area, which is covered with the long series of vegetation observations.
We used the vegetation index data for vegetation analysis. Being synthetic dimensionless indicator NDVI has strong correlations with the productivity and biomass of vegetation for various ecosystem types (Cherepanov at al., 2009).
For selection of the NDVI data source, we had to take into account the next three basic requirements. The first is that the spatial resolution of the data should allow us to detect the types of plant communities (Jia at al., 2006; Jia at al., 2009).
The second is that the observation series should be homogeneous and continuous in time and space during long periods, because of the high temporal variability of vegetation index. This variability depends on the intensive variability of the weather conditions at the region of interest. The changes in precipitation and temperature conditions during the calendar year cause an immediate response in vegetation characteristics. Additionally, Northern Eurasia characterized by the constant

* Corresponding author

presence of cloud cover that leads to the greater number of observation gaps. These gaps resume errors in the computation of long-term vegetation index trends. Then the observation technique should provide the opportunity to collect data with high frequency, which able to capture short cloud free periods.

Finally, the third is that the length of the data series should be at least 10 years. Within a shorter time interval, it is difficult to notice the changes that occur in the structure of vegetation cover.

Almost the only dataset that corresponds the above-mentioned requirements is the archive of MODIS imagery obtained from TERRA satellite with approximate spatial resolution of 250 meters per pixel. MODIS imagery cover the area of our interest every day and NDVI data are obtained from raw datasets as the standard product averaged for 16-day intervals. This technique increases the probability of cloudless vegetation data observation, and makes it possible to generate a continuous series of observations.

The vegetation types, vegetation density and condition, and also the surface slope and aspect are also influence NDVI magnitude (Cherepanov at al., 2009). Due to this context, we used the ground test area, where the terrestrial vegetation monitoring conducted repeatedly, for the correct interpretation of long-term NDVI trends (Figure 1). Additionally, one of meteorological stations of the Russian hydrometeorological observation network is located on the area, and allows the direct observation of air temperature and precipitation changes. Then the data collections of vegetation and meteorological monitoring are presented for this area for some past decades.

This test area located in the middle taiga zone of the Komi Republic (north-west region of Russia).

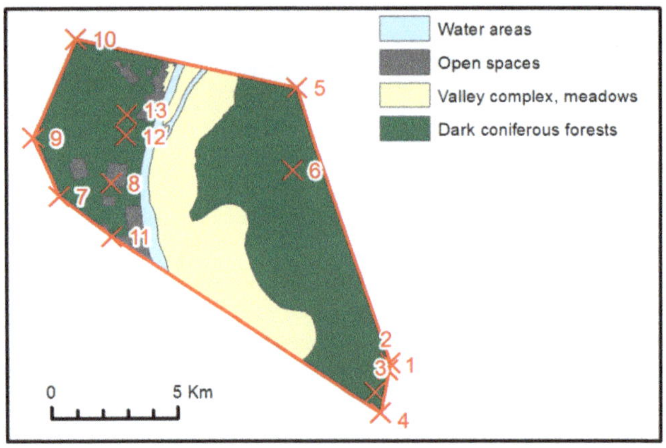


Figure 1. Distribution of the vegetation types on the ground test area. The numbers allocate sample plots positions on the ground test area that were used for direct terrestrial observations

## 3. DATA AND TECHNIQUES

### 3.1 Remote Sensing Data

The NDVI maps were produced using TERRA/MODIS imagery. These datasets preprocessed as standard product and distributed as public domain data[1]. Rasters of ~5 km, ~1 km and ~250 meters spatial resolution averaged per 1 month or 16 days and published online. In this project we used 16-day NDVI datasets of 250 meters spatial resolution, collected over the period from 2000 to 2013. These data allowed us to study NDVI change in connection to the change in temperature and precipitation during the full growing season.

### 3.2 Meteorological Data

For climate change investigation at the ground test area we used the observation data, which collected at the Syktyvkar meteorological station. The dataset range is also from 2000 to 2013.

To provide the statistical analysis we involved daily averaged air temperature and daily total precipitation. The data series of the mean daily temperature and total precipitation values were averaged for the 16-day periods equally to the NDVI data.

Furthermore, the secondary meteorological parameters were calculated additionally for the growing seasons:

1. Average air temperature
2. Accumulated amount of air temperatures
3. Average total precipitation
4. Accumulated total precipitation

### 3.3 Natural Conditions of the Study Area

The study area is located in the taiga zone in the European North-East (Russia, Komi Republic, Syktyvdinsky district). The ground polygon area covers the territory of Syktyvkar Forestry and forest territories administrated by Syktyvkar Timber Industry Complex and some agricultural companies.

The city of Syktyvkar with surrounding suburban areas, green areas and floodplain of river Vychegda are lying within the test area.

### 3.4 Vegetation of the Study Area

The vegetation cover types of the ground test area are presented on Figure 1. The area is located in the boreal forests (taiga) zone of the European North.

Forested part of the test area is located on the Syktyvkar Forestry territory. This forested part of Syktyvkar Forestry occupies approximately 92% of its total area.

The forested part combined by approximately ~77% of coniferous, which divided into pine forests (~42%) and spruce forests (~35%).

Additionally, the territory of aged wood stands is occupied by mature and overmature forests (~55% of total area) and the territory of the young stands is occupied by coniferous-deciduous (~38% of total area) (Forestry organization…, 1988).

### 3.5 Growing Seasons

The study of climate change was conducted for the growing seasons. We calculated the averaged beginning and ending dates of the growing seasons, in which some significant changes in vegetation index dynamics could be detected (Sadokov et al., 2012). These are spring, summer and autumn seasons. The boundary dates were calculated as the average dates for the period from 2000 to 2013 (Table 1).

The frames of the summer season are the dates of increasing air temperature transition through +10℃ (June 4) and the decreasing transition through +10℃ (August 21). The average beginning date of spring season is the date of increasing air temperature transition through +5℃ (May 13) and the end date of autumn season is the date of decreasing transition through +5℃ (September 29).

[1]http://modis.gsfc.nasa.gov/data/dataprod/index.php

| Period | Beginning | Ending | Duration |
|---|---|---|---|
| Spring | 13 May | 3 June | 22 days |
| Summer | 04 June | 21 August | 78 days |
| Autumn | 22 August | 29 September | 39 days |

Table 1. The averaged frames of the growing seasons during 2000-2013

### 3.6 Data Visualization

The ArcGIS[2] and ArcReader[3] software were used for processing of the raw data and for cartographic visualization of the spatial distribution of vegetation index values and climate parameters. The ArcGIS was used as the instrument of data processing and visualization. Additionally the ArcReader played the role of tool for data representation for visual analysis and detailed data examination by thematic specialists.

This two-stage data representation was used to provide single data view for all research group participants as well as to minimize the adaption efforts to new software for thematic specialists that generally use some another specific software for data visualization.

### 3.7 Comparative Analysis

We calculated the spatial distribution grids of linear trend coefficients for surface temperature, precipitation and NDVI for each growing season. Then we applied the Spearman's rank correlation method to calculate the spatial distribution grids of correlation coefficients between surface temperature and NDVI, as well as between precipitation and NDVI.

## 4. CLIMATE AND VEGETATION TRENDS

Air temperature and daily precipitation sums at the Syktyvkar meteorological station varied in a wide range during the study period in 2000-2013. The most dry and hot years were observed in the 2000, 2010 and 2011. The most cold and wet years were 2008, 2012.

The trend of 16-day averaged air temperature was about +1 ℃ for 14 years of the study period. Trend of 16-day averaged total precipitation was approximately +10 mm for 14 years.

Trends are significant at the 95% confidence level.

### 4.1 Air Temperature

The graphs of the air temperature averaged by growing seasons, show that the rise of temperature from 2000 to 2013 had different intensity (Figure 2).

Maximum positive coefficients of the air temperature linear trends were observed in spring. The average spring temperature during 14 years increased for more than 2 ℃. Rising of the summer average temperature did not exceed 1 ℃ during 14 years. Finally, in autumn season, the air temperature rose just about 1 ℃ during 14 years.

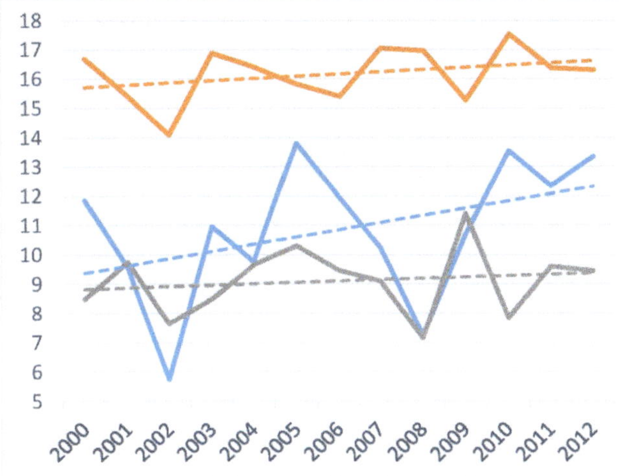


Figure 2. Graphs of the air temperature (℃) at the Syktyvkar station, from 2000 to 2013. The temperature is averaged by seasons (blue graph – spring, orange – summer, grey - autumn)

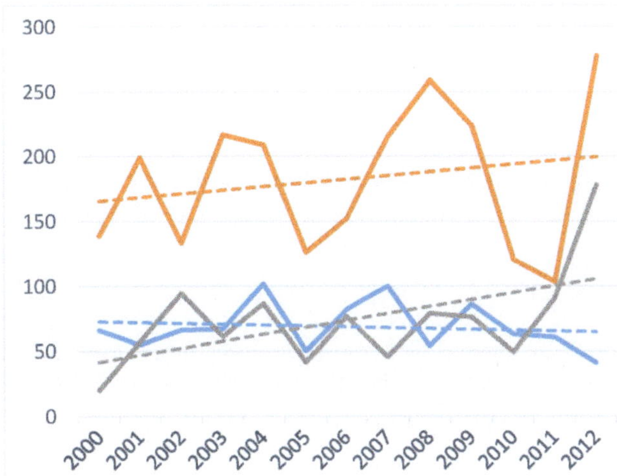


Figure 3. Graphs of the total precipitation (mm) at the Syktyvkar station, from 2000 to 2013. The precipitation is summarized by seasons (blue graph – spring, orange – summer, grey - autumn)

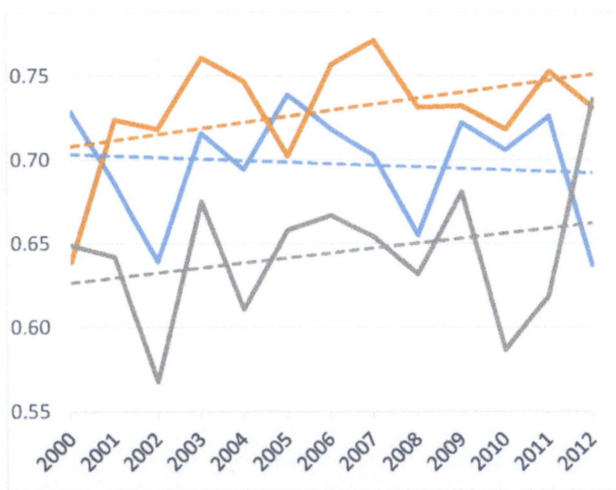


Figure 4. Graphs of the NDVI at the test area, from 2000 to 2013. The NDVI is averaged for test area and by seasons (blue graph – spring, orange – summer, grey - autumn)

---

[2]http://www.esri.com/software/arcgis/arcgis-for-desktop

[3]http://www.esri.com/software/arcgis/arcreader/download

### 4.2 Precipitation

The graphs of the summarized spring, summer and autumn precipitation shows that the precipitation sums rose with different intensity from 2000 to 2013 too.

Positive and statistically significant coefficients of the precipitation linear trends observed for summer. The summarized summer precipitation increased approximately for 40 mm during 14 years.

The precipitation, summarized for spring season, decreased slightly. Maximum increase of precipitation sums was observed in the autumn season and was about 60 mm during 14 years. (Figure 3).

### 4.3 Vegetation Index

Vegetation index, averaged for the ground test area was unstable from 2000 to 2013, due to the changing of temperature and precipitation conditions on the area. The main feature of the index is its slow rise in spring and summer and a sharp decrease in autumn. The highest NDVI magnitude observed in summer of 2007 (0.77). The lowest summer NDVI value observed in 2000 (0.64). Over the whole study period, the average NDVI value increased.

This NDVI increasing indicates the slow and gradual improvement of climate conditions for vegetation growth on the test area. The growth was consistent with the increasing of precipitation sums and air temperature.

The combinations of average index values for summer, spring and autumn periods show that the most favorable conditions for vegetation were in 2003 and 2006. The worst years for the vegetation were 2002, 2008 and 2010, 2012 (Figure 4).

NDVI variability was different in the different growing seasons. There was no significant change in the average NDVI magnitude in spring. The maximum increase occurred in summer (0.71 to 0.75). Additionally, the index rose from 0.63 to about 0.66 in autumn.

The trends are significant at the 95% confidence level.

### 4.4 Spatial Distribution of the Vegetation Index Trends

NDVI spatial distribution and spatial differentiation depend on the vegetation types, vegetation closeness and health. We calculated the full growing season NDVI trends for different sample plots of the ground test area, which are occupied by different vegetation types. Additionally we produced the NDVI trend maps of the test area, separately for each growing season (Figure 5).

Analysis of these trend maps shows us that most part of the area has the significant positive trend during all seasons. Sectors of the area, which are occupied by evergreen coniferous forests, preserves the positive NDVI trend also during all three seasons.

The sectors of the maximum trend values, which were observed in summer season, are occupied by shrub and meadow vegetation. In spring and autumn seasons the trends are largely neutral. The trend sign was positive in these seasons on scrubs and grasslands in floodplains.

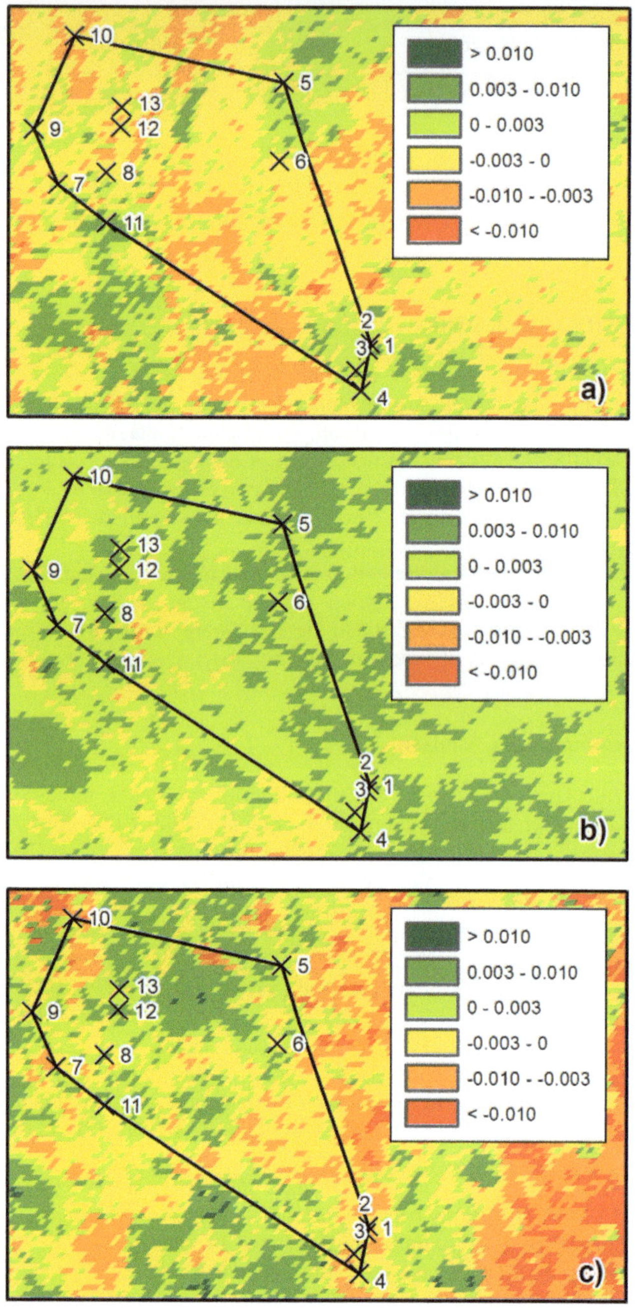


Figure 5. Spatial distribution of the vegetation index trends on the ground test area in the spring (a), summer (b) and autumn (c).

### 4.5 NDVI – Air Temperature and NDVI – Precipitation Relationships

The spatial distribution of coefficient values of the correlation between NDVI and air temperature, as well as between NDVI and precipitation sums, depends on growing season and vegetation type (Figure 6). The territory of significant (at the 95% confidence level) positive values of NDVI-air temperature correlation coefficient occupies the entire test area in spring season. Minimum values of the correlation coefficient were observed on the territories occupied by evergreen coniferous forest and non-forest floodplain lands.

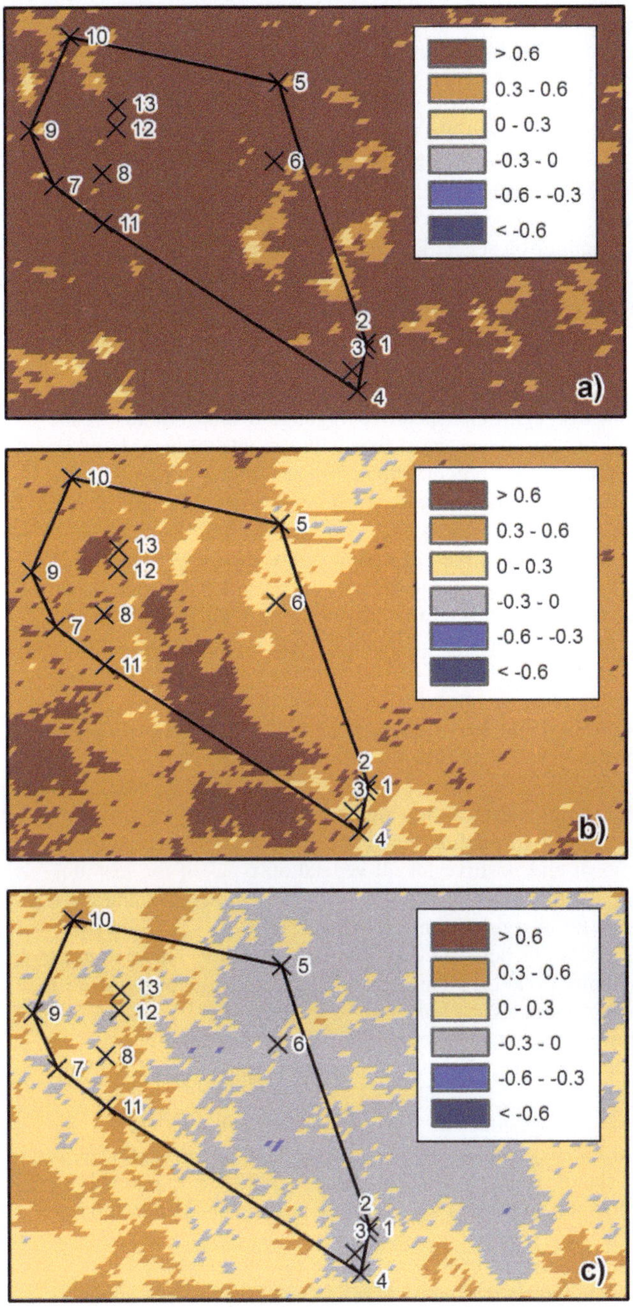


Figure 6. Spatial distribution of the correlation coefficients between NDVI and air temperature in the spring (a), summer (b) and autumn (c) seasons

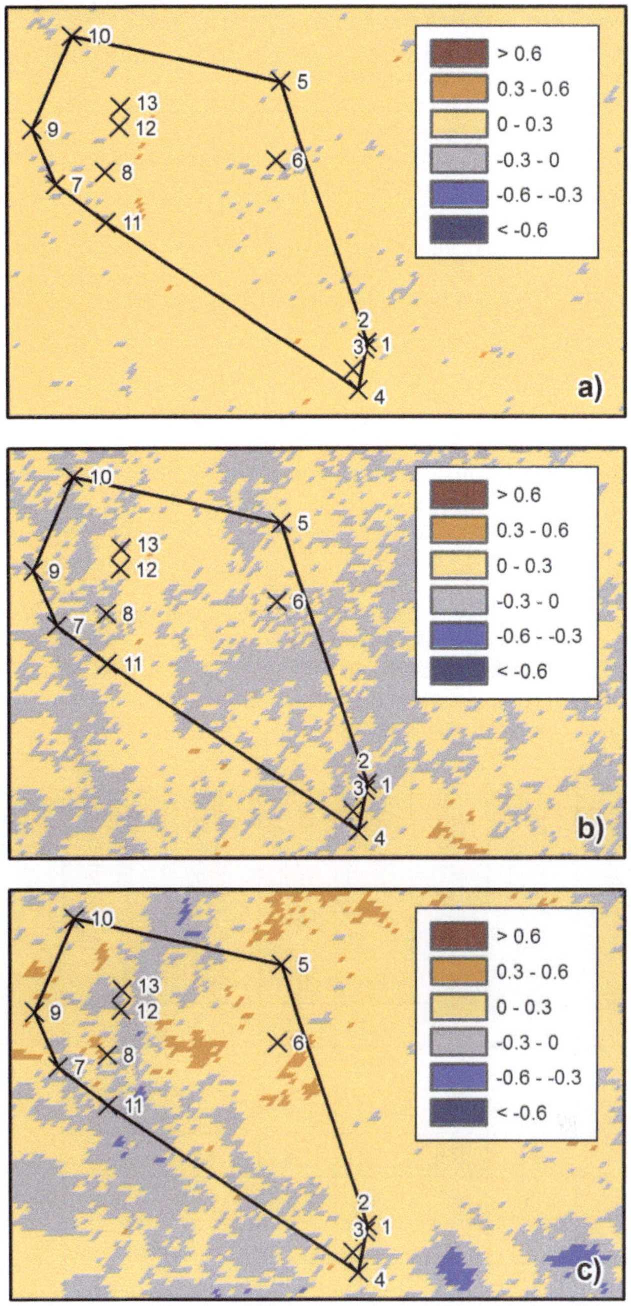


Figure 7. Spatial distribution of the correlation coefficients between NDVI and total precipitation in the spring (a), summer (b) and autumn (c) seasons

In summer season, the degree of correlation between NDVI and air temperature decreases. Maximum significant correlation coefficient values were observed on the territories occupied by non-forest lands. Minimum coefficient values were observed on the evergreen coniferous forests. In autumn, these territories have the negative values of correlation coefficient. The significant positive correlation values in autumn season remained only on floodplains.

The positive correlation coefficient values between NDVI and precipitation sums observed on significant part of the ground test area (Figure 7). In summer season, we can discover the significant positive correlation coefficient values mostly on non-forest lands in floodplains. On the territories of evergreen coniferous forests, we can allocate the significant negative values of correlation coefficient.

Also in autumn season the negative correlation values observed on non-forest lands. The greatest negative values of correlation coefficient are observed on the floodplains. The coefficient values of correlation between the precipitation sums and vegetation index magnitudes are always positive on the territories of evergreen coniferous forests.

### 4.6 Relationships of NDVI and Vegetation Characteristics of the Ground Test Area

There are most typical forest types for the North-West region of Russia grow on the territory of the ground test area, which was chosen for the project aims. The field observations, which are conducted every year at this test area, while the peak of the growing season, on 13 permanent sample plots. These observations provide such vegetation productivity parameters as stands density, the wood stock of living and dead trees and discoloration degree of the tree crowns.

The discoloration coefficients are calculated according to the method that implemented in the ICP-Forests manual (Manual on methods and criteria..., 1994). The discoloration of vegetation clearly reflects the state of tree vegetation. This parameter could be observed using satellite measurements as the value of chlorophyll index (vegetation index). However, the change of vegetation productivity could not be measured directly using remote sensing methods (Cherepanov at al., 2009).

We analyzed the change in NDVI on sample plots (Figure 1) on July 28. This date is the closest to the date of field observations. As we can see, the increase of NDVI values does not necessarily correlate with changes in the wood stock and discoloration degree of coniferous trees (Table 2).

The wood stock increases due to the growth of the tree, and positively correlates with an increase of NDVI. However, yellowing of pine and spruce needles at the six plots, which was observed on ground, is not associated with a decrease of NDVI. This discontinuity may be caused by the difference in size of the spatial resolution of NDVI maps that exceeded the size of the sample plots. The main reason of discontinuity is NDVI averaging over a large area that may lead to blurring of individual vegetation characteristics.

| Sample Plot | Vegetation Type | NDVI Increase | Discoloration Increase | Wood stock Increase | NDVI (28 July) / Wood Stock | NDVI (28 July) / Discoloration |
|---|---|---|---|---|---|---|
| 1 | A | 1,8 | 2,0 | 1,3 | + | + |
| 2 | B | 2,1 | 1,1 | 1,1 | + | + |
| 3 | A | 2,1 | 16,5 | 1,4 | + | + |
| 4 | B | 2,5 | 0,8 | 0,9 | - | - |
| 5 | A | 6,9 | 0,8 | 1,8 | + | - |
| 6 | A | 6,2 | 1,0 | 1,2 | + | - |
| 7 | C | 1,9 | 0,1 | 1,2 | + | - |
| 8 | C | 2,4 | 0,0 | 1,3 | + | - |
| 9 | C | 1,6 | 1,4 | 1,0 | + | + |
| 10 | C | 6,0 | 0,1 | 1,1 | + | - |
| 11 | D | 1,7 | 1,5 | 1,3 | + | + |
| 12 | B | 2,5 | 0,3 | | + | - |
| 13 | E | 3,3 | | 1,1 | + | + |

Table 2. Change of the NDVI, discoloration degree and wood stock. The positive signs mark the unidirectional changes, and the negative signs mark oppositely directed changes. Forest types marked as A – Pine-lichen, B – Pine-green moss, C – Spruce-green moss, D – Grassy pine, E – Pine-sphagnum

## 5. CONCLUSIONS

The study of long-term relationships and dynamics of climate characteristics and vegetation index at the ground test area leads to the following conclusions.

A gradual increase of the air temperature and precipitation was observed during the studied period at the Syktyvkar meteorological station. The most distinct positive and statistically significant trend of precipitation discovered in the autumn season, and the positive and statistically significant trend of air temperature observed in spring season.

The warmest year was in 2010, and the coldest in 2002. In addition, the wettest was in 2012, and the driest was in 2005. The dynamics of vegetation index was depended on significant volatility of the meteorological conditions in these years.

In accordance with the dates of the most cool and dry years, vegetation index reached the lowest values in 2002, 2008 and 2010. The most favorable conditions for plant growth were observed in 2003 and 2006.

Statistically significant positive vegetation index trends at the ground test area were consistent with the positive trends of air temperature and precipitation. These positive trends were preserved throughout three seasons on evergreen coniferous forest territories.

The maximum positive vegetation index trends were observed in summer season at the territories occupied by non-forest lands. In spring and autumn seasons the trend values remains positive only for non-forest lands located at floodplains.

The values of coefficient of the correlation between NDVI and air temperature is statistically significant and positive in the spring season for the whole ground test area. Rising of the air temperature and precipitation sums during the last 14 spring periods caused the corresponding improvement in vegetation conditions for all vegetation types.

The values of correlation between NDVI and air temperature is significant and positive for all vegetation types of the test area in the summer season, but the positive correlation values are minimal for evergreen coniferous forests.

The correlation between precipitation sums and NDVI during the summer period is significant and positive only for non-forest lands (floodplains). For the evergreen coniferous forest territories, the coefficient values of correlation between precipitation sums and NDVI are statistically significant and negative.

For autumn period, the correlation between NDVI and air temperature is statistically significant and positive for non-forest lands only. This dependence could be explained by prolongation of growth season for annual plants of floodplains. At the evergreen coniferous forest sectors, the autumn temperature regime does not affect on the vegetation state.

Negative correlation coefficient values for the precipitation sums and NDVI exist on the non-forest lands in autumn period. At the sectors occupied by evergreen coniferous forests, correlation of precipitation and NDVI is positive. The temperature growth improves vegetation conditions on non-forest lands only while decreasing of the precipitation sums.

The NDVI dynamics at the test area was verified using ground observation data produced on permanent sample plots. NDVI trends distribution at sample plots is mosaic like. This caused by heterogeneity of vegetation cover on the plots. Maximum positive linear trend coefficients were observed on the plots located at non-forest lands.

Negative NDVI values, which were observed at two plots, caused by the sanitary felling and general needles yellowing, which is due to the emissions from pulp and paper mill and forest diseases. Hence, we should consider in investigations the anthropogenic influence, which could distort natural NDVI pattern.

Increasing of NDVI values during the summer season does not always correlate with changes in the wood stock and discoloration degree of the crowns of coniferous trees. This finding requires more detailed investigations.

Basing on the above-mentioned information, we can conclude that the state of non-forest vegetation of floodplains, characterized by NDVI in the spring-summer period, could serve as an indicator of long-term changes in temperature and precipitation conditions in the region. The anthropogenic impact should be taken into account when assessing climate change. This impact can distort the natural pattern of vegetation.

## ACKNOWLEDGEMENTS

The study was partially supported by Russian Foundation for Basic Research (RFBR), research project No. 14-05-00858 a.

## REFERENCES

Cherepanov A., Druzhinina E., 2009. Spectral characteristics of vegetation and vegetation indexes. *Geomatics*, 3(4), pp. 28-32. (Published in Russian)

Forestry organization and development project of Syktyvkar mehleskhoz of the Komi ASSR Ministry of Forestry RSFSR Ministry of Forestry, 1988. Explanatory note, Part 1, Leningrad, pp.10-18. (Published in Russian)

Goetz S.J., Mack M.C., Gurney K.P., Randerson J.T., Houghton R. A., 2007. Ecosystem responses to recent climate change and fire disturbance at northern high latitudes: observations and model results contrasting northern Eurasia and North America. *Environmental Research Letters*, 2, 045031 (9pp). doi:10.1088/1748-9326/2/4/045031

Jia G.J., Epstein H.E., Walker D.A., 2006. Spatial heterogeneity of tundra vegetation response to recent temperature changes. *Global Change Biology*, 12, pp. 42-55.

Jia G.J., Epstein H.E., Walker D.A., 2009. Vegetation greening in the Canadian Arctic related to decadal warming. *Journal of Environmental Monitoring*, 11, pp. 2231-2238

Kattsov V.M., 2004. Arctic climate in the 21st century: Modelling and scenarios. Vulnerability to global environmental change. A summary of climate change impact assessments from the U.S. Country Studies Program. Cambridge University Press, pp.99-150

Krankina O. N., Pflugmacher D., Hayes D. J., McGuire A. D., Hansen M. C., Hдme T., Elsakov V., Nelson P., 2010. Vegetation Cover in the Eurasian Arctic: distribution, monitoring, and role in carbon cycling. Chapter 5. *Eurasian arctic land cover and land use in a changing climate.* Springer, 1st edition, pp. 79-108

Manual on methods and criteria for harmonised sampling, assessment, monitoring and analysis of the effects of air pollution on forests, 1994. Hamburg, 177 p.

Sadokov B. N., Kozeltseva C. F., Kuznetsova N. N., 2012. Definition of spring dates of steady transition of the average daily air temperature through 0, +5 °C, forecast and assessment. *Proceedings of the Hydrometeorological Center of Russia*, Vol. 347, pp. 162-172. (Published in Russian)

Stow D. A., Hopea A., McGuireb D., Verbylac D., Gamond J., Huemmriche F., Houstond S., Racinef C., Sturmg M., Tapeh K., Hinzman L., Yoshikawai K., Tweediej C., Noylek B., Silapaswanl C., Douglasm D., Griffithn B., Jiao G., Epsteino H., Walkerp D., Daeschnera S., Petersena A., Zhouq L., Myneni R., 2004. Remote sensing of vegetation and land-cover change in Arctic Tundra Ecosystems. *Remote Sensing of Environment*, 89, pp. 281-308

Walker M. D., Wahren C. H., Hollister R. D., Henry G. H. R., Ahlquist L. E., Alatalo J. M., Bret-Harte M. S., Calef M. P., Callaghan T. V., Carroll A. B., 2006. Plant community responses to experimental warming across the tundra biome. *Proceedings of the National Academy of Sciences of the United States of America*, 103(5), pp. 1342-1346

Wookey P. A., Aerts R., Bardgett R. D., Baptist F., Brathen K. A., Cornelissen J. H. C., Gough L., Hartley I. P., Hopkins D. W., Lavorel S., Shaver G. R., 2009. Ecosystem feedbacks and cascade processes: understanding their role in the responses of Arctic and alpine ecosystems to environmental change. *Global Change Biology*, Vol. 15, pp. 1153-1172

# RETRIEVAL OF OZONE TOTAL COLUMNS OVER ÉVORA-PORTUGAL USING REMOTE SENSING INSTRUMENTS DURING 2007-2011

A. F. Domingues [a,*], D. Bortoli [a,b], A.M.Silva [a,c], P. Kulkarni [a], R. Mendes [a]

[a] Geophysics Centre of Évora - University of Évora (CGE-UE), Rua Romão Ramalho, 59, 7000 Évora, Portugal – ana.filipa.domingues@gmail.com; pavannpl@yahoo.co.in; rrbmendes@hotmail.com
[b] Institute of Atmospheric Sciences and Climate (ISAC-CNR), Via Gobetti 101, 40129 Bologna Italy – d.bortoli@isac.cnr.it
[c] Department of Physics, University of Évora, Rua Romão Ramalho 59, 7000 Évora, Portugal – asilva@uevora.pt

**KEY WORDS:** Ozone, SPATRAM, Remote Sensing, spectromer, DOAS, TOC

**ABSTRACT:**

The present study deals with the retrieval of Ozone ($O_3$) Total Column (TOC) and analysis of its variability over the Observatory of the Geophysics Centre of Évora (CGE-UE) –Portugal (38.5°N; 7.9 °W, 300 m a.s.l.) for the period comprised 2007-2011. The data presented in this study are obtained applying the Differential Optical Absorption Spectroscopy (DOAS) methodology to the measurements of diffused spectral sky radiation carried out along the zenith direction performed with the multipurpose UV-Vis. Spectrometer for Atmospheric Tracers Monitoring (SPATRAM) in the 250-900 nm spectral range. This ground-based spectrometer is installed at CGE-UE performing daily and automatic measurements since 2004. Other products of this scanning spectrometer are the vertical profiles of some trace gases (e.g. $NO_2$) and monitoring of air quality. The $O_3$ retrieved with the SPATRAM instrument confirm the typical seasonal cycle for middle latitudes reaching the maximum during the spring and the minimum during the autumn. The ground-based results obtained for $O_3$ column are also compared with data from SCanning Imaging Absorption spectroMeter for Atmospheric CHartographY (SCIAMACHY) and Global Ozone Monitoring Experiment (GOME) instruments aboard ENVISAT and ERS-2 satellites, respectively. The results show a good agreement between the datasets. The main reasons for the observed differences are discussed.

## 1. INTRODUCTION

The rapid acceleration of the concern about a new range of global environmental problems led to real advances in related technologies for the acquisition and processing of environmental data in "real or near real-time", even from remote locations. This progress provides the assessment of data related to the atmosphere and climate, like trace gases and aerosol concentration and clouds, besides the dynamic and meteorological information of the atmosphere (Laj et al. 2009). Nowadays the monitoring of the atmosphere is carried on by means of 'in-situ' and Remote Sensing (RS) measurements obtained from ground-based instruments, balloons, airborne or satellite based equipments. The data-sets acquired with all these instruments and sensors enable the monitoring of the atmosphere and can be kept for present and future prospect studies. In a so complex system as the Earth's Atmosphere compared by Seinfeild et Pandis, 2006 "to an enormous chemical reactor in which a myriad of species are continually being introduced and removed over a vast array of spatial and temporal scales" the need of short and long term observations are evident since atmospheric processes are still far from being entirely understood.

The main motivation for accomplishing this study is the importance of monitoring the trace compounds, namely $O_3$, in the atmosphere due to their major role in the atmospheric processes despite their low concentrations. This work is focused in the retrieval of Total Ozone Columns (TOC) with remote sensing instruments (ground-based and satellite equipments) for the period of 2007-2011 in Évora-Portugal

## 2. MEASUREMENT SITE AND REMOTE SENSING INSTRUMENTS

Figure 1 a) shows the location of the Observatory of Geophysics Centre of Évora situated in Évora. Located in the interior of Portugal, this city can be considered as a continental rural site. Évora is placed at about 130 km eastward from the Lisbon area and about 100 km westward from Badajoz (Spain). Figure 1 b) c) illustrates the ground based instrument system used to perform the spectral measurements constituted by the Spectrometer for Atmospheric Tracers Monitoring (SPATRAM) instrument (Figure 1 b)), and the Vertical Looking Device (VELOD) ( Figure 1c)). This latter is a small telescope coupled to the spectrometer to collect solar radiation in the zenith direction. The SPATRAM system measures the zenith scattered radiation in the 200-900 nm spectral range. The main products are the total column and the vertical profiles of $NO_2$ and $O_3$. The core of SPATRAM is the monochromator which allows for the decomposition of light in several wavelengths using a grating by Jobin-Yvon of 1200 grooves/mm with a typical dispersion of 2.4 nm/mm at 300 nm ( Domingues, 2011). The complete description of the instrument and working method is presented in the works of Bortoli, 2005; Bortoli et al., 2009; Bortoli et al., 2009.

The ground-based results obtained were compared with data from SCanning Imaging Absorption spectroMeter for Atmospheric CHartographY (SCIAMACHY) and Global Ozone Monitoring Experiment (GOME) instruments aboard ENVISAT and ERS-2 satellites, respectively.

* Corresponding author.

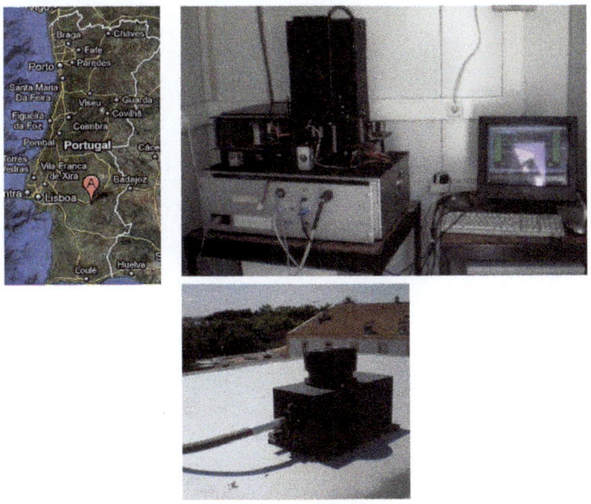


Figure 1. a) The location of Évora city in mainland Portugal. b) The SPATRAM instrument installed in the Observatory of the Geophysics Centre of Évora (38.5°N;7.9°W, 300 m a.s.l.). c) The VELOD telescope (source: Domingues, 2011).

## 3. METHODOLOGY

The Differential Optical Absorption Spectroscopy (DOAS) algorithms used for the retrieval of the mean trace gas concentrations along the optical path of measure (the Slant Column Density – SCD) are presented as follows.

First of all the SPATRAM high resolution spectral data suffers a pre-processing procedure. After that it is possible to apply the DOAS method to the SPATRAM datasets using the DOAS Master Equation (Equation 1), based on the Beer-Lambert Law.

$$\ln\left(\frac{I_0(\lambda,\theta_{min})}{I_s(\lambda,\theta)}\right) - \overline{\ln\left(\frac{I_0(\lambda,\theta_{min})}{I_s(\lambda,\theta)}\right)} = \Sigma_g\left(\sigma_g(\lambda) - \overline{\sigma_g(\lambda)}\right)SCD_g(\theta) \quad (1)$$

where

$I_0(\lambda,\theta_{min})$ = reference spectrum obtained at local noon

$I_s(\lambda,\theta)$ = spectrum measured for different condition of SZA

$\sigma_g(\lambda)$ = the absorber's cross sections

$\theta$ = solar zenith angle

$\theta_{min}$ = solar zenith angle at the local noon

$SCD_g(\theta)$ = slant column densities of the $g^{th}$ absorber

The superscript represents the high pass filter operator removing the low frequency features of the spectral series. Equation 1 is usually solved in a suitable wavelength interval that includes strong differential absorption structures of the trace gases of interest. To retrieve the $O_3$ SCD, the DOAS algorithms were applied in the 320-340 nm spectral range. And the data obtained with the SPATRAM instrument comprises all the SZA between the highest daily solar elevations (which corresponds to the minimum SZA) down to approximately 93° of SZA.

The final product resulting from the application of DOAS algorithms to SPATRAM data are the O3 SCD. Actually, from the definition of the DOAS Master Equation, the SCD are differences between the content of the absorber in the analysed spectrum and the amount of the considered compound in the reference spectrum. Consequently, the SCD are effectively the DSCD (Differential Slant Column Density), therefore the choice of the reference spectrum and the quantification of absorber's content in it, are crucial tasks. The reference spectrum - $I_0(\lambda,\theta_{min})$ - used for the DOAS analysis has to be measured in optimal atmospheric conditions (sunny day without clouds) gathered at local noon. At mid-latitudes, to estimate the content of the analyzed compound in the reference spectrum ($SCD_{I_0}(\lambda,\theta_{min})$) the Langley plots of the retrieved values during one day, before the local noon (AM) and after the local noon (PM) versus the series of the calculated AMF values are usually used. The $SCD_g(\lambda,\theta)$ can finally be calculated as follows in Equation 2.

$$SCD_g(\lambda,\theta) = DSCD_g(\lambda,\theta) + SCD_{I_0}(\lambda,\theta_{min}) \quad (2)$$

Since the SCD is dependent of the light path it is necssary to convert this quantity into a Vertical Column Density ( VCD) in order to remove the light path dependence. This procedure is important because allows for the comparison of datasets from different instruments. The SCD to VCD conversion is made by means of Equation 3, where is introduced the concept of Air Mass Factor ( AMF) . The AMF can be seen as a geometrical factor, function essentially of the Sun's elevation and of the wavelength, and it is defined as follows:

$$AMF_g(\theta,\lambda) = SCD_g(\theta,\lambda)/VCD_g(\theta,\lambda) \quad (3)$$

The AMF is computed by means of a Radiative Transfer Model (RTM), in this case the Atmospheric Model for Enhancement Factor Computation (AMEFCO). AMEFCO model was developed at ISAC – CNR and it is based on the Intensity Weighted Optical Path (IWOP) approach (Slusser et al.,1996). The AMEFCO is a single-scattering model using ray tracing in a spherical 2-D atmosphere (profile variation with SZA) with optical paths integrated over individual shells. Light refraction is not included. (Petrioli et al., 2002) .

## 4. RESULTS

### 4.1 Stratospheric $O_3$ seasonal variation

The typical distribution of TOC retrieved at Évora is in accordance with the literature for the mid-latitudes. From Figure 2 it is possible to point out a clear seasonal trend of the plotted values, following a sinusoidal pattern due to relation between solar activitiy and $O_3$ photochemical behaviour. The total ozone maximums are observed during Spring and the minimums in the Autumn due mainly to dynamic factors such as the Brewer Dobson circulation from tropical regions to the pole. This circulation is maximum during the Spring season, which correspond also to the higher $O_3$ formation in tropical regions leading to the enhancement of the $O_3$ concentration during that season. It is also visible the strong seasonal variability of the diurnal $O_3$ total columns. In the years 2008 and 2010 the Autumn and Spring TOC values are lower than the ones registered in the same season of the other analyzed period. This variability can be caused by dynamical and chemical factors such as the variability in the Brewer Dobson circulation, the possible intrusion of aerosols (e.g. dust from Saharan desert, or forest fire type, eruptions), the variation in solar cycle.

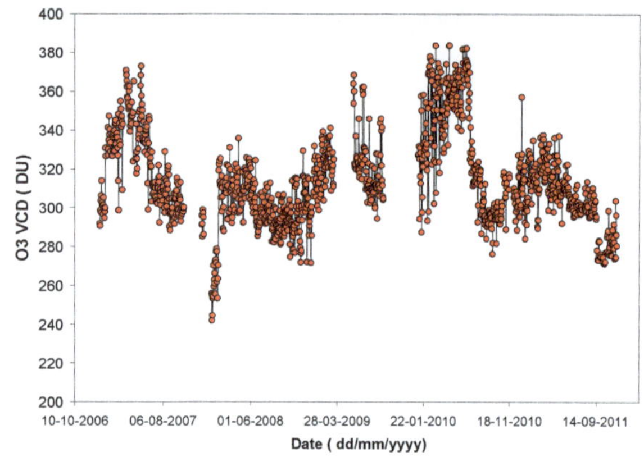


Figure 2. Time series of the $O_3$ VCD obtained with the SPATRAM equipment installed at Évora Observatory during 2007 -2011 period.

The time series obtained with SPATRAM and presented in this study are incomplete since the instrument participated in field campaigns outside Évora city .

### 4.2 SPATRAM *vs* Satellite data

The satellite data used for comparison with SPATRAM TOC are from SCIAMACHY and GOME. The description of the products used is presented as following:

1. The ESA usable algorithm for SCIAMACHY is called "Total Ozone retrieval scheme for SCIAMACHY based on the OMI DOAS algorithm" or "TOSOMI" (Eskes at al., 2006) and subsequently in GOME algorithm "TOGOMI" (Valks et al., 2004). This algorithm allows for the retrieval of Ozone columns from SCIAMACHY and was developed by the Royal Netherlands Meteorological Institute (KNMI). A complete description of the algorithm can be found at Eskes et al., 2006. TOSOMI product is freely available over several locations including Évora Observatory. Ozone data can be downloaded at: *http://www.temis.nl/protocols/o3col/data/scia/overpass_v2/Evora_Observatory_tosomi.dat.*
2. The GOME dataset is available from 1995 to 2011. For the Évora Observatory the data is freely available at *http://www.temis.nl/protocols/o3col/data/gome/overpass/Evora_Observatory_togomi.dat* for the considered period. The algorithm used for measuring trace gases that is current in operation is the GOME total ozone algorithm TOGOMI and was developed under ESA supervision by the German Aerospace Centre (DLR). More information about the algorithm can be found at literature (Valks et al., 2004).

In Figure 3 the two datasets (SPATRAM and SCIAMCAHY) are plotted versus time for the period of study.

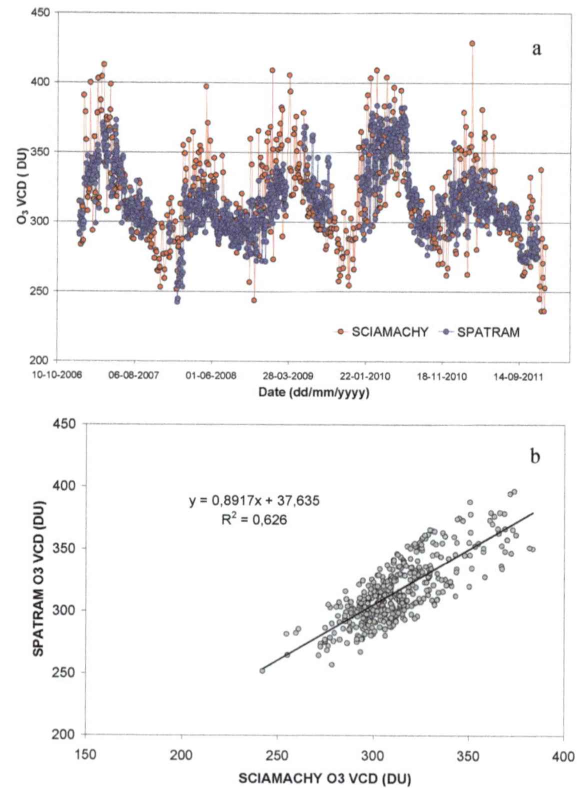


Figure 3. a) Time series of the $O_3$ VCD obtained with the SPATRAM equipment installed at Évora Observatory during 2007-2011, and the $O_3$ data from the SCIAMACHY instrument aboard the ENVISAT satellite. b) Scatter plot of the $O_3$ data from the SCIAMACHY instrument (TOSOMI product) *versus* SPATRAM data retrieved at Évora Observatory for the same period.

Figure 4 shows the comparison of SPATRAM and GOME datasets.

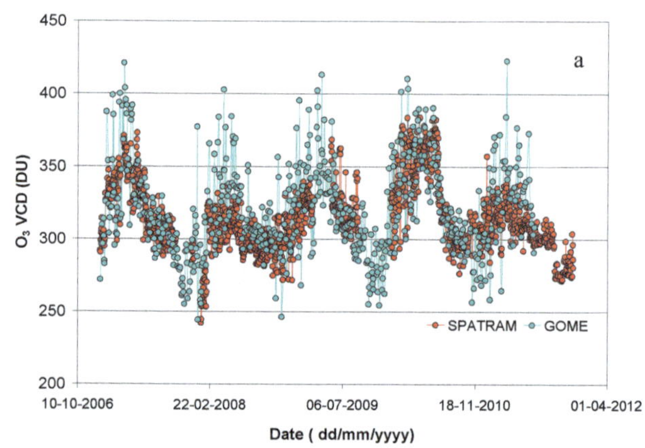

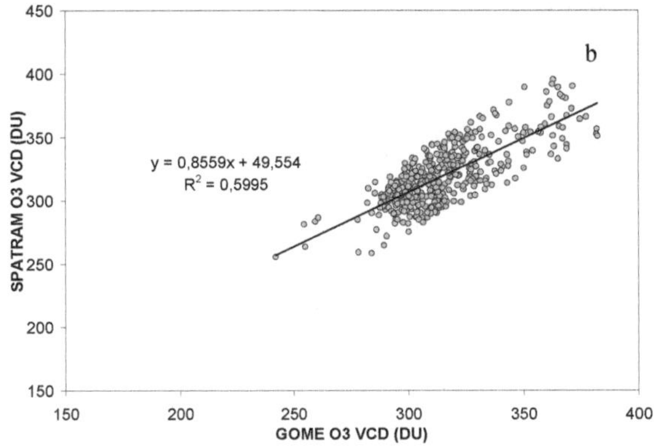


Figure 4. a) Time series of the $O_3$ VCD obtained with the SPATRAM equipment installed at Évora Observatory during 2007-2011, and the $O_3$ data from the GOME instrument aboard the ERS-2 satellite. b) The correspondent scatter plot of the $O_3$ data from the GOME instrument *versus* $O_3$ SPATRAM data retrieved at the same period including the regression line (black line).

From the analysis of Figures 3 and 4 and it is possible to infer that SPATRAM data-set is consistent with the SCIAMACHY and GOME data presenting the same seasonal pattern. By the analysis of the scatter plot (Figures 3b) and 4b)) one can see that the correlation between the two dataset is in fairly agreement with a correlation of $R^2$=0.63 and $R^2$=0.59, respectively. The deviation from the first satellite sensor associated is about 4,15% and for the second is 4.23% calculated using the relative error method.

### 4.3. Statistical analysis

The evaluation of some statistical parameters is made. The presented parameters are related to the comparison of SPATRAM data with data from each one of the satellite instruments. The values are presented in Table 1 and concern to the number of data (N), the slope and the correlation coefficient ($R^2$) of the scatter plots and the Root Mean Square Error (RMSE). This latter parameter is frequently used to represent the sample standard deviation of the differences between predicted values (in a model) and observed values (in the real atmosphere). The RMSE is a measure of the spread of data around the regression line and is calculated using Equation 4:

$$RMSE(\%) = \sqrt{\frac{100}{N}\frac{\sum_{j=1}^{N}(SPATRAM_{model,j} - SPATRAM_{obs,j})^2}{SPATRAM_{obs,j}}} \tag{4}$$

where
$SPATRAM_{obs}$ =observed values of TOC
$SPATRAM_{model}$ = modelled values calculated using the equation of regression line of the scatter plots (Equation 5)

$$SPATRAM_{model,j} = a + b \cdot Sat.Sensor_j \tag{5}$$

where $a$ = slope of scatter plot,
$b$ = intercept of scatter plot
$Sat.Sensor$ =values available from the sensors (SCIAMACHY and GOME) aboard the satellites.

In Table 1 the Mean Bias Error (MBE) and the Mean Absolute Bias Error (MABE) are also presented which are two statistical parameters computed using Equations 6 and 7.

$$MBE(\%) = \frac{100}{N}\sum_{j=1}^{N}\frac{SPATRAM_j - Sat.Sensor_j}{Sat.Sensor_j} \tag{6}$$

$$MABE(\%) = \frac{100}{N}\sum_{j=1}^{N}\frac{|SPATRAM_j - Sat.Sensor_j|}{Sat.Sensor_j} \tag{7}$$

| | N | Slope | $R^2$ | RMSE (%) | MBE(%) | MABE(%) |
|---|---|---|---|---|---|---|
| SCIAMACHY | 405 | 0.89 | 0.63 | 5.25 | -1.03 ± 0.24 | 4.15 ± 0.14 |
| GOME | 409 | 0.86 | 0.60 | 5.49 | -1.15 ± 0.24 | 4.24 ± 0.13 |

Table 1 . Correlation analysis of SPATRAM, GOME and SCIAMACHY total ozone data for the period of 2007-2011.

From the analysis of Table 1 it can be inferred that in most days the SPATRAM values are lower than the ones obtained with other sensors. Therefore there is an underestimation of the SPATRAM instrument that can be seen by the negative sign in MBE values.

The differences found between the results for TOC obtained with SPATRAM , SCIAMACHY and GOME instruments could be due to several facts such as:

- the satellite data presented above refers to the $O_3$ total column values corresponding to the daily overpass of the ENVISAT and ERS-2 satellites over a pixel of about 1800 $km^2$ (30x60 km) and 1600 $km^2$ (40x40 km) containing the Évora station while the SPATRAM Field Of View (FOV) covers a minor area;
- the differences of methodologies used for the retrieval of $O_3$ columns,
- the usage of different types of instruments i.e., the comparison between data from ground based spectrometer with instruments aboard satellites and the fact of SPATRAM "looks up" while the other instruments " look down" or, in other words, the SPATRAM measures the zenith sky scattered radiation and the satellite instruments carries out measurements of the radiation back-scattered by the Earth-atmosphere system,
- the employment of different AMF correction factors.

## 5. DISCUSSION

The analysis of the $O_3$ Total Columns retrieved with SPATRAM allowed for the study of the seasonal variation of that gas over Évora for the period 2007-2011.
The obtained TOCs were compared with the satellite data from SCIAMACHY and GOME instruments. The comparison revealed a fair agreement between the datasets proving that SPATRAM instrument (in zenith sky configuration) and the employed DOAS algorithms are a good tools for the retrieval of the concentration of these compounds.
With respect to the seasonal behavior of $O_3$ it is possible to point out that:
a) TOC shows a sinusoidal variation through the years;

b) the $O_3$ seasonal trend shows a maximum in Spring months and a minimum in Autumn as expected for the mid-latitudes;

c) it is possible to point out the inter-annual variability in $O_3$ values, in the years 2008 and 2010 the autumn and spring values are lower than the ones in the same season of the other analyzed period.

## ACKNOWLEDGEMENTS

The first author acknowledges a) the Fundação para a Ciência e a Tecnologia (FCT) which funded this work through the grant FCT- SFRH/BD/44920/2008 and project PTDC/AAC-CLI/114031/2009., b) To ISAC/CNR for the AMEFCO model c) the SCIAMACHY and GOME teams for the $O_3$ total columns data .

## REFERENCES

Bortoli, D., 2005. SPATRAM- Spectrometer for Atmospheric Tracers Measurements- A prototype Equipment for the monitoring of minor compounds of the atmosphere, PhD thesis, University of Évora.

Bortoli, D., Silva, A. M., Costa, M. J., Domingues, A. F., & G. Giovanelli, 2009a. Measurements of stratospheric ozone and nitrogen dioxide at Évora, Portugal, *Int. J. Remote Sens.*, 30 (15–16), pp.4209–4226.

Bortoli, D., Silva, A. M., Costa, M. J., Domingues, A. F., & G. Giovanelli, 2009b. Monitoring of atmospheric ozone and nitrogen dioxide over the south of Portugal by ground based and satellite observations. *Opt. Express*,17 (15),pp.12944-12959.

Bortoli, D., Silva, A. M., & Giovanelli, G., 2010. A new multipurpose UV-Vis spectrometer for air quality monitoring and climatic studies, *Int. J. Remote Sens.*, 31(3), pp.-725.

Domingues, A.F., Bortoli, D., Antón, M., & Silva, A.M., 2011. Ozone Seasonal Variation with Ground-Based and Satellite Equipments at Évora Observatory: Portugal During 2007-2010, In *Lecture Notes in Geoinformation and Cartography*, edited by Jukka M. Krisp, Liqiu Meng, Roland Pail, Uwe Stilla, Proceedings of Earth Observation of Global Changes (EOGC)(Springer-Verlag Berlin Heidelberg 2013), pp.137-146.

Eskes H. J., Van der A, R. J., & Brinksma, E., 2006. Algorithm Document TOSOMI . http://www.temis.nl/docs/AD_TOSOMI.pdf (20 august 2014).

Laj, P., et al., 2009. Measuring atmospheric composition change. *Atmos. Environ.* , 43(33), pp. 5351-5414.

Petritoli, A., Ravegnani, F., Giovanelli, G., Bortoli, D., Bonafe, U., Kostadinov, I., & Oulanovsky, A., 2002. Off-axis measurements of atmospheric trace gases by use of an airborne ultraviolet-visible spectrometer. *Appl. Opt.*, 41 (27), pp.5593–5599.

Seinfeld, J. H., Pandis, S. N., 2006. *Atmospheric Chemistry and Physics: From air pollution to climate change*. Wiley, New Jersey: 2$^{nd}$ edition, pp.33-38.

Slusser, J., et al., 1996. Comparison of air mass computations. *J. Geophys. Res.*,.101, pp.9315–9321.

Valks, P.J.M., de Haan, J.F., Veefkind, J.P., van Oss, R.F., & D.S. Balis, 2004. TOGOMI: An improved total ozone retrieval algorithm for GOME, XX Quadrennial Ozone Symposium, 1/6/2004-8/6/2004, C.S. Zerefos (Ed), Athens, University of Athens, pp. 129-130.

# RECOGNITION OF HUMAN POSE FROM IMAGES BASED ON GRAPH SPECTRA

A. A. Zakharov [a]*, A. E. Barinov [a], A. L. Zhiznyakov [a]

[a] Murom Institut Vladimir State University, CAD Department, , 602264, Orlovskaya 23, Murom, Russian Federation, aa-zaharov@ya.ru

**Commission VI, WG VI/4**

**KEY WORDS:** Image Recognition, Human Pose, Spectral Graph Matching

**ABSTRACT:**

Recognition of human pose is an actual problem in computer vision. To increase the reliability of the recognition it is proposed to use structured information in the form of graphs. The spectrum of graphs is applied for the comparison of the structures. Image skeletonization is used to construct graphs. Line segments are the nodes of the graph. The end point of line segments are the edges of the graph. The angles between adjacent segments are used to set the weights of the adjacency matrix. The Laplacian matrix is used to generate the spectrum graph. The algorithm consists of the following steps. The graph on the basis of the vectorized image is constructed. The angles between the adjacent segments are calculated. The Laplacian matrix on the basis of the linear graph is calculated. The eigenvalues and eigenvectors of the Laplacian matrix are calculated. The spectral matrix is calculated using its eigenvalues and eigenvectors of the Laplacian matrix. The principal component method is used for the data representation in the space of smaller dimensions. The results of the algorithm are given.

## 1. INTRODUCTION

Recognition of human pose is an actual problem in many areas: security, human-machine interface, medicine, transport, simulators. Marker and markerless approaches used for motion capture (Field, M., Stirling, D. A., Naghdy, F., Pan, Z., 2009). Marker motion capture system are optical passive, optical active, magnetic, mechanical, inertial, acoustic. Common disadvantage of marker motion capture systems is the impossible to use the sensors when viewed real scenes. This feature limits the use of marker techniques. Thus, the development of methods and software tools to capture and recognition of human pose using markerless methods is an important task. Questions of pose recognition based on optical methods insufficiently studied. Using computer vision is limited by the following factors: the small size of the scenes, the presence of noise and optical effects, varying textured background, overlapping objects.

To increase the reliability of the recognition it is proposed to use structured information in the form of graphs. The spectrum of graphs is applied for the comparison of the structures. The advantage of structural methods is that they give an opportunity to analyze a big set of elements on the basis of a small quantity of simple components and rules of forming the graphic model. Also structural methods allow to describe the characteristics of the object excluding its reference to another class that increases the reliability of the recognition.

## 2. SURVEY OF ALGORITHMS OF HUMAN POSE RECOVERY

Taxonomy of human pose recovery consists of the following: appearance, viewpoint, spatial relations, temporal relations and behavior (Perez-Sala, X., Escalera, S., Angulo C., Gonzalez, J, 2014). Appearance can be defined as image evidences related to human body and its possible poses. The following features are used to describe models: silhouettes and contours, intensity, color, texture, depth, optical flow, logical relations. The following classifiers are used to detect human pose: discriminative classifiers, generative classifiers, templates, interest points.

Viewpoint determines the number of images to pose recovery. Viewpoint approach is discrete and continuous. Discrete approach uses single image. Continuous approach uses multiple images.

Spatial models encode the configuration of the human body (Johnson, S., Everingham, M., 2011). Spatial models are of two types: hard (skeleton, bone lengths), a soft way (pictorial structures, grammars).

Temporal models use a sequence of images to reduce the the search space. Tracking is applied to ensure the coherence among poses over the time. Motion model is used to describe a variety of human actions (walking, running).

The term behavior includes actions and gestures. Recognition of actions is based on the images.

We use single image to recognize human pose. The spatial model is described as a skeleton. We use region-based approach to extract the skeleton of the image. We segment the image for the shock graph and shape matching (Thome, N., Merad, D., Miguet, S., 2008). The comparison of the graphs consists in searching the compliances between the structures of the graphs on the basis of more or less strict restrictions.

We propose to use the comparison methods on the basis of the spectral graph theory. The spectrum of the graph represents the vector of the eigenvalues of the connectivity matrix of the graph arranged in the decreasing or increasing order. The spectral methods are based on the following property: the

* Corresponding author

eigenvalues and the eigenvectors of the connectivity matrix of the graph are invariant concerning the permutation of the vertexes in a matrix (Chung, F. R. K., 1997). Therefore, if the two graphs are isomorphic, their connectivity matrices will have identical eigenvalues and eigenvectors. However, the opposite statement is not always right: there are graphs with different structures but with the identical spectrum (cospectral graphs).

The spectral representation of the graphs has the following advantages: it is invariant with respect to the order of the vertexes; it contains the information on the structure of the graph; it describes the structure of the graph in the form of numerical values, which is convenient for calculations; the calculation problem of the eigenvalues is well studied and is solved for the polynomial time (Zvetkovic, D., 2012).

Index based on spectrum graphs used to recognize human pose (Thome, N., Merad, D., Miguet, S., 2008). Topological signature vector is calculated as the sum of eigenvalues of the matrix (Shokoufandeh, A., Dickinson, S.J., Siddiqi, K., Zucker, S.W., 1999). Hermitian matrix is used to represent the human pose in the article (Haseeb, M., Hancock, E. R., 2012.).The angles between the joints are used to obtain the spectrum graphs.

We use the Laplacian matrix to obtain a spectrum graph. The principal component method is used for the data representation in the space of smaller dimensions.

## 3. RECOGNITION OF HUMAN POSE BASED ON GRAPH SPECTRA

### 3.1 A human pose model based on the graphs

The set of the segments located on the plane is obtained as a result of the image skeletonization. A segment is characterized by the coordinates of the end points. In the resulting graph the edges are represented by the segments and the vertices characterize the value of the angle between the edges. But it is difficult to represent the similar graph in the matrix form as it is impossible to reflect the weight of the vertices. Therefore, for the purpose of convenience it is proposed to transform the boundary representation into a linear graph, in which the vertices are represented by linear segments and edges connect adjacent segments of the graph and characterize the angle between them (Fig. 1). A Laplacian matrix is constructed using the values of the angles between adjacent segments.

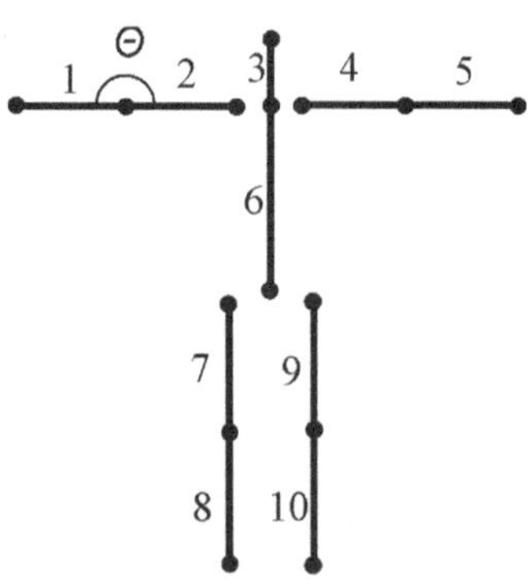


Figure 1. A Human Pose model based on the graphs

The adjacency matrix $A$ of a weighted graph $G=(V,E)$ is a $|V|\times|V|$ matrix whose element with row index $i$ and column index $j$ is

$$A(i,j)=\begin{cases} w_{i,j}, & \text{if } (i,j)\in E, \\ 0, & \text{otherwise.} \end{cases}$$

The degree matrix of a weighted graph $G$ will be denoted, and is the diagonal matrix such that

$$D(i,i)=\sum_j A(i,j).$$

The matrix $L=D\text{-}A$ is called the Laplacian matrix of graph $G$.

The set of the segments located on the plane is obtained as a result of the image skeletonization. A segment is characterized by the coordinates of the end points $v_i=(x_i,y_i)$.

Let us assume that there are two adjacent segments $e_i$ and $e_j$ with the nodes $v_{i1}$, $v_{i2}$ and $v_{j1}$, $v_{j2}$ respectively. The nearby nodes of segments in this case are $v_{i2}$ and $v_{j1}$. The angle between the vectors, on which the segments $e_i$ and $e_j$ are situated, is calculated using the following formula:

$$\theta_{ij}=arccos\left(\frac{\overline{v_{i2}v_{i1}}\cdot\overline{v_{j1}v_{j2}}}{|\overline{v_{i2}v_{i1}}|\cdot|\overline{v_{j1}v_{j2}}|}\right).$$

The adjacency matrix with the index of lines $i$ and the index of columns $j$ has the following form:

$$A_{i,j}=e^{\theta_{ij}},$$

where $\theta_{ij}$ – the angle between the vectors $\overline{v_{i2}v_{i1}}$, $\overline{v_{j1}v_{j2}}$.

### 3.2 Vector space embeddings of graphs via graph matching

Graphs from a some domain are mapped to vector spaces of predefined prototype graphs (Riesen, K., Jiang, X., Bunke., H., 2010).

**Definition (Graph Embedding).** Let us assume a graph domain $H$ is given. If $T=\{g_1,\ldots,g_N\}\subseteq H$ is a training set with $N$ graphs and $P=\{p_1,\ldots,p_n\}\subseteq T$ is a prototype set with $n$ graphs, the mapping $\varphi_n^P: H\rightarrow R^n$ is defined as the function $\varphi_n^P(g)=(d(g,p_1),\ldots,d(g,p_n))$ where $d(g,p_i)$ is any graph dissimilarity measure between graph $g$ and the $i$-th prototype graph.

We use for graph embedding a symmetric polynomial. A symmetric polynomial is a polynomial $S(x_1,x_2,\ldots,x_n)$ in $n$ variables, such that if any of the variables are interchanged, the same polynomial is obtained. For a set of variables $x_1,x_2,\ldots,x_n$ the elementary symmetric polynomials can be defined as:

$$S_n(x_1,x_2,\ldots,x_n)=\prod_{i=1}^{n}x_i.$$

A symmetric polynomial is invariant under permutation of the variable indices (Wilson, R. C., Hancock, E. R., & Luo, B., 2005).

The elementary symmetric polynomials can be computed from using the Newton-Girard formula

$$S_r = \frac{(-1)^{r+1}}{r}\sum_{k=1}^{r}(-1)^{k+r}P_r S_{r-k},$$

where $P_n(x_1, x_2, \ldots, x_n) = \sum_{i=1}^{n} x_i^n$ – the power symmetric polynomial functions.

The eigendecomposition of a Laplacian matrix is $L = U\Lambda U^T$, where $\Lambda = diag(\lambda_1, \lambda_1, \ldots, \lambda_{|V|})$ is the diagonal matrix with the eigenvalues in increasing, $U = (u_1, u_2, \ldots, u_{|V|})$ is the matrix with the ordered eigenvectors as columns. The Laplacian spectrum is the set of eigenvalues $(\lambda_1, \lambda_1, \ldots, \lambda_{|V|})$.

The spectra and eigenvectors are used to compare graphs.
We construct a complex spectral matrix

$$\Psi = (\sqrt{\lambda_1}U_1 \mid \sqrt{\lambda_2}U_2 \mid \ldots \mid \sqrt{\lambda_n}U_n).$$

Polynomial coefficients are calculated based on the Newton-Girard and variable of matrix $\Psi$ : $S_1(\psi_{1,i}, \ldots, \psi_{n,i})$, ..., $S_n(\psi_{1,i}, \ldots, \psi_{n,i})$.

Thus, the structures of each pose defined by the vector $S_n$. Space must be reduced to compare vectors. The principal component is used for this.

**3.3 An algorithm of human pose recognition based on graph spectra**

Step 1. The graph on the basis of the vectorized image is constructed. The angles between the adjacent segments are calculated.

Step 2. The Laplacian matrix on the basis of the linear graph is calculated:

$$L_{i,j} = e^{\theta_{ij}}.$$

Step 3: The eigenvalues and eigenvectors of the Laplacian matrix are calculated.

$$L = U\Lambda U^T.$$

Step 4. The spectral matrix is calculated using its eigenvalues and eigenvectors of the Laplacian matrix

$$\Psi = (\sqrt{\lambda_1}U_1 \mid \sqrt{\lambda_2}U_2 \mid \ldots \mid \sqrt{\lambda_n}U_n).$$

Step 5. Elementary symmetric polynoms are calculated on the basis of the $\Psi$ matrix. The columns form a vector which characterizes the structure of the model.

Step 6. The principal component method is used for the data representation in the space of smaller dimensions.

Step 7. The method of K-means is applied to clustering the points.

**3.4 The results of the algorithm operation**

Studies of the algorithm were performed using 30 images of human poses. As test example, the image in Figure 2 was used.

a) b) c)

Figure 2. Example of human poses: a)pose 1, b)pose 2, c)pose 3

The feature sets are used for the data representation in the 2D-space by means of the principal component analysis. The method of K-means is applied to clustering the points in the 2D-space. The result of recognition of views is shown in Figure 3.

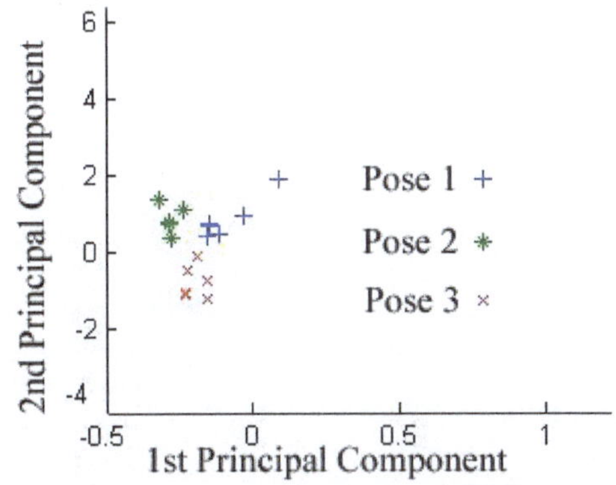


Figure 3. The data representation in 2D-dimensional space using the method of principal components

## 4. CONCLUSIONS

In this paper we proposed an approach to the recognition human pose using the spectral graph theory. The model of the image description is offered using the segments integrated in the graph. The information on the angle between them is used to create the graph. For obtaining the spectrum of the graph the Laplacian matrix is used. The results of the research showed that the proposed method will recognize the main types of the human poses. However, examples of simple poses are presented. Overlapping objects in the scene are often. Such human poses is difficult to recognize on the basis of single image. Further research will focus on the recognition of human pose by using multiple cameras on the basis of the presented approach.

## ACKNOWLEDGEMENTS

This work was supported by the RFBR grant 15-07-01612, project number 2918 in the framework of the basic tasks of the state of the Russian Ministry of Education.

## REFERENCES

Chung, F. R. K., 1997. Spectral graph theory, AMS, p. 207.

Field, M., Stirling, D. A., Naghdy, F., Pan, Z., 2009, Motion capture in robotics review, *IEEE International Conference on Control and Automation*, IEEE, Christchurch, New Zealand, pp. 1697-1702.

Haseeb, M., Hancock, E. R., 2012. Unsupervised Clustering of Human Pose using Spectral Embedding. In *Proceedings of Structural, Syntactic, and Statistical Pattern Recognition – Joint IAPR InternationalWorkshop*, SSPR/SPR, Berlin, Heidelberg. Springer-Verlag, pp. 467–473.

Johnson, S., Everingham, M., 2011. Clustered Pose and Nonlinear Appearance Models for Human Pose Estimation. In *Proceedings of the 21st British Machine Vision Conference (BMVC2010)*, pp. 1-11.

Perez-Sala, X., Escalera, S., Angulo C., Gonzalez, J, 2014. A Survey on Model Based Approaches for 2D and 3D Visual Human Pose Recovery. *Sensors*, 14(3), pp. 4189-4210.

Riesen, K., Jiang, X., Bunke., H., 2010. Exact and Inexact Graph Matching: Methodology and Applications. *Managing and Mining Graph Data Advances in Database Systems* Vol. 40, pp. 217-247.

Shokoufandeh, A., Dickinson, S.J., Siddiqi, K., Zucker, S.W., 1999. Indexing using a spectral encoding of topological structure. *Proc. Int'l Conf. Computer Vision and Pattern Recognition*, Vol. 2, pp. 491-497.

Thome, N., Merad, D., Miguet, S., 2008. Learning articulated appearance models for tracking humans: a spectral graph matching approach. *Signal Processing: Image Communication*, Vol. 23(10), pp. 769–787.

Wilson, R. C., Hancock, E. R., & Luo, B., 2005. Pattern Vectors from Algebraic Graph Theory. *IEEE Transactions on Pattern Analysis and Machine Intelligence*, *27*(7), pp. 1112–1124.

Zvetkovic, D., 2012. Spectral recognition of graphs. *Yugoslav journal of operations research*, no. 2, pp.145-161.

# The Impact of Multi-Sensor Data Assimilation on Plant Parameter Retrieval and Yield Estimation for Sugar Beet

Martina Hodrius [a], Silke Migdall [a], Heike Bach [a], Tobias Hank[b]

[a] Vista Geowissenschaftliche Fernerkundung GmbH, Gabelsbergerstraße 51, 80333 München, Germany, Email: hodrius@vista-geo.de, migdall@vista-geo.de, bach@vista-geo.de

[b]Ludwig-Maximilians-Universität München, Luisenstraße 37, 80333 München, Germany, Email: tobias.hank@lmu.de

**KEY WORDS:** yield modelling, crop growth model PROMET, sugar beet, multi-sensor data assimilation, SLC

**ABSTRACT:**

Yield Maps are a basic information source for site-specific farming. For sugar beet they are not available as in-situ measurements. This gap of information can be filled with Earth Observation (EO) data in combination with a plant growth model (PROMET) to improve farming and harvest management. The estimation of yield based on optical satellite imagery and crop growth modelling is more challenging for sugar beet than for other crop types since the plants' roots are harvested. These are not directly visible from EO. In this study, the impact of multi-sensor data assimilation on the yield estimation for sugar beet is evaluated. Yield and plant growth are modelled with PROMET. This multi-physics, raster-based model calculates photosynthesis and crop growth based on the physiological processes in the plant, including the distribution of biomass into the different plant organs (roots, stem, leaves and fruit) at different phenological stages.

The crop variable used in the assimilation is the green (photosynthetically active) leaf area, which is derived as spatially heterogeneous input from optical satellite imagery with the radiative transfer model SLC (Soil-Leaf-Canopy). Leaf area index was retrieved from RapidEye, Landsat 8 OLI and Landsat 7 ETM+ data. It could be shown that the used methods are very suitable to derive plant parameters time-series with different sensors. The LAI retrievals from different sensors are quantitatively compared to each other. Results for sugar beet yield estimation are shown for a test-site in Southern Germany. The validation of the yield estimation for the years 2012 to 2014 shows that the approach reproduced the measured yield on field level with high accuracy. Finally, it is demonstrated through comparison of different spatial resolutions that small-scale in-field variety is modelled with adequate results at 20 m raster size, but the results could be improved by recalculating the assimilation at a finer spatial resolution of 5 m.

## 1. INTRODUCTION

Yield and biomass maps are basic information sources for smart farming. These maps can be used for the daily assessment of plant development and site-specific fertilization measures. Sugar beet yield is usually not mapped during harvesting [Schmittmann 2002], hence, spatially distributed yield information is hard to obtain. Earth Observation (EO) data fill this lack of information and support smart farming by delivering up-to-date information on plant growth independent from in-situ data [Migdall et al. 2013]. However, operational agricultural application requires reliable information for any date during the vegetation period, whatever the weather conditions are. Thus, the combination of EO data with crop growth modelling is necessary to monitor the vegetation development continuously over the growing season [Bach & Angermair 2013, Hank et al. 2015].

In this study, multi-sensor EO data from 2012 to 2014 is used to derive plant parameters for sugar beet using an inversion of the SLC (Soil-Leaf-Canopy) model [Verhoef & Bach 2003, 2007]. The resulting green LAI maps are then assimilated into the crop growth model PROMET to model the plant development at different phenological stages and estimate yield. This approach is well-established and validated for winter wheat as shown in several studies [Bach & Angermair 2013; Hank et al. 2015, Migdall et al. 2013]. The same approach is now applied for sugar beet. The estimation of sugar beet yield based on optical satellite imagery is more challenging than for other crop types since the plants' roots are harvested, which are not directly visible from EO. However, PROMET calculates crop growth based on physiological processes, including the distribution of biomass into the different plant organs for each grid cell [Hank et al. 2015].

Three main issues are addressed by this paper: The first is to evaluate the impact of different sensor properties, e.g. of the spatial resolution, on plant parameter retrieval. Therefore, two data pairs have been selected, that are comparable in terms of acquisition time. The second question is, how accurate the yield estimation for sugar beet may become over 3 consecutive years using the multi-sensor approach (RapidEye, Landsat ETM+, OLI). This was done using field mean values provided by the farmer for validation. The third question is, how well small scale in-field heterogeneity is traced by the yield modelling approach. For this, yield was estimated at a 5 m and at a 20 m raster grid. The results were compared to evaluate the effect of spatial resolution on the results. As validation data, sampling points in two fields with a size of 4 $m^2$ were harvested manually to measure the spatial distribution of sugar beet yield.

## 2. DATA AND TEST SITE

Plant parameter retrieval and yield estimation for sugar beet have been conducted for a test site in Southern Germany. The test site is located near Straubing next to the river Danube, at the centre of the so called Gäuboden, a region in Lower Bavaria, which covers one of the largest loess regions in Southern Germany.

From 2012 to 2014 the development of the Leaf-Area-Index (LAI) during the vegetation period was monitored for several sugar beet fields. After assimilation into the crop growth model yield has been estimated. A multi-sensor approach has been used for the crop monitoring, since the whole vegetation period should be covered with satellite data to ensure an accurate yield model result.

The freely available data of the Landsat missions is well-suited for plant parameter retrieval. The spatial resolution of 30 m is sufficient for many applications in agriculture and the spectral range from Visible to SWIR (Short Wave Infrared), the latter of which is sensitive to plant water content, is ideal for agricultural analyses. Besides Landsat, also RapidEye data is used. The spectral configuration containing the red edge band and the spatial resolution of 5 meters make the sensors very suitable for vegetation monitoring, but the missing SWIR reduces the spectral information content. With the constellation of five satellites and pointing capabilities RapidEye allows to cover the region of interest every day. Accordingly, data can be acquired as soon as there is no cloud cover. Since the presented yield estimation methods are used in an operational mode, a multi-sensor approach is preferred for a reasonable compromise between data availability, data quality and economic interests (mix of cost-free and commercial data). For the test site, the availability of data has been varying in the different years. For 2012, only Landsat 7 ETM+ images have been used. Since 2013, also Landsat 8 OLI data is available. For 2014 the input data used are from Landsat 7 ETM+, Landsat 8 OLI and RapidEye. Table 1 shows a list of the used satellite data. For the analysis of these data, atmospheric correction was carried out to obtain surface spectral reflectance. Based on the calibrated data, the photosynthetically active leaf area is retrieved by using the SLC model (Section 3.2). The resulting maps are then assimilated into the crop growth model PROMET (Section 3.2) to model the plant development and resulting yield.

| | | | |
|---|---|---|---|
| 23 July 2012 | Landsat 7 ETM+ | 22 May 2014 | RapidEye |
| 9 Sept. 2012 | Landsat 7 ETM+ | 3 June 2014 | Landsat 8 OLI |
| 25 Sept. 2012 | Landsat 7 ETM+ | 10 June 2014 | Landsat 8 OLI |
| 11 Oct. 2012 | Landsat 7 ETM+ | 11 June 2014 | RapidEye |
| 16 June 2013 | Landsat 8 OLI | 19 June 2014 | Landsat 8 OLI |
| 2 July 2013 | Landsat 8 OLI | 19 July 2014 | RapidEye |
| 3 Aug.2013 | Landsat 8 OLI | 2 Aug.2014 | RapidEye |
| 28 Sept. 2013 | Landsat 7 ETM+ | 6 Aug.2014 | Landsat 8 OLI |
| 14 Oct. 2013 | Landsat 7 ETM+ | 7 Sept.2014 | Landsat 8 OLI |
| 7 Nov. 2013 | Landsat 7 ETM+ | 9 Oct.2014 | Landsat 8 OLI |

Table 1: Satellite data used for Plant Parameter Retrieval

For validation of the yield estimation, in-situ data was provided by the farmer and the Südzucker AG. Field mean values were available for the years 2012 to 2014. This data is used to validate the yield estimation for several years on field level. However, the assessment of the small-scale heterogeneity of yield requires spatially distributed data on smaller units than field level. For other crop types, which are harvested with a combine harvester, yield maps are potentially available. Such spatially distributed data have been used in previous studies to validate the in-field heterogeneity of the model results for winter wheat [Bach & Angermair 2013]. But root-crop harvesters as used for sugar beet do not map the yield distribution since the distinction between the roots and other plant components or soil is not possible during harvesting [Schmittmann 2002]. Thus, the validation of the yield estimation for sugar beet is more difficult since less data is available. A common way to estimate sugar beet yield before harvesting is destructive sampling at preselected points. This method was extended by defining several sampling points within one field to catch the spatial distribution of yield. The sampling points were defined with a size of 4m² from satellite-based biomass analyses, where high- and low-yield zones could be distinguished. The field sampling was conducted by Südzucker AG following their standard procedures and provided for two fields in 2014 (see Figure 1). This data set was used for the validation of the small-scale variability.

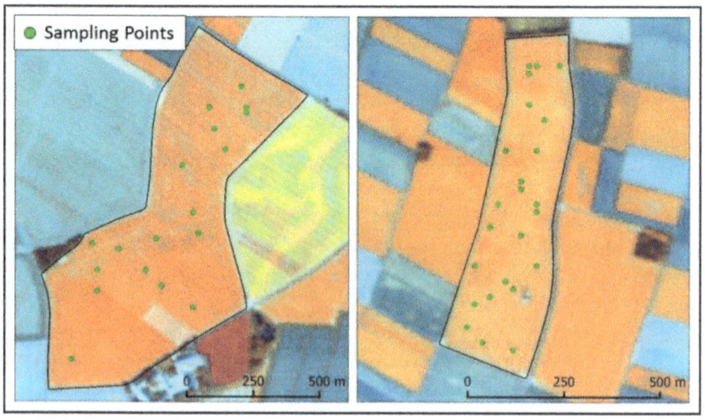


Figure 1: Sampling points for sugar beet yield (Background image: Rapid Eye image of Jul 19th 2014, UTM33N, colour combination: 656nm, 710nm, 804nm)

## 3. MODELS AND METHODS

### 3.1 Satellite Data Processing

To ensure the comparability of multi-temporal, multi-sensoral data and also the transferability of results, all used methods are based on physical processes and modelling. Hence, the raw satellite data has to be pre-processed extensively: it has to be geometrically fine-adjusted, radiometrically calibrated and atmospherically corrected. For this, the radiative transfer model MODTRAN, which was developed by the US Airforce, 2015 is used. This physically-based model calculates the path of the light from the sun through the atmosphere, the interactions with the surface and the path back through the atmosphere to the sensor. A model inversion delivers the surface spectral reflectance (MODTRAN interrogation technique [Verhoef & Bach 2007]). This way, the reflectance of each pixel can be calculated. Reflectance is a property of the surface that is independent of the atmospheric condition and allows for comparison between different images and hence for the analysis of time series.

The data processing and LAI retrieval was done for all three years at 20 m resolution. In 2014, for which the sampling points are available, it was also carried out at 5 m, which is the resolution of the RapidEye Level 3A Product.

For studying the impact of multi-sensoral data on yield estimation it would be desirable to have two complete coverages of each sensor type following the whole growing cycle. This, of course, is not realistic (see Table. 1). RapidEye and Landsat vary not only in their spectral configuration but also in spatial resolution. In order to mimic these differences and to allow for a similar temporal LAI monitoring [Hank et al. 2013], an alternative approach was chosen. The "Landsat 20m" based assimilation uses spatially degraded RapidEye scenes from 22 May and 19 July to fill temporal gaps at the beginning and peak of the LAI development. The "RapidEye 5m" based assimilation is complemented with Landsat derived LAI values during crop maturity (7 Sep and 9 Oct). Table 2 lists the data used for the respective yield estimations.

| Date in 2014 | DOY | Sensor | AZA[°] | GSD [m] | "Land-sat 20 m" | "RapidEye 5 m" |
|---|---|---|---|---|---|---|
| 22 May | 142 | RE | 2.95 | 6.5 | x | x |
| 10 June | 161 | OLI | Nadir | 30 | x | |
| 11 June | 162 | RE | 9.75 | 6.5 | | x |
| 19 June | 170 | OLI | Nadir | 30 | x | x |
| 19 July | 200 | RE | 6.42 | 6.5 | x | x |
| 2 Aug. | 214 | RE | 0.27 | 6.5 | | x |
| 6 Aug. | 218 | OLI | Nadir | 30 | x | |
| 7 Sept. | 250 | OLI | Nadir | 30 | x | x |
| 9 Oct. | 250 | OLI | Nadir | 30 | x | x |

Table 2: Satellite data used for 5m and 20m yield estimation; light green indicates that these dates have been used for inter-sensor comparison (RE= RapidEye, OLI = Landsat 8 Operational Land Imager, AZA= Acquisition Zenith Angle, GSD = Ground Sampling Distance of original satellite image, DOY=Day of Year)

### 3.2 The Radiative Transfer Model SLC

For the retrieval of the plant parameters, an inversion of the SLC model, an extended version of the SAIL model family, is applied [Verhoef & Bach 2003, 2007]. Based on a four-stream concept, the radiative transfer between soil, canopy and single leaves is modelled. The PROSPECT [Jacquemoud & Baret 1990] sub-model is used to describe the transmittance of green and brown leaves [Migdall et al. 2009]. The input parameters for the forward modelling of reflectance describe the structural and physiological properties of the soil and the vegetation canopy, among them Leaf-Area-Index (LAI) and the fraction of brown leaves, characteristics of the leaves (e.g. chlorophyll content and plant water content) as well as the sun-observer geometry [Verhoef & Bach 2012].

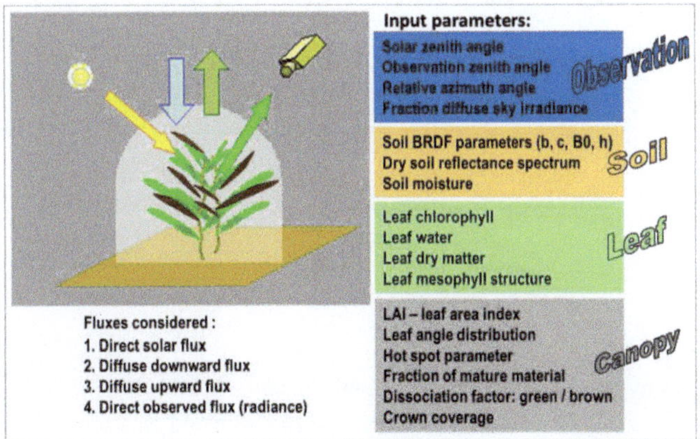


Figure 2: The SLC model and its input parameters [Verhoef & Bach 2003, 2007]

Figure 2 on the right-hand side lists the required input parameters. On the left-hand side it shows the four radiation fluxes considered in the SLC model. The parametrization of the SLC model takes the Spectral-Response-Function of the sensors into account. Therefore, it can be adapted to any sensor according to the sensor's spectral and geometrical properties [Migdall et al. 2009]. The sun-observer geometry is recorded during the image acquisition and therefore is known. The soil reflectance and its variation with moisture are described by a soil BRDF (bi-directional reflectance distribution function) sub-model, based on the soil model by Hapke, 1981. Some of the leaf and canopy parameters are assumed to be constant for one crop type or within one specific phenological stage and are either obtained from literature or were determined using hyperspectral and in-situ data. The remaining parameters, which are highly variable (e.g. LAI, chlorophyll), can be retrieved by model inversion using the RMS error between the simulated and the measured spectra as criterion for the best fit [Migdall et al. 2009].

### 3.3 The Crop Growth Model PROMET

While optical remote sensing data can retrieve accurate information on the developed leaf area, it cannot see the absolute biomass or its distribution into the different plant compartments. It can definitely not directly observe the root biomass, which in case of sugar beets makes up the actual yield.

Therefore, the green leaf area serves as spatially distributed input for crop growth and yield modelling with PROMET [Mauser & Bach 2009, Hank et al. 2015]. This multi-physics, raster-based model calculates crop growth based on the physiological processes in the plant, including the distribution of biomass into the different plant organs (roots, stem, leaves and fruit) at different phenological stages. The model calculates the plant growth in hourly time-steps for the whole growing period, using background data such as a Digital Terrain Model and Soil Maps as well as up-to-date meteorological data. The model generates in an ensemble mode different scenarios for varying soil conditions [Hank et al. 2015].

Green LAI maps, retrieved with the SLC model, are used to find the scenario that fits the current growth conditions best. For this, the LAI maps are assimilated into the plant growth simulation as raster data sets. Small-scale soil variations due to e.g. different water holding capacity, which cannot be included in the more generalized background data, will thus be considered in the modelling [Migdall et al. 2009]. Since the unknown spatial heterogeneity of soil conditions is considered a major cause for in-field variations of plant development, the assimilation of remotely sensed data into the model improves the model outcomes significantly [Hank et al. 2015]. The concept of assimilating multi-sensoral EO data into PROMET is shown in Figure 3.

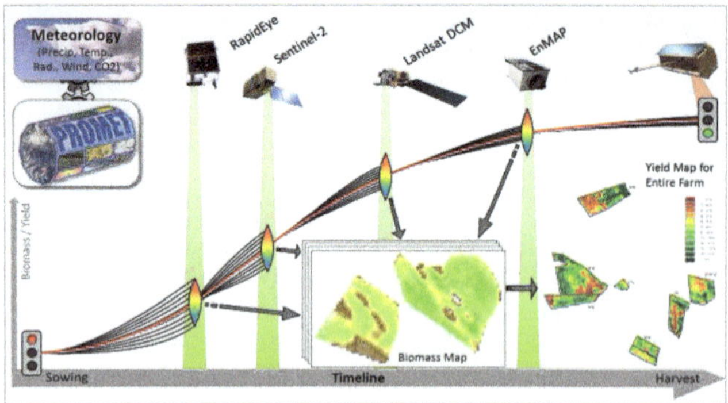


Figure 3: Concept of using PROMET model ensembles for assimilating multi-sensoral remote sensing data, modified after Hank et al. 2015

## 4. RESULTS

Based on the pre-processed data, the inversion of the SLC model was applied to retrieve the green LAI. Then, the green LAI was assimilated into the crop growth model at 5m and 20m resolution to estimate yield. The following section shows the results.

### 4.1 Impact of different Sensors on Plant Parameter Retrieval

Figure 4 shows the results for the green LAI retrieval for all available scenes in 2014 over the whole sugar beet growing period. Two different effects can be observed in this figure. For one, the LAI time series derived from different sensors is consistent. There is no large deviation due to sensor characteristics, only small differences of about 0.15 LAI occur at Day of Year 218. Secondly, true temporal coverage of the crop growing period is only possible through the combination of multi-sensor data. From the four RapidEye acquisitions, just two are at about the same time as the Landsat data. For these data pairs, quantitative comparisons of atmospherically corrected spectral signatures and retrieved LAI values are given.

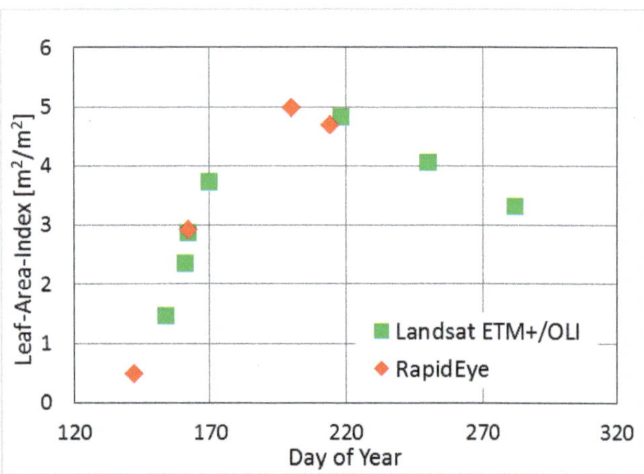


Figure 4: LAI retrieval for 2014 based on Landsat OLI/ETM+ and RapidEye data

Figure 5 shows the comparison of the average reflectance measured for sugar beet with RapidEye and Landsat 8 OLI for the DOY 214/218.

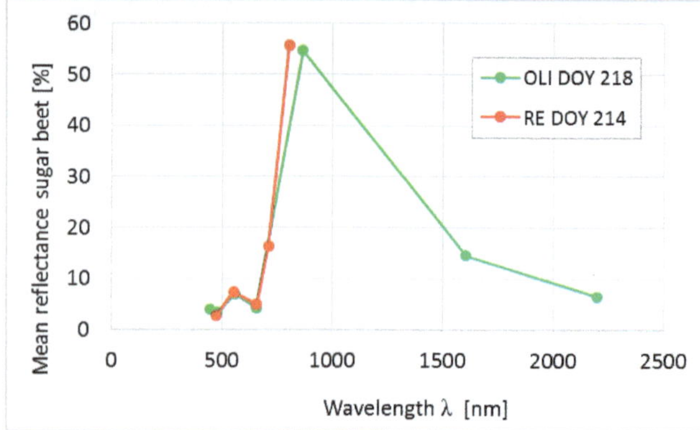


Figure 5: Reflectance measured for Sugar Beet with Landsat 8 OLI and RapidEye

While the visible bands of Landsat and RapidEye correspond to a large extent in their spectral response function (SRF), the NIR bands differ substantially [Blackbridge 2012, NASA 2015]. This can also be recognized in the shift of central wavelength of the NIR band in Figure 5. The challenge of the multi-sensor approach is to take these different spectral configurations for the plant parameter retrieval into account. The parametrization of the SLC model uses the SRF of the sensors [Migdall et al. 2009]. Thus, the influence of the spectral configurations should not lead to differences in the retrieved plant parameter. Whether those requirements are fulfilled is analysed by comparing the LAI derived from different sensors acquired during the same timeframe as described above.

Figure 6 shows the LAI retrieved with RapidEye compared to the LAI retrieved with Landsat 8 OLI for each 20m pixel. The discrete steps visible in this scatter plot are caused by the applied Look-Up-Table inversion. The steps of the tables can be recognized and the non-linear stepping is visible (smaller steps for lower LAI values where higher accuracies are targeted). The absolute values of LAI retrieval show a high congruency, as the gain with 0.97 is very close to 1. The RMSE between the two sensor retrievals amounts to 0.6 $m^2/m^2$. The scattering of the values increases with increasing LAI values. This is caused by a saturation effect that occurs at very high LAI values and makes a distinction between LAI 5 and 6 much more difficult than between 2 and 3 [Bach et al. 2012].

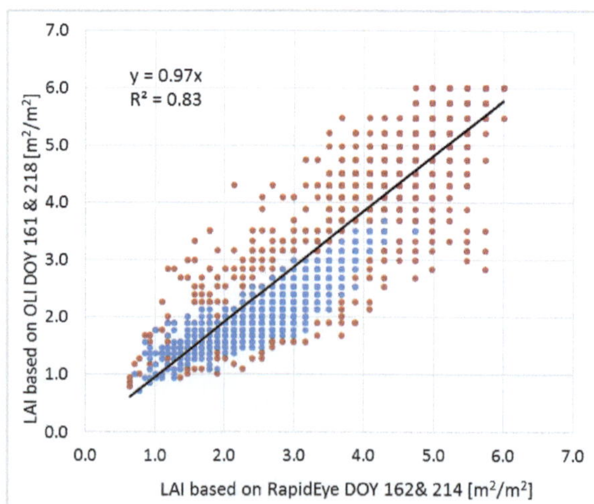


Figure 6: Comparison of LAI retrieval based on Landsat OLI/ETM+ and RapidEye data for the two dates analysed (blue: DOY 161, red: DOY 218)

Table 3 compares the averages of the retrieved LAI. The disagreement of the LAI is 0.4 and 0.1 for the two analysed dates. There is no systematic offset observable, as the deviation occurs in both directions.

| DOY | Plant Parameter | OLI | RE |
|---|---|---|---|
| 161/162 | Green LAI [$m^2/m^2$] | 2.4 | 2.8 |
| 214/218 | Green LAI [$m^2/m^2$] | 4.8 | 4.7 |

Table 3: Plant Parameter retrieved from OLI and RE

### 4.2 Multi-Sensor-based Yield Estimation for Sugar Beet 2012 – 2014

Sugar beet plant parameter retrieval was done for three consecutive years. The next question to answer is how accurate this multi-sensor approach is. In the years 2012 and 2013, only Landsat ETM+/OLI data was available, whereas in 2014 all three sensors have been used.

Figure 7 shows the mean LAI development of all three observed years. Sugar beet development has differed significantly in those years. The peaks of the LAI development are in 2012 and 2014 much earlier and higher in absolute value than in 2013. Furthermore, the increase of the LAI was slower in 2013. These differences are mainly caused by the weather conditions. In 2013, seeding took place almost one month later then in 2014 due to snow cover, rain and wet soil conditions. Extreme weather conditions were dominant during the whole year. A phase with heavy rainfall in June was followed by drought in July. Altogether, the growing season was four weeks shorter in 2013 than in 2014. In contrast, the conditions in 2014 were ideal. The seeding took place very early and the weather conditions were optimal for sugar beet growing during the whole season.

The varying LAI development leads to a different amount of accumulated biomass in the roots and thus to yield differences. This results in a very high modelled yield in 2014, where the LAI is constantly higher than in the other years. In contrast, the modelled yield in 2013 is very low which is expected due to the late start and slow LAI increase.

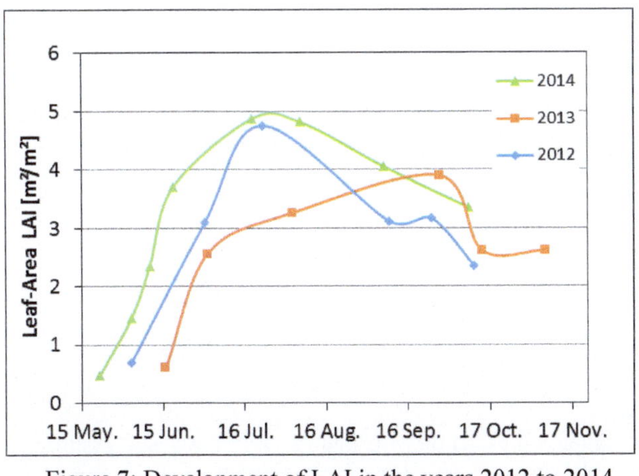


Figure 7: Development of LAI in the years 2012 to 2014

These modelled results are validated with available in-situ data as shown in figure 8. The different colours represent the three years. The linear regression, which is very close to the 1-to-1-line, proofs that the approach is very suitable for sugar beet yield estimation. Both, the absolute values of yield and variations on field level are well reproduced by the model. Not only the variation between different years but also the spatial variations between fields in each year are well represented. A gain value very close to one shows that there is no offset in the modelling results, neither in very low nor in very high yield ranges. This is also indicated by the low RMSE of 4.4 t/ha, which is only 4.5% of the yield mean over all 3 years.

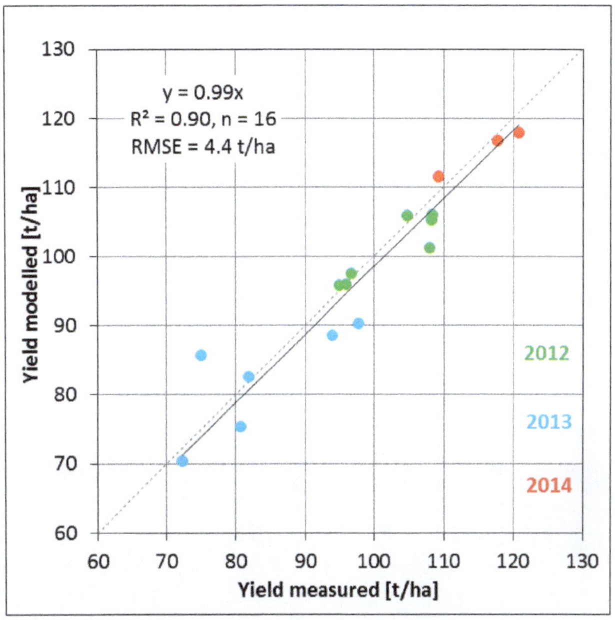


Figure 8: Yield validation on field level

Concluding, it could be shown that with the multi-sensor approach it is possible to retrieve plant parameters and to model yield with a high accuracy on field level.

### 4.3 Validation of Modelled In-Field Heterogeneity

The validation of the multi-sensor yield modelling should not be limited to field averages, but also consider the accuracy of the modelling of the in-field heterogeneity of yield. Sampling points within the sugar beet fields were harvested by hand to assess the spatial distribution of the yield. This was done for two fields in 2014 and the collected data was used for validation of the modelled yield maps (see Figure 1). There is usually some loss of yield during harvesting with a root-crop harvester. This loss is calculated as 7% by comparing the mean of the sampling points with effective yield of the fields. Accordingly, the yield samples were multiplied with the factor 0.93 for comparison with the model results.

The LAI retrieval and yield modelling was performed at 5 m ("RapidEye like") and at 20 m ("Landsat like") resolution. The average yield of the "Landsat-like 20m" and "RapidEye-like 5m" yield results vary only slightly (see Table 4). This supports the conclusion that the multi-sensoral approach is reliable and produces comparable results.

| Field-number | Mean Yield "LS 20 m" [t/ha] | Mean Yield "RE 5 m" [t/ha] |
|---|---|---|
| 2808 | 119.3 | 122.6 |
| 2877 | 114.3 | 116.8 |
| 2920 | 108.6 | 110.6 |
| 2961 | 113.0 | 115.7 |

Table 4: Results of the yield estimation for the "Landsat like 20m" and the "RapidEye like 5m" estimation

On the other hand, this approach also allows to compare the effect of sensor resolution in terms of spatial accuracy of yield. Figure 9 shows the yield maps in 20 m and 5 m resolution for one field in comparison with the sampling points for both fields. While the same overall structures can be seen in both resolutions, the additional detail in the 5m version is visible. In the 5 m yield map even row structure becomes visible.

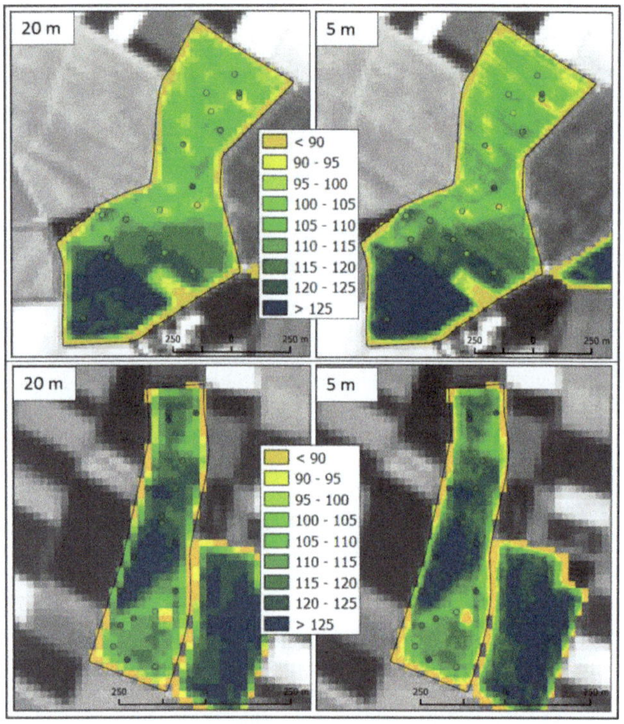


Figure 9: Results of yield modelling in 20 m (left) and 5 m (right) resolution in comparison with sample points

Figure 10 shows the result of the validation for 20 m and Figure 11 for 5 m. The higher resolution could improve the coefficient of determination ($R^2$), which means that the spatial variance is better reproduced. But also it could be shown that with only the Landsat data small scale variability can be modelled with adequate results.

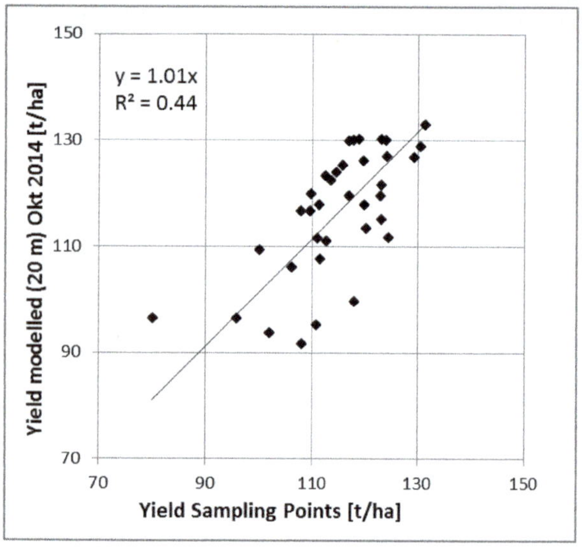


Figure 10: Yield validation sampling points - yield modelled at 20m resolution

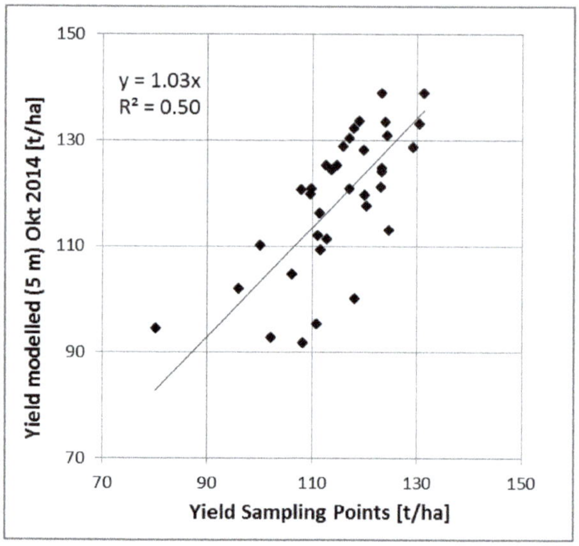


Figure 11: Yield validation sampling points - yield modelled in 5 m resolution

## 5. CONCLUSIONS

For precision farming applications, the management unit is not the whole field. Management is rather conducted on smaller units depending on the spatial distribution of site-characteristics and the working width of the machinery (e.g. 24 m or 36 m). With sugar beet, site-specific applications are not common yet (in opposition to e.g. wheat), because site-specific information about sugar beet growth and especially yield is hard to come by, as the main biomass is under ground and there is no technology for site-specific harvesting of sugar beets available on the market yet. Therefore, information derived from EO data and crop growth modelling is a new and exciting spatial data source for new site-specific sugar beet applications.

Yield estimation based on SLC and PROMET was successfully conducted for sugar beet during 3 consecutive years. It could be shown that the accuracy of the yield estimation is very high on field level. Additionally, the small-scale in-field variety is modelled with adequate results at 20 m raster size, but even better results are achieved at a 5 m raster. The comparison of the LAI retrieval based on RapidEye and OLI shows that the SLC model and the data assimilation concept in PROMET is very suitable for the multi-sensor approach, since it is physically based and SLC takes the individual spectral configurations of the different sensors into account.

The demonstration that satellite data of variable spectral and spatial characteristics can be successfully used in crop yield estimation is of special importance, since using only one sensor often does not allow monitoring the LAI development very well. Thus, the multi-sensor approach can improve the accuracy of the yield estimation by increasing the number of assimilated LAI maps.

The presented methods can be used in an operational mode to support site-specific farming for sugar beet. The up-to date plant monitoring based on satellite imagery can be used for the daily assessment of the sugar beet crop. Thus, the occurrence of plant diseases, pests and other challenges can be detected early and the necessary measures can be conducted. Some phenological stages are very important for the vegetation development, e.g. the phase of row closure. This information can be provided spatially distributed. Additionally, the crop growth model delivers information on the root development, which is not observable from above. Using this information, the harvesting can be optimised in terms of logistics and time planning [Angermair & Bach 2015].

## ACKNOWLEDGEMENTS

This work was supported by ESA within the IAP Demonstration Project TalkingFields, funded under contract number 4000100879/10/NL/US (ARTES 20 Programme) and the EU FP7 project MELODIES (Grant Agreement 603525) coordinated by the University of Reading.
Thank you to our TalkingFields partners Wolfgang Angermair and Stefan Burgstaller from PC Agrar GmbH for organising the field trial. Thanks to the Poschinger-Bray'sche Güterverwaltung in Irlbach, namely Jürgen Schwarzensteiner, for the cooperation and providing validation data for our work. Thank you also to the Südzucker AG for supporting the field trial and harvesting of sampling points.

## REFERENCES

Angermair, W.; Bach, H., 2015. Fernerkundung: Teilflächenbewirtschaftung Zuckerrübe. *Agrarmanager 4/2015,* pp. 81-83.

Bach, H.; Angermair, W., 2013. Ein großer Schritt voran; *DLG Mitteilungen* 2/2013, pp. 76 – 78.

Bach, H.; Migdall, S.; Spannraft, K.; Hank, T; Mauser, W. , 2012. Potential and challenges of using Sentinel-2 for Smart Farming; *Sentinel-2 Preparatory Symposium*, ESA-ESRIN, Frascati, Italy, 23 to 27 April 2012.

Blackbridge: Spectral Response Curves of the RapidEye Sensors, 2012. http://blackbridge.com/rapideye/upload /Spectral_ Response_ Curves.pdf (20.02.2015).

Hank, T., Frank, T., Bach, H., Spannraft, K., & Mauser, W. , 2013. Assessing the required temporal frequency of optical EO acquisitions for the modelling of winter wheat yield in Northern Germany , *In: Borg, E., Daedelow, H. & Johnson, R. (Hrsg.): RapidEye Science Archive From the Basics to the* Service ISBN 978-3-95545-002-1 , GITO Berlin , Reviewed, published. pp. 39 – 55.

Hank, T., Bach, H., Mauser, W. (2015): Using a remote sensing supported hydro-agroecological model for field-scale simulation of heterogeneous crop growth and yield: application for wheat in Central Europe. *Remote Sens.* 2015, 7, doi:10.3390/rs70403934, pp. 3934-3965.

Hapke, B.W., 1981. Bi-directional reflectance spectroscopy 1. Theory. *Journal of Geophysical Research*, 86, pp. 3039-3054.

Jacquemoud, S.; Baret, F.(1990): PROSPECT: A model of leaf optical properties spectra. *Remote Sensing of Environment*, 34, pp. 75-91.

Mauser, W.; Bach, H., 2009. PROMET - large scale distributed hydrological modelling to study the impact of climate change on the water flows of mountain watersheds. *Journal of Hydrology* 2009, 376, pp. 362-377.

Migdall, S.; Bach, H.; Bobert, J.; Wehrhan, M.; Mauser, W., 2009. Inversion of a canopy reflectance model using hyperspectral imagery for monitoring wheat growth and estimating yield. *Prec. Agr,* 2009, 10, pp. 508–524.

Migdall, S., Spannraft, K., Bach, H., Hank, T., Frank, T., Mauser, W., Burgstaller, S., Tüller, G. & Angermair, W. (2013. On-Farm Application of Operational Integrated Satellite Services, *ESA Living Planet Symposium 2013, Special Publication SP-722* , Edinburgh (UK) , Proceeding, published.

NASA, 2015. Spectral Response of the Operational Land Imager In-Band, Band-Average Relative Spectral Response. http://landsat.gsfc.nasa.gov/?p=5779 (20.02.2015).

Schmittmann, O., 2002. Teilflächenspezifische Ertrags-messung von Zuckerrüben in Echtzeit unter besonderer Berücksichtigung der Einzelrübenmasse. *Dissertation*, Institut für Landtechnik, Universität Bonn.

US Airforce: http://modtran5.com (05.01.2015) .

Verhoef, W., Bach, H., 2003. Simulation of hyperspectral and directional radiance images using coupled biophysical and atmospheric radiative transfer models. *Remote Sensing of Environment*, 87, pp. 23-41.

Verhoef, W., Bach, H., 2007. Coupled soil–leaf-canopy and atmosphere radiative transfer modeling to simulate hyperspectral multi-angular surface reflectance and TOA radiance data. *Remote Sensing of Environment*, 109, pp. 166-182.

Verhoef, W.; Bach, H., 2012. Simulation of Sentinel-3 images by four stream surface atmosphere radiative transfer modeling in the optical and thermal domains. *Remote Sensing of Environment*, 120, pp. 197-207.

# AN APPROACH FOR INDOOR WAYFINDING REPLICATING MAIN PRINCIPLES OF AN OUTDOOR NAVIGATION SYSTEM FOR CYCLISTS

A. Makri [a], S. Zlatanova [b] E. Verbree [c]

[a] MSc Geomatics, Faculty of Architecture and the Built Environment, Delft University of Technology, Julianalaan 134, 2628 BL, Delft, The Netherlands, A.Makri@student.tudelft.nl
[b] 3D Geoinformation, Faculty of Architecture and the Built Environment, Department of Urbanism, Delft University of Technology, Julianalaan 134, 2628 BL, Delft, The Netherlands, s.zlatanova@tudelft.nl
[c] GIS Technology, Faculty of Architecture and the Built Environment, OTB Research Institute for the Built Environment, Delft University of Technology, Julianalaan 134, 2628 BL, Delft, The Netherlands, E.Verbree@tudelft.nl

**Commissions WG IV/7 and WG V/5**

**KEY WORDS:** Indoor wayfinding, landmark, signpost, decision point, node network, route graph, Constrained Delaunay Triangulation

**ABSTRACT:**

This work presents an approach to enhance navigation in indoor environments based on a landmark concept. It has already been proved by empirical research that by using landmarks the wayfinding task can be significantly simplified. Navigation based on landmarks relies on the presence of landmarks at each point along a route where wayfinders might need assistance. The approach presented here is based on the Dutch system for navigation of cyclists. The landmarks that are used in the proposed approach are special signposts containing the necessary directional information in order to guide the wayfinder in the space. The system is quite simple, efficient and satisfactory in providing navigational assistance in indoor space. An important contribution of this research is the generation of an approach to automatically determine the decision points in indoor environments, which makes it possible to apply it to navigational assistance systems in any building. The proposed system is verified by placing numbered landmark-signs in a specific building. Several tests are performed and the results are analysed. The findings of the experiment are very promising, showing that participants reach the destinations without detours.

## 1. INTRODUCTION

In recent years, navigation has become a very active research area with a wide range of application fields. Navigation, also called finding one's way, whether in real or electronic world, is a fundamental but also complex human activity. Despite the fact that navigation systems have primarily been developed for vehicles, supporting wayfinding in indoor environments has emerged to be a significant field of interest as people spend most of their times indoors. Especially in public buildings, such as airports and train stations, hospitals, offices or university buildings, individuals often fail to find their way immediately under time pressure and would, therefore, benefit from a well-established system offering navigational assistance.

Without the support of external navigation aids people have to rely on common knowledge about the structure of buildings, on their previous experience and on the visual input they encounter while moving in the building (Holscher et al., 2007). However, this is a complex task. The conceptualization of wayfinding, as a spatial problem solving, incorporates information processing, decision making or planning and decision execution (Passini, 1992). Wayfinding requires explicit decision-making, such as selecting routes to follow, orienting towards non-perceptible landmarks and scheduling trips (Montello & Sas, 2006). The sense of orientation in indoor spaces is affected by several parameters. Firstly, indoor space is characterized by the existence of the third dimension which is expressed though the different floor levels (Brunner-Friedrich & Radoczky, 2006). Vertical movement can have a serious impact on the wayfinding performance. Moreover, indoor environments are composed of smaller, fragmented spaces, with a limited field of view and change of direction is imposed more often than outdoors. Consequently, people encounter significant difficulties in the pursuit of their destination target when they are navigating in unfamiliar indoor environments. Humans depend mainly on wayfinding directions, either provided by other people or given by maps or other wayfinding services. People use landmarks when they give route directions to anchor actions in space or to provide confirmation that the right track is still being followed (Michon & Denis, 2001). However, while it is proven that directions provided by people are based on using landmarks as references, especially at decision points, the directions given by wayfinding services are generated based on the geometry of the space.

Empirical research has established the importance of landmarks in our understanding of and communication about space (Lynch, 1960; Michon & Denis, 2001; Raubal & Winter, 2002; Snowdon & Kray, 2009). Landmark-based navigation is the most natural concept for humans to navigate themselves through their environment, especially in new and unfamiliar environments where wayfinding may be a time consuming and even challenging task, as in this type of navigation people choose a readily-identifiable feature of the general landscape and use it as a base. They can move out from the base to explore, return to the base whenever they get lost or whenever they want to start over and explore in a different direction. When they become familiar with an area, they may move from one landmark to another in order to navigate through larger spaces. It has already been proved by empirical research that landmarks simplify significantly the wayfinding task (May et al., 2003).

This paper proposes a wayfinding assistance system based on specially designed indoor landmarks. The term landmark is used here to indicate prominent signposts, which have the potential of serving as distinctive, recognizable and salient landmarks and at the same time can provide additional information (Fontaine & Denis, 1999; Millonig & Schechtner, 2007). Using such signposts, a network is created, which guides the users though the building.

The remainder of this paper is structured as followed. The second section provides a background information on human wayfinding process. Section 3 discusses the original outdoor Junction Network System for cyclists, which main principles are used to build the indoor network. Section 4 presents the proposed Junction Network System for Indoor Environments. The final section discusses the validation tests and concludes on the findings.

## 2. COMPONENETS OF HUMAN WAYFINDING

Wayfinding in indoor environments is a complex process, which requires knowledge about human orientation and movement indoors as well as the type of supporting information humans need to locate themselves and find their way.

### 2.1 Human orientation and navigation

Substantial research have been completed to investigate the processes that take place when people orient themselves and navigate through space (Raubal & Egenhofer, 1998). Several theories have been developed in order to address issues such as how people find their ways in the physical world, what they need in order to find their way, how they communicate directions and how their abilities affect wayfinding. Successful wayfinding means that a goal or destination is being reached efficiently. This requires spatial knowledge and various cognitive abilities (Raubal & Egenhofer, 1998). McNight et al (1993) summarize the three primary ways that people employ in order to find their way: 1) landmark-navigation, in which people select easily identifiable points of reference in the environment and use them as a base, 2) route-navigation, in which they put the landmarks in a sequence creating navigation paths and they navigate in the space by learning routes between locations, and 3) map-navigation in which people create a general frame of reference (mental/cognitive map) containing the spatial relationships between objects and use it to navigate. People's perception of the real world develops gradually through these three levels by recording information about the environment. Accordingly, for successful wayfinding information about the environment - what is in the environment and where it is - are required.

Weisman (1981) distinguished four classes of environmental variables which influence the wayfinding process: 1) visual access to familiar cues or landmarks; 2) the degree of architectural differentiation between different parts of a building; 3) use of signs or room numbers; and d) plan configuration. These variables have been examined in several studies (Weisman, 1981; O' Neill, 1991; Montello and Sas, 2006; Hölscher & Brösamle, 2007). These studies reveal that orientation becomes difficult and wayfinding performance decreased when the visual access is restricted. However, wayfinding success increases with the presence of signage. Floor plan complexity decreases the people success in wayfinding as well. Inexperienced users prefer as much as possible central parts of the building even if it entails more detours (Hölscher et al., 2006). Users familiar with the environment use the direction strategy (following a direction) or the floor strategy (finding the floor first) in order to find their way (Hölscher & Brösamle, 2007).

Another aspect that influence wayfinding is orientation. Orientation refers to the ability of humans to be aware of their location relative to their destination and to other places or objects (Montello & Sas, 2006). The orientation requirements for a wayfinder can be quite coarse; it is important to know what is essential in order to reach the destination effectively and to avoid getting lost. Geographic disorientation can generate anxiety and frustration to the wayfinder. Humans can orient themselves in an environment through landmark-based and dead reckoning processes (Montello & Sas, 2006).

Several approaches can be identified for wayfinding assistance. The most common way of navigating in indoor spaces is by presenting information about the building on maps or by signs.

Maps are the most commonly used tools in the wayfinding process. Replacing the traditional paper maps/instructions, digital maps and route descriptions are becoming wide-spread mean for wayfinding. Variety of tools have been developed, such as verbal navigational instructions, static and interactive maps, 3D visualizations and animations. While well-developed for outdoor, indoor applications are relatively limited and not well adapted to user expectations. User unfamiliar with the building or with the systems can still get lost.

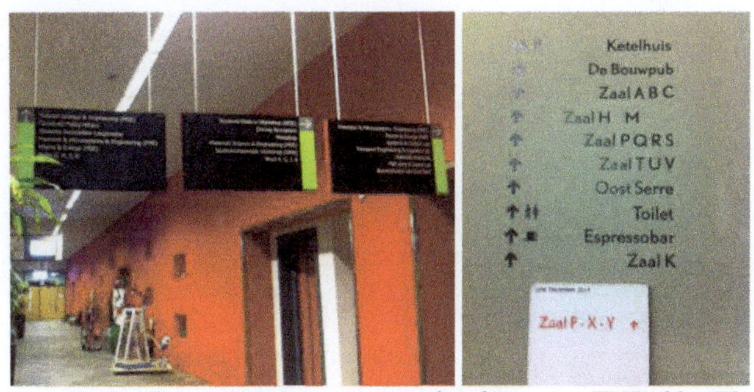


*Figure 1: Sign systems in two faculties at TUDelft, The Netherlands (3EM and Architecture)*

Signage is the most commonly employed physical means of enhancing wayfinding efficiency in indoor environments (Figure 1). Most building complexes offer well developed strategy for wayfinding taking into account the room names, directional signs or other graphical elements. Several studies (O' Neill, 1991; Hölscher & Brösamle, 2007) indicate the positive relationship between signage and wayfinding performance. Signs can create confusion as well. People may have problems to understand the signs in case they are not clear or they have too much information or even confusion is generated when there are too many signs (Montello & Sas, 2006). In this paper we argue that signs are the better approach for wayfinding indoors. Eventual difficulties with understanding and following the signs can be attributed to lack of a common reference system, i.e. language, sign nomenclature and location of the signs.

### 2.2 Landmarks as source of information

Signs are well-defined and distinguishable landmarks, which provide information about the indoor environment and the options for moving. Presson & Montello, 1998 define landmarks as features that are relatively well-known and which define the location of other points. Landmarks are stationary, distinct, and salient objects or places, which serve as cues for structuring and building a mental representation of the surrounding area (Millonig & Schechtner, 2007).

A landmark is structurally attractive if it plays a major role or has a prominent location in the structure of the spatial environment (Raubal & Winter, 2002). From a structural point of view, landmarks can be categorized into global and local landmarks. In a wayfinding context, global landmarks are typically used for conveying directional information; they are at a distance, or off the route. Local landmarks are typically used for conveying positional information; they are close to the route, and are sometimes further categorized into landmarks at decision points, landmarks at potential decision points, and on-route landmarks along segments (Klippel et al., 2005; Lovelace et al, 1999).

Most approaches to include landmarks in wayfinding instructions focus either on landmark identification (Nothegger, Winter and Raubal, 2004; Elias, 2003) by specifying the area in which landmarks has to be sought and then identifying the features that act as outliers in the area, or on the integration of landmarks into the generated instructions (Klippel et al, 2005; Caduff & Timpf, 2005, Ruso & Zlatanova 2014). The failure of incorporating landmarks in commercial applications can be attributed to the costs associated with the acquisition of the required data and the highly skewed distribution of landmark candidates in available spatial data (Richter, 2013)

### 2.3 Giving directions

Route directions are important part of navigation process. These are the instructions on how to follow a route, i.e. alerting at an appropriate location (decision point) of the route which action has to be taken (Richter & Klippel, 2005; Richter et al, 2008). The action can be related to changing the direction of movement or investigating possible options for movement or completing a task (e.g. counting doors) (Richter et al, 2008). A variety of details can be given when providing directions for wayfinding, such as landmarks, cardinal directions, street names, distances and turn descriptions (Hund & Padgitt, 2010). Human route directions cannot make use of numerical references such as distances or angles. Instead people very often use landmarks (doors, stairs, plants) to anchor actions in space or to provide confirmation that the right track is still being followed (Michon & Denis, 2001; Richter, 2013).

*Figure 2: Junction network in the Netherlands*

Specifying a direction is critical for guiding. Directional relations state the location of entities encountered along the route (like landmarks) with respect to the human or other entities; they announce a change of heading at decision points, i.e. represent turning actions; and they may relate these actions to an entity's location to better anchor them in space. In general, it can be stated that wayfinding can be characterized as following a route segment up to a decision point, making a directional choice, following the next route segment up to the next decision point, making a directional choice, and so on (Klippel et al., 2004).

## 3. WAYFINDING JUNCTION SYSTEM

### 3.1 Outdoor

The 'Junction Network System' is a special approach of providing direction instructions to cyclists, used in the Netherlands and Belgium (Figure 2). The main concept of the system is a network of nodes, which are placed at junctions of cyclist paths. The nodes are marked with a signpost (with map overview of the area) containing a unique number and directional information to the next-encountered nodes/intersections. This network has two important characteristics: 1) the number at the sign gives a unique identity to the junction and provides relative location in the space, and 2) remembering (recording) a sequence of such unique numbers allows a user-specified route to be determined. The most important principle of the system is that the assistance is provided at the point when a user has to take a decision how to go on. The signage system is a green sign with the node numbers in a white circle, which has emerged to be a special type of landmark as it is easily identified by cyclists (Figure 3).

*Figure 3: Signs of the system*

### 3.2 Indoor Junction system

The proposed system, 'Junction Network System for Indoor Environments', is the first attempt of adapting the main principles of the original 'Junction Network System' for indoor and focusing at pedestrians as a target group. The indoor concept follows closely the main principles of the original Junction Network System. The indoor variant of the system should be also able to respond to the three major questions of wayfinding communication: what information should be presented, where the information will be provided, and in what form.

Similar to outdoor, the proposed approach is based on the creation of a network of locations, equipped with a special type of signpost, the landmark-sign, containing a unique number for every location and directional information of the other numbered locations in the vicinity of it.

In this research, floor mounted signs are designed as they attract people's caution and they are well visible from a distance. Moreover, there are no important limitations regarding their size, they do not interfere in the configuration of space and do not intervene in people's movement (Figure 4).

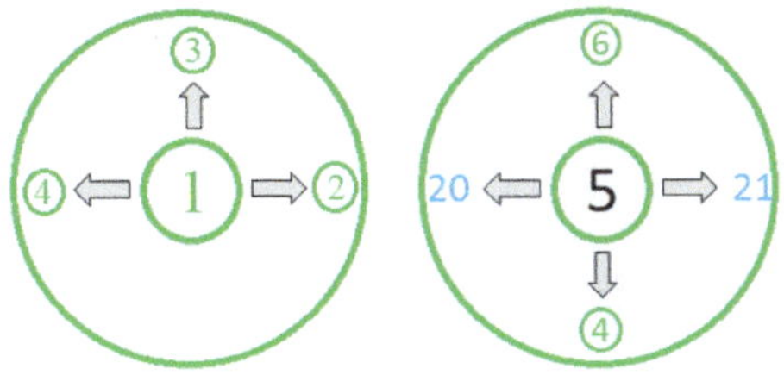


*Figure 4: Designed floor signs*

The landmark-signs are placed at decision points, where people need assistance in order to proceed further (i.e. the junctions of the original system). The location and directional information are revealed through the landmark-signs. By supplying every landmark-sign with a unique number the closest destination spaces to this sign are automatically registered to it. For example, when the user is located at number 5, he/she can infer his relative location in the building (Figure 4, right). A directional assistance is provided by the signs that contain directional arrows indicating the closest numbered decision points or the arrival at a destination space in all possible directions of moving. The destination spaces are also marked with a unique number. It is suggested that the numbering of destination spaces (e.g. rooms) is completed after the numbering of the decision points. This facilitates following a specific route to a destination point. The location of the signs should support the conceptualization of turning instructions, enhancing the understanding of direction instructions.

The connections between the landmark-signs and the attached destinations, create the whole network of nodes which provides all possible combinations of routes between starting points and destinations. In this way, the basic representation of the system is a sequence of decision points with their accompanying actions. Guidance is given by referring to numbers of landmark-signs and anchoring actions to them. The network of landmark-signs has the same twofold role as outdoor: 1) it acts as a routing system that enhances the successfully reach of destination and 2) it plays the role of a referencing system which provides location information to the users.

There are several advantages of this system:

- It is based on an already existing and recognized system (in the Netherlands and Belgium) meaning that people can easily get used to it.
- It provides an indication of the relative position of the user in the environment. It acts not only as a wayfinding/routing system but also as a referencing system.
- The landmark-sign is well-defined object, which expresses the critical points where people need to choose direction of moving and provides location and directional information.
- It is a solution that can be applied to different building cases.
- It can be used for seamless indoor-outdoor navigation system since the principles are the same.
- No special devices (indoor positioning and digital maps) are required.

One of the most critical difference between indoor and outdoor junction system is locations of the landmark-signs. Outdoors the signs are placed at the junctions as they are assumed to be the decision points. Indoors, the decision points might be much more and located not only at places of crossing corridors. In contrast to outdoor, indoor environments do not have well-defines paths and humans can move freely in the open space. Therefore it is critical to decide which the possible routes are and where the landmark-signs can be placed. In this research, an automatic approach is followed to derive a network for movement and landmark-sign location placement.

The network (Figure 5) to represent the possible movement structure of the indoor environment is based on a Medial Axis Transformation (MAT) generated from Constrained Delaunay Triangulation (CDT). MAT is a good approximation of depicting the human movement in these spaces (Mortari et al., 2014). The MAT represents the connectivity of the (triangulated) spaces. Spaces are mapped to nodes of the graph, which provide access to all possible parts of the building. All building spaces are identified and represented by their central point, applying the concept of duality of spaces (Zlatanova et al, 2013). All the possible destination points and functional areas are also indicated and included in the graph (Kruminaite and Zlatanova, 2014). The connections between the spaces can represent adjacency and connectivity relationships. Connectivity ensures that humans can move further in the space. Connectivity can be established on the basis of sematic information (e.g. knowing that certain object is a door) or on the basis of geometry. For example, knowing that a floor of a room is triangulated, all the adjacent triangles have a connectivity relationship if they are not neighbouring to a non-navigable spaces such as walls. In that way a network of nodes and edges is created which represents all the possible routes that humans can employ in a certain environment. The idea of a route skeleton corresponds to certain degree to the central-point wayfinding strategy that people usually employ in unfamiliar environments.

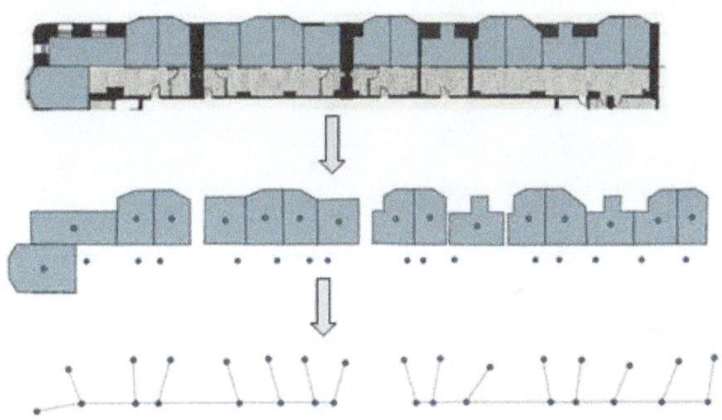

*Figure 5: Graph model of part of a building*

To define the location of the landmark-signs, the triangles of the CDT are investigated. Two types of generated triangular facets can be distinguished: 1) those that are built from edges that consist part of the user-specified constraints and 2) others that are free of these constraints. The last type of triangles are adjacent to other generated triangles from all three sides. Thus, humans standing at these sub-spaces are allowed to walk to three possible directions - there are no constraints to limit them to any direction. This type of generated sub-spaces corresponds in reality to areas that two or more navigable connecting spaces intersect. Consequently, in these sub-spaces decision points are located. By classifying the triangles the decision points are determined. The distinction between the two triangle types is performed based on their topological overlay relations with the originating polygon. Therefore, the topological spatial relations between the generated sub-spaces and the originating polygon are examined by using the Dimensionally Extended Nine-Intersections Model (DE-9IM) (Strobl, 2008). If the result of the intersection of the boundaries of the two geometries is a point set then the triangle contains zero constrained edges. Otherwise if it results in a line set the triangle has at least one constrained edge. In that way the triangles without constrained edges are selected.

Concluding, the steps to be followed in designing Indoor Junction system are as follows:

1. Define destinations spaces (rooms) and transition spaces (doors) of interest.
2. Determine connection spaces, i.e. the paths where movement occurs - main corridors of buildings or well-structured paths.
3. Define decision points, i.e. locations where two or more connecting spaces intersect or locations where gaps in the boundary of the connecting spaces occur.
4. Place of a landmark-sign at the decision points.
5. Provide location indication and directional information through the numbered landmark-signs.
6. Place maps of the building at the entrance and floor change points for getting overview.
7. Construct network by connecting the decision points using the triangulation analysis approach.
8. Decide on numbering approach for large multi-storey buildings, e.g. include the number of the floor in the numbering.

## 4. IMPLEMENTATION

The proposed concept was developed and tested for two buildings. This paper presents the results of the building of the Faculty of Architecture and the Built Environment, Delft University of Technology (Figure 6). The building is quite complex, contains large open spaces and elaborated existing wayfinding system. The tests were facilitated by the large number of users (students), which had the opportunity to compare the proposed with the existing wayfinding system.

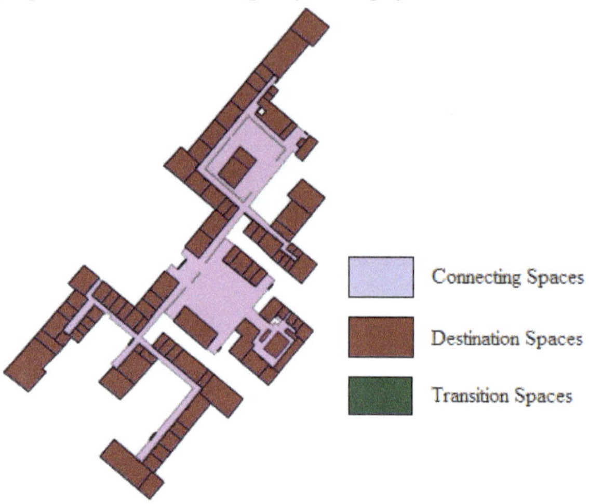


*Figure 6: Identification of all spaces*

The spaces of interest where identified and marked on the map of the ground floor of the building. As visible on Figure 6, the ground floor contains two large connecting spaces, which apparently require several landmark-sings.

To establish the network and the locations of the landmark-signs, the indoor environment is processed in two steps. Firstly, in the more detailed level a CDT is performed on the connecting spaces. This method subdivides the floor plan into a number of triangular-shaped non-overlapping facets, while it retains the boundary information (Figure 7). Constrained segments of the triangulation can be perceived as entities blocking the movement in the space, e.g. walls, tables, etc. Secondly, analysis of the triangulated space is performed to establish decision points and nodes for the network for human movement.

As described above, the locations of the landmark-signs is done by estimating the triangles and adjacency between them. The centroids of the triangles without constrained edges are calculated as they are the most representative points in order to act as the precise locations of decision points.

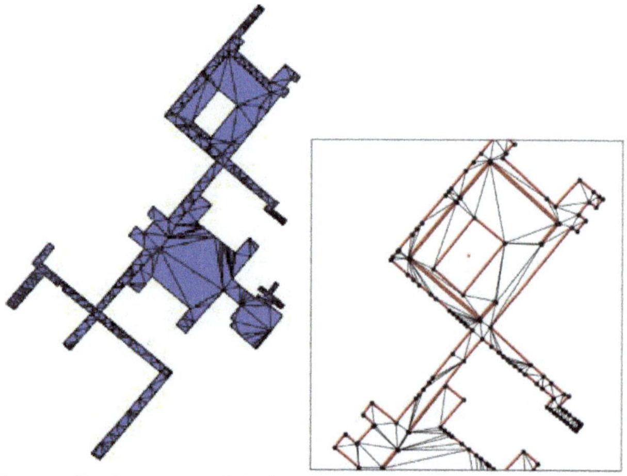

*Figure 7: Constrained Delaunay Triangulation: ground floor (left) and fragment of the floor (right)*

However, in some cases, some of these possible locations are very close to each other, as it is illustrated in (Figure 8). This creates redundancy of landmark-sign locations. The problem arising from this situation is that: 1) two or more signs can be installed very close to each other or even partially overlap depending on the sizes of the selected signs and 2) this entails the risk of creating confusion or misinterpretation of the assistance. Moreover, from a practical point of view, the presence of a higher number of landmark-signs requires more effort and maybe higher costs depending on the materials used for the fabrication of the landmark-signs.

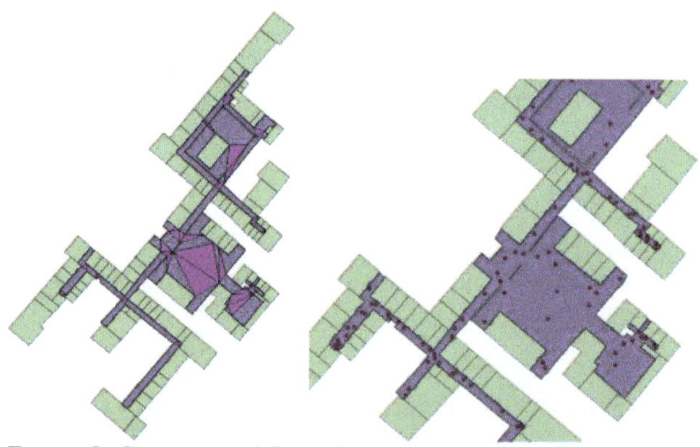

*Figure 8: Spaces candidates for landmark-sign placement (left) and the final locations (right)*

Another detected issue was that the locations of the landmark-signs can sometimes be quite before or even after the real location where a direction change is imposed. The cause of this circumstance is that the CDT generates large triangles with irregular shapes and small angles. In order to solve this problem, the Steiner points were inserted in order to split the constrained segments and create a more symmetrical and more regular space subdivision result (Figure 9). The Conforming Delaunay Triangulation is then used, while further area and angle constraints can be imposed to generate an even better outcome containing triangles as equilateral as possible.

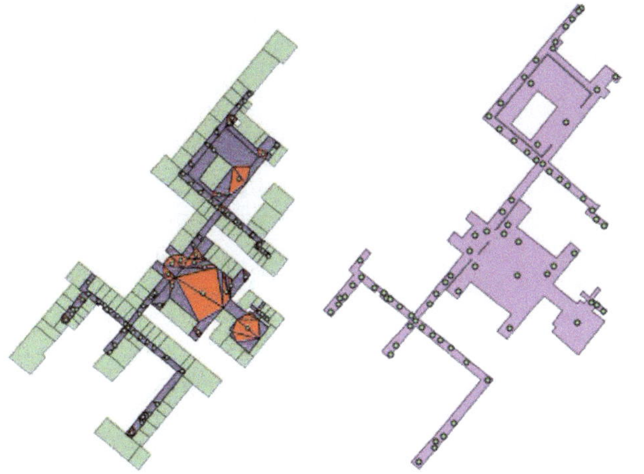

*Figure 9: Final spaces (left) and locations (right) of landmark-signs.*

For the generation of the network for the entire floor, the central points of polygons representing destination spaces and transition spaces are calculated and linked to the closest decision point node. In this way, all the possible routes of the building are represented and distances between the nodes can be estimated as the graph reflects not only the topological relationships between spaces but also the geometry of the building. The fact that at least three possible directions occur at the decision points is certified by the degree of the nodes. The decision points have a degree of three or more as at least three edges are incident to these vertices. All other nodes of the graph representing connecting spaces have only two links. The same is for most of the nodes representing destination spaces.

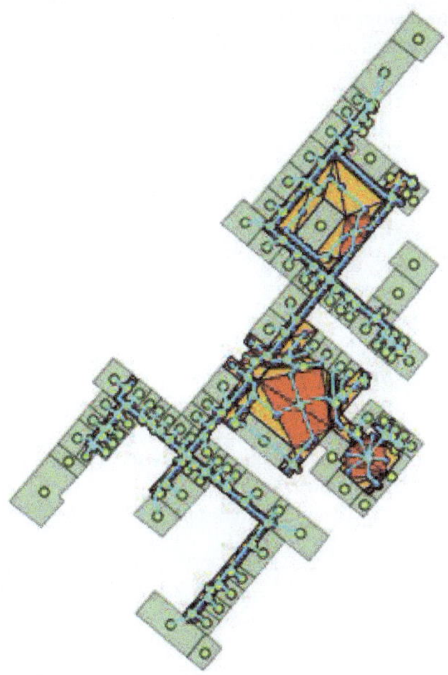

*Figure 10: Final network of ground floor to be used in the Indoor Junction System*

The last step, after determined all the decision points and their links in the network, is numbering the points and fabricating the landmark-signs. The number of possible directions of moving and consequently, the amount of directional arrows in the landmark-sign is derived by the degree of each one of the decision points. For the test building, the maximum degree of node encountered is four. Therefore, only landmarks-signs with three and four indications of the next numbered node were used.

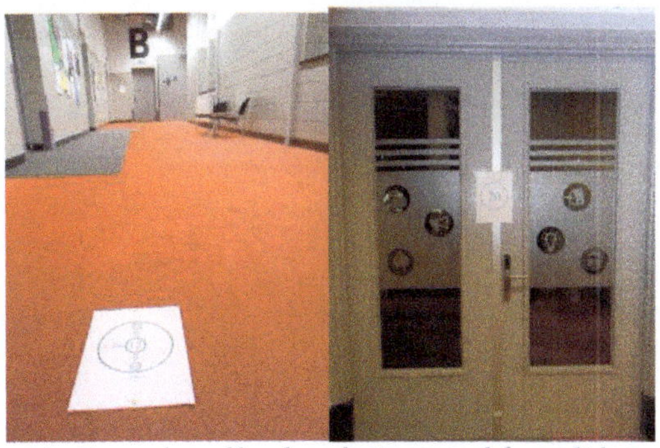

*Figure 11: Examples of landmark-signs used for validation*

## 5. DISCUSSION

This paper presented a new approach for indoor navigation based on a landmark-sign system, which adopts the principles of a well-known outdoor system. The network and landmark-sign locations are automatically defined. Specific landmark-signs are proposed to indicate the decision points.

An important advantage of this system is that it provides a global solution for facilitating the wayfinding process in indoor settings. Another benefit is that the same system exists for outdoor navigation purposes, making it easier to gain people's recognition and acceptance and creating the potential of combining the two systems for providing assistance even in the case that a transition from outdoor to indoor environment is involved. Additionally, by enumerating the signposts the system can be combined not only to the outdoor Junction Network System but also with other outdoor systems based on numbers, e.g. a system numbering the individual buildings of a complex of buildings such as a university campus. Then the number of the building can be linked to the numbers of the decision points, especially to those at the entrances and exits as this is the transition area between indoor and outdoor environment.

The approach has been tested with several groups of students (familiar and not familiar with the building). They were asked to follow different routes. It was observed that the users got quicker to their destination using the proposed system, compared to the existing wayfinding system (Figure 1, right). The satisfaction with the system was quite high, the landmarks were perceived as understandable and appropriate. The feedback received after the test of the system at the Faculty of Architecture and the Built Environment reveals high peoples acceptance

The promising result of the human-based survey shows the potential of the system to act as a solution for the wayfinding indoors. People are able to follow it and that it is simple in its perception and usability. However, further testing in different building cases is definitely necessary in order to accept the system as a reliable solution.

There are also several aspects that need further attention. As mentioned above, in some cases the decision points do not coincide with the actual location where the choice of direction is imposed, so that the human should modify the automatically generated points according to the common sense during the installation phase. The main source of this error is identified at the manual work phase, where the boundaries of the polygons were digitized. Test with more buildings should further investigated how to improve the analysis of the triangulated space.

The way of numbering the landmark-signs should be carefully decided. The numbers should follow a logical sequence, so that people to be able to remember and follow them easier, and to estimate their relative location at the building and the travelled distance. A uniform numbering system that reflects the structure of all indoor environments requires thorough investigation of the diversity of indoor space configurations.

It should be taken into consideration that in very crowded environment, detecting the landmark-signs (on the floor) from a distance might become problematic. Crowded environments were not examined. The validation test took place when the building was relatively empty. However, this problem might be overcome by the fact that the density of existing landmark-signs creates a security that humans follow the right route until they come to the next in the order landmark-sign. Nevertheless, the investigation of other locations of landmark-signs such as free-standing or ceiling-hung should be investigated as well.

Finally, one of the important conclusions of the validation tests is that some people need to be aware of the meaning of the numbers and to have an overview of the numbering system for the whole environment. This was not investigated in the context of this experiment as the focus was on testing people's ability to perceive the principles of the system. Further studies should concentrate on designing appropriate landmark-sings and a numbering system.

## 6. REFERENCES


Brunner-Friedrich, B., & Radoczky, V. (2006). Active landmarks in indoor environments. In Visual Information and Information Systems, Springer, Berlin Heidelberg, pp. 203-215.

Caduff, D., & Timpf, S. (2005). The Landmark Spider: Representing Landmark Knowledge for Wayfinding Tasks. In AAAI Spring Symposium: Reasoning with Mental and External Diagrams: Computational Modeling and Spatial Assistance, pp. 30-35.

Elias, B. (2003). Extracting landmarks with data mining methods. In Spatial Information Theory. Foundations of Geographic Information Science, Springer, Berlin Heidelberg, pp. 375-389

Fontaine, S., & Denis, M. (1999). The production of route instructions in underground and urban environments. In Spatial Information Theory. Cognitive and Computational Foundations of Geographic Information Science, Springer, Berlin Heidelberg, pp. 83-94.

Hölscher, C., Meilinger, T., Vrachliotis, G., Brösamle, M., & Knauff, M. (2006). Up the down staircase: Wayfinding strategies in multi-level buildings. Journal of Environmental Psychology, 26(4), pp.284-299.

Hölscher, C., Büchner, S. J., Brösamle, M., Meilinger, T., & Strube, G. (2007). Signs and maps–cognitive economy in the use of external aids for indoor navigation. In Proceedings of the 29th annual cognitive science society, Austin, TX: Cognitive Science Society, pp. 377-382.

Hund, A. M., Padgitt, A.J. (210). Direction giving and following in the service of wayfinding in a complex indoor environment. Journal of Environmental Psychology, 30(4), pp. 553-564.

Klippel, A., Dewey, C., Knauff, M., Richter, K. F., Montello, D. R., Freksa, C., & Loeliger, E. A. (2004). Direction concepts in wayfinding assistance systems. In Workshop on Artificial Intelligence in Mobile Systems, pp. 1-8.

Kruminaite, M. & Zlatanova, S. (2014) Indoor Space Subdivision for Indoor Navigation, ISA'14, Proceedings of the Six ACM SIGSPATIAL International Workshop on Indoor Spatial Awareness, pp. 25–31.

Lynch, K. (1960). The image of the city (Vol. 11). MIT press.

May, A. J., Ross, T., Bayer, S. H., & Tarkiainen, M. J. (2003). Pedestrian navigation aids: information requirements and design implications. Personal and Ubiquitous Computing, 7(6), pp. 331-338.

McKnight, C., Dillon, A. and Richardson, J. (1993). Space -- the final chapter: Or why physical representations are not semantic intentions

Michon, P. E., & Denis, M. (2001). When and why are visual landmarks used in giving directions?. In Spatial information theory, Springer, Berlin Heidelberg, pp. 292-300.

Millonig, A., & Schechtner, K. (2007). Developing landmark-based pedestrian-navigation systems. Intelligent Transportation Systems, IEEE Transactions on,8(1), 43-49.

Montello, D. R., & Sas, C. (2006). Human factors of wayfinding in navigation.

Mortari, F., Zlatanova, S., Liu, L., & Clementini, E. (2014) Improved Geometric Network Model (IGNM): A novel approach for deriving connectivity graphs for indoor navigation. ISPRS Annals of Photogrammetry, Remote Sensing and Spatial Information Sciences,4, pp. 45-51.

O'Neill, M. J. (1991). Effects of signage and floor plan configuration on wayfinding accuracy. Environment and Behavior, 23(5), 553-574.
Presson, C. C., & Montello, D. R. (1988). Points of reference in spatial cognition: Stalking the elusive landmark*. British Journal of Developmental Psychology, 6(4), 378-381.

Raubal, M., & Egenhofer, M. J. (1998). Comparing the complexity of wayfinding tasks in built environments. Environment and planning B, 25, pp. 895-914.

Raubal, M., & Winter, S. (2002). Enriching wayfinding instructions with local landmarks, Springer, Berlin Heidelberg, pp. 243-259.

Richter, K. F., & Klippel, A. (2005). A model for context-specific route directions. In Spatial Cognition IV. Reasoning, Action, Interaction, Springer, Berlin Heidelberg, pp. 58-78.

Richter, K. F., Tomko, M., & Winter, S. (2008). A dialog-driven process of generating route directions. Computers, Environment and Urban Systems,32(3), pp. 233-245.

Richter, K. F. (2013). Prospects and Challenges of Landmarks in Navigation Services. In Cognitive and Linguistic Aspects of Geographic Space, Springer, Berlin Heidelberg, pp. 83-97.

Ruso, D., Zlatanova S. & Clementini E. (2014) Route directions generation using visible Landmarks, ISA'14, Proceedings of the Six ACM SIGSPATIAL International Workshop on Indoor Spatial Awareness, pp. 1–8.

Snowdon, C., & Kray, C. (2009). Exploring the use of landmarks for mobile navigation support in natural environments. In Proceedings of the 11th International Conference on Human-Computer Interaction with Mobile Devices and Services, ACM, p. 13.

Strobl, C. (2008). Dimensionally Extended Nine-Intersection Model (DE-9IM). Encyclopedia of GIS, Springer, Berlin Heidelberg, pp. 240-245.

Weisman, J. (1981). Evaluating Architectural Legibility Way-Finding in the Built Environment. Environment and behaviour, 13(2), pp. 189-204.

Zlatanova, S., Liu L & Sithole G. (2013) A Conceptual Framework of Space Subdivision for Indoor Navigation. ISA '13 Proceedings of the Fifth ACM SIGSPATIAL International Workshop on Indoor Spatial Awareness, ACM New York, NY, USA. pp. 44-48.

# CLASSIFICATION OF FEATURES OF IMAGE SEQUENCES

E. V. Pugin[a] A. L. Zhiznyakov[a]

[a] Murom Institute (branch) Vladimir State University, Information Technologies Department, 602264 Orlovskaya str. 23, Murom, Russia - egor.pugin@gmail.com, lvovich@newmail.com

**Commission V, WG III/3**

**KEY WORDS:** image sequences, feature extraction, feature classification, semantic features, feature clarification

**ABSTRACT:**

Processing of image sequences is a very actual trend now. This is confirmed with a vast amount of researches in that area. The possibility of an image sequence processing and pattern recognition became available because of increased computer capabilities and better photo and video cameras. The feature extraction is one of the main steps during image processing and pattern recognition. This paper presents a novel classification of features of image sequences. The proposed classification has three groups: 1) features of a single image, 2) features of an image sequence, 3) semantic features of an observed scene. The first group includes features extracted from a single image. The second group consists of features of any kinds of image sequences. The third group contains semantic features. Reverse feature clarification method is the iterative method when on each iteration we use higher level features to extract lower level features more precisely. The proposed classification of features of image sequences solves a problem of decomposition of the source feature space into several groups. Reverse feature clarification method allows to increase the quality of image processing during iterative process.

## INTRODUCTION

Today we can see the growing popularity of methods of processing of image sequences. This is due to several factors. At first, the increasing capabilities of computer systems allow to process more images for the constant time. Secondly, more powerful sensors can provide several images at the same time. For example, during multispectral photography cameras give several image of a single place in different frequency spectrums. Also we can consider here video recording instead of simple photography. Thirdly, complex methods and algorithms of image processing evolve. Some of them requires creating or generating new images based on current image during their work. This is done for extracting more image features from a single image. For example, during wavelet analysis three additional images are generated. Forthly, in the most cases an image sequence contain more information than a single image. Therefore, processing of the whole sequence should give us more precise results. A lot of publications in this area only confirms its actuality.

Feature extraction is one of the main steps during image processing and pattern recognition. It can also be performed during processing of image sequences. While single image features and ways of their extraction are well-known, features of image sequences is not studied well enough. With the growth of the number of features there is a problem of their ordering. This paper presents a general mathematical model of feature extraction from digital image sequences and a novel classification of features of image sequences. The reverse feature clarification method is described. It allows to increase results of image processing and pattern recognition during an iterative process.

In chapter 1 we show the evolution of methods of image sequences processing. Chapter 2 briefly describes the main kinds of image sequences. The general mathematical models of image sequences are given in chapter 3. The classification of image sequences features and the reverse feature clarification method are described in chapter 4.

## 1. THE EVOLUTION OF IMAGE SEQUENCES PROCESSING

The first references of image sequences appeared almost at the same time after the digital image processing was born in the 1950-1960s (Pratt, 1974, Meyer-Eppler and Darius, 1956, Anuta, 1969, Barnea and Silverman, 1972). In that time there was no term of image sequences, because computers were very limited and the scientists were focused on processing of a single image. With the growth of capabilities of computers and cameras the number of publications related to image sequences increased.

In (Pratt, 1974, Meyer-Eppler and Darius, 1956, Anuta, 1969) a problem of comparing shifted objects on two images was described. Authors used a correlation link between objects. The images had the same object came from different sensors or from single sensor but with time delay. In (Barnea and Silverman, 1972) authors used the sequential approach instead of correlational. The optimization of the given methods were described. The correlational approach uses the finding of a maximum of a correlation function. The sequential approach uses the iteration over the image pixels and their comparison with the neighborhood of the first object.

In (Jain and Jain, 1981, Jain, 1981, Jane et al., 1995) the methods that uses motion in spatial area for segmentation were given. Authors used the approaches of building of image differences. Mathematical model of images of a sequence looks like $f(x, y, t_i)$, where we can see the third argument — the time of the frame. This description is suitable for describing video sequences. Authors of (Rajala et al., 1983, Shariat and Price, 1990, Cumani et al., 1991, Dellen and Wrgtter, 2011) also studied the problems of using motion during segmentation, but in spectral area using the Fourier transform. Moreover the motion estimation is described in (Shapiro and Stockman, 2001, Sonka et al., 1999).

At the end of 1980s — beginning of 1990s the first publications related to wavelet analysis were published (Mallat, 1987, Meyer, 1992, Daubechies, 1992). Many image processing (Castleman,

1996) and compression (Antonini et al., 1992) methods were based on wavelets.

## 2. TYPES OF IMAGE SEQUENCES

There are a lot of kinds of image sequences. The main are video sequences and image sequences created during multiscale transforms (Zhiznyakov, 2007). They are image pyramids and images created during wavelet analysis. Besides there are some more types of image sequences. They are processed with the same methods as used with main kinds of image sequences. The other types are:

1. Images from multispectral photography.
2. Color images. During processing of color images the sources pixels are intensity values of the each color.
3. Image sequences generated with some processing operators (filters etc.).

Detailed review of different types of image sequences can be found in (Pugin, 2014).

## 3. MATHEMATICAL MODELS OF IMAGE SEQUENCES

There are a lot of mathematical models of image sequences today. Because the number of problems in this area is great, authors often choose well-suited models for their purposes or simply do a verbal description of the image sequence model. In this case they assume that image sequence is a set of two or more images related to each other by some criteria, operator or semantics. For example, during video sequences processing the next formula is used (Shapiro and Stockman, 2001):

$$\mathbf{F} = f(x, y, t_i), \tag{1}$$

where $\mathbf{F}$ — the image sequence, $f(x, y, t_i)$ — the frame in time $t_i$ $(i = 1, 2, ..., n)$.

Color image in RGB format can be described as

$$\mathbf{F} = \{f_R, f_G, f_B\}. \tag{2}$$

The sequence of generated images during processing can be presented as

$$\mathbf{F} = \{T_1[f], T_2[f], ..., T_n[f]\}, \tag{3}$$

where $T_i[\bullet]$ — some operator $(i = (1, 2, ..., n))$.

Another approach of mathematical model of image sequences is given in (Zhiznyakov, 2008). An image sequence $\{f_n\}$ is an ordered set of images $\mathbf{f}$. Every image has an index

$$\{f_n\} = \{f_0, f_1, ..., f_n\}. \tag{4}$$

The expression (4) is well suited for many kinds of image sequences, so we will be using it in the next chapters.

Each image $f_i(x, y)$ of the sequence has its own set of features $\mathbf{S}_i$. So, we can build the next sequence

$$\{\mathbf{S}_n\} = \{\mathbf{S}_0, \mathbf{S}_1, ..., \mathbf{S}_n\}. \tag{5}$$

## 4. CLASSIFICATION OF FEATURES OF IMAGE SEQUENCES

The proposed classification has three groups:

1. "classic" image features or features of a single image,
2. features of an image sequence,
3. semantic features of an observed scene.

The first group includes features extracted from a single image $f(x, y)$. Let $O_i[\cdot]$ be an operator of feature extraction. And the feature extraction process can be described as

$$\mathbf{X_1} = O_1[f(x, y)], \tag{6}$$

where $\mathbf{X_1}$ — the set of features of the image of the first group. This group includes statistical, fractal (Zhiznyakov et al., 2014), geometric, topological, color, spectral features.

The second group consists of features of any kinds of image sequences. The feature extraction can be done as

$$\mathbf{X_2} = O_2[\mathbf{f_n}(x, y)] \tag{7}$$

or

$$\mathbf{X_2'} = O_2'[\{\mathbf{X_1}\}, \mathbf{X_2}], \tag{8}$$

where $\mathbf{X_2}$, $\mathbf{X_2'}$ — extracted features, $O_2[\cdot]$, $O_2'[\cdot]$ — the extraction operators, $\mathbf{f_n}(x, y)$ — the set of images or the image sequence, $\{\mathbf{X_1}\}$ — the set of first group features of the image sequence. The first form is used for extracting image features from the image sequence $\mathbf{f_n}(x, y)$. The extraction can use only several images of the sequence. The second form extracts features using first and second feature groups. Image sequence features are used in motion and change detection, motion analysis, object detection and tracking and in some other areas. Feature examples of video sequences are the focus of expansion, the focus of contraction, the motion field, speed of changing of features of the first group during the sequence.

The third group contains semantic features. These features are usually selected empirically and based on the features of previous groups. Such features are used for understanding of 3D-scenes and objects, their analysis, behavior estimation. The extraction of semantic features is

$$\mathbf{X_3} = O_3[\{\mathbf{X_1}\}, \mathbf{X_2}], \tag{9}$$

where $\mathbf{X_3}$ — a set of semantic features. Sometimes it is possible to extract semantic features from each other

$$\mathbf{X_3'} = O_{3'}[\mathbf{X_3}] \tag{10}$$

or from all groups of features

$$\mathbf{X_3'} = O_{3'}[\{\mathbf{X_1}\}, \mathbf{X_2}, \mathbf{X_3}]. \tag{11}$$

Examples of semantic features are speed of objects on the scene, movement vector in 3D-space, other scene characteristics.

The general algorithm of features extraction from an image sequence can be described as:

1. Extract features from every single image in sequence and add them to common multidimensional array.
2. Extract sequence features from the given array.

3. Extract semantic features from the features of two first steps.

To increase the quality of image processing and pattern recognition it is necessary to study the influence of proposed groups of features on each other. Obviously that the features of the lower groups directly affect the higher level feature. Many research papers aims to study such influence. But in the same time it is interesting to investigate the backward influence of higher level features on the lower level features. There is a lack of research papers in that area. This task can solve the following question: "If we know semantics of the scene, can we improve the quality of recognition of the separate objects on it?"

Let's call such technique as "reverse feature clarification method". It is the iterative method when on each iteration we use higher level features to extract lower level features more precisely. For example this method can be used in reconstruction of the location of the hidden object on early frames of video sequence in case if this object has predictable speed and stable forms (find a car in the car stream).

In general case we can present the reverse feature clarification method as

$$\mathbf{X}_{\mathbf{i}}^{\mathbf{k+1}} = Q[\mathbf{X}_{\mathbf{j}}^{\mathbf{k}}], \quad 1 \leq i < j \leq 3, \qquad (12)$$

where $\mathbf{X}_{\mathbf{i}}^{\mathbf{k}}$ — features of $i$ group of image sequence on iteration $k$; $i, j$ — numbers of features groups, $k$ — the iteration number, $Q[\bullet]$ — the operator of clarification.

## CONCLUSIONS

In this paper we did a review of evolution, types and mathematical models of image sequences. We presented a novel classification of image features extracted from image sequences. The proposed classification of features of image sequences solves a problem of decomposition of the source feature space into several well defined groups. The reverse feature clarification method is given. It is based on presented classification. Reverse feature clarification method allows us to increase the quality of image processing during an iterative process.

## REFERENCES


Antonini, M., Barlaud, M. and Mathieu, P., 1992. Image coding using wavelet transform. IEEE Trans. Image Processing 1(2), pp. 205–220.

Anuta, P. F., 1969. Digital registration of multispectral video imagery. Photo-Optical Instr. Engineers J. pp. 168–178.

Barnea, D. I. and Silverman, H. F., 1972. A class of algorithms for fast image registration. IEEE Trans. Computers pp. 179–186.

Castleman, K. R., 1996. Digital Image Processing. Prentice Hall.

Cumani, A., Guiducci, A. and Grattoni, P., 1991. Image description of dynamic scenes. Pattern Recognition 24(7), pp. 661–674.

Daubechies, I., 1992. Ten Lectures on Wavelets. Society for Industrial and Applied Mathematics.

Dellen, B. and Wrgtter, F., 2011. A local algorithm for the computation of image velocity via constructive interference of global fourier components. International Journal of Computer Vision 92, pp. 53–70.

Jain, J. R., 1981. Dynamic scene analysis using pixel-based processes. Computer 14(8), pp. 12–18.

Jain, J. R. and Jain, A. K., 1981. Displacement measurement and its application in interframe image coding. IEEE Trans. Comm. COM-29, pp. 1799–1808.

Jane, R., Kasturi, R. and Schunk, B., 1995. Computer Vision. McGraw-Hill.

Mallat, S., 1987. A compact multiresolution representation: The wavelet model. Proc. IEEE Computer Society Workshop on Computer Vision pp. 2–7.

Meyer-Eppler, W. and Darius, G., 1956. Two-dimensional photographic autocorrelation of pictures and alphabet letters. Proceedings 3rd London Symposium on Information Theory pp. 34–36.

Meyer, Y., 1992. Wavelets and applications. Proceedings of the International Conference, Marseille, France.

Pratt, W. K., 1974. Correlation techniques of image registration. IEEE Trans. Aerospace and Electronic Systems pp. 353–358.

Pugin, E. V., 2014. Review of methods and algorithms of processing of sequences of digital images. Algorithms, methods and systems of data processing (3), pp. 50–59.

Rajala, S. A., Riddle, A. N. and Snyder, W. E., 1983. Application of one-dimensional fourier transform for tracking moving objects in noisy environments. Computer vision, Image processing 21, pp. 280–293.

Shapiro, L. G. and Stockman, G., 2001. Computer Vision. Prentice Hall.

Shariat, H. and Price, K. E., 1990. Motion estimation with more than two frames. IEEE Trans. Pattern Anal. Machine Intell. 12(5), pp. 417–434.

Sonka, M., Hlavac, V. and Boyle, R., 1999. Image Processing, Analysis, and Machine Vision. PWS Publishing.

Zhiznyakov, A. L., 2007. Formation and analysis of feature sets of multiscale sequences of digital images. Program products and systems (4), pp. 24.

Zhiznyakov, A. L., 2008. Theoretical foundations of processing of multiscale image sequences. Vladimir State University.

Zhiznyakov, A., Privezentsev, D. and Pugin, E., 2014. Use of fractal signs of digital images for detection of surface defects. Institute of Electrical and Electronics Engineers Inc., pp. 391–392. Conference of 2014 24th International Crimean Conference Microwave and Telecommunication Technology, CriMiCo 2014.

# MUSIC-ELICITED EMOTION IDENTIFICATION USING OPTICAL FLOW ANALYSIS OF HUMAN FACE

V. V. Kniaz [a] *, Z. N. Smirnova [b]

[a] St. Res. Institute of Aviation Systems (GOSNIIAS), Moscow, Russia – vl.kniaz@gosniias.ru
[b] Gnessin Russian Academy of Music, Moscow, Russia – art.vivus@gmail.com

**Commission V, WG V/5**

**KEY WORDS:** optical flow, musical psychology, emotion identification, human face recognition

**ABSTRACT:**

Human emotion identification from image sequences is highly demanded nowadays. The range of possible applications can vary from an automatic smile shutter function of consumer grade digital cameras to Biofied Building technologies, which enables communication between building space and residents. The highly perceptual nature of human emotions leads to the complexity of their classification and identification. The main question arises from the subjective quality of emotional classification of events that elicit human emotions. A variety of methods for formal classification of emotions were developed in musical psychology. This work is focused on identification of human emotions evoked by musical pieces using human face tracking and optical flow analysis.

Facial feature tracking algorithm used for facial feature speed and position estimation is presented. Facial features were extracted from each image sequence using human face tracking with local binary patterns (LBP) features. Accurate relative speeds of facial features were estimated using optical flow analysis. Obtained relative positions and speeds were used as the output facial emotion vector. The algorithm was tested using original software and recorded image sequences. The proposed technique proves to give a robust identification of human emotions elicited by musical pieces. The estimated models could be used for human emotion identification from image sequences in such fields as emotion based musical background or mood dependent radio.

## 1. INTRODUCTION

Emotions are one of the most complex aspects of human behaviour. They deeply influence the way humans perceive the world in everyday life. The problem of human emotion monitoring, widely known as emotional intelligence (EI), is a rapidly developing field of modern psychology. Nowadays due to wide use of imaging devices and the rapid progress in image processing techniques vision based methods for emotion monitoring are becoming one of the main tools of EI. Moreover EI provides a basis for a range of brilliant new technologies based on human behaviour such as Biofied Building (Kita, 2015), which enables communication between building space and residents.

### 1.1 Emotions monitoring using facial expressions

The first systematic framework for emotional intelligence is commonly credited to Salovey and Mayer who define EI as 'the ability to monitor one's own and others feelings and emotions, to discriminate among them and to use this information to guide one's thinking and actions' (Salovey, 1990a). The authors outline two main kinds of expression of emotion: verbal and nonverbal. Nonverbal expression of emotion, being more instinctive and deeply rooted in human feelings, provides the way to express human mood and state of mind via facial expressions and gestures.

Though more complex in interpretation nonverbal expression of emotions could often provide more accurate information about human mood and state of mind, especially in case when verbal expression is unavailable (e.g. children nonverbal communication, cinema viewers, etc.). Being one of the most prominent indicators of emotions, human's face have attracted interests of many researchers for more than a century since Darwin's now classic study of facial expression (Darwin, 1872). However due to a highly dynamical nature of facial expressions their stable recording and monitoring could be done only by means of video recording system.

### 1.2 Computer vision based human emotion identification

Though first attempts to record human emotions by means of photography could be found as early as 1975 (Boucher, 1975), the rapid development of facial expression monitoring had begun only when consumer grade digital cameras became available in 1990s. The development MPEG-4 Face and Body Animation (FBA) International Standard (ISO/IEC 14496-1) in 1999 was a major keystone in the field of facial emotion modeling and classification. This standard defines 84 facial feature points (FP) on neutral face, that provide spatial references to general elements of the human's face (eyes, nose, lips, etc.) In addition MPEG-4 FBA defines Face Animation Parameters (FAP) including visemes, expressions, head rotation and low-level parameters to move all parts of the face. They could be used to either represent recorded facial expressions or to create facial expressions using deformable 3D-model of a face. The facial animation parameters are based on the study of minimal perceptible actions (MPA) and are closely related to muscle actions (Pandzic, 2002).

The problem of emotion identification using facial features and computer vision could be decomposed into three main steps:

* Corresponding author

1. Face detection and tracking
2. Estimation of the facial animation parameters – relative position and speeds of facial feature points in spatial space
3. Transformation from the facial animation parameters-space to emotion-space using emotion model

Usually the third step presents most of the problems due to dependence on the formal model of human emotions. Human emotions are highly subjective and nonlinear in nature. It is quite hard to describe them even using informal language. Thus the creation of an accurate model representing the transformation from facial animation parameters-space to emotion-space is an ill-defined problem. The design of such model requires the selection of a representative emotion-space and high-quality training data.

The base facial feature point set called Facial Action Coding System (FACS) proposed by Ekman et. al in (Ekman, 1978) has given a rise to series of publications on human emotion monitoring. The general approach is to collect training data by recording facial expressions a volunteer. Some specially selected images, video or dialogues are used to elicit a known set of emotions during the recording session. Esau et. al proposed to use a fuzzy emotion model to recognize facial expression based represented in FACS (Esau, 19. The training data was obtained by telling volunteers to express a single emotion. The proposed model showed 87% accuracy in recognition of 3D emotion space with axes representing anger, happiness and surprise.

Speech Analysis and Interpretation Laboratory (SAIL) developed an interactive emotional dyadic motion capture database (IEMOCAP) with twelve hours of training data. This database was recorded from ten actors in dyadic sessions with markers representing MPEG-4 facial feature points, which provide detailed information about their facial expression during scripted and spontaneous spoken communication scenarios (Busso, 2008). The actors played specially selected scripts to elicit specific types of emotions (happiness, anger, sadness, frustration and neutral state).

### 1.3 Music-elicited emotion identification

The use of scripted scenarios requires subjective labeling of emotions for model estimation. Moreover verbal communication may lead to rapid changes in emotions, which are quite hard to describe using simple models. This difficulty could be overcome by using music to elicit the desired emotions. Music being smooth and continuous in nature is composed to be emotionally expressive. Thus specially selected musical pieces could be used to evoke a broad range of continuously changing emotions.

In this paper we present the results of music-elicited emotion identification using optical-flow analysis of human face. A set of specially selected songs from '1000 Songs for Emotional Analysis of Music' dataset (Soleymani, 2013) and custom selected classical music pieces were used to record training data. We propose a modified face tracker based on LBP features for initial face position estimation and optical flow for accurate sub-pixel tracking of face center.

The same method is used for accurate tracking of facial feature points and estimation of their facial animation parameters. Both linear and nonlinear autoregressive exogenous models (ARX) were used to relate the extracted facial animation parameters with the music labeled in emotion-space. The estimated models were used to reconstruct elicited emotions in evaluation data. The second part of this paper presents the music-elicited emotion model. In the third part data collection method is presented. Part four refers to modified LBP-optical flow face tracker. The fifth part presents the results of model estimation.

## 2. MUSICAL-ELICTED EMOTION MODEL

A number of state-space emotion representations were developed in musical psychology. A good balance between the flexibility and simplicity of the model provides valence-arousal emotion representation.

### 2.1 Valence-arousal emotion-space

One of the most well known emotion-space is so called 'Circumplex model' or valence-arousal (V-A) emotion representation proposed in (Russel, 1980). Being easy to use for both researcher and volunteer valence-arousal model represent all emotions in two-dimensional space. The first axes called valence represent the human's attitude to some object that could be positive or negative. The second axes called arousal indicates the strength of the emotion. Typical emotion-space vectors positions for affective words are shown in figure 1.

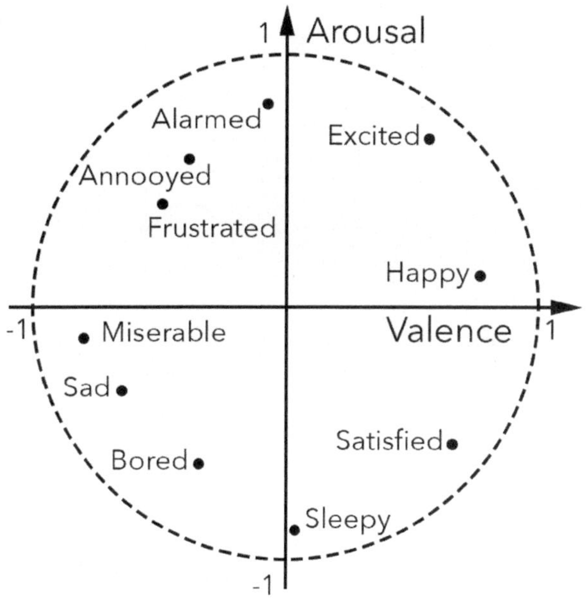


Figure 1. Valence-arousal emotion-space

### 2.2 The facial animation parameters-space

A number of features from original facial feature points set presented in MPEG-4 FBA were used to extract emotional expressions from images of human face. The selected feature points were used to estimate facial animation parameters that represent facial expressions (table 1).

An original two-dimensional state space was used to represent facial expressions:

$$\boldsymbol{Y}(t) = \begin{bmatrix} v_h \\ v_l \end{bmatrix} \quad (1)$$

$$v_h = |\varphi| + |\psi|$$

$$v_l = |v_{24}| + |v_{25}|$$

Here $v_h$ – modulus of sum of head yaw $\psi$ and roll $\varphi$ speeds, $v_l$ – modulus of sum of lip corners speeds $v_{24}, v_{25}$. Allowable yaw and roll angles for training data are restricted to ±30°. Therefore corresponding yaw and roll speeds can be calculated by dividing pixel speeds of feature point #9.12 (nose bone) by visible head radius. The configuration of the used facial feature points is presented on figure 2.

| FAP # | FAP name | FP # | FP name |
|---|---|---|---|
| 49 | Head yaw | 9.12 | Middle lower edge of nose bone |
| 50 | Head roll | 9.12 | Middle lower edge of nose bone |
| 6 | Horizontal displacement of left lip corner | 2.4 | Left corner of inner lip contour |
| 7 | Horizontal displacement of left lip corner | 2.5 | Right corner of inner lip contour |

Table 1. Facial animation parameters and corresponding facial feature points used for representations of facial expression

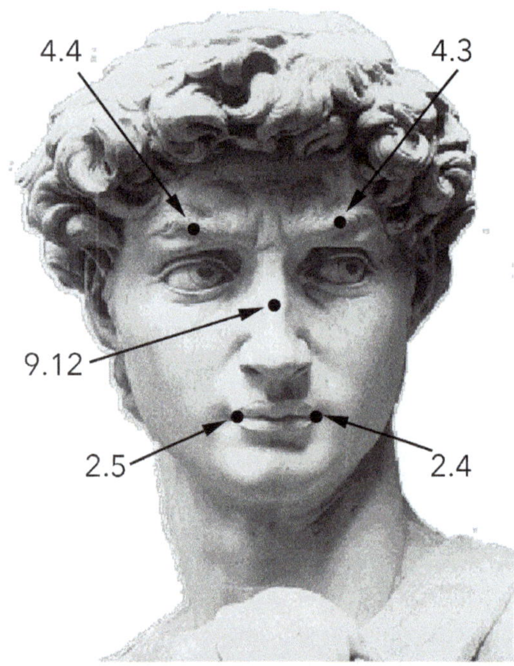


Figure 2. Configuration of the used facial feature points

### 2.3 Musical-elicited facial expression dynamical model

As the valence-arousal emotion representation can provide a smooth and continuous representation of emotional flow a linear time invariant (LTI) model of facial expressions elicited by emotions could be used:

$$\dot{\boldsymbol{X}}(t) = \boldsymbol{AX}(t) + \boldsymbol{BU}(t) \qquad (2)$$
$$\boldsymbol{Y}(t) = \boldsymbol{CX}(t) + \boldsymbol{DU}(t)$$

Here output vector $\boldsymbol{Y(t)}$ is given by equation (1), input vector $\boldsymbol{U(t)}$ is composed of valence value $v$ and arousal value $a$:

$$\boldsymbol{U}(t) = \begin{bmatrix} a \\ v \end{bmatrix} \qquad (3)$$

However due to highly nonlinear nature of human emotions only rough similarity of the LTI model output with the real emotions could be expected. To estimate more accurate emotion model a nonlinear autoregressive exogenous models (NARX) with wavenet network was used. The original model structure proposed in (Zhang, 1997) was used.

## 3. DATA COLLECTION METHOD

Training and evaluation data was recorded from ten volunteers listening to musical pieces. A number of specially selected musical pieces with various emotional features were used. To record image sequences and label the musical pieces using V-A emotion-space an original emotion capture software was developed.

### 3.1 Emotion capture software

To benefit from distributed data collection from many volunteers a mobile platform was selected for software development. The following functions were implemented in the software:

- Video recording at 30FPS synchronized with the music playback
- Valence and arousal recording using user interface (dynamic slider) at 30FPS synchronized with the music playback
- Database keeping

The user interface of 'EmotionGrabber' software is presented on figure 3.

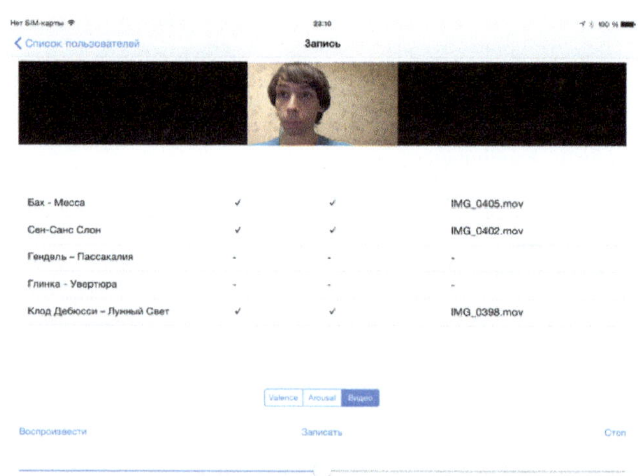


Figure 3. 'EmotionGrabber' software user interface

### 3.2 Musical pieces selection

Music is composed to elicit a variety of emotions and often the composer state the proposed emotion in the name of the musical piece. A number of classical music pieces were carefully selected to provide a variety of emotions. The most prominent emotional fragment of each musical piece with the length of 45s was extracted and uploaded to 'EmotionGrabber' software. The list of musical pieces, their emotional labels, and averaged valence and arousal labeled by volunteers are presented in table 2.

### 3.3 Data recording workflow

Each volunteer has listened to musical pieces three times. During the first listening session the video recording was done. During the second and the thirds session arousal and valence labelling was performed. To avoid loss of concentration on the music a five-minute pause was done between after every third recording session. First fifteen seconds of recording were discarded, as they are distorted due to initial concentration on 'EmotionGrabber' software user interface.

| # | Musical piece | Composer | Emotion Label | V | A |
|---|---|---|---|---|---|
| 1 | Passacaille (No.6), Suite in G minor, HWV 432 | Handel G. F. | Sad | -0.8 | -0.3 |
| 2 | Mass in B minor, BWV 232 | Bach J. S. | Miserable | -0.9 | -0.1 |
| 3 | Lacrimosa (Requiem) | Mozart W.A. | Sad | -0.8 | -0.2 |
| 4 | Mephisto Waltz No. 1, S.514 | Liszt F. | Alarmed | -0.1 | 0.8 |
| 5 | Overture from opera 'Ruslan and Lyudmila' | Glinka M. I. | Excited | 0.6 | 0.6 |
| 6 | The Poem of Ecstasy, Op. 54 | Scriabin A. N. | Aroused | 0.1 | 0.9 |
| 7 | Isle of the Dead, Op. 29 | Rachmaninoff S. V. | Tense | -0.1 | 0.8 |
| 8 | The Carnival of the Animals, Part V 'The elephant' | Saint-Saëns C. | Happy | 0.8 | 0.2 |
| 9 | Clair de Lune | Debussy C. | Serene | 0.3 | -0.5 |
| 10 | Danse sacrale from ballet 'The Rite of Spring' | Stravinsky I. F. | Afraid | -0.3 | 0.7 |

Table 2. List of musical pieces

## 4. DATA PROCESSING

To process the recorded data a modified LPB-optical flow face tracker algorithm was proposed. The algorithm was implemented in original software for off-line processing of the recorded data.

### 4.1 Modified LBP-optical flow face tracker

LBP feature trackers are commonly used to track human face in image sequences. However the accuracy of such tracker is usually not good enough to track subtle sub-pixel motions. The LBP-tracker also tends to have random false face detections, which should be detected and filtered.

An optical flow based tracking of human face could be used to detect sub-pixel motion of the face. However due to momentary errors in optical flow calculation such trackers usually have an accumulated error. A modified LBP-optical flow face tracker is proposed to overcome drawbacks of both trackers. The LBP-optical flow trackers uses LPB features to detect the face position on the first frame. After that an optical flow is calculated for the head region for each consequent frame. Non-zero regions of optical flow are averaged across the face region to obtain accurate sub-pixel movement of the frame. The calculated movement is compared with the difference LBP-tracker estimations of head position. If the difference is smaller than maximum allowable head movement between two frames, the optical flow field estimated movement is fused with the movement estimated by LBP-tracker. The fusion between two motion estimates is done using given ratio *k*. The algorithm of the proposed tracker is summarized in figure 4. The algorithm was implemented using optical flow estimation method proposed in (Farnebäck, 2003).

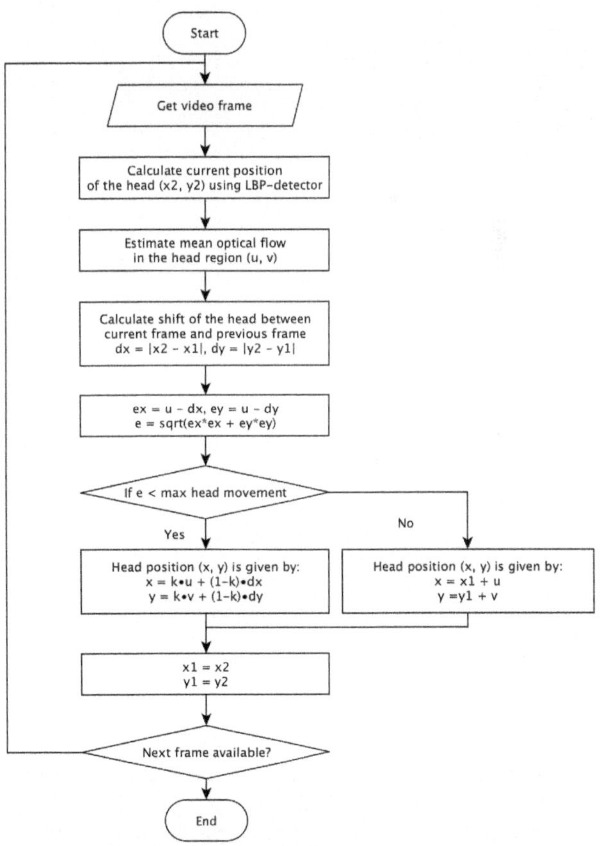


Figure 4. LBP-optical flow face tracker algorithm

### 4.2 Facial feature points trackers

A general structure of facial feature points' locations was taken from MPEG-4 FBA standard. To correct the locations of abstract facial feature points to their locations on particular face an LBP-classifier for each facial feature point was generated. Real locations of the feature points on the first frame were found by search in the local windows centered at general feature point location. After that modified LBP-optical flow tracker was used to track position of each feature relative to the head's position on consequent frames.

The process of facial feature points tracking is presented on figure 5.a. Position of feature point #9.12 (nose bone) estimated using basic optical flow tracker is marked with an ellipse. Due to accumulated error in optical flow estimation the tracked feature position has moved away from the actual position (up and toward left eye). Position of feature point #9.12 estimated using LBP-optical flow tracking is marked with a rectangle. Static error of the feature point position estimated using the proposed tracker doesn't exceed 1 pixel. Positions of feature points #2.5, 2.4 (lip corners) are marked with black circles.

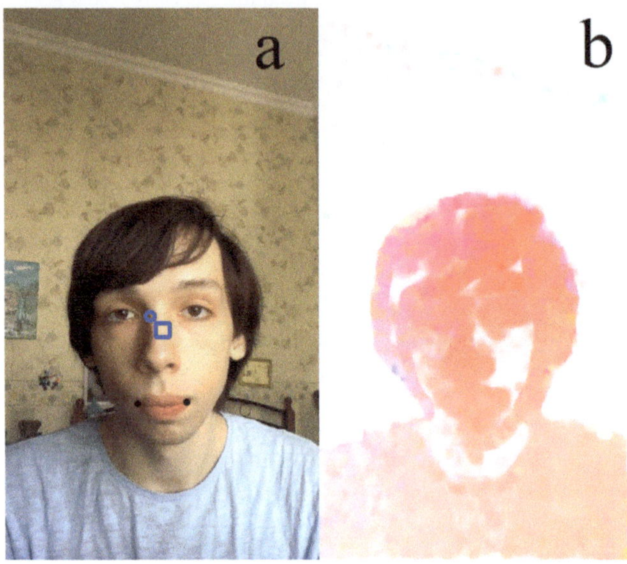


Figure 5. Facial expression extraction using LBP-optical flow tracker. (a) Extracted facial feature points. (b) Corresponding optical flow in Middlebury color-coding.

## 5. MUSIC-ELICITED EMOTION MODEL ESTIMATION

Two kinds of models were used to represent the connection between labelled emotions and facial expressions LTI models and NARX models.

### 5.1 LTI model estimation

A state-space model of the system given by (2) was identified from the recorded data using Matlab System Identification Toolbox (Ljung, 2003). The normalized root-mean-square error (NRMSE) of the estimated model was 75%. The comparison of original output data and output of the estimated model is shown figure 7.

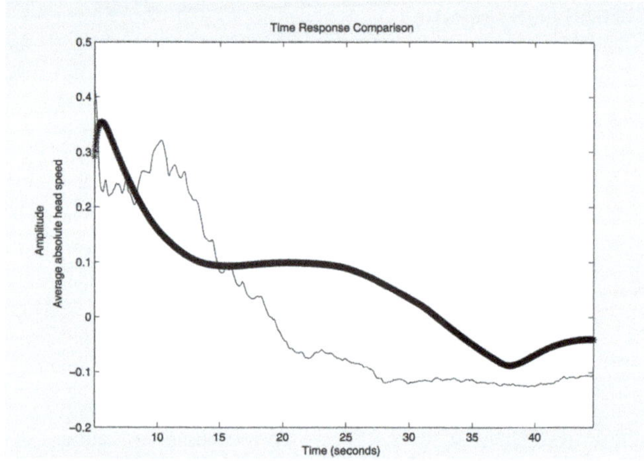


Figure 7. Original output and the output of the LTI model

### 5.2 NARX model estimation

The NRMSE of the estimated model was 17% (figure 6). The significant reduction of NRMSE of the NARX model could be explained by a better representation of nonlinear nature of emotional behavior by NARX model.

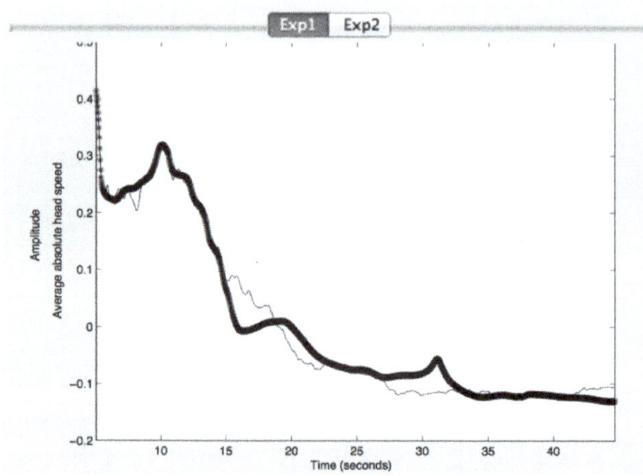


Figure 7. Original output and the output of the NARX model

## 6. CONCLUSIONS

A method for musical-elicited emotion model estimation was proposed. The model is estimated using the black-box model identification method. The input vector for model identification represents emotion-space, the output vector – facial expression space. The estimated model could be used to reconstruct musical-elicited emotion in valence-arousal emotion-space from recorded image sequence.

The proposed method was tested using specially selected list of classical musical pieces with various emotional content. A training data was recorded from ten volunteers listening to musical pieces. Dedicated software for a mobile tablet computer was designed to automate the recording workflow.

A modified LBP-optical flow optical was proposed to track face and facial features on the recorded video. The proposed tracker fuses data from optical flow motion estimation and LBP-tracker to achieve robust performance and sub-pixel feature tracking.

Two kinds of model were used to represent the connection between elicited emotions and facial expressions. LTI model failed to reconstruct dynamic nonlinear changes in facial expressions. NARX model proved to be accurate enough to reconstruct emotions on the evaluation data with 17% NRMSE.

## REFERENCES


Boucher, J. D., Ekman P. 1975. Facial Areas and Emotional Information. *Journal of Communication*, 25 (2).

Busso C., Bulut M., Lee C., Kazemzadeh A., Mower E., Kim S., Chang J. N., Lee S., Narayanan S. S., 2008. IEMOCAP: Interactive emotional dyadic motion capture database. *Language Resources and Evaluation*, 42 (4), pp. 335-359.

Darwin, C., 1872/1955. Expression of the Emotions in Man and Animals. *Philosophical Library*, New York.

Ekman P., Friesen. W. V. 1978. *Facial Action Coding System.* Consulting Psychologists Press Inc., California,

Esau, N., Wetzel, E., Kleinjohann, L., Kleinjohann, B. 2007. *Real-Time Facial Expression Recognition Using a Fuzzy Emotion Model*, Fuzzy Systems Conference, FUZZ-IEEE 2007. IEEE International, pp. 1-6

Farnebäck G., 2003. Two-Frame Motion Estimation Based on Polynomial Expansion, *Image Analysis, 13th Scandinavian Conference, SCIA 2003 Halmstad, Sweden*, pp. 363-370

Kita S., Mita A., 2015. Emotion Identification Method using RGB information of Human Face, *Proc. SPIE 9435, Sensors and Smart Structures Technologies for Civil, Mechanical, and Aerospace Systems 2015*

Ljung, L., 2003. *System Identification Toolbox*, The MathWorks, Inc

Pandzic, I., Forchheimer R. 2002. MPEG-4 Facial Animation - The standard, implementations and applications. John Wiley & Sons.

Russell. J. A., 1980. A circumplex model of affect. *Journal Personality Social Psychology*, 39, pp. 1161–1178.

Salovey, P., Mayer, J. D, 1990. Emotional intelligence. *Imagination, Cognition, and Personality*, 9, pp 185-211.

Soleymani, M., Caro M. N, Schmidt E. M., Sha C.-Y., Yang Y.-H. 2013. 1000 Songs for Emotional Analysis of Music, *CrowdMM '13 Proceedings of the 2nd ACM international workshop on Crowdsourcing for multimedia*, pp. 1-6.

Zhang, Q. 1997. Using wavelet network in nonparametric estimation, *IEEE Transactions on Neural Network*, Vol. 8, No. 2

# RICE CROP MONITORING AND YIELD ASSESSMENT WITH MODIS 250m GRIDDED VEGETATION PRODUCT: A case study in Sa Kaeo Province, Thailand

J. S. J. Wijesingha[a, *], N. L. Deshapriya[b], L. Samarakoon[b]

[a] Department of Physical Geography and Ecosystem Science, Lund University, Sölvegatan 12, S-223 62 Lund, Sweden - jayansrijeewantha@gmail.com

[b] Geoinformatics Center (GIC), Asian Institute of Technology (AIT), PO Box 04, Klong Luang, Pathumthani, Thailand - lakmal@ait.ac.th, lalsamarakoon@gmail.com

**KEY WORDS:** Rice Phenology, MODIS, NDVI, EVI, Whittaker Smoother

**ABSTRACT:**

Billions of people in the world depend on rice as a staple food and as an income-generating crop. Asia is the leader in rice cultivation and it is necessary to maintain an up-to-date rice-related database to ensure food security as well as economic development. This study investigates general applicability of high temporal resolution Moderate Resolution Imaging Spectroradiometer (MODIS) 250m gridded vegetation product for monitoring rice crop growth, mapping rice crop acreage and analyzing crop yield, at the province-level. The MODIS 250m Normalized Difference Vegetation Index (NDVI) and Enhanced Vegetation Index (EVI) time series data, field data and crop calendar information were utilized in this research in Sa Kaeo Province, Thailand. The following methodology was used: (1) data pre-processing and rice plant growth analysis using Vegetation Indices (VI) (2) extraction of rice acreage and start-of-season dates from VI time series data (3) accuracy assessment, and (4) yield analysis with MODIS VI. The results show a direct relationship between rice plant height and MODIS VI. The crop calendar information and the smoothed NDVI time series with Whittaker Smoother gave high rice acreage estimation (with 86% area accuracy and 75% classification accuracy). Point level yield analysis showed that the MODIS EVI is highly correlated with rice yield and yield prediction using maximum EVI in the rice cycle predicted yield with an average prediction error 4.2%. This study shows the immense potential of MODIS gridded vegetation product for keeping an up-to-date Geographic Information System of rice cultivation.

## 1. INTRODUCTION

### 1.1 Background

Rice is the second most harvested staple food in the world and the leading staple food in Asia. Rice would be contributes heavily to the alleviation of hunger and poverty since millions of small farmers grow millions of hectares of rice in the Asian region. In addition, there are millions of landless workers who generate some income by working on these farms. Moreover, 60% of the global population and 90% of the world rice production is derived from the Asian continent (Claessens, 2013).

Rice monitoring and mapping are thus very important for optimizing food security, environmental sustainability, water security, greenhouse gas emission reduction and general economic prosperity. Most countries in the Asian region use limited survey methods to collect rice paddy data from community level to national level. These data sources do not meet the needs of science and policy researchers, who need geo-spatial databases of rice agriculture with updated spatial and temporal resolution (Xiao et al., 2006).

Remote Sensing is becoming an essential tool for monitoring and mapping rice growing over large areas, at repeated time intervals (Son et al., 2012). According to a review article of Remote Sensing of rice crop areas, Remote Sensing combined with *Geographical Information System* (GIS) (Kuenzer and Knauer, 2013), can provide reliable information for the following purposes related to rice farming:

- Mapping and monitoring the extent of rice growing ecosystems
- Monitoring and assessment of rice growth and health status
- Assessment of crop pattern and crop system efficiency
- Estimation of crop growth related parameters
- Precision agriculture
- Improvement, and extension of, crop growth and yield models

*Corresponding author.

High temporal and medium spatial resolution optical Remote Sensing *Vegetation Indices* (VI) known as *Normalized Vegetation Index* (NDVI) and *Enhanced Vegetation Index* (EVI) and *Land Surface Water Index* (LSWI) are utilized to map and monitor rice agriculture on a small scale, and analyzed in a global, continental, and subcontinental context (Sakamoto et al., 2005); (Xiao et al., 2006). *Moderate Resolution Imaging Spectroradiometer* (MODIS) sensor data was the major data source for those studies. Various time series image analysis techniques, such as wavelet analysis, Artificial Neural Network (ANN), Time Series Regression and Time Series Image Classification, were used to derive rice crop acreage and phenology. (Kuenzer and Knauer, 2013);(Jonai and Takeuchi, 2012);(Sakamoto et al., 2005);(Son et al., 2012);(Xiao et al., 2006).

### 1.2 MODIS Gridded Vegetation Product

The MODIS land science developed Global MODIS VI to provide consistent spatial and temporal comparisons of vegetation conditions. Visible red (620 670nm), *near infrared* (NIR) (841 876nm) and visible blue (459 479nm) surface reflectance of MODIS sensor data are used to make global NDVI and EVI products with 250m, 500m, 1000m, and 5600m spatial resolution, and sixteen-day, or one-month temporal resolution. MODIS sensor contains both a Terra and Aqua satellite platform, which were used to make vegetation products MOD13XX and MYD13XX. (The O refers to Terra platform, the Y refers to Aqua platform, and XX refers to one letter and one number. For example, if XX is Q1, then a sixteen-day composite image with 250m resolution is shown. If XX is A3, then a one-month composite with 1000m resolution is presented). Each year, one satellite sensor provides 23 images of sixteen-day composites. Production of Terra and Aqua products can be improved using the following temporal phase shifts: Terra 16-day period starting day 001, with Aqua 16-day period starting day 009) (Didan et al., 2010).

One *Hierarchical Data Format* (HDF) data-set in MXD13Q1 provides twelve layers which are NDVI, EVI, Surface reflectance of visible blue, visible red, NIR, *Mid-Infrared* (MIR) (2105-2155nm), VI *Quality Assessment* (QA) layer, composite day-of-year layer and four other layers.

MODIS VI data composite algorithm was based on Advanced Very High Resolution Radiometer-NDVI composite algorithm. *Constrained View Angle-Maximum Value Composite* (CV-MVC) and *Bidirectional Reflectance Distribution Function Composite* (BEDF-C) were developed by MODIS science team to design MODIS VI. Currently, CV-MVC is used as the primary compositing algorithm for MODIS VI Products (Shao et al., 2011).

NDVI defined by equation (1) is a normalized difference measure, comparing the near infrared and visible red bands defend by equation 1 where $_{\rho}NIR$ (841-876nm) and $_{\rho}Red$ (620-670nm) are the bi-directional surface reflectance for their respective MODIS bands. The NDVI is closely correlated with paddy area Leaf Area Index (LAI) (Xiao et al., 2006).

$$NDVI = \frac{_{\rho}NIR -_{\rho} Red}{_{\rho}NIR +_{\rho} Red} \qquad (1)$$

EVI was developed as an alternative vegetation index to NDVI, and is defined in the following equation (2), where $_{\rho}NIR$, $_{\rho}Red$, and $_{\rho}Blue$ (458-479nm) are atmospherically corrected (or partially atmospherically corrected) surface reflectance, and $C_1$, $C_2$, and $L$ are coefficients to correct for atmospheric condition (i.e., aerosol resistance). For the standard MODIS EVI product, $L$=1, $C_1$=6, and $C_2$=7.5 (Didan et al., 2010).

$$EVI = G\frac{_{\rho}NIR -_{\rho} Red}{_{\rho}NIR + C_1{}_{\rho}Red - C_2{}_{\rho}Blue + L} \qquad (2)$$

The MODIS Land VI QA provides quality assurance data for each pixel of the product as a bit field value. Composite day-of-year layer offers information for each pixel, representing the Julian day during the sixteen-day compositing period when reflectance was recorded (Pringle et al., 2012).

The noticeable difference between MODIS NDVI and EVI is the dynamic range. In high bio mass areas, MODIS NDVI appears to be saturated, whereas EVI shows more sensitivity (Shao et al., 2011).

### 1.3 Research Problem and Objectives

National and regional level rice crop monitoring and mapping is not enough to address food security, water management, and environmental stability problems in detail. Provincial or district level crop monitoring and mapping will be an effective way to address the above problems.

A number of studies have involved high level image processing techniques to rice crop monitoring, using satellite data. However, these methodologies were not applied to develop an active up-to-date rice crop monitoring system at the provincial- or district-level, which can be easily used by government officials as well as administrators to manage and plan rice cultivation.

The specific objectives of this study were to investigate general applicability of high temporal resolution MODIS 250m gridded vegetation product for rice crop growth monitoring, rice crop acreage mapping, and yield analysis at the province level. This study addresses the following three primary research questions concerning rice crop monitoring and mapping:

I. How is the rice plant growth illustrated by the Remote Sensing VI?
II. How can you apply Threshold (Beurs and Henebry, 2010) method to determine rice crop information (area, phenology) using MODIS data?
III. How do MODIS NDVI and EVI time series data represent a rice crop and its yield?

To address these questions, a Remote Sensing image analysis was performed combination with statistical analysis using MODIS 250m gridded vegetation product data from May to December 2013 across the Sa Kaeo Province in Thailand.

## 2. MATERIALS AND METHODS

### 2.1 Study Area

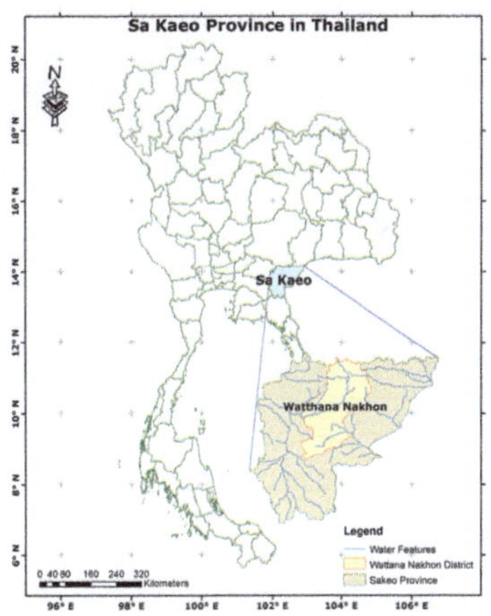


Figure 1: Map of the Sa Kaeo Province and Watthana Nakhon District

The study area for this research work was Watthana Nakhon District (102.4° $E$, 13.8° $N$), Sa Keao Province, Thailand (Figure 1). The northern part of the area includes part of the Thaplan National Park. Paddy, cassava, and permanent crops such as rubber and eucalyptus are cultivated in the central and the southern parts of the district. The paddy agriculture area is flat land with an elevation varying between 10m and 100m. According to 2012/13 agricultural statistics published by the District Agriculture Office, the total agricultural land area is 445053rai (1rai = 0.16ha) and the total area of rice cultivation is 207338rai. The main rice cultivation season is the wet season, which lasts from May to November and consists of rain-fed water. Dry season rice cultivation is less significant due to lack of water and cultivation of optional crops such as cassava and maze.

## 2.2 Data

**2.2.1 Remote Sensing / GIS Data:** MODIS gridded VI product with 250m resolution (MXD13Q1) h27v07 tile was used as the main Remote Sensing data. The VI product data from both Aqua and Terra Satellites (31 images from each platform) was taken over an eight month period (from 2013-05-01 to 2013-12-27). *Advance Land Observation Satellite* (ALOS) global forest/non forest data was used to identify the forest area in the study.

**2.2.2 Field Statistics:** Filed visit and field data collection was performed at four paddy plots in Watthana Nakhon District over the course of one complete rice cycle, during which time the MODIS satellites were observing the crops. A summary of the field sites is presented in Table 1. Plant height, water level in the field and canals and yield are collected during the field visit. Geotag photographs were taken in every visit to record the change of paddy growth. Province-level agriculture statistics were obtained from the Office of Agriculture Economics and district-level agriculture statistics were provided by the Watthana Nakhon District Agriculture Office.

| Site No | Village | Cultivation started | Rice variety |
|---|---|---|---|
| 1 | M6 - Ban Nongkum | 10 July | Leuang Awn |
| 2 | M3 - Ban Noongik | 06 July | Khao Dawk Mali 105 |
| 3 | M14 - Ban Wang Siaw | 07 July | Khao-Tah Haeng 17 |
| 4 | M9 - Ban Tahg Luang | 04 July | Khao Dawk Mali 105 |

Table 1: Summary of the field data collection sites

## 2.3 Step Processing

The whole study can be divided into the following four parts: (1) data pre-processing, (2) rice plant growth analysis using VI, (3) extracting rice crop parameters (e.g. acreage, start-of-season mapping from VI time series and accuracy assessment), and (4) rice yield analysis with MODIS VI.

**2.3.1 Data pre-processing:** Vegetation index layers (NDVI and EVI), VI QA layer, spectral bands (red, NIR, and MIR), and composite day-of-year layer were extracted from each MODIS HDF product as Geotiff images and then projected to WGS84 geographic projection. Eight days of NDVI and EVI images were stacked as a time series and detailed beyond Sa Kaeo province were masked out. The main element of pre-processing was the smoothing of the time series image stack. The VI time series was smoothed by using modified Whittaker Smoother with negatively biased noise of VI QA values and composite day-of-year values (Atzberger and Eilers, 2011);(Mattiuzzi and Lobo, 2013). Figure 2 shows selected time series before and after smoothing.

**2.3.2 Rice plant growth monitoring with RS VI:** Rice plant height data from field observation was plotted against extracted corresponding pixel values in VI (NDVI, EVI) data. Rice plant growth and VI pattern was observed.

**2.3.3 Extracting Rice Cultivation Parameters:** There are several spatio-temporal methods for extracting rice acreage and phenology from VI time series (Kuenzer and Knauer, 2013). This research work utilizes the simple threshold method (Beurs and Henebry, 2010). First, VI time series graphs were plotted for each land cover type (forest, water, urban, and paddy) to identify patterns and determine threshold values. After analyzing each land cover time series graph, an unique pattern was observed in

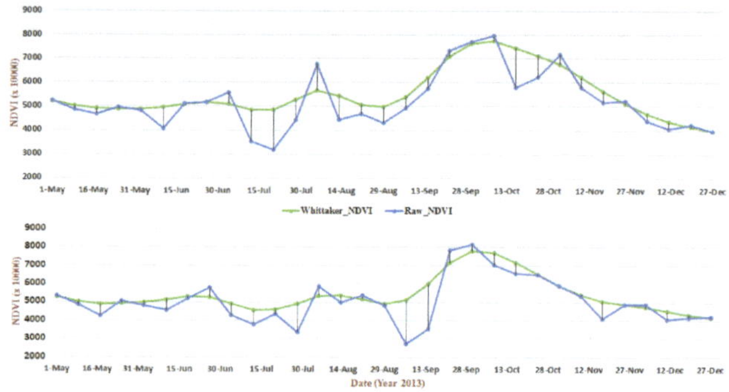


Figure 2: VI time series before and after applying smoothing

the paddy land cover data. Figure 3 and 4 show the how each land cover was interpreted by NDVI and EVI time series. When the rice paddy transitions from vegetative phase to reproductive phase, VI values increase up to a maximum. During the transition from reproductive phase to harvesting, VI values decrease. Paddy rice NDVI and EVI time series profiles therefore exhibit one major peak throughout the season. The date of this maximum is referred to as the heading date. By plotting several paddy pixel time series graphs, the threshold values for extracting paddy area were determined to be 0.6 and 0.47, for NDVI and EVI respectively.

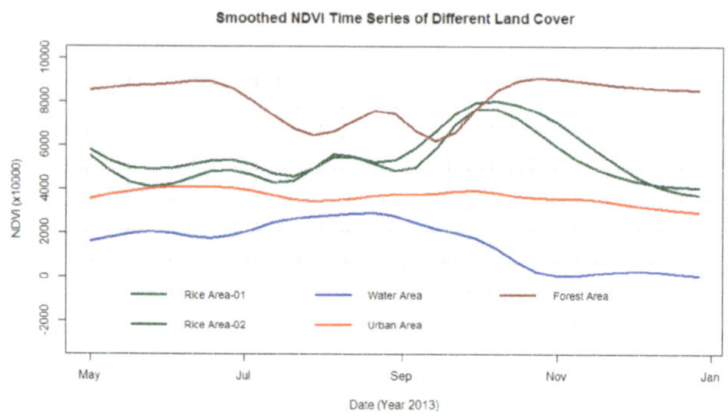


Figure 3: NDVI Time series graphs of different land cover

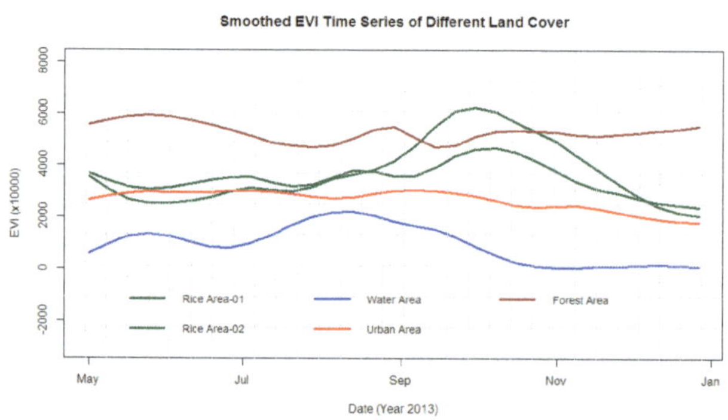


Figure 4: EVI Time series graphs of different land cover

Each time series pixel was analyzed to determine how many peaks exceeded the threshold values for given VI time series. The following information was recorded: number of peak images, first peak date image, second peak date image, and third peak date image. Single peak pixels were considered as single rice crop areas. Null and triple peak pixels reflect non-paddy or non-agriculture areas. Double peak values were further analyzed by calculating the difference between the dates of the first and second peaks. If the time difference was more than 80 days, those pixels were considered as representing a double rice paddy. There are two reasons for this condition: (1) a typical rice cycle is about 120 days, and thus cannot be much less than that, and (2) cassava and maze are cultivated with rice as a double crop pattern in some areas of the Sa Kaeo province. ALOS forest cover data was used to mask out the forest area in the study area.

In general, Sa Kaeo province rice cultivation cycle varies from 120 days 140 days. Therefore the rice planting date was occurred 60-70 days before the heading date. The start-of-season for each rice cultivation pixel was calculated by subtracting 70 days from heading date, and start dates were grouped monthly.

Accuracy of rice acreage was measured by comparing province-level and district-level rice cultivation area statistics and calculating the confusion matrix and kappa coefficient using ground truth from the field data.

**2.3.4 Rice yield analysis with Remote Sensing VI:** Chlorophyll content is related to the production of yield of the rice plant and it is reflected by Remote Sensing VI. The VI data was therefore used to determine if there is a correlation between MODIS VI and rice yield, and how the correlation changes with NDVI and EVI. Yield assessment was done using field data gathered from Site 2 and Site 4. The correlation coefficient and liner regression analysis of yield was evaluated against maximum VI of the season, cumulative mean VI of the season, and VI of the flowering date (for both NDVI and EVI). Regression analysis results from each VI were compared to determine which VI was associated with the highest rice yield. The yield of the Site 3 was predicted from the most highly correlated VI model and compared to the actual yield.

## 3. RESULTS & DISCUSSION

Figure 5: Geo-tag photograph time series of rice paddy cultivation

Rice plant growth observation from the field work is shown in Figure 5. The time series geo-tag photographs clearly show how the plant color changes from light-green to green, and then to yellow.

Figure 6 illustrates rice plant height time series in each field site. According to the time series graph, rice plants grew more than 1.2m height and plants grew with approximately uniform speed.

### 3.1 Plant height and vegetation indices

We can observe how VI changes with plant height from the graphs of Figure 7. When plant height reached between 0.7m and 1.0m,

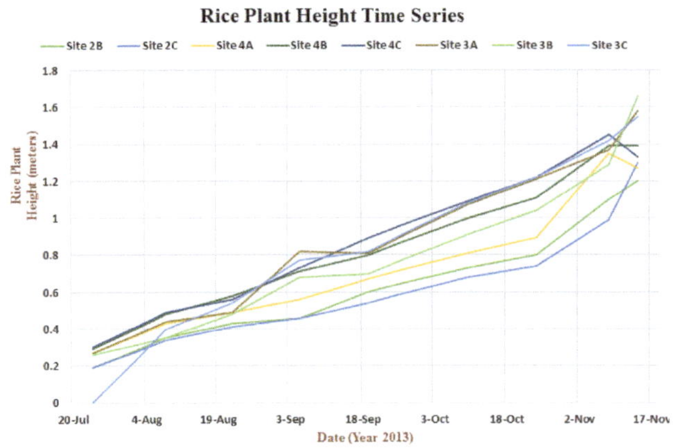


Figure 6: Rice Plant Time Series

NDVI and EVI show their peak values during the plant cycle. However, both NDVI and EVI show a similar pattern in plant height. At the start-of-the season, VI increases with plant height and reaches to peak level, and then it decreases while plant height is increasing.

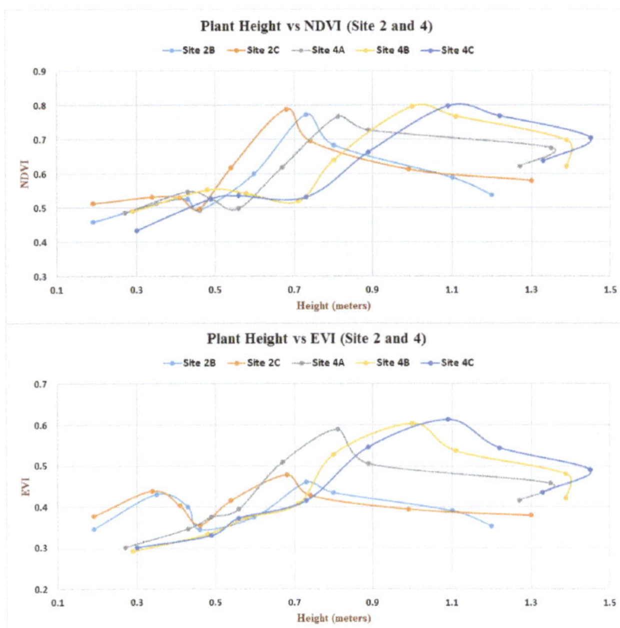


Figure 7: Graph of Rice plant height vs VI (NDVI and EVI)

### 3.2 Rice acreage and start-of-season Mapping

Figure 8 and Figure 9 exhibit NDVI and EVI maps of the rice crop distribution Sa Kaeo Province. Total rice cultivation area in Sa Kaeo province is calculated from both analyses. They are 138795.16ha (NDVI) and 152944.08ha (EVI).

The starting month for single crop areas are shown in Figure 10 and Figure 11. Both NDVI and EVI analyses show more than 80% of the paddy fields in single crop areas were started in June, July, or August in year 2013.

For both the NDVI and EVI time series analyses of double crop areas, April and May are the starting months for the first cultivation season, and July and August are the starting months for the second cultivation season.

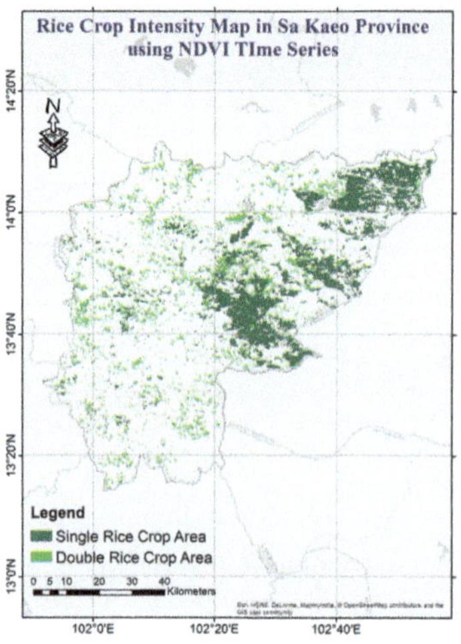


Figure 8: Rice Crop Intensity map from NDVI in Sa Kaeo Province, Thailand)

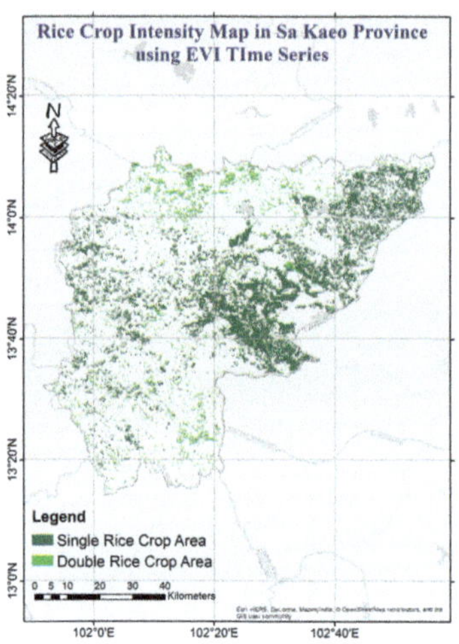


Figure 9: Rice Crop Intensity map from EVI in Sa Kaeo Province, Thailand)

Overall accuracy of NDVI-based rice paddy area mapping is 77.43% and the Kappa coefficient is 0.55. The EVI-based rice acreage mapping overall accuracy is 76.47%, with a the Kappa coefficient of 0.52. The associated confusion matrices from both analyses are shown in Table 2 and Table 3.

| NDVI | Non Paddy | Paddy | Total | P. Acc. (%) |
|---|---|---|---|---|
| Non Paddy | 89 | 19 | 108 | 82.41 |
| Paddy | 32 | 86 | 118 | 72.88 |
| Total | 121 | 105 | 226 | |
| U. Acc. (%) | 73.55 | 81.90 | | |
| **O. Acc. (%)** | **77.43** | | | |
| **Kappa** | **0.55** | | | |

Table 2: Accuracy assessment of NDVI-based rice acreage mapping

Province-level rice acreage was calculated by comparing the VI-determined acreage with the statistics provided by the Office of Agriculture Economics. Total acreage determined by NDVI was

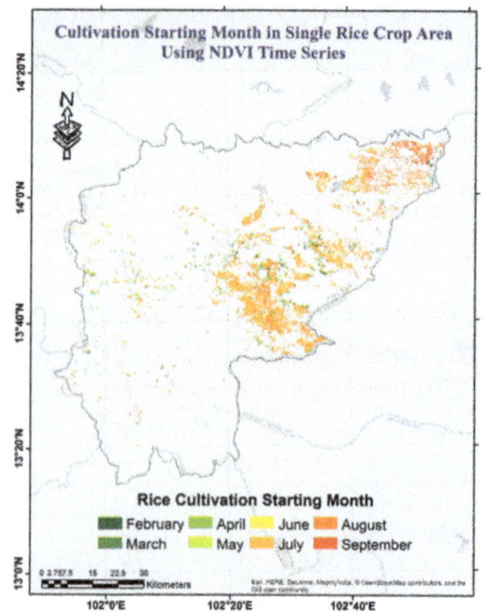


Figure 10: Single rice crop area start-of-season in Sa Kaeo Province, using NDVI

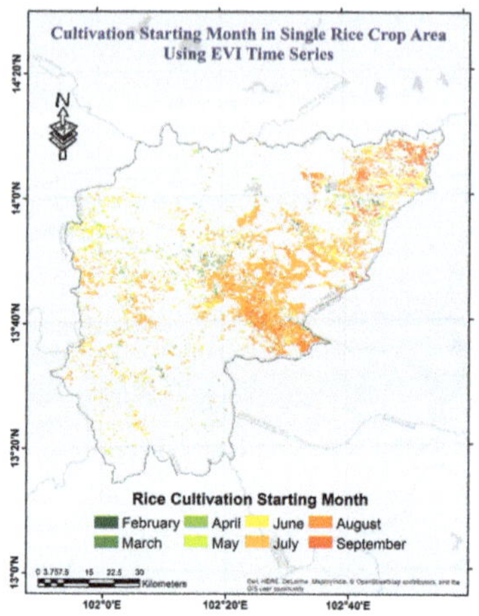


Figure 11: Single rice crop area start-of-season in Sa Kaeo Province, using EVI

| EVI | Non Paddy | Paddy | Total | P. Acc. (%) |
|---|---|---|---|---|
| Non Paddy | 100 | 34 | 134 | 74.63 |
| Paddy | 18 | 69 | 87 | 79.31 |
| Total | 118 | 103 | 221 | |
| U. Acc. (%) | 84.75 | 66.99 | | |
| **O. Acc. (%)** | **76.47** | | | |
| **Kappa** | **0.52** | | | |

Table 3: Accuracy assessment of EVI-based rice acreage mapping

138795ha, and total acreage determined by EVI was 152944ha. These estimates deviate from the published statistics by +3529ha (NDVI) and +17678ha (EVI), producing relative error of 2.6% (NDVI), and 13.1% (EVI). At the district-level, rice acreage accuracy was calculated by comparing the VI-determined acreage with the statistics reported by Watthana Nakhon District Agriculture Office. Rice cultivation area within Watthana Nakhon District is 33174ha. However, the district rice acreage determined by NDVI was only 29491ha, and the district acreage determined by EVI was only 31916ha, producing relative error of 11.1% (NDVI) and 3.8% (EVI). These statistics suggests rice acreage can be estimated by NDVI and EVI with in 14% maximum deviation. Confusion matrix analysis shows that rice acreage estimation using NDVI and EVI has an accuracy of more than 75%. Table 4 summarizes the area-based accuracy.

| Level | Area (ha) | Method | Estimated (ha) | RE (%) |
|---|---|---|---|---|
| Province | 135265.60 | NDVI | 138795.16 | -2.61 |
| | | EVI | 152944.08 | -13.07 |
| District | 33174.08 | NDVI | 29490.81 | +11.10 |
| | | EVI | 31916.44 | +3.79 |

Table 4: Summary of the area based accuracy assessment

### 3.3 Yeild Analysis

Rice yield and NDVI/EVI correlation analysis was done in three different ways. First, a correlation and linear regression analysis was performed using maximum VI of the time series and the yield. For the yield with maximum NDVI and maximum EVI, the correlation coefficient were 0.45 and 0.98, respectively. Analysis shows that the maximum EVI has the highest correlation with yield. The R-square values from linear regression analysis are 0.2 (NDVI) and 0.95 (EVI).

Second, the correlation between yield and flowering date VI was investigated. The average flowering date was founded by a survey done on farmers and it was on the 14th October, 2013. The $R^2$ values for linear regression are 0.72 (NDVI) and 0.92 (EVI). Table 5 summarizes regression analysis results of the above two methods.

| Method | VI | $R^2$ Value |
|---|---|---|
| Maximum VI | NDVI | 0.20 |
| | EVI | 0.95 |
| Flowering Date VI | NDVI | 0.72 |
| | EVI | 0.93 |

Table 5: Regression for yield with maximum VI and flowering date VI

A third yield analysis compared yield and cumulative mean of VI. Cumulative VI mean was calculated for the eight days after the start-of-season. The correlation coefficient of each cumulative VI mean and yield was also calculated. The cumulative VI with highest correlation coefficient was used in a regression analysis of the yield. Figure 12 illustrates how the correlation coefficient changes with cumulative NDVI/EVI and yield. Both NDVI and EVI exhibits a similar pattern of correlation coefficient; both VI are negatively correlated with yield at the start of the season, and then become positively correlated, with high correlation coefficient values.

Regression analysis results of the highest correlation between cumulative mean VI and yield are described in Table 6. $R^2$ values from linear regression analysis are 0.71 (NDVI) and 0.87 (EVI).

| VI | Highest CC | $R^2$ Value |
|---|---|---|
| NDVI | 0.84 | 0.71 |
| EVI | 0.94 | 0.88 |

Table 6: Regression analysis results of highest correlated cumulative mean VI with yield

The results of all three yield analyses show that yield has a higher correlation with EVI than NDVI. The $R^2$ values of regression analysis with yield are 0.95 (for maximum EVI), 0.93 (for flowering dates EVI), and 0.88 (for cumulative mean EVI).

The highest correlated EVI was utilized to predict yield for Site 3, which was then compared with actual yield. The most precise

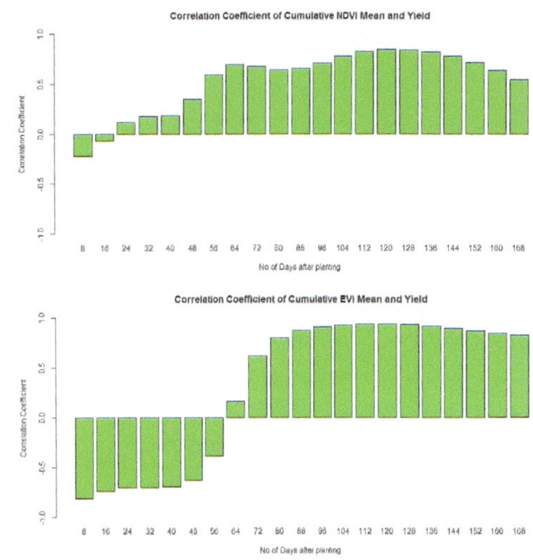


Figure 12: Correlation coefficient changing graph of the Cumulative VI mean and yield)

prediction is given by the maximum EVI of the time series, with an average error of 4.21%. The flowering data EVI also give better results (5.62% error) than cumulative mean EVI (10.87%). Figure 13 compares the actual yield with EVI-predicted yield.

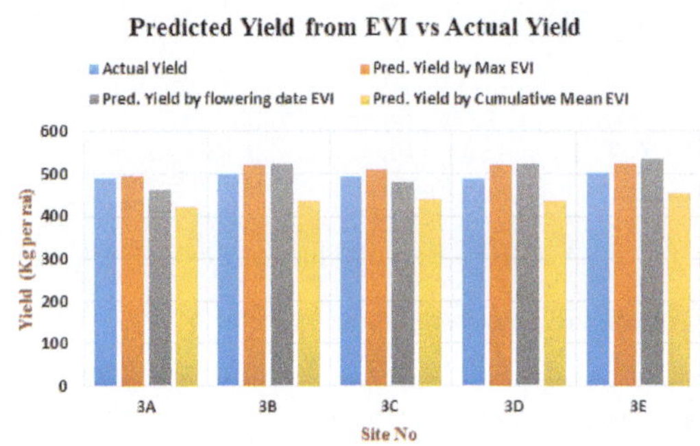


Figure 13: Actual yield and predicted yield from EVI by different method in Site 3)

## 4. CONCLUSION

This study illustrates the significant potential of using MODIS data for rice crop monitoring. MODIS 250m gridded vegetation NDVI and EVI data from both sensors (Aqua and Terra) increase the frequency of observation and improve rice monitoring. The use of the Whittaker smoothing technique, which includes a combination of quality assessment data and R program library, is recommended (Atzberger and Eilers, 2011).

The results obtained by this research confirm that there is an observable relationship with rice plant growth and VI changes in rice paddy land cover. MODIS NDVI and EVI rice paddy acreage mapping (using the Threshold method and crop calendar data) gives a minimum of 86% area accuracy and 75% classification accuracy. Rice crop cultivation starting month extracted via MODIS NDVI data matched the field data. MODIS NDVI data is better than EVI for rice acreage mapping and start-of-season mapping.

Pixel level yield analysis shows yield and VI have a noteworthy relationship. EVI is better for yield analyzing and Maximum VI throughout the crop season which indicates the highest correlation with yield. Yield was predicted for site 03 with 4.2% error using maximum EVI of the season.

It was proved that time series MODIS 250m vegetation index product data has immense potential for monitoring rice crops. This study shows the usefulness of MODIS gridded vegetation product for the following techniques: (1) mapping rice cultivation area at the province-level, (2) generating an up-to-date database of rice acreage and start-of-season dates, and (3) predicting yield of rice crop before harvesting. Applying this novel methodology for implementing up-to-date GIS rice database is recommended for other areas where rice cultivation is prominent.

## ACKNOWLEDGEMENTS

The authors thank Ms. Udompna Promnart of the Prachinburi Rice Research Center, Thailand, whose enormous support made this field work possible. The authors are also grateful for assistance from Watthana Nakhon District Agriculture Office staff, and from the village leaders and farmers.

## REFERENCES

Atzberger, C. and Eilers, P., 2011. A time series for monitoring vegetation activity and phenology at 10-daily time steps covering large parts of South America. *International Journal of Digital Earth* 4(5/5), pp. 365–386.

Beurs, K. D. and Henebry, G., 2010. Spatio-Temporal Statistical Methods for Modelling Land Surface Phenology. In: Hudson I.L. and M. R. Keatley (eds), Phenological Research, Springer, Netherlands.

Claessens, G., 2013. IRRI's Rice Production Course. Online.

Didan, K., Solano, R., Jacobson, A. and Huete, A., 2010. MODIS Vegetation Index User's Guide - Version 2.00.

Jonai, H. and Takeuchi, W., 2012. Development of new water index with modis and amsr-e for global rice paddy field mapping. In: Proceedings of 34th Asian conference on remote sensing (ACRS), AARS.

Kuenzer, C. and Knauer, K., 2013. Remote Sensing of rice crop areas. *International Journal of Remote Sensing* 34(6/12), pp. 2012–2139.

Mattiuzzi, M. and Lobo, A., 2013. Filter vegetation index time series imagery with the modified whittaker approach.

Pringle, M., Denham, R. and Devadas, R., 2012. Identification of cropping activity in central and southern Queensland, Australia, with the aid of MODIS MOD13Q1 imagery. *International Journal of Applied Earth Observation and Geoinformation* 19, pp. 276–285.

Sakamoto, T., Yokozawa, M., Toritani, H., Shibayama, M., Ishitsuka, N. and Ohno, H., 2005. *Remote Sensing of Environment* (3/4), pp. 366–374.

Shao, Y., Taff, G. and Lunetta, R., 2011. A Review of Selected MODIS Algorithms, Data Products, and Applications. In: Q. Weng (ed.), *Advances in Remote Sensing*, CRC Press LLC, Boca Raton, Florida.

Son, N., Chen, C. and Cru, C., 2012. Mapping major cropping patterns in Southeast Asia from MODIS data using wavelet transform and artificial neural networks. *International Archives Photogramm. Remote Sensing* XXXIX-B3, pp. 421–425.

Xiao, X., Boles, S., Frolking, S., Li, C., Babu, J., Salas, W. and Moore, B., 2006. A crop phenology detection method using time-series modis data. *Remote Sensing of Environment* 96(1), pp. 95–113.

22

# FACIAL FEATURES DETECTION USING TEXTURE HOUGH TRANSFORM


V. S. Gorbatsevich

The Federal State Unitary Enterprise «State Research Institute of Aviation Systems» Russia, Moscow - gvs@gosniias.ru



**Commission V, WG V/5**

**KEY WORDS:** Local Binary Pattern, Object detection, facial features detection, Hough transform



**ABSTRACT:**

The paper presents an original method for object detection. The "texture" Hough transform is used as the main tool in the search. Unlike classical generalized Hough transform, this variation uses texture LBP descriptor as a primitive for voting. The voting weight of each primitive is assumed by learning at a training set. This paper gives an overview of an original method for weights learning, and a number of ways to get the maximum searching algorithm speed on practice.


## 1. INTRODUCTION

One of the most popular issue in technical vision is the issue of identification of certain points on the image of a face (mouth, nose, eyes, eyebrows, etc). These special points which align with different part of the face are used, more or less, in almost every application developed in the digital face image processing sphere. For example, there is an abundance of methods for comparing images of faces, pose evaluation algorithms, algorithms for composing 3D-models of faces and many others. This being said, besides the quality, the computation speed is very important for the methods of finding the special points. Currently, there are a lot of ways to find the points, from already classical Voila-Jones(Viola 2001) method, to Hough Forest(Gall 2010). All these methods give us proper means to deal with the identification issue; however the problem of quality/speed ratio is still relevant.

This work makes an attempt to look into a matter of objects identification from the perspective of the concept of generalized Hough transform and voting weights training. It also suggests modification of Hough transform, as well as follow-up methods of evidence analysis.

## 2. TEXTURE HOUGH TRANSFORM

### 2.1 Hough transform

Nowadays Hough transform is one of the most popular methods of analytically defined objects identification. Its popularity arises from the excellent ratio of quality and speed. The central idea of Hough transform is voting of some primitives for hypotheses about the position of the sought object. The hypotheses that have gathered the votes above a certain threshold are considered true. To illustrate the central idea of Hough transform, let's take a look at the simplest equation for identification of a black line on a binary image. As is known, the equation of line takes the form of:

$$y = k * x + b$$

where K – slope of the line
X, y – coordinates
B - is the y-intercept of the line

While in actual practice the following parameterization is more commonly used:

$$r = x \cos\theta + y \sin\theta$$

where R - radius vector
Θ – angle between the line and the horizontal axis

In this case every hypothesis appears as a pair of parameter values (R, Θ). A range of all these hypotheses are combined into a so-called Hough space. Therefore every point in the Hough space corresponds to the certain line. Points different from the background (in this case – black) are used as primitives. For every fixed point at the coordinates ($x_p$, $y_p$), there are infinite straight lines coming through it. In Hough space these lines are assigned via the sinewave, derived from:

$$r(\theta) = x_P \cos\theta + y_P \sin\theta \qquad (1)$$

So, set of points of some line (r, Θ) corresponds to the set sinewaves of the form of (1), which intercross at the point [r, Θ]. Thus, the line identification reduces to the voting of every point for all hypotheses in the line of equation (1). In actual practice a special accumulator-array is used for hypotheses storage, and the process of voting is reduced to summing.

Figure 1 shows an example of a line and a corresponding accumulator.

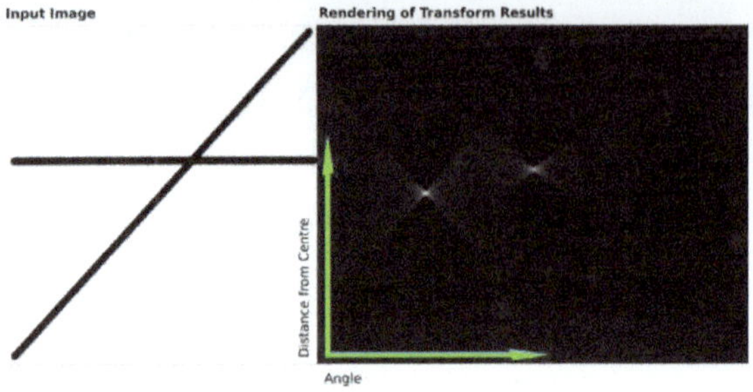


Figure 1. An example of Hough transform for lines identification. Initial image (on the left) and accumulator (or the right).

Hough transform is used the same way for identification of any curves defined by analytical equations.

## 2.2 Generalized Hough Transform

Generalized Hough Transform (GHT) is used for detection binary objects of an arbitrary shape. Unlike the classical variation of the transform, the generalized Hough transform is used in situations when the desired objects cannot be defined by any analytical equations. However, if the center is set manually, a convex curve can be defined as a look-up table. Therefore, it is possible to make a process of voting of edge points for possible center for simple distortion models simple. In this case a Hough space is a space of possible positions of center, and an accumulator-array is an array aligned with the size of an image. Figure 2 shows an example of a voting process for identification of an object of an arbitrary shape.

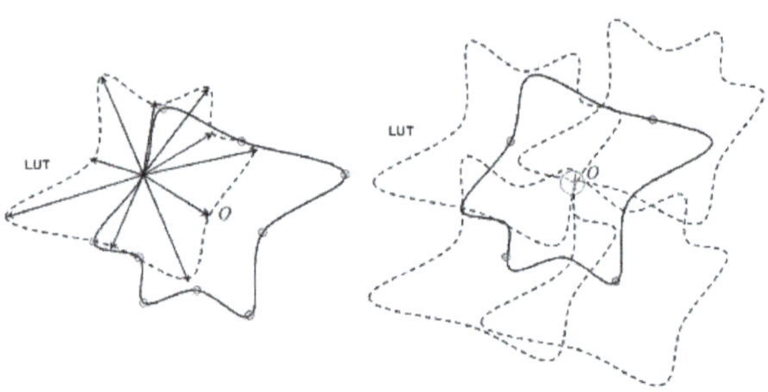


Figure 2. An example of Hough transform for lines identification. Initial image (on the left) and accumulator (or the right).

This example of a process of identification of some object in the image does not account for possible distortions although it does demonstrate the basic ideas of generalized Hough transform. Currently there are a lot of GHT modifications, which are capable to work with different distortions.

## 2.3 Texture Hough Transform

And now we are back to facial features identification. Unlike the previous cases, our current example is greyscale, namely, a greyscale image of an eye. A figure 3 shows a greyscale image of an eye.

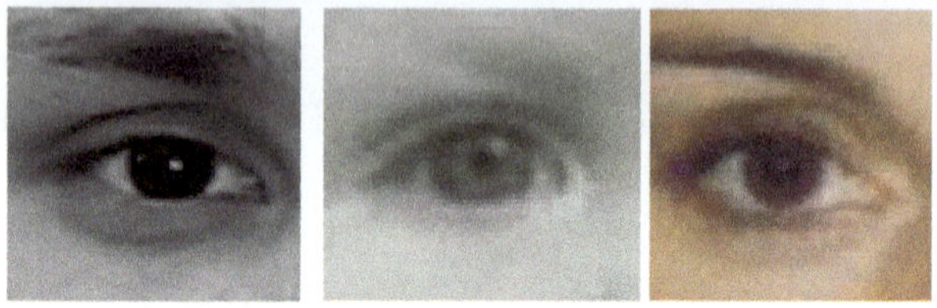

Figure 3. A greyscale image of an eye.

Unfortunately, it is impossible to apply Hough transform directly because:

**Firstly**, in this case we work not with just an object, but with the whole class of objects, the appearance of which can vary depending on race, gender, emotional state, and other conditions;

**Secondly**, preliminary appliance of highly-accurate segmentation and edge detection algorithms is needed;

**Thirdly**, there is no "object" – "background" image mapping. However, the ideas of generalized Hough transform can still be applied. Let's take a look at some texture descriptor. Let this descriptor be able to assume a limited set of values. Then we can take some specific value of this descriptor as a primitive. Then, just like with GHT, the Hough space consists of many possible object centers. Thus, the accumulator is a rectangular array, the size of which corresponds to the size of the image.

Let the size of an object be fixed at the [W*H], so that for some primitive with coordinates [X, Y] the area of possible center positions of the object takes a rectangular shape with the size of [W*H] and the center at the [X, Y], as shown in the figure 4.

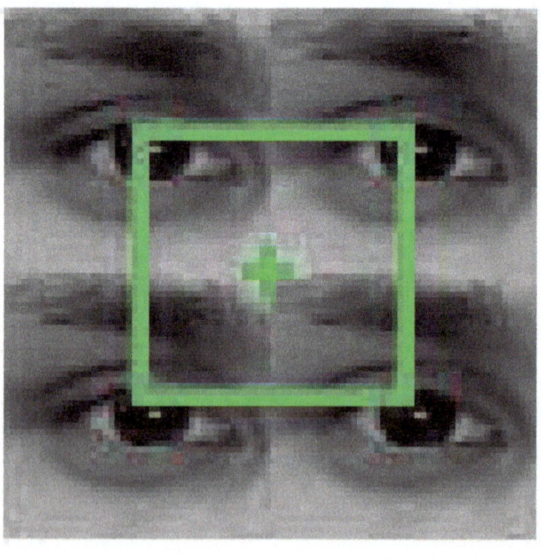

Figure 4. An area possible object center positions (green rectangular) for some primitive position (green cross).

Therefore, just a small voting weights map is required for every descriptor value, describing possible center positions of the desired object relative to the descriptor position.

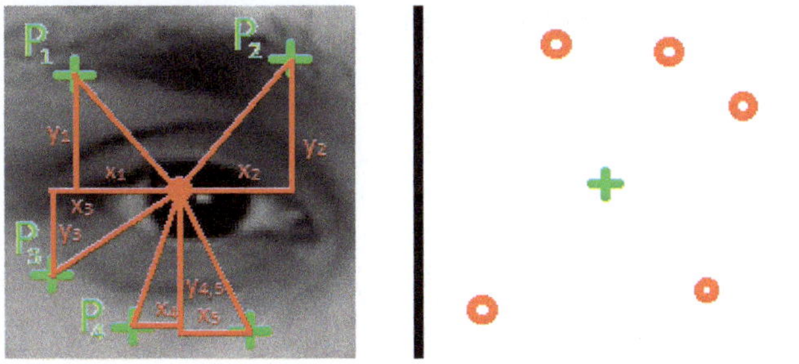

Figure 5. An example of weights array for some descriptor value. Descriptor with some fixed value (on the left). Weight map with possible center positions (on the right).

The voting algorithm looks like this:

*For every point in the image (x,y):*

*for dx = -W/2 to W/2*
*for dy = -H/2 to H/2* (2)

$$Akk_{x+dx,y+dy} = Akk_{x+dx,y+dy} + Weigth_{Diskr_{x,y},dx,dy}$$

where $Akk_{x,y}$ – accumulator array
$Weight_{VAL,dx,dy}$ – weight mapping for VAL descriptor value
$Diskr_{x,y}$ – value of descriptor in a point [x,y]

Detected objects are defined by:

$$\text{Centers} = \{(x,y): Akk_{x,y} > Thr\}$$

where Thr – threshold

Therefore, the accumulator filling occurs just only with summing operations. This will be discussed in the following section in more detail.

Albeit being really similar to the classical GHT, the described method of texture Hough transform has a number of important differences. All points in the image are used in the voting, and not just those on the edges. On one hand it makes the identification of objects with any texture and shape possible, but in the other this method causes the problem of assigning voting weights and, consequently, forming the corresponding weight maps.

## 3. LBP DESCRIPTOR

This work uses the LBP descriptor as a texture descriptor. This descriptor is one of the most well-known texture descriptors. Currently, there is an abundance of texture descriptors, like SIFT(Lowe 2001), SURF(Bay 2006), G-SURF(Pablo 2013), and many others. And yet the LBP has a number of attributes which make it more preferable for texture Hough transform:

1. Resistance towards additive and multiplicative brightness distortion
2. Very high speed of composition
3. There are only 256 descriptor value units
4. Every LBP descriptor matches with only one pixel

The process of getting a LBP descriptor can be briefly described as follows: a circular neighborhood of a certain radius is considered for every point in the image, and then every point is compared to a value of a current point of the image:

$$LBP(p) = \sum_{i=0}^{n} 2^i \begin{matrix} 1: I(p) \geq I(o_i) \\ 0: I(p) < I(o_i) \end{matrix}$$

where N – number of neighborhood points
$O_i$ – neighborhood points
P – current point of the image

The result of comparison is recorded as binary vector (see Figure 4).

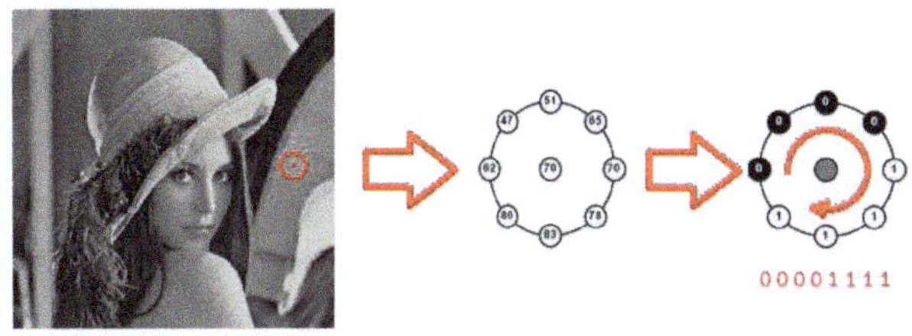


Figure 4. An example of LBP descriptor calculation for a point in the image

## 4. WEIGHTS MAPS LEARNING

The texture Hough transform described in the section 2.3 implies that there are preset weights maps for every value of a texture descriptor. It is the weights values that assign search object model and background model, and that's why the task of weights assigning is really important in using texture Hough transform. This work suggests using original learning procedure for weights value calculation. In order to do that we would need to rewrite the earlier mentioned texture transform as Viola-Jones style strong binary classifier:

$$C(x,y) = \begin{cases} \textbf{object}: val(x,y) \geq Thr \\ \textbf{background}: val(x,y) < Thr \end{cases}$$

$$val(x,y) = \sum_{dx=-W/2}^{+W/2} \sum_{dy=-H/2}^{+H/2} \sum_{LBPVAL=0}^{255} Weigth_{LBPVAL,dx,dy}\, S(x+dx, y+dy, LBPval) \quad (3)$$

Where

$$S(x,y,LBPval) = \begin{cases} 1: Diskr_{x,y} = LBPval \\ 0: Diskr_{x,y} \neq LBPval \end{cases}$$

*W,H* – object sizes

$Weight_{LBPVAL,dx,dy}$ – weight map for descriptor value = *LBPval*
$Diskr_{x,y}$ – $LBP_{val}$ value weight map
Thr – some threshold

It is easy to verify that using such classifier in a sliding window is identical to the voting process from 2.3. Therefore, the assigning/setting the voting weights mirrors the training of such classifier. The usual method for training such classifiers is boosting (Rojas 2009). However, in this case the situation is slightly different from usual process. This not something for Haar-like characteristics. The main difference is that there is just a small number of classifiers that can be used for every position, namely, only W*H classifiers. This gives us an opportunity to use new unexpected algorithm for weights calculation.

Imagine a small subset Base. It is possible to use quality quadratic functional for evaluation of classifier's quality:

$$Q(\mathbf{Base}, \boldsymbol{Weigths}) = \frac{1}{2}\sum_{a\epsilon\mathbf{Base}} (val(a, \boldsymbol{Weigths}) - Ans(a))^2$$

where val(a) – classifier's response for example a (see 2)
Ans(a) – trainer's response:

$$\boldsymbol{Ans(a)} = \begin{cases} \mathbf{1}: \boldsymbol{a - object} \\ \mathbf{0}: \boldsymbol{a - background} \end{cases}$$

$\boldsymbol{Weigths}$ –weight maps

The learning procedure can be represented as minimization problem:

$$\boldsymbol{ResW} = \mathbf{argmin}_{\boldsymbol{Weigth}} \boldsymbol{Q}(\mathbf{Base}, \boldsymbol{Weights})$$

For minimization we would use a stochastic gradient descend. E. g. for every training example $a_i \in$ Base, changes are iteratively made in the weights value, of the form:

$$\Delta = -\alpha \frac{\partial q(a, Weigths)}{\partial Weights}|_{a=a_i} \quad (3)$$

where $\mathbf{q}(\boldsymbol{a}, \boldsymbol{Weigths})=(\boldsymbol{val}(\boldsymbol{a}, \boldsymbol{Weigths}) - \boldsymbol{Ans}(\boldsymbol{a}))^2$

$\boldsymbol{\alpha} - \textit{speed parameter}$

Nevertheless it should be noted that the number of nonzero elements at Δ for every training example is equal to W*H, which allows for a high speed of the process of weights correction. All in all this approach is basically similar to back propagation, commonly used in neural networks training.

As a first approximation, the probability estimate is used in a form of:

$$\boldsymbol{Weigth_{LBPVAL,dx,dy}} = \boldsymbol{TP/(TP+FP)} \quad \mathbf{(4)}$$

where TP – true positives
FP – false positives

Then the $\boldsymbol{Weigth_{LBPVAL,dx,dy}}$ must be normalized, in such a way as to val (3) function highest value is no more than 1.0. Such approximation allows for getting a stable result for most real objects.

On completion of weights values learning, the value of strong classifier threshold is calculated. The point corresponding to the threshold is determined by a FAR/FRR graph, made for validation frame based on specific standards for the final classifier.

## 5. IMPLEMENTATION SPECIFICS

### 5.1 Learning

During the implementation of the described method in actual practice, a number of specifics should be used for the best results.

Since the assumed descriptors are less resistant towards the scale change than classical Haar-like features, the training selection should include artificial scale and rotation angle distortion.

As "background" examples should be used a small random subset of real images cuts. This set iteratively grown during learning process trough adding wrong classified examples from huge dataset of real data.

### 5.2 Scale invariance

As aforesaid, some small scale fluctuations are recorded directly into the training selection. A classical method provided for an image pyramid is used to record more tangible distortions.

### 5.3 Cascades

It is possible to use cascades technique using the discussed classifiers in a similar way to Viola-Jones (Viola 2001) cascade classifiers. The cascades tuned to different scales and working with different level of the pyramid are the most appropriate.

### 5.4 Voting process

The voting is the most resource-hungry task in Hough texture transform operation. Let us consider the formula (2). The sum operation is the only operation used there. The algorithm structure allows for the full use of SSE2 and AVX vector commands. It should be noted that it is possible to use 16-bit целое представление for weights maps and for accumulator with little to no losses of detection quality, which makes using vector commands to the full extent possible.

Another opportunity for huge performance improvement is using weights maps for two nearby descriptors. Such weights maps can be easily extracted from the initials. Although it leads to the huge increase in the used memory, it makes the voting really faster.

## 6. TESTING RESULTS

A number of classifiers was trained for the identification of 24 basic points on the digital image of a face (see figure 5) with the help of such method. "Faces in the Wild" database (LFW from "http://vis-www.cs.umass.edu/lfw/") was used in this testing. The special property of this database is that all its images were taken in unregulated conditions. The manually labeled images subset were used as a reference. The response is considered true if identified coordinates value's deviation on from the reference image is no more that 5% of the space between the eyes. The testing results are shown in the table 1.

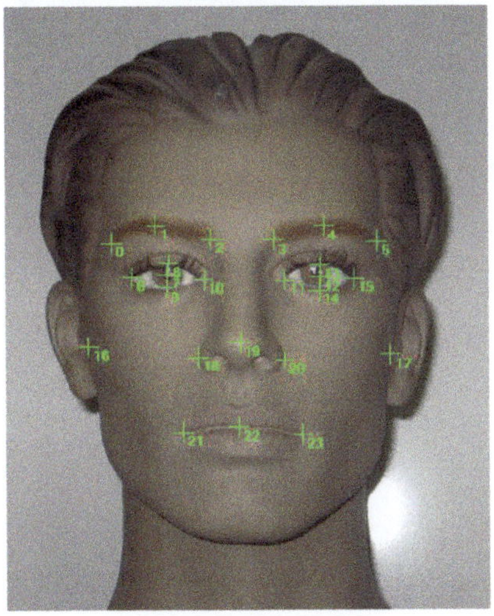

Figure 5. Facial features.

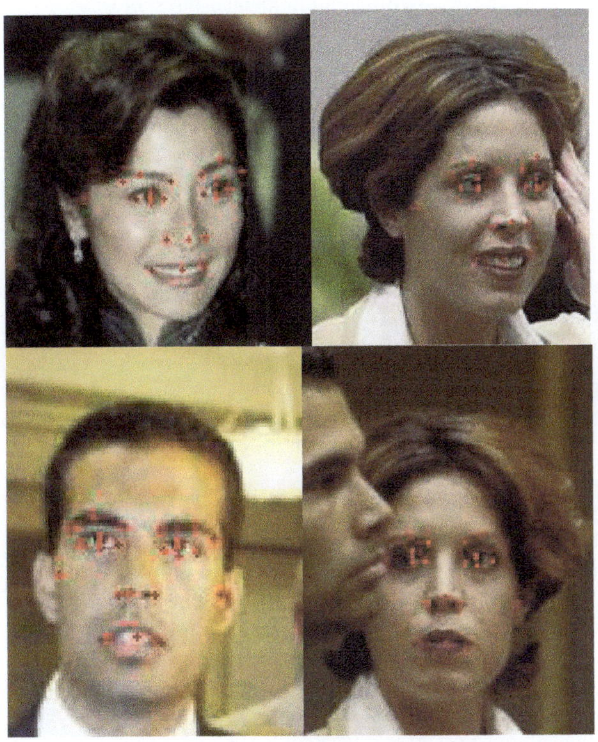

Figure 6. Detection results.

| Feature point id (see Fig 5) | TPR(FPR = 0.99) |
|---|---|
| 0/5 | 0,81 |
| 1/4 | 0,83 |
| 2/3 | 0,92 |
| 6/15 | 0,88 |
| 7/12 | 0.99 |
| 8/13 | 0,95 |
| 9/14 | 0,95 |
| 10/11 | 0.96 |
| 16/17 | 0,72 |
| 18/20 | 0,97 |
| 19 | 0,96 |
| 21/23 | 0,97 |
| 22 | 0,95 |

Table 1. Results on LFW Dataset

All in all the results of classifiers' work are on a par with modern methods of identification of the facial features.

## 7. CONCLUSIONS

This article discusses the unique method of the objects identification in the image. The main feature of this meth is using Hough-like voting process. Texture LPB-descriptors are used as voting primitives. Voting weights are defined by training in some training selection. The work describes the training algorithm in detail, and gives recommendations on its implementation. The method was tested by the identification of the special points on the faces from the LFW database. The results confirm the efficiency of the method and its actionability.

## REFERENCES

Viola, Jones: Robust Real-time Object Detection, //IJCV 2001 p 1-25

Juergen Gall, Victor Lempitsky: Class-Specific Hough Forests for Object Detection, //CVPR 2009 pp. 1022-1029

Herbert Bay, Tinne Tuytelaars, and Luc Van Gool: SURF: Speeded Up Robust Features // Computer Vision and Image Understanding (CVIU).Vol. 110, No. 3, pp. 346–359, 2008.

Lowe, David G. "Object recognition from local scale-invariant features"//International Conference on Computer Vision - ICCV , vol. 2, pp. 1150-1157, 1999

Pablo F. Alcantarillaa, Luis M. Bergasab, Andrew J. Davisonc: Gauge-SURF Descriptors // Image and Vision Computing 2013

Rojas, R. AdaBoost and the super bowl of classifiers a tutorial introduction to adaptive boosting. //Freie University, Berlin, Tech. Rep. 2009

# 23

# RICE-PLANDTED AREA EXTRACTION BY TIME SERIES ANALYSIS OF ENVISAT ASAR WS DATA USING A PHENOLOGY-BASED CLASSIFICATION APPROACH: A CASE STUDY FOR RED RIVER DELTA, VIETNAM

D. Nguyen [a,b,] *, W. Wagner [a], V. Naeimi [a], S.Cao[a]

[a] Department of Geodesy and Geoinformation, Vienna University of Technology, Austria – nguyenbaduy@humg.edu.vn , Wolfgang.Wagner@geo.tuwien.ac.at, Vahid.Naeimi@geo.tuwien.ac.at, senmao.cao@geo.tuwien.ac.at
[b] Department of Photogrammetry and Remote Sensing, Hanoi University of Mining and Geology, Vietnam – nguyenbaduy@humg.edu.vn

**KEY WORDS:** Rice phenology, Time series, ENVISAR ASAR WS, SAR, Red River delta, Classification

**ABSTRACT:**

Recent studies have shown the potential of Synthetic Aperture Radars (SAR) for mapping of rice fields and some other vegetation types. For rice field classification, conventional classification techniques have been mostly used including manual threshold-based and supervised classification approaches. The challenge of the threshold-based approach is to find acceptable thresholds to be used for each individual SAR scene. Furthermore, the influence of local incidence angle on backscatter hinders using a single threshold for the entire scene. Similarly, the supervised classification approach requires different training samples for different output classes. In case of rice crop, supervised classification using temporal data requires different training datasets to perform classification procedure which might lead to inconsistent mapping results. In this study we present an automatic method to identify rice crop areas by extracting phonological parameters after performing an empirical regression-based normalization of the backscatter to a reference incidence angle. The method is evaluated in the Red River Delta (RRD), Vietnam using the time series of ENVISAT Advanced SAR (ASAR) Wide Swath (WS) mode data. The results of rice mapping algorithm compared to the reference data indicate the Completeness (User accuracy), Correctness (Producer accuracy) and Quality (Overall accuracies) of 88.8%, 92.5 % and 83.9 % respectively. The total area of the classified rice fields corresponds to the total rice cultivation areas given by the official statistics in Vietnam ($R^2 \approx 0.96$). The results indicates that applying a phenology-based classification approach using backscatter time series in optimal incidence angle normalization can achieve high classification accuracies. In addition, the method is not only useful for large scale early mapping of rice fields in the Red River Delta using the current and future C-band Sentinal-1A&B backscatter data but also might be applied for other rice cultivation areas.

## 1. INTRODUCTION

*"RICE is life, for most people living in Asia"*(Gnanamanickam, 2009). Rice has shaped the cultures, diets and economies of millions of people. Rice is the primary main food supply for more than half of the world's population (N.V.Nguyen, n.d.), providing livelihood and employment for the most of rural populations (Khush, 2005). In many countries of East and Southeast Asia, rice plays a very significant role in society, the agrarian system and livelihoods of a majority of farmers, being a leading user of land and water resources, and featuring heavily in local cultures and traditions (Gnanamanickam, 2009). In the East and Southeast Asia, Vietnam is known as one of the most important rice producers and exporters. It is also the second biggest rice exporters country (after Thailand) (FAO, 2012).

RRD is known as the second most important rice producers and exporters in Vietnam (after Lower Mekong River delta) with the sown areas of 1.139 million hectares for the year 2012 (*Statistical Yearbook of Vietnam 2013*). The area is the country's most densely populated region: 20.439 million inhabitants are crowded onto slightly more than 15.000 km$^2$, 13.881 million of those live in rural areas and only 6.558 million of those inhabitants live in cities. More than 8.100 million people are working in agricultural activities (*Statistical Yearbook of Vietnam 2013*). Rapid population growth, along with intensive rice cultivation, has caused degradation of land in many parts of this region, consequently influencing rice yields and land productivity. An effective rice crop mapping method in the RRD is essential to provide policymakers with accurate information on rice-growing areas so that they can devise better economic development plans, as well as ensure security of the food supply for the region.

Synthetic aperture radar (SAR) imaging systems have demonstrated great potential for the rice fields mapping and monitoring, because of these advantages, such as all time (day /night), weather capacity and wide coverage. The radar backscatter signatures of each frequency band usually have unique patterns that depend on the growth of the rice canopy so that it well adapted to distinguish rice from other land cover types (Toan et al, 1989). However, at least three images should available during the growth cycle and the optimum acquisition dates are during the flooded vegetative phase, at the end of reproductive phase and shortly before harvest (Aschbacher et al., 1995) that sometimes impossible for available SAR sensors. Moreover, previous studies on the assessment of classification methods based on temporal change using ENVISAT ASAR data (A Bouvet et al, 2009; Alexandre Bouvet & Le Toan, 2011; Chen et al., 2007) and studied using RADARSAT (Ribbes & Le Toan, 1999) have emphasized the necessary of a high temporal observation frequency (e.g, around every ten days) to achieve acceptable classification accuracy. To improve number of

* Corresponding author.

measurement per location we proposed a method that combining SAR images from different tracks and different year into a single assimilation of year (Figure 1); however, this leads to a difference in local incidence angle and potentially reduces the accuracy of classification results. This study, we performed local incidence angle normalization by using empirical, regression based approaches to the data to minimize impacts on the mapping results prior to analysis. For classification task, we proposed an automatic approach identifying double cropped-rice used the phonological information of rice crops that we called temporal crop parameters identify, such as, the relative minimum and relative maximum backscatter values, their difference, minimum and maximum increment values between two subsequent acquisitions.

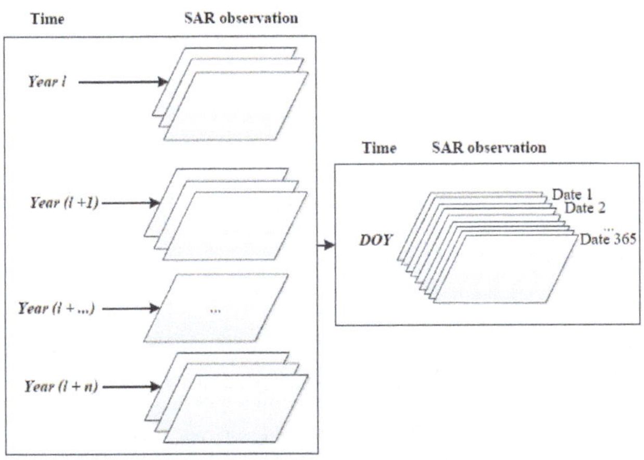


Figure 1. Time series data combination strategy

The aim of this paper is to present a novel method of time series ENVISAT ASAR WS mode data analysis as well as develop a phenology-based classification approach for mapping rice-cropping systems in in the RRD, Vietnam using Gaussian filter algorithm and phenological information of rice crop phenology. The main objects of this study are as follows: (1) local incidence angle normalization by using empirical, regression based approaches, (2) time series data combining and filtering, and (3) a phenology based method to derive rice cropping systematic in RRD. Section 2 describes the test site and rice crop phenology. Section 3 describes the data used in the study. The mapping method is developed in Section 4. Section 5 presents the mapping results and validation. Conclusion and discussion are presented in Section 6.

## 2. STUDY AREA & RICE CROP PHENOLOGY

### 2.1 Study area

Study area locates in the northern part of Vietnam (Lat. 19.5° – 21.5° N, Long. 105.3° – 107.1° E), western coastal zone of the Gulf of Tonkin, includes 10 provinces and covers approximately 15,000 km$^2$. RRD is the second largest rice-producing area of the country and one of the five largest rivers on the of East Asia coast, almost flat plain with most of the land below 20 m (Figure 2).

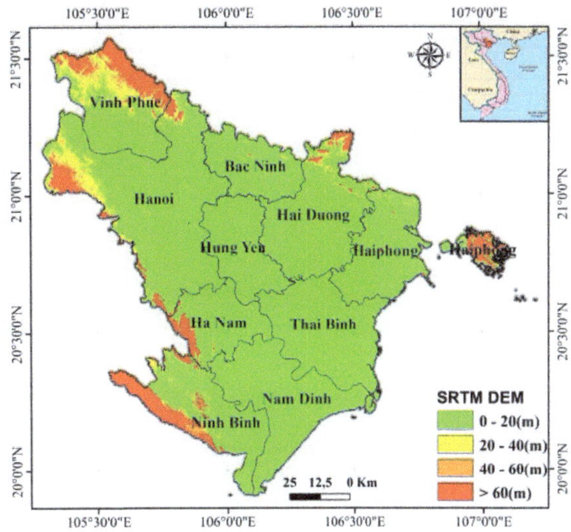


Figure 2. Map of 10 provinces in the RRD & topography from SRTM

Cultivated rice is primarily activity and is considered as the major source of income for most of rural households in this study area. Farmers typically grow three crops per year, double-cropped + winter crop (such as maize, soybean, or sweet potato). Spring or early rice (ER) from February/March to Jun/July, summer or late rice (LR) from Jun/July to September, and a winter crop from October to February/March (because it is often affected by heavy rainfall or storms) (Denning & Vo, 1995; Devienne, 2006). Two major zones could be distinguished in the delta, depending on whether there was tidal action or not. Generally, throughout the delta, the highland rice fields supported one rice cycle during the rainy season, the lowland rice fields one dry season cycle, and the midlands all supported one rainy season cycle and if they could be irrigated during the dry season via manual scooping or by the ebb and flow of the tide, a second rice cycle. These seasons form three main types of rice crop: single-cropped rice, double-cropped rice and double-cropped rice + main crop. Depending on the location and irrigation conditions, the double-cropped rice has two types: long-duration varieties and short-duration varieties (Devienne, 2006).

### 2.2 Rice crop phenology

In RRD, although direct seeding is practiced less in the northern part than in the southern part of the area, it is becoming popular as irrigation systems are improved and as herbicides become available. Firstly, the sowing period when rice is sown directly into some fields flooded soils, other fields are kept flooded with 2-5 cm water depth. In this period radar backscatters from rice fields are actual low dues to specular reflectance. Secondly, the growing period that consists of the vegetative stage, reproductive stage, and ripening stage. In this period the vegetative stage is characterized by a speedy increase in rice plants height, leaf and tiller number, and also the cause of moisture content increase. In addition, free space between rice stems is decreased, and therefore, radar backscatter increases speedy during this period. In microware remote sensing applied for rice mapping point of view, the vegetative phase and reproductive phase are the two most important times, because of in these stages the significant development of tillers, leaves and a gradual increase in plant height, then reaching heading date when the leaves begin to whither and die. In this time period, radar backscatter intensity also begins levelling off. Finally, in the fallow period, when the rice fields are almost bare or flooded, hence radar backscatter responses from rice fields are very low again (Figure 3).

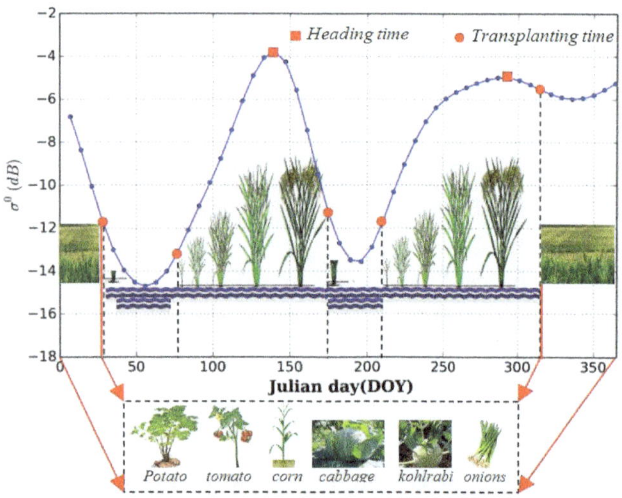


Figure 3. Double-cropped rice + winter main crop pattern plot based on 5 years ENVISAT ASAR WS backscatter data combination (rice plant image from the International Rice Research Institute (IRRI)-Rice Knowledge Bank)

## 3. DATA COLLECTION AND PREPROCESSING

### 3.1 Satellite Data

#### 3.1.1 ENVISAT ASAR Wide swath mode data

This paper uses data acquired by the Wide Swath mode of the ASAR on board of the European Environmental Satellite ENVISAT. In Wide Swath (WS) mode it covers a swath of 405 km, with a spatial resolution of 150 m and incidence angle in each image ranges from 17° to 42°. A total of 120 ASAR WS images which completely or partially cover the Red River delta were taken from European Space Agency (ESA) between 2007 and 2011 (Figure 4).

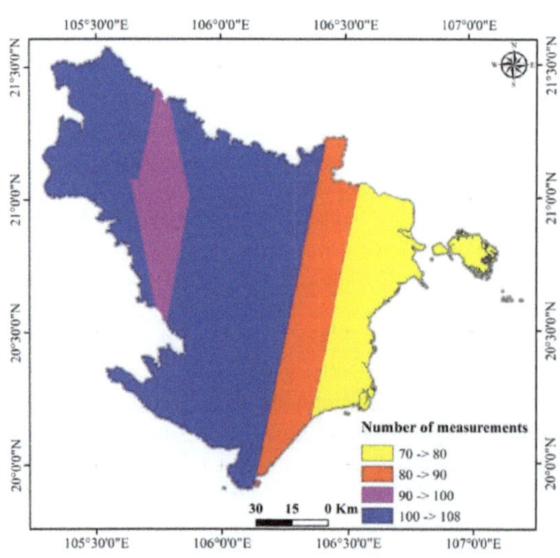


Figure 4. ENVISAT ASAR WSM images over study area for a period of 5 years (2007-2011)

The ENVISAT ASAR WS data were pre-processed by using the open source NEST (Next ESA SAR Toolbox) software. Pre-processing steps consist of geocoding, radiometric calibration, and resampling. In the geocoding step, DORIS precise orbit information and the other ENVISAT auxiliary files were automatically downloaded from ESA via FTP. The Range-Doppler Range -Approach was carried out by using the 30 arc-seconds DEM (SRTM) and bilinear interpolation as the image and DEM resampling method. The resampling of these images to a fixed grid (cover all the study area) in a database was carried out in order to allow efficient time series analysis, which was required for the extraction of the backscatter parameters.

#### 3.1.2 LANDSAT Data

Because of the difficulty of collecting simultaneous ground-truth data for rice cropped types from historical land use maps, the simultaneous or recent available LANDSAT images can be used as a reference parameter to generate training data. By this means, obvious stability and changes in land use land cover could be reliably detected by image interpretation. In this study, due to the "time gap" between the establishments dates of the land use map in 2007 and 2011, all the free and low-cloud cover LANDSAT 7 Enhanced Thematic Mapper Plus ($ETM^+$) and LANDSAT 5 Thematic Mapper (TM) data downloaded for a period of 5 years (2007-2011) through the USGS website http://earthexplorer.usgs.gov/ were used. Then, with the assistance of reference maps (land use map for the year 2005 and 2010), an interpretation key that describes the spectral reflectance and spectral reflectance changed for each rice cropped type was developed and carefully evaluated. When the definition of representative training areas is known, selecting training sites on the image of false-colour composite 4-3-2 (FCC) has the advantage of easily visually distinguishing the classes.

### 3.2 Land Use Land Cover and Rice Area Statistic Data

Ancillary maps, including the Land use & land cover map of 10 provinces in the study at 1/50,000 (2010) collected from the General Department of Land Administration of Vietnam (Figure 5a). Rice area statistics were also provided by the General Statistics Office of Vietnam (Figure 5b). Because the land use land cover map just have been updated every 5 years and are recorded in vector format, thus we used the land use map in combine with LANDSAT time series data to digitize sampling sites vector layers. These vector layers were then overlaid upon the 2007 and 2011 time series Landsat images, and the polygons were modified wherever stability rice fields were found. Digitizing sampling sites then were converted to a raster file (75 m resolution) which used for rice cropped classification and validation (70 percent and 30 percent of this training data were used, respectively. In addition, the accuracy of classification result was verified by comparing with the government rice statistics data at province level. In addition, because of the absent of single crop information in both land use land cover map (2010) and government statistics data, in this study, a new shape file was created from time series LANDSAT data.

## 4. MAPPING METHOD

The classification algorithm developed for identification two types of rice cropping systems in the RRD (from time series SAR backscattering data) includes five main steps, as following:

i) SAR backscatter coefficient normalization (empirical regression-based approach) and combination (Nyquist–Shannon sampling theorem) : $\sigma^{o} \rightarrow \sigma^{o}_{NOR} \rightarrow \sigma^{o}_{NOR\_average}$ ;

ii) The time series data filtering using Gaussian filter (moving window) method: $\sigma^{o}_{NOR\_average} \rightarrow \sigma^{o}_{NOR\_average\_filtered}$ ;

iii) Filtered backscattering profiles of different rice crops and other land cover types are then analysed to selected thresholds for classification;

iv) Classification time series data using decision tree approach;

v) Accuracy assessment using ground truth data, rice statistics at province level, and time series Landsat images.

This algorithm might be extended and applied to identify rice-based cropping or other types of crop in the other regions that have distinct growing seasons.

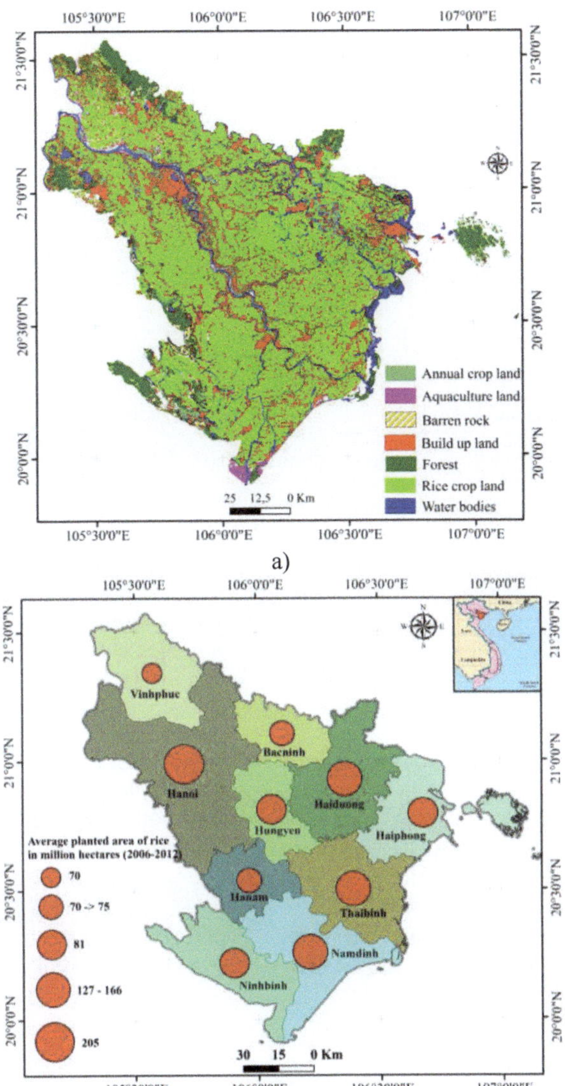


Figure 5. (a) Land use & land cover map of the RRD in 2010 (b) Average of planted area of rice by province and by rice seasons for the period 2007-2011

## 4.1 Backscatter coefficient normalization, combination and filtering.

Following (Loew et al, 2006; Pathe et al, 2009), the local incidence angle and slope parameter were used to normalize backscatter measurements acquired over the study area at different incidence angles, to a single reference incidence angle ($\theta = 30°$). Moreover, in our case, data acquisition sequence (day of year sequence) over the study area is as follow: the maximum and minimum in intervals of respectively 16 days and one day. Hence to guarantee that the information from original datasets are not excluded at and to minimize the impact speckle noise or other environmental conditions, the Nyquist–Shannon sampling theorem was applied to generate average of $\sigma^{o}_{NOR}$ images (Nyquist, 1928). The 7 days SAR backscatter coefficient data was combined to determine the mean value of backscatter, it is denoted by $\sigma^{o}_{NOR\_average}$ . As the result, 52 $\sigma^{o}_{NOR\_average}$ scenes that were then stacked into a composite scene (52 bands). Although averaging time series backscatter coefficient value compositing ($\sigma^{o}_{NOR\_average}$) increases ability to identify double rice crop, data analysis remains affected by the residual effects of cultivated time difference over period 5 years and other environmental conditions. Therefore, a smoothing of temporal $\sigma^{o}_{NOR\_average}$ signal is required. There are many methods of data smoothing. Considering the robust issues, maintain the integrity of signals, and characteristic of noise in $\sigma^{o}_{NOR\_average}$ data, Gaussian filter was applied to smooth time series $\sigma^{o}_{NOR\_average}$ data. The smooth time-series $\sigma^{o}_{NOR\_average\_filtered}$ data was then used for characterizing the temporal characteristics of double rice crop patterns throughout the year.

## 4.2 Thresholds selection and Classification

A straightforward decision tree approach was used to discriminate different type of rice cropped. This approach is based on the unique seasonal patterns of double rice crop in the magnitude of SAR backscatter coefficient throughout the growth cycle. For each pixel, the maximum backscatter coefficient value, minimum backscatter coefficient value, the different between the minimum of local maximum backscatter coefficient value and the maximum of local minimum backscatter coefficient value, and the different between the two consecutive local maximum backscatter coefficient value were identified and calculated from $\sigma^{o}_{NOR\_average\_filtered}$ temporal profile. Accordingly, by scanning the smoothed time profile of $\sigma^{o}_{NOR\_average\_filtered}$ data to find characteristic points including maximum points and inflection points, it is possible to detect phenological stages of rice such as the time of planting, heading and harvesting.

According to time series SAR data for backscatter from rice fields, maximum $\sigma^{o}_{NOR\_average\_filtered}$ values are obtained around heading date. We therefore employed a $\sigma^{o}_{NOR\_average\_filtered}$ time profile on rice field data as the local maximal point would correspond to each heading date. The local maxima algorithm was applied for $\sigma^{o}_{NOR\_average\_filtered}$ to identify heading dates (peaks) of rice as the maximum values that extracted from $\sigma^{o}_{NOR\_average\_filtered}$ time profile. However, because of double-cropped rice is employed on a wide scale study area, it is difficult to identify the appearance of the heading date using only local maximum values from the $\sigma^{o}_{NOR\_average\_filtered}$ time profile. Moreover, other land cover types might have the same number of heading date, like grassland, shrub or small scale mixed horticulture, for example. Consequently several thresholds of feature parameters had to be employed to identify type of rice crop for each pixel. Along with the number of heading time (number of local maxima in time-series backscatter profile that greater than a threshold), two other feature parameters are also used, the first parameter is "absolute value of minimum backscatter different of rice" (the different between the minimum of local maximum and the maximum of local minimum), and the second parameter is "minimum length of rice cultivation" (the different between the two probable dates). Based on training data, a group of SAR backscatter coefficient profile were plotted to select thresholds for corresponding parameters, as can be seen from Figure 6. The first, local maximal points with backscatter coefficient values greater than -8 dB, the second, the different between the minimum of local maximum and the maximum of local minimum with absolute value greater than 2.5 dB, and the third, the "minimum length of rice cultivation" is 77 days.

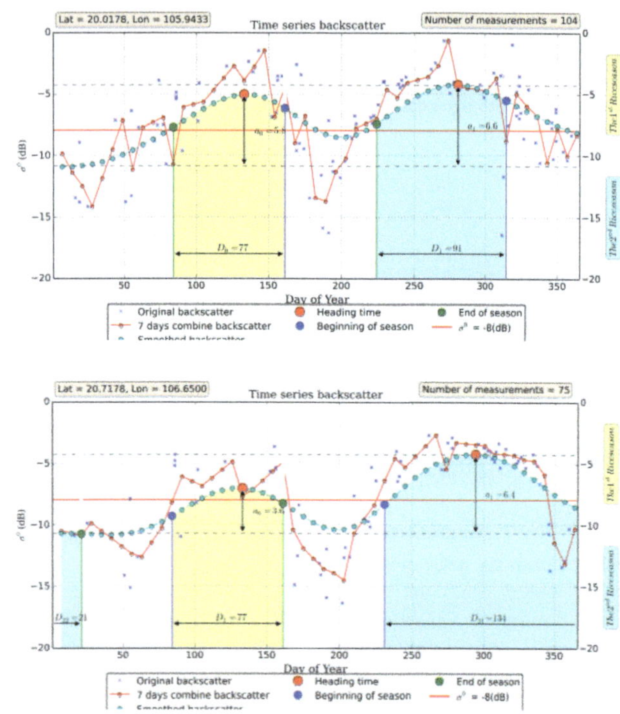


Figure 6. Rice crop phenology parameters retrieval from time series SAR-C band backscatter coefficient (Beginning of rice crop season, End of rice crop season, Heading time, Rice crop amplitude, Length of season)

### 4.3 Validation

In order to validate the accuracy of the classification result, shape files of double-cropped rice was generated based on land use map of the year 2010 and time series LANDSAT data, also called knowledge-base digitizing datasets. These shape files then were converted into binary raster format and used to evaluate the quality of former classification results. True positive pixels are all the pixels in both datasets that have the value of 1. False Negative pixels are the pixels which have a value of 1 in knowledge-base digitizing dataset and a value of 0 in classification results.

## 5. RESULT AND DISCUSSION

### 5.1 Classification result & Validation

Double-cropped rice fields estimated from multi-year and multi-track ENVISAT WS SAR data of the study area is shown in Figure 7. The overestimation in the number of crops per year may have been due to mixed-pixel effects caused by adjacent land surfaces. The accuracy of classification is presented in the Table 1. The former provides the Correctness of 88.8% and Completeness of 92.5% and Quality (Overall accuracy) of 83.9%.

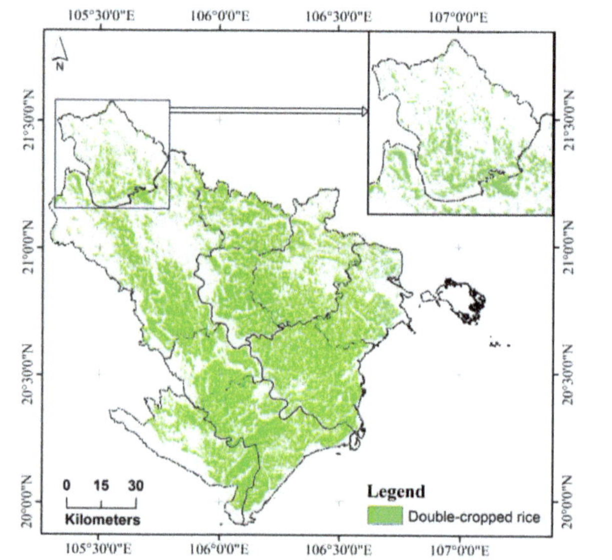


Figure 7. Rice crop map derived from five year time series ENVISAT WS mode SAR data in the RRD

| Paremeter / Crop types | Corr (%) | Comp (%) | Quality (%) |
|---|---|---|---|
| Double rice crop | 92.5 | 88.8 | 83.9 |

Table 1. The result of the classification accuracy assessment

The relationship between the ENVISAT WS time series data derived rice areas and those from the government's rice area statistic for each year was also examined. The results showed that the correlation between two datasets was vigorous. The correlation coefficient ($R^2$) used to measured the adequacy of regression model between two datasets, for rice planted area (Figure 8) indicated that the models could explain almost exactly the variability in the data ($R^2 \approx 0.96$) with a root mean square error (RMSE) of 3993.98 hectares per province.

### 5.2 Discussion

As can be seen from Table 2 and Figure 8, the planted area estimation is excellent for most of the provinces in study area; only in Vinhphuc provinces are the area estimates somewhat less satisfactory. These sources of error include the effect of mixed pixels, omission errors (rice area classified as non-rice). The effect of mixed pixels is more important and is directly related to spatial resolution of the input data that is clearly represented in the case of Vinhphuc province, another factor may be involved (Figure 9).

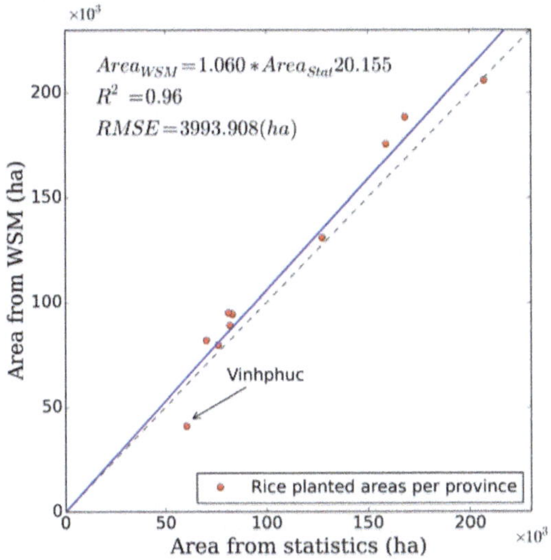


Figure 8. Retrieved rice planted areas per province (in hectares) for the sum of double cropped rice vs. statistical rice planted areas. The thick green line represents the linear regression between the two datasets.

| Province name \ Data | WSM data ( ×10³ ha) | Statistical database (×10³ ha) | Diff (×10³ ha) |
|---|---|---|---|
| Hanoi | 206.0 | 206.9 | -0.9 |
| Vinhphuc | 41.1 | 60.5 | -19.4 |
| Bacninh | 79.6 | 76.2 | 3.4 |
| Haiduong | 131.0 | 127.5 | 3.5 |
| Haiphong | 94.5 | 83.1 | 11.4 |
| Hungyen | 89.1 | 81.9 | 7.2 |
| Thaibinh | 188.7 | 168.3 | 20.4 |
| Hanam | 81.8 | 70.4 | 11.4 |
| Namdinh | 175.8 | 159.0 | 16.8 |
| Ninhbinh | 95.0 | 81.3 | 13.7 |

Table 2. Maximum planted area of rice by province (in ha) for all Season in period 5 year (2007-2011) from the statistical database and from the WSM data

Three pixels were selected in this area to describe the time-series backscatter behaviour (after smoothing with Gaussian filter algorithm) for double-cropped rice and plotted in Figure 10. This figure shows the patterns (the amplitudes of backscatter coefficient after filtered) of double-cropping. The crop season patterns of the same type of rice crop (bur they locate at different location over the study area) are represented in red, green and blue profile. It is worth noting here that the amplitude value of blue profile differs from others, with the value equalling 5.2 dB (bigger than 2.5 dB that was used for double-cropped rice classification), meanwhile, two other profiles (red and green colour) of $\sigma^{o}_{NOR\_average\_filtered}$ have amplitude value quite small (less than 2.0 dB). In addition, it is difficult to observe two heading time characteristic of double-cropped rice from green colour profile. This is a minor issue that led to the double-cropped rice areas classified as non-rice areas. This effect can explain a significant part of the difference between the rice area extent derive from ENVISAT WS data and the rice area extent from the government statistics data. And reasons for this issue come from the fields size, terrain effect, types of rice, as well as cultivation activities in upland areas. It really is available in Vinhphuc province (Figure 10).

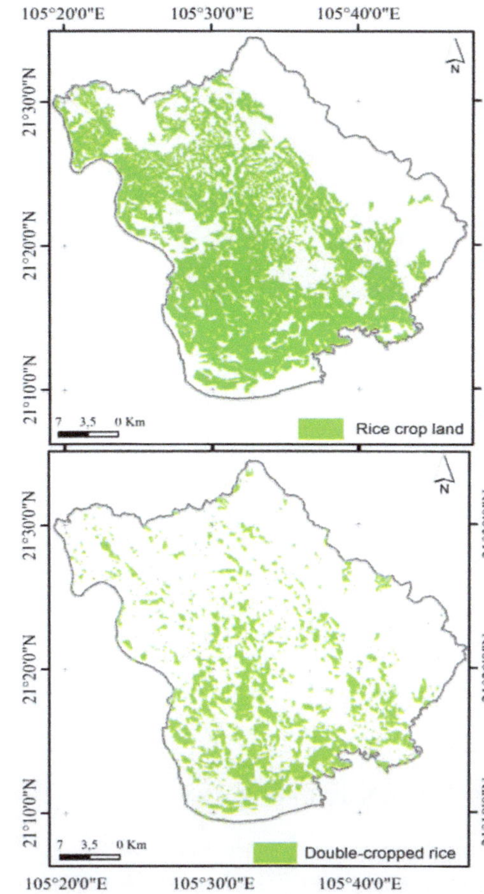


Figure 9. Rice map derived from the WSM dataset (below) and land cover map at scale 1:50000 (above) in Vinhphuc

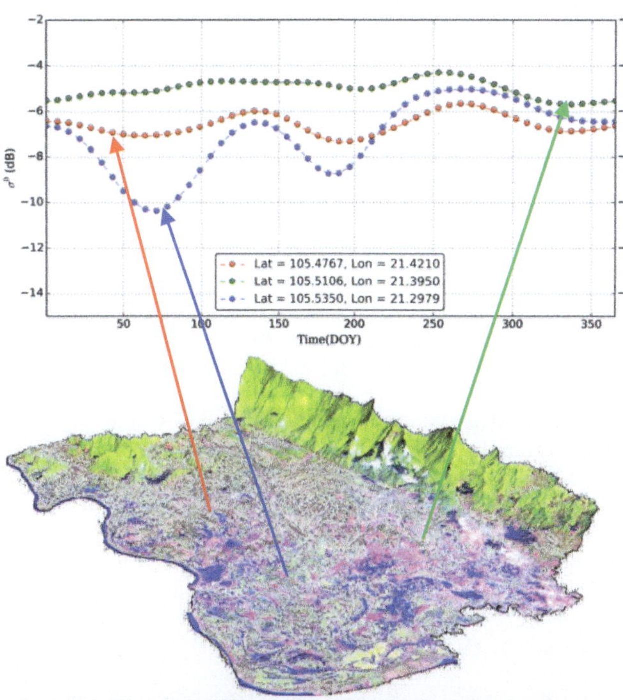


Figure 10. The time series backscatter pattern from Double rice at different location of Vinhphuc province. Red-dot line: Double rice at the open slope areas (wide fields area extend), Green-dot line: Double rice at the complex terrain areas (small fields area extent), Blue-dot line: Double rice at the flat areas.

## 6. CONCLUSION

A novel rice mapping method has been developed, based on former methods using the temporal backscatter time profile as a classification feature, and an adapted method to normalize multi-year and multi-track wide-swath datasets. The effect of the spatial and temporal variation of the incidence angle within the dataset is tackled by using linear relationship coefficients, which slope parameter is derived from multi-temporal ENVISAT WSM data. Five years data composite allows increasing significantly the observation frequency and the size of the mapped areas compared to the former methods using single-track narrow- swath datasets or single year data.

The first rice planted map of double-cropped rice for the entire RRD was generated using ENVISAT ASAR WS mode data. Although several errors causes-notably topography issues resolution bias, as well as mixed-pixel issues, between the classification map and the reference data contributed to the accuracy level of classification results, the Correctness (Corr) and Quality achieved by comparing the classification map and the reference data were approximately 89% and 84% for double-cropped rice, respectively. These results were also reaffirmed by close correlation between the ENVISAT WS mode SAR-derived rice area and the rice area from statistics data ($R^2 \approx 0.96$).

## ACKNOWLEDGEMENTS

The authors would like to thank Department of Geodesy and Geoinformation, Vienna University of Technology for support us the data and software to do this research. The financial support of the ESA funded TIGERNET project (ESA Contract No 4000105732/12/I-NB).

## REFERENCES

Aschbacher, J., Pongsrihadulchai, A., Karnchanasutham, S., Rodprom, C., Paudya, D. R., & Toan, T. Le. (1995). Assessment of ERS-1 SAR data for rice crop mapping and monitoring. *IEEE Geoscience and Remote Sensing Letters*, 83–85.

Bouvet, A., & Le Toan, T. (2011). Use of ENVISAT/ASAR wide-swath data for timely rice fields mapping in the Mekong River Delta. *Remote Sensing of Environment*, *115*(4), 1090–1101. doi:10.1016/j.rse.2010.12.014

Bouvet, A., Le Toan, T., & Lam-Dao, N. (2009). Monitoring of the rice cropping system in the Mekong Delta using ENVISAT/ASAR dual polarization data. *IEEE Transactions on Geoscience and Remote Sensing*, *47*(2), 517–526.

Chen, E. X., Li, Z. Y., Tan, B. X., Pang, Y., Tian, X., & Li, B. B. (2007). Supervised wishart classifier for rice mapping using multi-temporal ENVISAT ASAR APS data. In *Envisat Symposium 2007* (Vol. 2007, pp. 2–7).

Denning, G. L., & Vo, T. X. (1995). Vietnam and IRRI : a partnership in rice research : proceedings of a conference held in Hanoi, Vietnam, 4-7 May 1994. Manila, Philippines; Hanoi, Vietnam: International Rice Research Institute ; Ministry of Agriculture and Food Industry.

Devienne, S. (2006). *Red River Delta: Fifty Years of Change. Moussons.*

FAO. (2012). http://www.fao.org/docrep/016/ap288e/ap288e.pdf.

Gnanamanickam, S. (2009). Rice and Its Importance to Human Life. In *Biological Control of Rice Diseases SE - 1* (Vol. 8, pp. 1–11). Springer Netherlands. doi:10.1007/978-90-481-2465-7_1

Khush, G. (2005). What it will take to Feed 5.0 Billion Rice consumers in 2030. *Plant Molecular Biology*, *59*(1), 1–6. doi:10.1007/s11103-005-2159-5

Loew, A., Ludwig, R., & Mauser, W. (2006). Derivation of surface soil moisture from ENVISAT ASAR wide swath and image mode data in agricultural areas. *Geoscience and Remote Sensing, IEEE Transactions on*, *44*(4), 889–899.

N.V.Nguyen. (n.d.). Global climate changes and rice food security. Retrieved July 29, 2014, from http://www.fao.org/forestry/15526-03ecb62366f779d1ed45287e698a44d2e.pdf

Nyquist, H. (1928). Certain topics in telegraph transmission theory. *AIEE Trans*, *47*, 617–644. doi:10.1109/5.989875

Pathe, C., Wagner, W., Sabel, D., Doubkova, M., & Basara, J. B. (2009). Using ENVISAT ASAR Global Mode Data for Surface Soil Moisture Retrieval Over Oklahoma, USA. *IEEE Transactions on Geoscience and Remote Sensing*, *47*(2), 468–480. doi:10.1109/TGRS.2008.2004711

Ribbes, F., & Le Toan, T. (1999). Rice field mapping and monitoring with RADARSAT data. *International Journal of Remote Sensing*, *20*(2-4), 745–765.

*Statistical Yearbook of Vietnam 2013*. (2012) (pp. 1–7). General Statistics Office of Vietnam.

*Statistical Yearbook of Vietnam 2013*. (2013) (pp. 1–7). General Statistics Office of Vietnam.

Toan, T. Le, Laur, H., Mougin, E., & Lopes, A. (1989). Multitemporal and dual-polarization observations of agricultural vegetation covers by X-band SAR images. *IEEE Transactions on Geoscience and Remote Sensing*, *27*(6), 709–718.

# DEVELOPMENT OF A PEDESTRIAN INDOOR NAVIGATION SYSTEM BASED ON MULTI-SENSOR FUSION AND FUZZY LOGIC ESTIMATION ALGORITHMS

Y. C. Lai [a,] *, C. C. Chang [b], C. M. Tsai [b], S. Y. Lin [c], S. C. Huang [b]

[a] Dept. of Aerospace and Systems Engineering, Feng Chia University, Taiwan - yingclai@fcu.edu.tw
[b] Microsystem Technology Center, Industrial Technology Research Institute, Taiwan - (jinja, scart)@itri.org.tw
[c] Cloud Service Technology Center, Industrial Technology Research Institute, Taiwan - yun049@itri.org.tw

**Commission WG IV/7 and WG V/4**

**KEY WORDS:** Dead Reckoning, Fuzzy Logic, Multi-Sensor Fusion, Inertial Measure Unit, Wearable Device, Portable Device

**ABSTRACT:**

This paper presents a pedestrian indoor navigation system based on the multi-sensor fusion and fuzzy logic estimation algorithms. The proposed navigation system is a self-contained dead reckoning navigation that means no other outside signal is demanded. In order to achieve the self-contained capability, a portable and wearable inertial measure unit (IMU) has been developed. Its adopted sensors are the low-cost inertial sensors, accelerometer and gyroscope, based on the micro electro-mechanical system (MEMS). There are two types of the IMU modules, handheld and waist-mounted. The low-cost MEMS sensors suffer from various errors due to the results of manufacturing imperfections and other effects. Therefore, a sensor calibration procedure based on the scalar calibration and the least squares methods has been induced in this study to improve the accuracy of the inertial sensors. With the calibrated data acquired from the inertial sensors, the step length and strength of the pedestrian are estimated by multi-sensor fusion and fuzzy logic estimation algorithms. The developed multi-sensor fusion algorithm provides the amount of the walking steps and the strength of each steps in real-time. Consequently, the estimated walking amount and strength per step are taken into the proposed fuzzy logic estimation algorithm to estimates the step lengths of the user. Since the walking length and direction are both the required information of the dead reckoning navigation, the walking direction is calculated by integrating the angular rate acquired by the gyroscope of the developed IMU module. Both the walking length and direction are calculated on the IMU module and transmit to a smartphone with Bluetooth to perform the dead reckoning navigation which is run on a self-developed APP. Due to the error accumulating of dead reckoning navigation, a particle filter and a pre-loaded map of indoor environment have been applied to the APP of the proposed navigation system to extend its usability. The experiment results of the proposed navigation system demonstrate good navigation performance in indoor environment with the accurate initial location and direction.

## 1. INTRODUCTION

With the enormous progress in electronic and communication technologies, portable or wearable devices have become popular and affordable nowadays, and they have become widely used in various applications, such as sport monitoring and management, healthcare, visitor navigation, etc. The Global Navigation Satellite Systems (GNSS) such as Global Position System (GPS) can provide accurate location information in open outdoor spaces, but they are not always available especially in GPS-denied environments such as indoors, undergrounds, and urban environments (Hemin et al., 2015).

To compensate the drawback of GNSS, some solutions, such as assisted GPS (A-GPS) and cellular-based positioning, but their accuracy is not enough to meet the demand of most indoor applications (Hui et al., 2007). Therefore, the development of indoor positioning systems, which are categorized into two types: device-free and device-based, has been a hot topic (Harle, 2013). Since the device-free type is still in its infancy and has a lot of limitations, the device-based becomes the main stream of indoor positioning (Youssef et al., 2007). Typical device-based location sensing methods include proximity(Hightower, 2004), triangulation (Harle, 2013), fingerprinting (Chintalapudi, et al., 2010), dead reckoning (Woodman and Harle, 2008), and so on.

Due to the progress in low-cost sensors based on micro electro-mechanical system (MEMS), Pedestrian Dead Reckoning (PDR) systems are becoming feasible options for indoor navigation and tracking. Low-cost sensors mean lower performance and high noisy compared with the expensive sensors. Therefore, the integration and combination of different sensors and algorithms are required to enhance the accuracy and usability of the PDR with low-cost sensors.

In this study, a sensor calibration procedure based on the scalar calibration and the least squares methods are induced in this study to improve the accuracy of the inertial sensors. With the calibrated inertial data, a PDR system based on multi-sensor fusion and fuzzy logic estimation algorithms is proposed. It is a self-contained dead reckoning navigation that means no other outside signal is demanded.

In order to achieve the self-contained capability, a portable and wear able inertial measure unit (IMU) were developed. The developed multi-sensor fusion algorithm provides the amount of the walking steps and the strength of each steps in real-time. Consequently, the estimated walking amount and strength per step are taken into the proposed fuzzy logic estimation algorithm to estimates the walking length of the user. Since the walking length and direction are both the required information of the dead reckoning navigation, the walking direction is

* Corresponding author. This is useful to know for communication with the appropriate person in cases with more than one author.

calculated by integrating the angular rate acquired by the gyroscope of the developed IMU module. Both the walking length and direction are calculated on the IMU module and transmit to a smartphone with Bluetooth to perform the dead reckoning navigation which is run on a self-developed APP.

## 2. HARDWARE DEVELOPMENT AND INERTIAL SENSOR CALIBRATION

### 2.1 Wearable Inertial Sensor Module Development

The sensors of the proposed pedestrian navigation system are low-cost accelerometer and gyroscope that measure the signals from the inertial motion of the target objective. They are also termed inertial sensors. These low-cost sensors based on MEMS technology are commonly found in various applications such as human body motion tracking and flight control of a consumer drone. The inertial sensors transduce the inertial motion into analog or digital signals. In regard to acceleration motion, the accelerometer transfers the objective acceleration into physical value with unit in $m/s^2$ or G. For the rotation motion, the gyroscope transfers the objection rotation into physical value with unit in deg/s. In this study, both the step length and moving direction are estimated from the outputs of the inertial sensors.

The features of MEMS sensors are their light weight and small size. However, the low-cost sensor implies lower performance compared to an expensive and highly accurate one. The major drawbacks of the low-cost sensors are large noise and measurement errors. This is the reason why the calibration and verification of the sensor module are critical and necessary procedures to enhance its accuracy and performance before its application (Lai et al., 2010). Therefore, in order to acquire the stable and reliable data, a low-cost and wearable sensor module has been developed in this study.

As shown in Figure 1, the developed module consists of one 3-axis accelerometer, one 2-axis X-Y gyroscope, one single-axis Z gyroscope, and one 3-axis e-compass. The applied 2-axis IDG-500 and single-axis ISZ-500 gyroscopes could be replaced by one 3-axis ITG-3050 gyroscope, and the adopted inertial sensors could also be updated to one 6-axis sensor containing a 3-axis gyroscope and a 3-axis accelerometer such as a MPU-6050 device, made by InvenSense, Inc. All the sensing data are collected and processed in a micro-controller, which is C8051F500 made by Silicon Labs. The estimated price of the adopted sensors and controller is under $20. In addition, a Bluetooth module was utilized to allow communication with other devices, such as smartphones or laptops, to execute the pedestrian navigation algorithm.

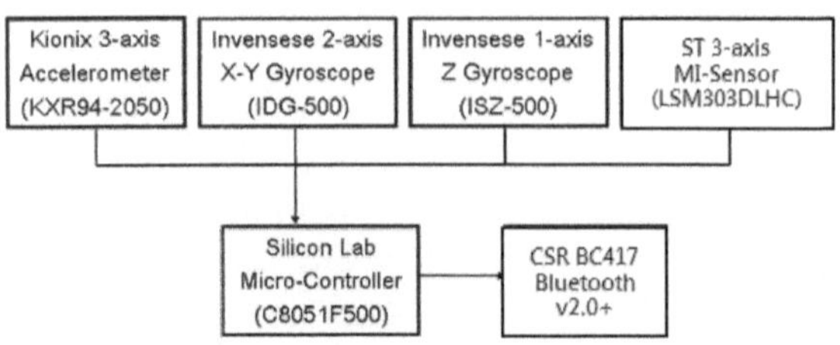


**Figure 1.** The block diagram of the developed inertial sensor module

### 2.2 Sensor Error Model and Calibration Method

As mentioned above, the low-cost MEMS sensors suffer from high noise and errors which are the results of manufacturing imperfections and other side effects. Basically, these errors can be divided into two categories: random constant and time-correlated random process errors (Rogers 2003). Since the dynamical representation of random constant errors is simple and have a significant influence to the performance of the sensor, the selected calibration parameters are random constant errors which could result from scale factors, biases, and orthogonalization angles. The applied sensor error model of the specified sensor error triad is as follows:

$$\vec{y} = \boldsymbol{K}\mathbf{T}\vec{u} + \vec{\boldsymbol{b}} \tag{1}$$

where $\vec{y}$ : ***sensor outputs***
$\vec{u}$ : ***observed physical quantity***
$\boldsymbol{K}$ : scale factor matrix
$\mathbf{T}$ : orthogonalization angles matrix

Figure 2 shows the block diagram of the applied sensor error model. The equations of ***K***, **T**, and ***b*** parameters are as follows:

$$\mathbf{K} = \begin{bmatrix} k_x & 0 & 0 \\ 0 & k_y & 0 \\ 0 & 0 & k_z \end{bmatrix}, \quad \vec{b} = \begin{bmatrix} b_x \\ b_y \\ b_z \end{bmatrix} \tag{2}$$

$$\mathbf{T}_s^p = \begin{bmatrix} 1 & 0 & 0 \\ \alpha_z & 1 & 0 \\ -\alpha_y & \alpha_x & 1 \end{bmatrix} \tag{3}$$

where $\alpha_x, \alpha_y, \alpha_z$ nonorthogonal angles

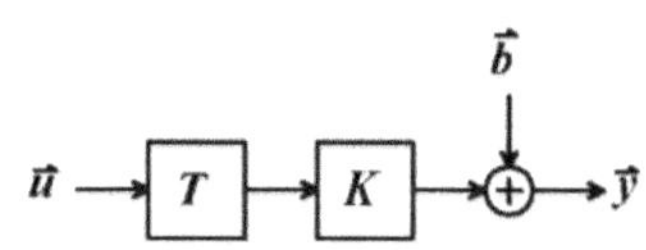


**Figure 2.** Sensor error model of specified sensor triad

The sensor model is applied to different sensor triads, which are the accelerometer, gyroscope triads in this study. With the developed sensor module, the calibration procedures for each sensor triad can be accomplished by rotating the sensors to various random orientations or specified rotations. For each sensor triad, there are nine parameters, three scale factors, three biases and three nonorthogonal angles, to be determined, so that nine or more orientations and rotations are required to determine these parameters. The calibrations for each sensor triad are accomplished by using the scalar calibration and the least squares methods. More detail of the applied calibration methods can be referred to the study in (Lai et al, 2010).

### 2.3 Accelerometer Calibration

The calibration of the acceleration triad can be accomplished by using the magnitude of the Earth's gravity field, which is constant in a specified location without the influence of other disturbances. Therefore, the reference value of the accelerometer triad is the gravitational constant represented by G. It is an empirical physical constant and independent of the orientations of the sensor module when rested. For this reason, the exact orientation of the sensor triad is not required, but it is necessary to maintain static when acquiring the data.

The calibration of accelerometer triad was performed on a six-axis translation and rotation platform as shown in Figure 3. By fixing the sensor triad to at least nine different orientations on the platform and applying the least squares optimization, the nine parameter vector can be determined. The calibration results of the accelerometer triad are presented in Table 1. Figure 4 shows that the errors of the accelerometer triad lower 0.3% of the gravitational gravity G.

**Figure 3.** Six-axis translation and rotation platform

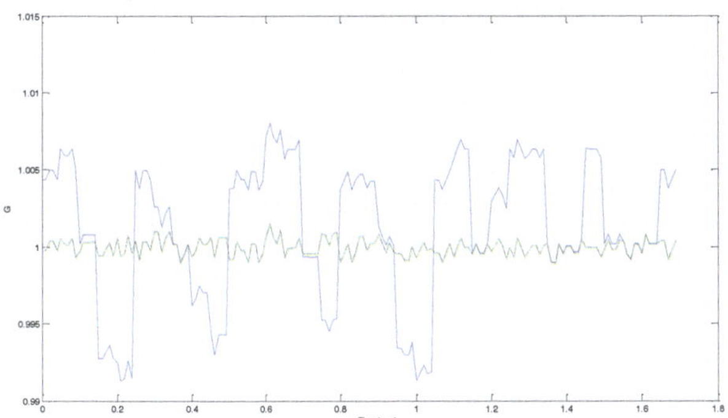

**Figure 4.** The calibration results of acceleration triad

| Axis | Scale factor | Bias | Nonorthogonal angle |
|---|---|---|---|
| x | 0.9968 | 0.0506 | -0.0207 |
| y | 0.9991 | 0.0006 | 0.0004 |
| z | 1.0025 | 0.0019 | -0.0505 |

**Table 1.** Calibrated parameters of the accelerometer triad

### 2.4 Gyroscope Calibration

The calibration of the gyroscope triad also can be accomplished by applying the scalar calibration with a rotation platform. In this study, a single-axis rotation platform, as presented in Figure 5, was utilized to perform the calibration of the gyroscope triad. The platform can hold the sensor triad in desired orientations and perform the specified rotation with constant angular rate while the reference value of the rotating sensor triad can be determined. The specified angular rate was set to 300deg/sec during the calibration. Moreover, the accuracy biases of the gyroscope triad can be first determined by keeping the platform in static condition. After the biases are determined, the remaining unknown parameters are reduced to six, and it requires at least six different rotations of the platform with constant angular rate to apply the least squares optimization.

The result of the gyroscope calibration with this procedure is shown in Table 2. Figure 6 shows that the errors of the gyroscope triad with an error percentage lower than 0.2% of the reference angular rate.

| Axis | Scale factor | Bias | Nonorthogonal angle |
|---|---|---|---|
| x | 1.0139 | 0.0506 | -0.0050 |
| y | 0.9704 | 0.0021 | 0.0007 |
| z | 1.0025 | 0.0006 | 0.0054 |

**Table 2.** Calibrated parameters of the gyroscope triad

**Figure 5.** Single-axis Rotation Platform

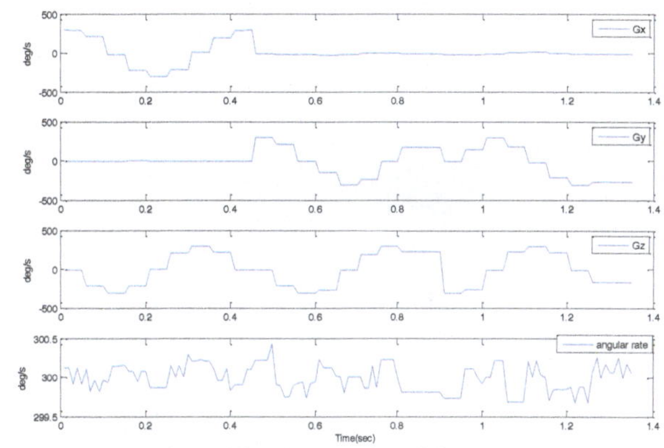


**Figure 6.** The calibration results of the gyroscope triad

## 3. PEDESTRIAN NAVIGATION ALGORITHM

### 3.1 Pedestrian Navigation Scheme

The developed wearable sensor module is designed to be waist-mounted or handheld as shown in Figure 7 and 8. With some preliminary tests, the waist-mount type was selected to implement the pedestrian navigation in this study, due to the higher accuracy and reliability than the handheld type. The waist-mounted type has the capability to describe the motion and action of lower limb more precise. In contrast, the handheld type is affected by how the user holds the device and the swinging motion of the hand while moving.

The proposed pedestrian navigation system is based on the step-and-heading-based dead reckoning and utilizes the multi-sensor fusion and fuzzy logic estimation algorithms. The scheme of this pedestrian navigation system is presented in Figure 9. The process of the multi-sensor fusion algorithm acquires the raw data from the sensor module to perform the sensor calibration and attitude estimation with a second order complementary filter. It then outputs the estimated roll and pitch angles to the Step Strength and Frequency and the estimated heading angle to the Heading Estimation blocks. The Step Strength and Frequency block applies the roll angle, pitch angle, and the velocity in world x-axis, which is the integration of the acceleration in world x-axis, to calculate the step counting, step frequency, and step strength simultaneously. Figure 10 shows the roll angle, pitch angle, and the velocity in world x-axis, which are the forward/backward velocity and denoted as Vx. The red line in the top plot is moving detection index, and the values greater than zero are recorded to be moving period. The red asterisks in the bottom plot indicate the step counting results by utilizing the velocity in world x-axis to perform the step detection. Each asterisk in the plot represents a detected step.

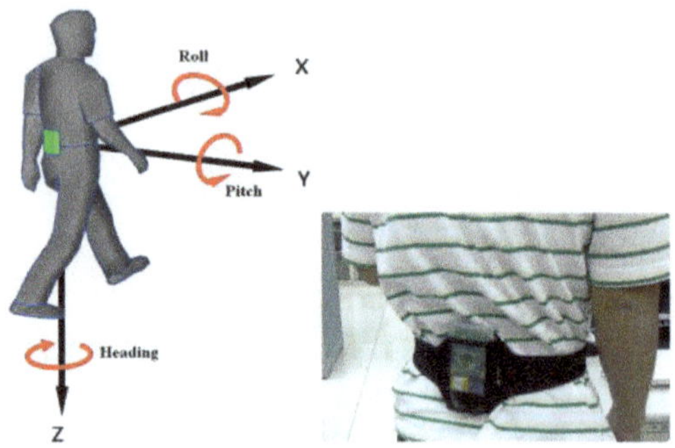


**Figure 7.** Waist-mounted type

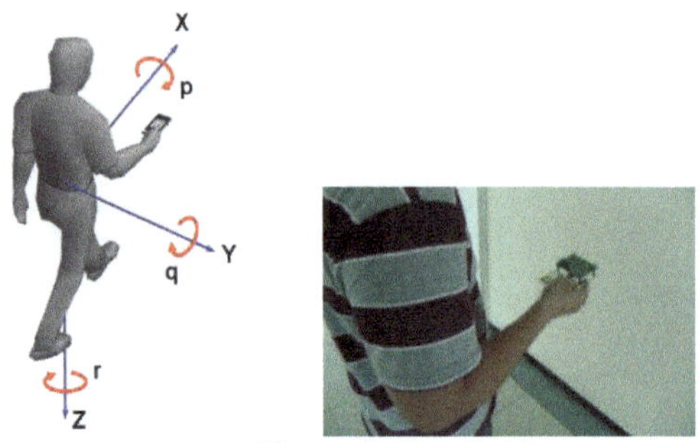


**Figure 8.** Handheld type

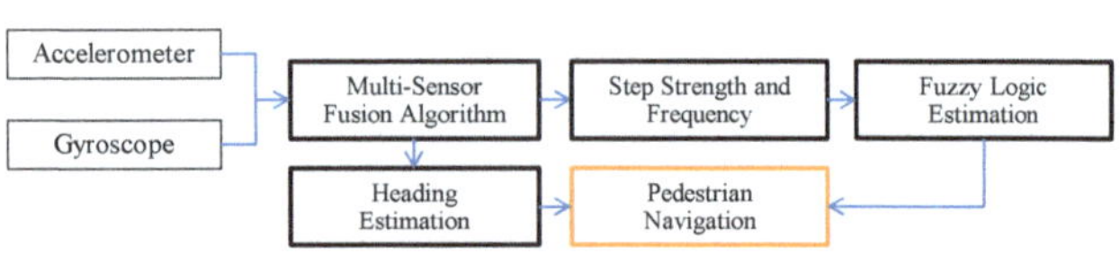


**Figure 9.** The scheme of the proposed pedestrian navigation system

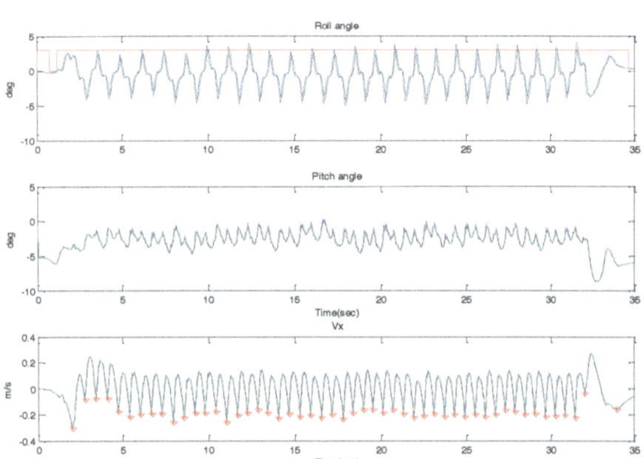


**Figure 10.** The roll angle, pitch angle, and the velocity in world x-axis while walking

### 3.2 Multi-Sensor Fusion Algorithm

After completing the sensor calibration, the calibrated parameters of each sensor triad are implemented to the micro-controller to calibrate the data in real-time. Hence the performance and accuracy of the developed sensor module are improved. The direct outputs of the sensors are acceleration and angular rate of the target objective, but these signals are not enough to describe the motion of human body while walking. In order to extract more information of pedestrian activities, the multi-sensor fusion algorithm is one of the methods to achieve this goal. In this study, the multi-sensor fusion algorithm is used to derive the orientation information of the sensor module. The orientation is useful information to indicate the spatial condition of the target objective when performing various daily activities such as walking, jogging, cycling, etc. In this study, the Euler angles, namely roll, pitch and yaw angles, are adopted to be the orientation representation. Sometimes the yaw angle is termed the heading angle.

The roll and pitch angles are estimated by fusing the outputs of the accelerometer and gyroscopes with a second-order complementary filter (Aggarwal et al., 2006). The basic idea of the complementary filter is to fuse the roll and pitch angles derived from the gyroscope outputs through a high-pass filter and from the accelerometer through a low-pass filter to obtain the estimated attitude, thus compensating for the drift on the gyroscope and for the slow dynamics of the accelerometer. Consequently, the estimated attitudes would have both short-term and long-term accuracies.

Figure 11 presents the estimated roll and pitch angles from different methods when the subject is walking. The data denoted as "acc", ''gyro", and "CF" are the roll and pitch angles derived from the accelerometer, the gyroscope, and the complementary filter methods, respectively. The data reveal that the accelerometer suffers from high noise during high dynamic motion, the gyroscope suffers from drift problem, and only the complementary filter can perform well during both short-term and long-term.

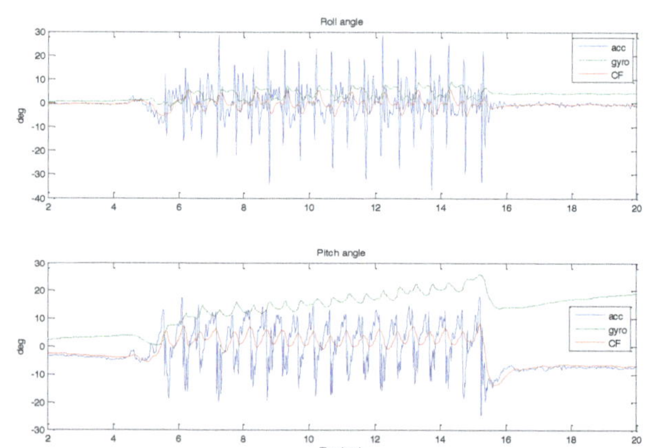


**Figure 11.** The roll and pitch angles from different methods

### 3.3 Fuzzy Logic Estimation Algorithm

Fuzzy logic is applied to be the step length estimation of this study since it is a nonlinear and rule-based method, which means that no complex model is required, and it is comprised of linguistic rules, which implies that it can be easy realized by most computer programming languages. It has been applied to many fields, from control theory to artificial intelligence. Basically, a fuzzy logic consists of three components, the fuzzifier, fuzzy inference engine and defuzzifier. The fuzzifier transforms a crisp input into some fuzzy sets, which are represented by membership functions. The fuzzy inference engine uses the fuzzy rules to reason for the fuzzy output. Then, the fuzzy output is transformed to crisp output by the defuzzifier.

In this study, the adopted fuzzy rules are Mamdani-type (Tzafestas and Venetsanopoulos, 1994), also called max-min controllers, and the variables are two inputs and one output. The defuzzification strategy is the center of gravity method (Tzafestas and Venetsanopoulos, 1994). The input variables of the adopted fuzzy logic are step frequency and acceleration strength, and its output value is the estimated step length. The acceleration strength is calculated from the integration of resultant acceleration of every step. Figure 12 shows the design of the rule table in this study. Due to the difference of height, weight, and leg length between every subject, the design of the rule table varies with different individuals. For every subject, the fuzzy logic design needs to be adjusted from the testing data.

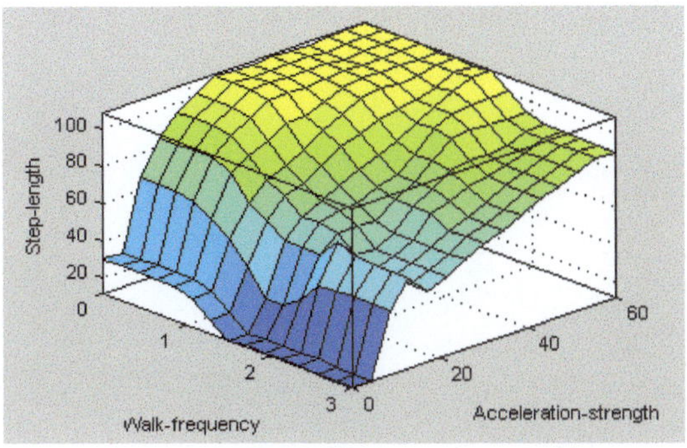


**Figure 12.** Rule table design

## 4. PEDESTRIAN NAVIGATION ALGORITHM

In order to verify the performance and accuracy of the proposed methods, an experiment was designed with a subject to walk in a straight line. The subject was asked to walk fifty steps naturally with different step lengths that are 30cm, 60cm, and 90cm. The walk direction and step lengths are referred to the arrangement and size of floor tiles, respectively. The applied multi-sensor fusion and fuzzy logic estimation algorithms were implemented to the MCU in real-time. The estimated walk heading and step length were transmitted to another device, such as a laptop or a smartphone, to perform pedestrian navigation. In this experiment, a laptop was applied to record the data and to monitor the execution of the system health. The experiment results are presented in Table 3. It shows that the accuracy of the step counting is up to 99% and the maximum error of the estimated step length is under 2cm.

| Experiment | Step counting | | Estimated length(cm) | | Walking distance(m) | |
|---|---|---|---|---|---|---|
| | detected | Reference | Mean | Standard Deviation | Estimated | reference |
| 30cm | 50 | 50 | 29.54 | 1.69 | 14.47 | 15.00 |
| 30cm | 49 | 50 | 28.81 | 2.07 | 14.11 | 15.00 |
| 90cm | 49 | 50 | 91.56 | 2.35 | 44.86 | 45.00 |
| 90cm | 50 | 50 | 90.67 | 1.88 | 45.33 | 45.00 |
| 60cm | 50 | 50 | 59.20 | 4.22 | 29.60 | 30.00 |
| 60cm | 50 | 50 | 58.35 | 1.49 | 28.59 | 30.00 |
| Total | 298 | 300 | | | | |

**Table 3.** The experiment results of step length estimation

With the estimated heading and step lengths, the proposed methods were implemented on smartphone to perform pedestrian navigation in an exhibition hall. In this demonstration, the estimated heading and step lengths are transmitted to a smartphone with Bluetooth. This smartphone has the capability to run a self-developed APP with the map matching and particle filter algorithms. A subject wore the developed IMU module and held the smartphone. After setting the initial location and direction, the subject walked into this exhibition hall and return to the exit as shown in Figure 13. The result shows that the subject is able to return to the same exit with the help of the proposed pedestrian navigation system.

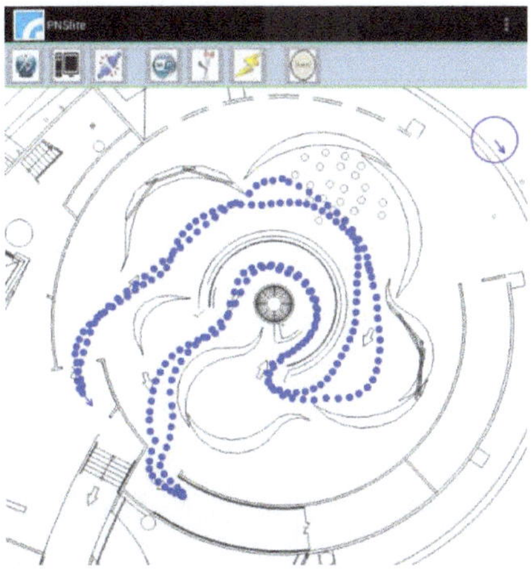

**Figure 13.** Demonstration of the proposed pedestrian navigation in an exhibition hall

## 5. CONCLUSION

A pedestrian indoor navigation system based on the multi-sensor fusion and fuzzy logic estimation algorithms has been successfully developed in this study. The proposed pedestrian navigation system is based on the step-and-heading-based dead reckoning. It applied a self-developed portable and wearable inertial measure unit (IMU). Its adopted sensors are the low-cost inertial sensors, accelerometer and gyroscope, based on the micro electro-mechanical system (MEMS). There are two types of the developed IMU modules, handheld and waist-mounted. Since the low-cost MEMS sensors suffer from various errors, a sensor calibration procedure based on the scalar calibration and the least squares methods has been induced in this study to improve the accuracy of the inertial sensors.

With the calibrated data acquired from the inertial sensors, the step length and strength of the pedestrian are estimated by multi-sensor fusion and fuzzy logic estimation algorithms. The developed multi-sensor fusion algorithm provides the amount of the walking steps and the strength of each steps in real-time. The experiment results show that the accuracy of the step counting is up to 99% and the maximum error of the estimated step length is under 2cm. Due to the error accumulating of dead reckoning navigation, a particle filter and a pre-loaded map of indoor environment have been applied to a self-developed APP to demonstrate the proposed navigation system. This experiment result demonstrated good navigation performance of the proposed system in indoor environment with the accurate initial location and direction.

## ACKNOWLEDGEMENTS

This study was supported in part by the Ministry of Science and Technology of Taiwan under grant number MOST 103 - 2221 - E - 035 - 067.

## REFERENCES

Aggarwal, Priyanka, Zainab Syed, Xiaoji Niu, and Naser El-Sheimy. 2006. Cost-effective testing and calibration of low cost MEMS sensors for integrated positioning, navigation and mapping systems. *Proceedings of XIII FIG Conference.*

Chintalapudi, Krishna, Anand Padmanabha Iyer, and Venkata N Padmanabhan. 2010. Indoor localization without the pain. *Proceedings of the sixteenth annual international conference on Mobile computing and networking.*

Harle, R. 2013. A Survey of Indoor Inertial Positioning Systems for Pedestrians. *Communications Surveys & Tutorials, IEEE 15 (3):1281-1293.*

Hemin, Zhang, Yuan Weizheng, Shen Qiang, Li Tai, and Chang Honglong. 2015. A Handheld Inertial Pedestrian Navigation System With Accurate Step Modes and Device Poses Recognition. *Sensors Journal, IEEE 15 (3):1421-1429. doi: 10.1109/JSEN.2014.2363157.*

Hightower, Jeffrey. 2004. The location stack. *PhD Dissertation Citeseer.*

Hui, Liu, H. Darabi, P. Banerjee, and Liu Jing. 2007. Survey of Wireless Indoor Positioning Techniques and Systems. *Systems, Man, and Cybernetics, Part C: Applications and Reviews, IEEE Transactions on 37 (6):1067-1080.*

Lai, Ying-Chih; Jan, Shau-Shiun; Hsiao, Fei-Bin. 2010. Development of a Low-Cost Attitude and Heading Reference System Using a Three-Axis Rotating Platform. *Sensors 10, no. 4: 2472-2491.*

Rogers, Robert M. 2003. *Applied mathematics in integrated navigation systems*. Vol. 1: Aiaa.

Tzafestas S.G., Venetsanopoulos A.N., 1994, *Fuzzy reasoning in information, decision, and control systems*, Kluwer Academic, Dordrecht ; Boston.

Woodman, Oliver, and Robert Harle. 2008. Pedestrian localisation for indoor environments. *Proceedings of the 10th international conference on Ubiquitous computing, Seoul, Korea.*

Youssef, Moustafa, Matthew Mah, and Ashok Agrawala. 2007. Challenges: device-free passive localization for wireless environments. *Proceedings of the 13th annual ACM international conference on Mobile computing and networking, Canada.*

# FRAMEWORK FOR COMPARING SEGMENTATION ALGORITHMS

George Sithole[a], Langalethu Majola[b]

[a]Geomatics Division, School of Architecture, Planning and Geomatics, University of Cape Town, Private Bag X3, Rondebosch, 7701, South Africa, Email: george.sithole@uct.ac.za
[b]Geomatics Division, School of Architecture, Planning and Geomatics, University of Cape Town, Private Bag X3, Rondebosch, 7701, South Africa, Email: mjllan001@myuct.ac.za

**Commission IV/WG IV/7**

**KEY WORDS:** Segmentation, Algorithms, Mapping, Point Clouds

**ABSTRACT:**

The notion of a 'Best' segmentation does not exist. A segmentation algorithm is chosen based on the features it yields, the properties of the segments (point sets) it generates, and the complexity of its algorithm. The segmentation is then assessed based on a variety of metrics such as homogeneity, heterogeneity, fragmentation, etc. Even after an algorithm is chosen its performance is still uncertain because the landscape/scenarios represented in a point cloud have a strong influence on the eventual segmentation. Thus selecting an appropriate segmentation algorithm is a process of trial and error.
Automating the selection of segmentation algorithms and their parameters first requires methods to evaluate segmentations. Three common approaches for evaluating segmentation algorithms are 'goodness methods', 'discrepancy methods' and 'benchmarks'. Benchmarks are considered the most comprehensive method of evaluation. This paper shortcomings in current benchmark methods are identified and a framework is proposed that permits both a visual and numerical evaluation of segmentations for different algorithms, algorithm parameters and evaluation metrics. The concept of the framework is demonstrated on a real point cloud. Current results are promising and suggest that it can be used to predict the performance of segmentation algorithms

## 1. INTRODUCTION

### 1.1 Introduction

Mass data acquisition techniques that produce a cloud of points for indoor modelling are now widely used. The acquisition techniques used include photogrammetry and laser scanning deployed on mobile or static platforms. The point clouds generated are very large and pre-processing is required before they are in a form useful for modelling or reduction to a wireframe or parameterised model. Pre-processing includes the cleaning, thinning (decimation) and meshing of point clouds.

**1.1.1 Segmentation:** Often before pre-processing a point cloud has to be segmented, i.e., dividing the point cloud into disjoint point sets (called segments) that have homogeneous geometric properties, such as surface curvature.

Let V be a point cloud, then segmentation is an operation, F, on the point cloud which partitions it into n segments, S1, S2… $S_n$. Alternatively it can be thought of as a membership function that generates labels for points in the cloud such that points in the same segment receive the same label.

$$F(V) = \{f(v \in \mathbb{R}3) \mid \forall v \ \in V\}$$

Here f is a function, in this paper called a similarity function that generates a label for a point as a function of the neighbourhood of the point. Definitions of the quality of a segmentation vary; however the general definition given by (Hoover et al. 1996) is used here. According to them a segmentation has to satisfy the following conditions:

1. $V = \bigcup_{i=1}^{n} S_i$
2. $S_i \text{ is spatially connected} \qquad for\ i = 1, 2, 3 \ldots n$
3. $S_i \cap S_j = \{\emptyset\} \qquad for\ i \neq j$
4. $P(S) = true \qquad for\ i = 1, 2, 3 \ldots n$
5. $P(S_i \cup S_j) = false \qquad for\ i \neq j$

Conditions 1 and 3 respectively state that all segments combined yield the original point cloud and no two segments have points in common. Condition 2 states that spatial connectivity must be maintained after the segmentation, i.e. points that are spatially connected are most likely to be on the same surface thus should fall in the same segment. Condition 4 defines a predicate, P, on S that requires points in a segment lie on or near the same geometric body (plane, surface etc.). Finally Condition 5 states that P applied to the combination of points from any two segments should be false, i.e., points from two or more geometric bodies should not be captured in the same segment.

Segmentation algorithms differ according to (a) the features they yield, (b) their similarity/dissimilarity functions, (c) the parameters used in the functions, and (d) the metrics used to describe their performance.

**1.1.2 Segmentation Algorithms:** The features that are often sought in a point cloud are, edges, surfaces, and objects. These can be further distinguished based on whether they are regular or irregular, e.g., a straight line vs a curve. This paper restricts itself to surface extracting segmentation algorithms, e.g. connected components, region growing, and clustering. These methods of segmentation will not be discussed further as they are beyond the scope of the paper.

## 2. THE PROBLEM

Because segmentation is a fit for purpose partitioning of a point cloud into homogeneous point sets (segments) with user defined geometric properties (e.g., surface curvature), the notion of a 'Best' segmentation does not exist. Therefore, no segmentation can be said to be better than the other.

However, if segmentation is to be included in a point cloud processing work flow, then (1) the selection of the ideal segmentation algorithm has to be automated, (2) the selection of optimal parameters of the chosen segmentation algorithm has to be automated, and (3) the quality of the result of a segmentation has to be quantified/evaluated to test for fitness for purpose.

The purpose of this paper is to provide a framework capable of answering these three questions. The problem is non-trivial because the optimality of segmentation is specific to the scene and its geometric qualities. For example a high density point cloud of a scene will likely provide a greater number of segments than its low density equivalent. The next section looks at how others have attempted to answer the above questions.

### 2.1 Previous work

Most of the work done on this subject resides in the image processing and computer vision domain. Nonetheless the concepts are transferrable to point cloud segmentation.

**2.1.1 Automated algorithm selection:** The selection of ideal segmentation algorithms depends on the features being extracted (e.g., surfaces, edges, etc.,) and the properties of the point cloud (e.g., point density, noise, point attributes, etc.). The selection of an algorithm is thus not as problematic as the selection of algorithm parameters and the evaluation of the performance of the algorithm.

**2.1.2 Parameter selection:** A common approach for automatically selecting an algorithm's optimal parameter values is to segment a data set for many different values of the algorithm's parameters. For each segmentation the number of segments generated is counted. Typically, small segments i.e., those that contain less than a user defined number of points, are not counted. An assumption is then made that the segmentation that yields the greatest number of segments is the ideal segmentation.

**2.1.3 Evaluation of Segmentation Algorithms:** Segmentation evaluation methods fall into two general categories, Analytical methods and the Empirical methods.

**Analytical evaluation methods:** These assess segmentation algorithms by analysing their fundamental principles; this is done without actually implementing these algorithms. In this approach the requirements, utilities and complexities of each algorithm are studied without implementing these algorithms (Zhang 1996). Due to the lack of a general theory for segmentation, the analytical approach can fail to obtain all the required properties of the segmentation algorithm (Zhang 1996). For the purpose of this research analytical methods will not be further considered because they do not adequately account for the variability and complexities of different scenes.

**Empirical methods:** These methods quantify the performance of a segmentation algorithm by directly applying them to test point clouds and then evaluating their result. Empirical methods fall into two categories: goodness methods, discrepancy methods and benchmarking.

- **Goodness methods**: This approach evaluates the internal quality of the segmentation algorithm. Statistical information within and between segments is evaluated. These methods do not require prior knowledge about the ideal segmentation. These techniques evaluate performance by quantifying homogeneity, heterogeneity or both in the segments. The first measure evaluates the local variance within a segment, ideally a "good" segmentation should maintain high intra-region uniformity (inversely proportional to variance) and the second measure evaluates the variance of the segments with respect to adjacent or all other segments, ideally the segmentation results should have high inter-region heterogeneity. Examples of goodness methods are presented by Levine & Nazif (1985), Liu & Yang (1994), Huang & Dom (1995), and Borsotti et al. (1998). The application of these measures will be discussed in section 3. Figure 1 exemplifies the concept of goodness evaluation methods. An N-dimensional space is defined for parameters, their values and different metrics. Ideal segmentations exist for certain regions of this space. If a segmentation maps into the 'Ideal' regions, it is then considered optimal. The problem with this method is that it assumes that the definition of 'Ideal' applies for all types of scenes.

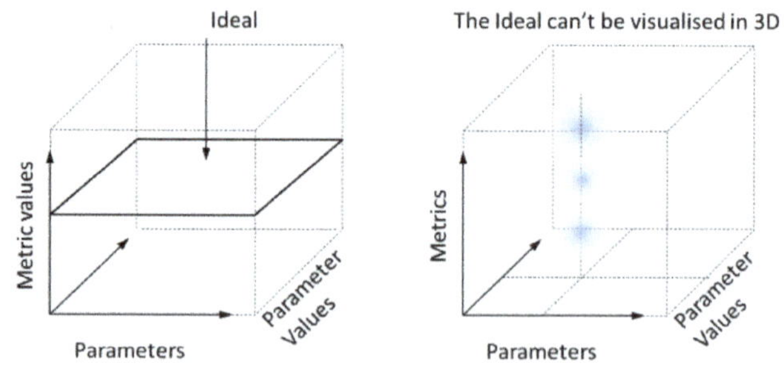


**Evaluation of A single Segmentation**: A single metric tested against a spectrum of parameters. The result is a histogram

**Evaluation of A single Segmentation**: Multiple metrics tested against a spectrum of parameters. The result is a scalar field

**Figure 1** Goodness Methods, two possible variations. Parameters, their values and metrics are defined in an N-dimensional space. The ideal segmentation exists in particular regions of this space.

- **Discrepancy methods**: This approach evaluates the disparity between computed segments and ground-truth segmentations, with the ground-truth being representative of an "ideal" segmentation. The larger the disparity the lower the performance of the segmentation algorithm. A number of measures are used in these methods for example, number of mis-segmented points, position of mis-segmented points, feature values of segmented regions etc. Examples of discrepancy methods are presented by Yasnoff et al. (1977), Huang & Dom (1995), and Chen et al. (2009).

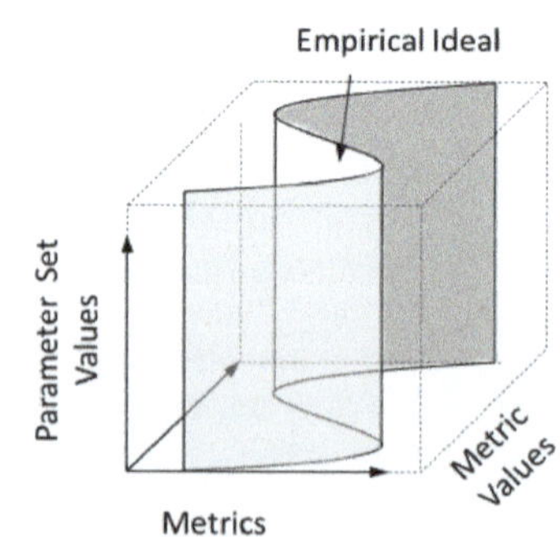


**Evaluation of A single Segmentation:**
A single metric tested against the values of multiple parameter sets

**Figure 2** Discrepancy methods. Metrics, their values and parameter sets for a single segmentation are defined in an N-dimensional space. The metrics for a segmented reference point cloud (the empirical ideal) is mapped into this space. The parameter sets that yield segmentations that map into the neighbourhood of the empirical ideal are treated as being optimal.

Figure 2 shows the concept of the discrepancy method. The metrics for a segmented reference point cloud (the empirical ideal) is mapped into an N-dimensional space defined by the Metrics, their values and parameter sets for a single segmentation. Parameter sets that yield segmentations that map into the neighbourhood of the empirical ideal are treated as being optimal. The limitation of this method is that typically a single algorithm is tested against a single metric at a time.

- **Benchmarking**: The results of segmentation can be quantified using a variety of metrics. Typically metrics have to be considered together because a single metric on its own is not enough for an evaluation. For this reason evaluation methods use benchmarks. Benchmarking expands on the principle of empirical discrepancy evaluation, by testing algorithms against a range of 'ideal' segmentations and a range of metrics. Benchmarks are used to rank segmentation algorithms according to performance metrics and the types of scenes being segmented. Examples of benchmarking for segmentation are presented by Hoover et al. (1996), Chen et al. (2009), Estrada & Jepson (2009), Zhao et al. (2011) and Li et al. (2013).

Mersmann et al. (2010) describe the process of benchmarking as follows:

1. Define the problem domain; restrict the domain of the benchmarking process to a specific application, e.g., benchmarking of point cloud segmentation algorithms.
2. Define the algorithms and input parameters to be benchmarked. Also select a set of performance indicators.
3. Select a standard algorithm, method or data set.
4. Lastly, rank the algorithms according to their performance, against the standard algorithm or data set.

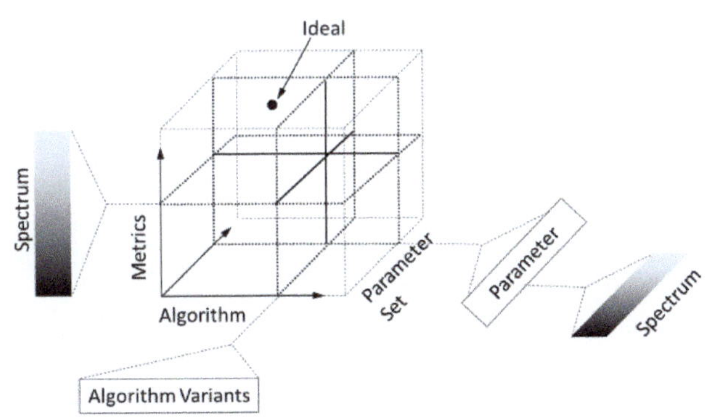


**Figure 3** Benchmarking. Many algorithms are compared against many parameter sets, many metrics and reference segmentations (the empirical ideal).

Figure 3 shows the concept of benchmarking. For a single problem domain, many algorithms are compared against many parameter sets and many metrics. The problem here is that because of the multiplicity of algorithms, parameter sets, and metrics, finding the ideal segmentation within this N-dimensional space is not obvious. Moreover, as already mentioned, the benchmark has to be applicable across multiple domains (scene types). In the next section a benchmarking framework is proposed to address this problem.

## 3. PROPOSED BENCHMARKING FRAMEWORK

The benchmarking framework proposed here is designed with the intention of answering the following question: "Given point clouds of different scenes, what are ideal segmentation algorithms to use and what are their ideal parameters?"

The solution proposed here is to devise a benchmarking frame work for comparing the performance of segmentation algorithms. The framework is composed of three elements (1) a parameter spectrum, (2) a performance metric spectrum, and (3) a set of segmented reference point clouds. The framework provides both a quantitative and visual comparison of segmentation algorithms.

### 3.1 Elements of the framework

**3.1.1 Parameter Spectrum:** In a segmentation algorithm, a similarity function is used to test whether two neighbouring points belong to the same segment. If the function yields a value less than a given threshold the points are judged to belong to the same surface. The similarity function is here given by f(v, $\phi$), where v is a point in the cloud and $\phi = \{\phi_1, \phi_2, \phi_3, \dots \phi_p\}$ is the parameter set, Domain:{$\phi \in \Re$ }, of the function f. The parameter set, $\phi$, contains p number of parameters, $\phi_i$. Typical parameters used in segmentation algorithms are:

- **Proximity:** Points that belong to the same surface are often assumed to be close to each other. Proximity is often computed using the Euclidean distance between two points.
- **Curvature**: Curvature is the amount by which the surface of an object deviates from being planar. The curvature at a point is computed using a number of methods, e.g., mean curvature and Gaussian curvature. An example of computing curvature is presented by Rabbani et al. (2006) who estimate surface curvature using the variation of point normals.
- **Colour difference**: Neighbouring points on the same surface are often expected to have the same colour, i.e., the difference in their colours is expected to be small. Colour difference is typically computed as the Euclidean distance between the colour of two points in an RGB colour cube.

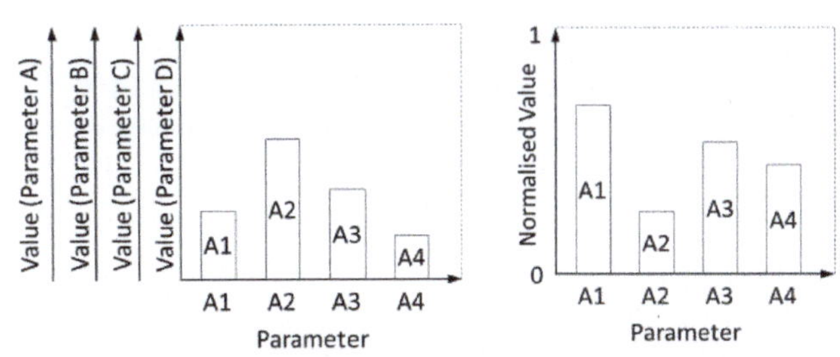


**Figure 4** A segmentation done with four parameters A1 to A4 (e.g., curvature, etc). (a) The parameters have different useable ranges. (b) For visual convenience the parameters are normalised against their useable ranges.

Associated with every parameter is a useable range, $\phi_{i.a} < \phi_i < \phi_{i.b}$, i.e., that range for which a parameter yields segments (see Figure 4a). The values $\phi_{i.a}$ and $\phi_{i.b}$ are the lower and upper bounds of this range respectively. For example if curvature is expressed as an absolute angle, then this angle has a range from 0 to $\pi$, i.e., $\phi_{i.a} = 0$ and $\phi_{i.b.} = \pi$. Because the parameters have different useable ranges it becomes convenient to normalise them over the useable range, i.e., re-range them between 0 and 1 (see Figure 4b).

The p-dimensional space formed by the p parameters and their normalised ranges now constitute the parameter spectrum. For visual evaluation it is convenient to represent this p dimensional space in 2 dimensions. The approach favoured here is a 100% stacked bar chart of the normalised parameter sets as shown in Figure 4b. An example of this is shown in Figure 5. Here multiple parameter sets of a segmentation algorithm can be shown (in the case of Figure 5 it's four, namely A, B, C and D). The relative size of the parameter values in a parameter set indicate the relative strength of each parameter used for that segmentation.

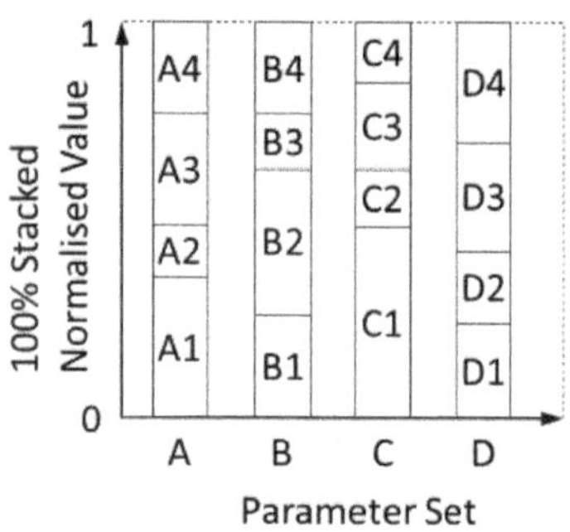


**Figure 5** Parameter spectrum as 100% stacked normalised parameter sets.

**3.1.2 Performance Metric Spectrum:** As mentioned in section 2 algorithms have to be compared across a spectrum of metrics. Typical Metrics are:

- **Segment count**: This is the number of segments generated after segmentation. Typically segments containing less than a user defined number of points are not counted.

- **Homogeneity**: This is a function of the similarity of points within a segment across a segmentation. Two coomon forms of homogeneity are Intra-segment homogeneity and Inter-segment homogeneity. Intra-segment homogeneity is a measure of the disparity within a segment. Inter-segment homogeneity is a measure of the disparity between neighbouring segments. More on this can be found in Levine & Nazif (1985), Liu & Yang (1994), Huang & Dom (1995), Borsotti et al. (1998) and Radoux and Defourny (2008).

- **Heterogeneity**: This is a function of the di-similarity of points within a segment across a segmentation

- **Fragmentation metric**: After segmentation points belonging to the same surface maybe segmented into $n_f$ number of segments. Here $n_f$ is called the fragmentation of a segment. Statics of the distribution of $n_f$ across all the segments yields a set of metrics for fragmentation, e.g., mean, standard deviation, percentile of $n_f$.

- **Deviations from a modelled surface**: The points in a segment can be used to model a surface. The distribution of the orthogonal distance/deviation of the segment points from this surface can be determined. For ideal segments the range of this distribution should be small.

Each segmentation algorithm has a parameter set. Each segmentation is performed for given parameter set values. A given segmentation is also evaluated by a set of metrics. Therefore, associated with each segmentation is a set of metrics and corresponding parameter set values (see Figure 6). Figure 6a shows the metric value for four parameter sets. This however is restrictive because it does not permit the visual comparison of multiple metric values against multiple parameter sets. To overcome this restriction the multiple metrics are represented as a 100% stacked bar chart of the normalised metric sets as shown in Figure 6b. Note here that by visualising Figure 5 and Figure 6b side by side it is now possible to compare segmentation parameter sets against their corresponding metrics.

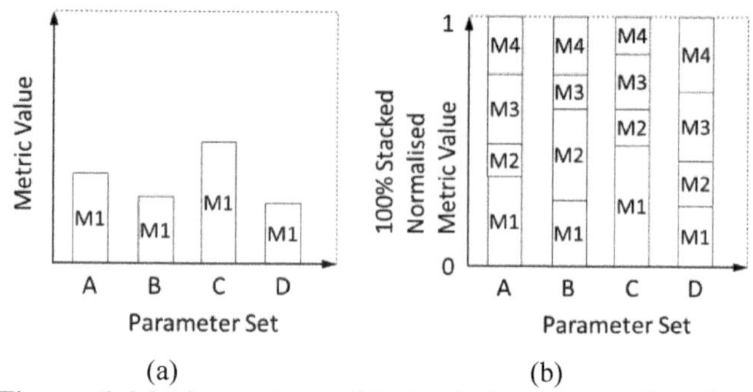


**Figure 6** Metric spectrum. (a) A single metric value for multiple parameter sets A, B, C and D. (b) Metric spectrum as 100% stacked normalised metric sets.

**3.1.3 Segmented Reference Point Clouds:** As mentioned already segmentations are fit for purpose. The purpose changes with the scene and the intended application. The segmented reference point clouds are fit for purpose segmentations for different types of application (e.g., extracting planes, edges, etc.). Segmented reference point clouds can be generated manually or they can be points that have been segmented by trial and error and found to be optimal.
By calculating metrics for each scene type the optimal metrics for each scene type are obtained, Figure 7.

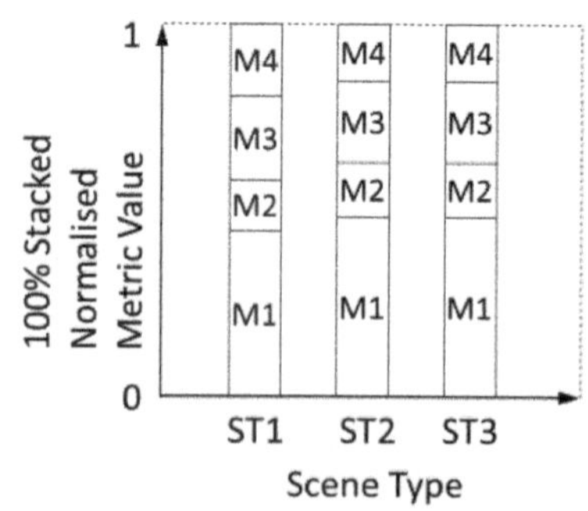


**Figure 7** Metric spectrum for different scene types.

**3.1.4 Using the Benchmark**: By viewing the different charts side by side (Figure 5, Figure 6b, and Figure 7) it is now possible to compare segmentation parameters against their corresponding metrics and optimal metric sets. The problem of visually evaluating the parameter sets, parameter values, metrics and metric values in N-dimensional space has been reduced to 2 dimensions.

Figure 8 shows an example of one way of using the framework. In this example the best segmentation parameters for a scene of Type ST1(e.g., urban-residential) are being sought. The metrics for the reference ST1 are compared to the metric spectrum for the parameter sets of the segmentation algorithm. The closest match is parameter C. Therefore, this is chosen as the 'Algorithm – Parameter Set' combination likely to yield the best segmentation result.

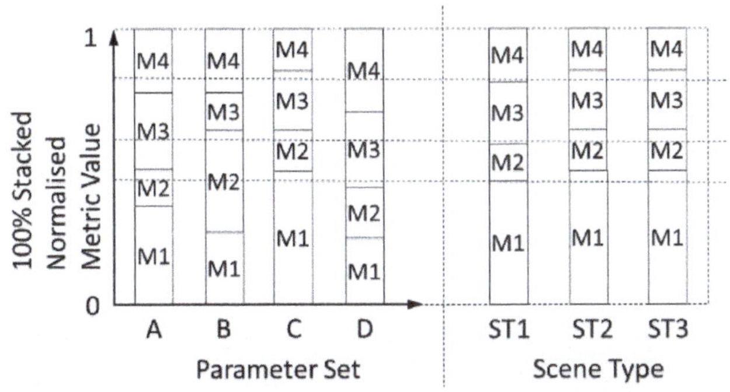


**Figure 8** Example of finding the optimal parameter set for a scene type. Here the parameter set C has the metrics that best correspond to scene ST1. Therefore if a scene is of type ST1, for this algorithm the parameter set C is the optimal choice.

One final problem remains, and this is to quantify and rank parameter sets and segmentation algorithms for given scene types. The concept behind the proposed ranking is shown in Figure 9. The absolute difference of corresponding metric values for the reference scene and a parameter set are summed. The sum for a given parameter set, $r_P$, is given in equation 1, where m is the number of metrics in the metric spectrum.

$$r_P = \sum_{i=0}^{m} |M_{i,P} - M_{i,ST}| \quad (1)$$

If the parameter set matches the scene type, then $r_P$ should be small. The value $r_P$ can be calculated for a number of parameter sets. The values can then be ranked from smallest to largest. The best match is the parameter set with the smallest $r_P$.

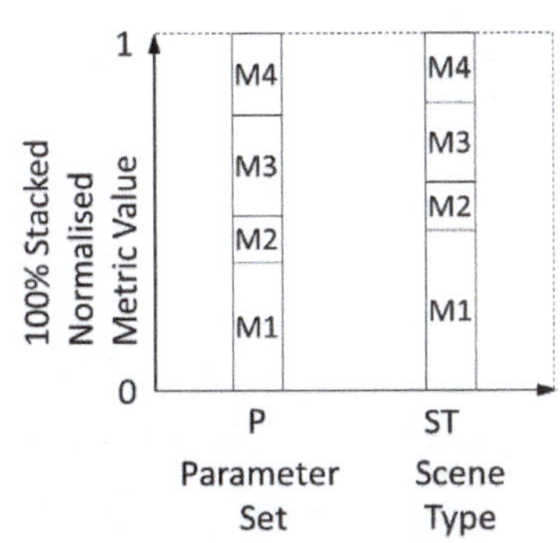


**Figure 9** Comparison of two sets of metrics. A parameter set and a scene type match if the difference between the values of their corresponding metric values is small.

# 4. RESULTS

To test the concept of the framework a point cloud of a part of the Cape Town harbour was used. The point cloud has an average spacing of about 25 cm. This point cloud was manually segmented. This segmentation yielded 157 surfaces which included roof facets, domes and the ground, shown in Figure 10.

The point cloud was segmented using a three parameter (angle between neighbouring point normal, neighbourhood size, and minim segment size) region growing algorithm. Twenty seven different segmentations were done using different combinations of values (parameter sets). For each segmentation five metrics were calculated, absolute deviation, homogeneity (three variants, namely homogeneity, NPSSD and ratio p), and heterogeneity. Figure 11 shows the chart in which the parameter sets and their corresponding metric spectrum (compare this to Figure 8) are mapped. Also show in the figure is the metrics for the reference segmentation (ST, this is at the far right of the chart).

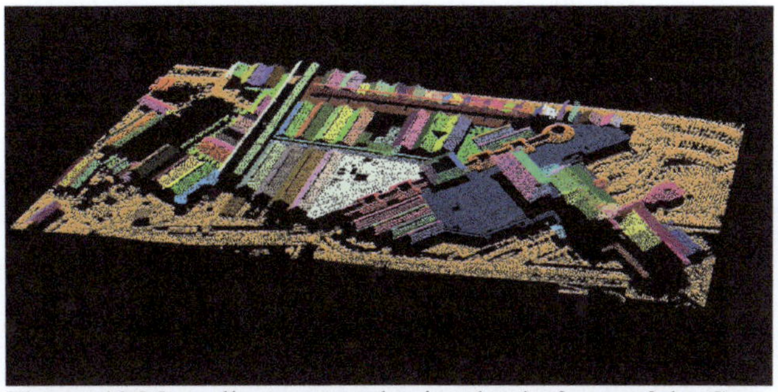

**Figure 10** Manually segmented point cloud of part of the Cape Town harbour.

In Figure 11 it is apparent that parameter sets 1 to 9 are comparable to the reference segmentation. It's therefore expected that a region growing segmentation done with parameter sets 1 to 9 should yield fairly acceptable segments. Visual inspection showed that parameter set 3 yielded the best results while the other eight segmentations where tolerable.

Identification of the optimal parameter set should improve by using more metrics. Here only five were used and this may explain why the other eight parameter sets identified as ideal.

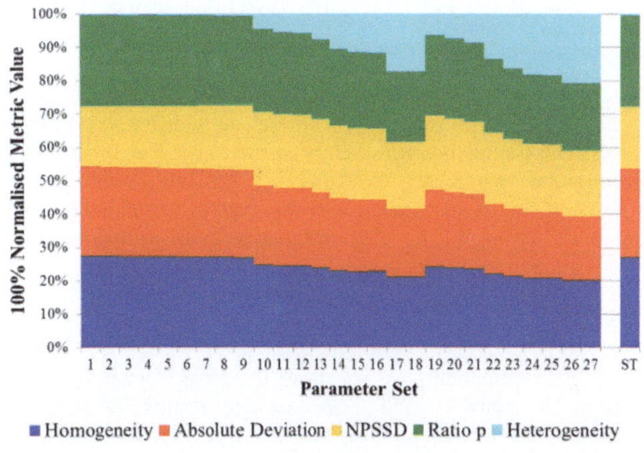


**Figure 11** Performance Metric Spectrum for a particular algorithm. The metrics used to assess the algorithm are absolute

deviation, homogeneity (three variants, namely homogeneity, NPSSD and ratio p), and heterogeneity. The right most column is the metric spectrum for the reference scene (ST). The vertical-axis shows the 100% stacked normalised metrics values and the horizontal-axis shows the different parameter sets (here 27 different sets of parameters sets were used)

## 5. DISCUSSION

Early results are promising and suggest that the framework is sound. However, the framework does have shortcomings:

- **Too few metrics**: At the moment only five metrics were tested in the framework. Two important metrics were not considered in the tests, and these are metrics for over segmentation (fragmentation), and under segmentation (fusion). As the framework is developed these metrics will be included as metrics in the framework.
- **Geometry of the point cloud**: The framework doesn't account for variations in the geometry of a point cloud, e.g., point spacing. How this will be built into the framework will have to be considered.
- **Surface Extraction:** The framework has only been tested for surface extraction. Segmentation algorithms also extract edges, objects and so forth. The framework will have to be tested for edge and object extraction algorithms.

## 6. CONCLUSION

As more feature extraction algorithms use segmentation, so too will automated segmentation gain in importance. This paper proposes a framework for evaluating segmentation algorithms with a view to providing a mechanism for automatically selecting both optimal algorithms and optimal parameters for different types of scenes.

The framework extends the concept of benchmarking and attempts to reduce the selection problem from an N-dimensional domain to a 2-dimensional domain. The purpose of this is to permit visual inspection.

The framework has been tested on a real data set and early results are promising although much more work needs to be done. The framework still needs to be expanded to account for varying geometries of point clouds, additional metrics, and the extraction of other types of features (e.g., edges, objects).

When the framework is complete an attempt will be made to build a database of reference segments for many different types of scenes. The framework will then be tested to see if automated algorithm and parameter selection is possible with the framework. If this is successful, then this type of algorithm and parameter selection will be tried on other types of problems to see if it is general in application.

### 6.1 References

Borsotti, M., Campadelli, P., Schettini, R., 1998: Quantitative evaluation of color image segmentation results. *Pattern recognition letters*. 19(8), pp. 741-747.

Chen, X., Golovinskiy, A., Funkhouser, T., 2009. A benchmark for 3d mesh segmentation. *ACM Transactions on Graphics (TOG)*. Vol. 28, p. 73.

Estrada, F. J., Jepson, A. D., 2009. Benchmarking image segmentation algorithms. *International Journal of Computer Vision*. 85(2), pp. 167-181.

Hoover, A., Jean-Baptiste, G., Jiang, X., Flynn, P. J., Bunke, H., Goldgof, D. B., Bowyer, K., Eggert, D. W., Fitzgibbon, A., Fisher, R. B., 1996. An experimental comparison of range image segmentation algorithms. *IEEE Transactions on Pattern Analysis and Machine Intelligence*. 18(7), pp. 673-689.

Huang, Q., Dom, B., 1995. Quantitative methods of evaluating image segmentation. *Proceedings, International Conference on Image Processing, IEEE*. 3. pp. 53-56.

Levine, M. D., Nazif, A. M., 1985. Dynamic measurement of computer generated image segmentations. *IEEE Transactions on Pattern Analysis and Machine Intelligence*. (2), pp. 155-164.

Li, H., Cai, J., Nguyen, T. N. A., Zheng, J., 2013. A benchmark for semantic image segmentation. *IEEE International Conference on Multimedia and Expo (ICME), IEEE*. pp. 1-6.

Liu, J. & Yang, Y. H., 1994. Multiresolution color image segmentation. *IEEE Transactions on Pattern Analysis and Machine Intelligence*. 16(7), pp. 689-700.

Mersmann, O., Preuss, M., Trautmann, H., 2010. Benchmarking evolutionary algorithms: Towards exploratory landscape analysis, Springer.

Rabbani, T., van den Heuvel, F. & Vosselmann, G., 2006: Segmentation of point clouds using smoothness constraint. *International Archives of Photogrammetry, Remote Sensing and Spatial Information Sciences*. 36(5), pp. 248-253.

Radoux, J., Defourny, P., 2008. Quality assessment of segmentation results devoted to object-based classification. *Object-Based Image Analysis, Lecture Notes in Geoinformation and Cartography, Springer Berlin Heidelberg. Blaschke, T., Lang, S., Hay, G.* pp. 257-271.

Yasnoff, W. A., Mui, J. K., Bacus, J. W., 1977. Error measures for scene segmentation. *Pattern Recognition*. 9(4), pp. 217-231.

Zhang, Y. J. 1996. A survey on evaluation methods for image segmentation. *Pattern recognition*. 29(8), pp. 1335-1346.

Zhao, Y., Nie, X., Duan, Y., Huang, Y, Luo, S., 2011, A benchmark for interactive image segmentation algorithms. *IEEE Workshop on Person-Oriented Vision (POV), IEEE*. pp. 33-38.

# AUTOMATIC EXTRACTION OF ROAD SURFACE AND CURBSTONE EDGES FROM MOBILE LASER SCANNING DATA

A.Miraliakbari, M. Hahn, S. Sok

Department of Geomatics, Computer Science and Mathematics, University of Applied Sciences Stuttgart, Schellingstraße 24, 70174 Stuttgart, Germany, - (alvand.miraliakbari, michael.hahn, sokunthet.sok)@hft-stuttgart.de

**WG IV/7 and WG V/4**

**KEY WORDS:** Automation, Extraction, Laser scanning, Mapping, Mobile, Surface, Urban

**ABSTRACT:**

We present a procedure for automatic extraction of the road surface from geo-referenced mobile laser scanning data. The basic assumption of the procedure is that the road surface is smooth and limited by curbstones. Two variants of jump detection are investigated for detecting curbstone edges, one based on height differences the other one based on histograms of the height data. Region growing algorithms are proposed which use the irregular laser point cloud. Two- and four-neighbourhood growing strategies utilize the two height criteria for examining the neighborhood. Both height criteria rely on an assumption about the minimum height of a low curbstone. Road boundaries with lower or no jumps will not stop the region growing process. In contrast to this objects on the road can terminate the process. Therefore further processing such as bridging gaps between detected road boundary points and the removal of wrongly detected curbstone edges is necessary. Road boundaries are finally approximated by splines.
Experiments are carried out with a ca. 2 km network of smalls streets located in the neighbourhood of University of Applied Sciences in Stuttgart. For accuracy assessment of the extracted road surfaces, ground truth measurements are digitized manually from the laser scanner data. For completeness and correctness of the region growing result values between 92% and 95% are achieved.

## 1. INTRODUCTION

Road extraction from aerial images, airborne laser scanning data, and high-resolution satellite images has a long research tradition. The main purpose of those developments was to generate or update roads maps. However, many details of a road surface are not visible in those data. With mobile laser scanning this limitation is overcome. These systems capture the road corridor and generate huge point clouds that describe a road scene in detail.
Road surface extraction from laser scanning data is an important preparatory step for many tasks related to road maintenance and infrastructure planning. The extraction of a road surface is a first step for road condition mapping which is the background of the developments in this paper.
Mobile laser scanning covers sensors and procedures for point cloud acquisition from moving platforms. By mounting the system on road vehicles and driving the vehicles through the city, the recorded point cloud covers road surfaces, curbstones, sidewalks, traffic signs, facades etc. By mounting the scanner such that scan direction is perpendicular to the driving direction, typically a high point density is achieved in transverse direction on the road. This can favourably be used to separate road surfaces from the sidewalks. Often used criteria are height jumps or slope changes at the vicinity of the region boundary. The curbstone height above ground varies typically from 10 cm to 20 cm (Wikipedia, 2014). In practice, however, there are many exceptions as shown for example in Figure 1 side of a road protrude around 3 cm above the road surface.

Figure 1. A example of a low curb on the left side of the imaged road

Several related researches have been published over the past decade. (Vosselman and Liang, 2009) investigated the detection of curbstones in airborne laser scanning data. Locations with 'height jumps' are detected which are potential curbstone positions. Moreover they reconstructed the line topology and bridged the gaps in the road boundary outlines. (Samadzadegan et al., 2009) investigated automatic road extraction from airborne Lidar data based on classifier fusion. They employed minimum distance and maximum likelihood classifiers and combined them using weighted majority voting and selective Naïve Bays technique. The authors stated that the Naïve Bays method leads to better outcomes. (Siegemund et al., 2010) investigated curb reconstruction using conditional random field (CRF). The input data they used is a dense 3D point cloud generated from images by stereo matching techniques. These images were taken from the camera mounted in the vehicle windshield. Their suggested workflow operates online. The 3D traffic area with street floor, curbstones and sidewalks is captured. The proposed method is able to reconstruct the curbstones that have different heights from adjacent points on the street. The authors report that they fit straight lines but also higher order polynomials through the detected curbstone edges. (Michalke et al., 2010) examined a self-adaptive method for curbstone detection. Their method is activated by processing of 'human neural signal processing'. They fused a multi-sensor which is like human characteristics of observation for online adaptation of road boundaries and curbstones position. Gabor filter kernels are employed for edge (curb line) detection. (Gallo et al., 2008) studied curbstone detection by means of time-of-flight (TOF) camera. They employed a camera called Canesta. Their plane-based approach follows a RANSAC version called CC-RANSAC. They stated that contrary to the ordinary RANSAC which considers all inliers, the modified version takes only the largest associated components of inliers into the consideration. They claimed that the proposed approach prevents RANSAC failure if shallow curbs exist in the search region. Also the cost efficiency of their method due to the less required iterations is addressed. (El-Halawany et al., 2011) studied curb line detection using mobile laser scanning data. They first separate ground and non-ground depending on parameters such as direction of the 'surface normal and the size of normalized eigenvalues'. Undesired objects are eliminated, and finally the curb lines are detected based on some 3D segmentation and edge detection techniques.

In the following we pick up ideas on region growing using irregular point clouds and analyse the point cloud locally to develop criteria for the region growing process. The growing results in the street floor region. The region boundaries are mostly but not always curbstone edges which are finally approximated by splines to define the road boundaries.

The concept of our work is outlined in the next section. We present and discuss results in section 3. Summary and an outlook are given in the last section.

## 2. CONCEPT

We consider this study as an initial step for our main goal which is automatic detection of road surface depressions such as ruts, potholes and general unevenness. Readers can find further details in (Miraliakbari et. al., 2014).

In this study, we assume that the road surface has a lower height compared to its environment. So we extract the road surface by detecting height jumps, which we assume to coincide with the street borders (curbstone edges). We examine two variants of jump detection, namely: height difference-based (HDB) and histogram-based (HB) jump detection.

Detected jumps are marked in a symbolic road image. Starting from seed pixels on the road, a region growing is performed for which the detected height jumps serve as stop criteria. However, not in all cases the road border is characterized by a height jump. Junctions of bikeways and streets feature frequently smooth transitions for the convenience of bikers. Such shallow curbs imply difficulties: due to the region growing condition, side walk and bikeways might be detected partially as street floor, because the region growing only stops at facades or other obstacles far outside the road. This causes the detection of false curbstones, while the real road border is missing or interrupted. Similar effects are induced by occlusions of the laser rays by cars etc., which also cause gaps in the curb lines; additionally, in such cases the road is not completely covered by laser points. To overcome these effects, the detected height jumps have to be filtered for outliers and have to be completed to smooth continuous lines. Once the final continuous road borders are found, the road surface is identified as the area between these borders.

The results of the above-mentioned variants for jump detection are compared with each other and with the hand-operated outcomes which we use as ground truth for accuracy assessment.

### 2.1 Pre-processing, gross height filtering

The majority of objects represented by the mobile laser scanning data are non-ground objects which must be removed as much as possible. The scanner trajectory of the dataset is divided in equidistant intervals of 2 meters in order to generate point cloud segments. As the point cloud is denser on the road surface than at other locations, especially close to the foot points of the laser scanner positions on the ground, the most frequent height values are encountered at points located on the street. Figure 2 shows a fraction of the unprocessed point cloud. Pre-processing results are presented in blue. The majority of the points, in each section, are located on the street floor and the pedestrian area. Points located higher than 0.5m compared to the location of the highest point density are removed. In Figure 2, only the points within the blue band are retained.

Figure 2. Cross section of a fraction of raw and pre-processed laser scanning data

### 2.2 Principle of height jump detection

In order to find jumps as indications of curbstone edges, we investigate local features of the points. In particular, we surround each point by a vertical cylinder or by cuboid and consider the vertical point distribution within either body. The corresponding histogram represents the frequency per length unit. We introduce two criterion functionals in order to decide whether the neighbourhood contains a jump or not. We call these two variants height difference-based (HDB) and histogram-based (HB) jump detection, respectively.

#### 2.2.1 HDB

In this variant, the target function is the difference between maximum minus minimum height of the points located inside the vertical cylinder. Figure 3 shows a schematic representation of the criterion. If the maximal height difference exceeds a certain threshold, the laser points inside the cylinder represent the border between road surface and the sidewalk.

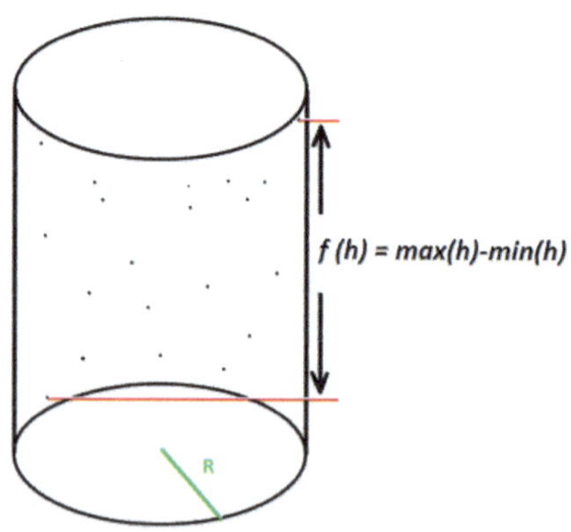


Figure 3. Schematic view of the criterion used in the HDB variant

In this variant, we have considered only the magnitude of the highest and lowest point along the z-axis without considering the distribution within the cylinder.

#### 2.2.2 HB

HB variant replaces the cylinder shown in Figure 3 by equidistant-voxels. The height of the voxels is the same as the predefined threshold. Due to the different longitudinal and transverse intervals of the laser points, the edges of the voxel are not equal. Following algorithm is implemented for this variant;

A) Input the pre-processed point cloud segments. Each segment contains two parts, namely right-scan and left-scan. We call them R-section and L-section.
B) Rotate the scan-lines to be parallel to the Y-axis of the new coordinate system.
C) Calculate the bounding box of each subsection. $\Delta X = maxX - minX$ and $\Delta Y = maxY - minY$.
D) Voxelize the subsections: Magnitude of edges of voxel elements depends on the regular curbstone height and density of the point cloud. In our research we have specified: $dx_v = 10$cm, $dy_v = 5$cm, $dz_v = 2$ cm.
E) Associate binary values with the voxel elements. If voxel contains one or more than one laser scanning point, the associated value amounts to one otherwise zero.
F) Sum up the values of each vertical pile above the pixel. The result is a 2D grid. We classify the grid into three classes; 'zero', 'one' and 'two'. If the result of the summation is more than one, the respective grid cell belongs to class 'two'.

Two constraints must be considered during the region growing process in this variant:
a- If within the region growing, a voxel with value zero is reached, the process will end.
b- In each vertical pile, if there is an empty voxels between non-empty voxels (right above the non-empty voxel), the region growing will be continued. The reason is to avoid confusion of car bumpers with curbstones. Of course one should consider the extra data which remains in the road surface class. The extra data which is in such case part of the car bumpers must be filtered in the post-processing.
Figure 4 shows the schematic presentation of the HB variant. Although as explained in part (F), the summation of the non-empty voxels along this are more than one, but due to having empty voxel in between, the class value amounts to one.

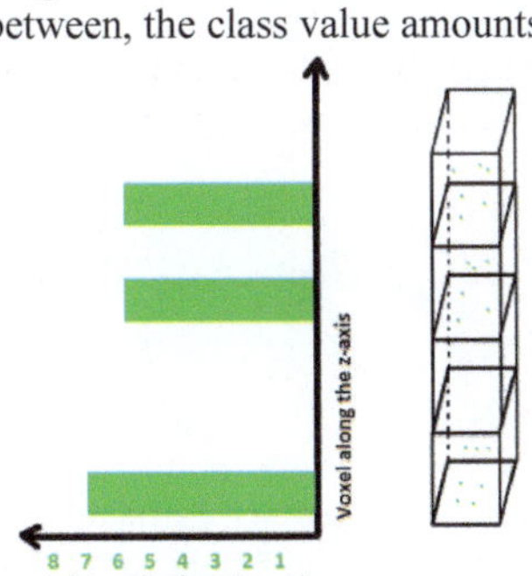


Figure 4. Schematic view of the HB variant

### 2.3 Region growing

We have proposed two region growing variants, namely: four-neighbourhood and two-neighbourhood.

#### 2.3.1 Four-neighbourhood region growing

In this region growing, as seed points we employ the grid cells which are closest to the projected scanner trajectory on the ground. Each cell (candidate grid) of the 2D grid overlaid on the 2D coordinates of the point cloud shall be the centre of a circle with a certain radius (used in HDB) or centre of the rectangle (used in HB) surrounded by the respective 2D irregular points. If the criterion of the target function discussed in previous subsections is fulfilled, the region growing will be continued; otherwise the candidate grid will be inactivated. For testing the fulfilment of the criterion, Z coordinate of the point cloud is taken into account.

#### 2.3.2 Two-neighbourhood region growing

**2.3.3** The major difference of this region growing with the former one is the dimension of region growing. This option offers two-neighbourhood region growing. In four-neighbourhood region growing, if an arbitrary small part of a data segment has shallow curbstone, the four-neighbourhood region growing contaminates the whole sidewalk. Alternatively we shall not obtain such dramatic results in two-neighbourhood variant. In Two-neighbourhood region growing, seed points are again pixels located nearby the 2D laser scanner positions. The search starts from the lowest row of every column of the R-section and goes as long as the criterion of either HDB or HB is fulfilled. Same condition applied for L-section.

### 2.4 Detection and vectorisation of the curbstone edges

In order to achieve two aims with one algorithm, we used the result of road surface extraction as an input to detect and vectorise curbstone edges. There are always some gaps in the detected curbstone edges which occur due to the parked cars or nonappearance of curbstones in some special area such as carriage drive or wheelchair locations (Vosselman and Liang, 2009). The gaps must be filled and all positions of the curbstones must be connected to each other.
Results of HDB and HB variants are employed for street boundary detection. Using result of the HDB, surrounding laser points of the inactivated candidate cell will be considered as potential curbstone positions. In this variant, the obstacles like car bumpers will wrongly be considered as curbstone edges. In HB variant, after a 'zero' value is encountered (e.g. occlusion), or the grid value is more than 'one' and there is no empty voxel between the pile, the process ends. The grid cell where the region growing comes to halt is considered as potential curbstone edge.
Elimination of the obstacles and refinement of the curbstone edges is the next task. Obstacles, mainly parked vehicles, degrade the process of road boundary detection. Following algorithm shows how to get rid of wrongly detected objects:

A) Convert detected potential curbstones to a binary image. The grid increment must not be lower than the longitudinal laser point distance.
B) Perform a connectivity analysis for the curbstone pixels.
C) Outlier removal such as wrongly detected potential curbstones in binary image using their geometric properties. For such filtering we have considered first: length of the segments (connected cells of the binary image which show the potential position of curbstones) and secondly the parallelism of the segments. The vectorisation is applied by fitting proper spline on the potential curbstone positions.

## 3. RESULTS AND DISCUSSION

The study area contains avenues and main streets with bikeways, park places and shallow curbstones. Such condition motivates us to examine road surface and boundary of the entire study area with our proposed methods. Two test areas (A and B) around the University of Applied Sciences in Stuttgart were selected and laser scanning test drives were carried out using FARO Focus3D X 330 laser scanner and Applanix POS LV 420 GNSS/INS system. Table 1 presents the characteristics of the laser scanner datasets.

| Dataset | Length [m] | Longitudinal point distance [m] | Transverse point distance [m] |
|---|---|---|---|
| A | 670 | 0.05 | 0.0025 |
| B | 1100 | 0.04 | 0.004 |

Table 1. Characteristics of the datasets

### 3.1 Comparison of results of HDB and HB road surface detection

Both HDB and HB variants together with the four-neighbourhood and two-neighbourhood region growing have their own pros and cons. In this study we have considered the results of HDB with four-neighbourhood and HB with two-neighbourhood. As advantage of four-neighbourhood region growing used in HDB variant, we noticed if there are cracks and potholes inside the road surface, the region growing shall not stop the process even if the height difference of these features is more than the threshold. The drawback of using this option is that if curbstone height from adjacent road surface is smaller than the predefined threshold, the region growing may count the sidewalks as road surface. Luckily using 2 m segment, less pedestrian area will be contaminated. Moreover because of the HDB functional which considers the absolute height difference, existence of objects above street floor such as car bumpers will terminate the process.

The superiority of HB variant with two-neighbourhood region growing is that due to the region growing along Y-axis, less pedestrian area will be detected as street floor (see Figure 5). Moreover street floor portion below car bumpers and bushes will be separated correctly from non-road surface area. Contrary to HDB variant with four-neighbourhood region growing, the existence of deep cracks, potholes and any other road surface depression terminates the process in the respective area. This is because of two neighbourhood region growing which is chosen in HD variant. Figure 5 shows the result of road surface detection using both conditions located in one of the most critical area. As presented in the Figure 5 (top), many parts of the sidewalks have been wrongly detected as street floor. On the other hand nearby both sides of zebra cross area, less sidewalks are wrongly detected as road surface in Figure 5 (down). This is because of two neighbourhood region growing. As nearby zebra cross we have shallow curbstones, it is more probable that the street floor extraction process fails.

### 3.2 Comparison of the results with different thresholds

The threshold in either of the variants (HDB and HD) plays an important role in the road surface detection process. Selecting small threshold value causes more time for region growing but it is worthwhile.

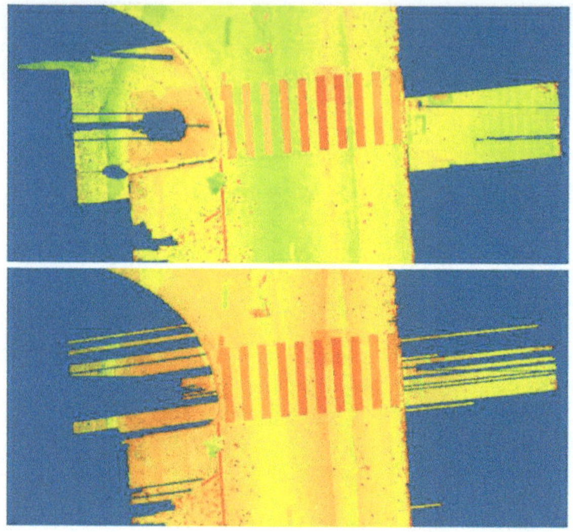

Figure 5. Comparison of HDB with four-neighbourhood region growing (top) and HD with two-neighbourhood region growing (down)

Figure 6 shows the result of road surface extraction using HD variant with two-neighbourhood region growing. The upper part of this figure looks neat compare to the lower sample. The reason is using smaller value for threshold. The area nearby the wheelchair sign is not properly separated and some sidewalks have been wrongly categorized as road surface. This is because of the shallow curbstones in that area. By reducing the voxel height (see Figure 6, top) we shall decrease the problem.

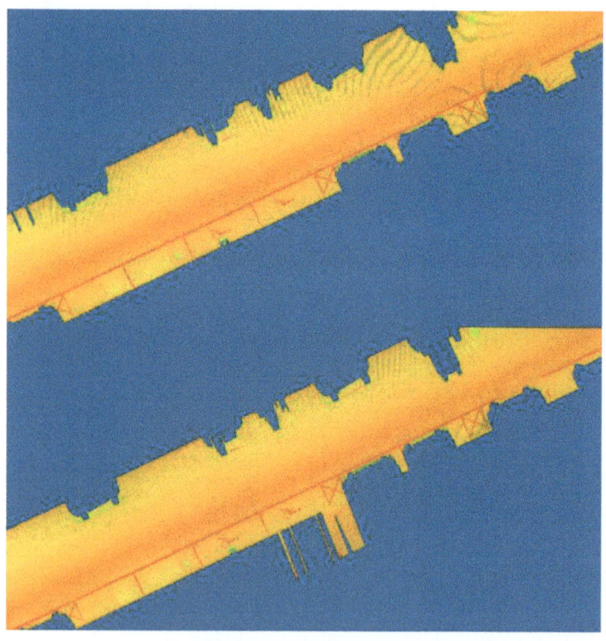

Figure 6. Comparison of road surface extraction using 2 cm (top) and 4 cm (down) voxel heights as threshold

### 3.3 Vectorising the curb line position

Generally it is important to apply post-processing to the result of potential curbstone detection. The post-processing aims to remove the wrongly selected regions such as car wheels (see Figure 7). Additionally due to the existence of obstacles like cars, the process of curbstone detection ends up having some gaps in the curbstone regions. As discussed, the obstacle removal is done by connectivity analysis. The gaps are filled and vectorisation is applied by fitting splines. The experimental investigations related to the curbstone detection show that apart from parking area and junctions, most of the vectorisations are done by fitting straight-lines to the potential curbstone positions.

Figure 7. Extracted potential curbstone positions including outlier shown mostly as car wheels.

Figure 8 presents result of curbstone edge detection and vectorisation (green straight-lines) using 3D laser scanning point cloud. Figure 9 shows a situation in which the park place is located beside the zebra cross area. We have utilized the 4th order spline to detect the curbstone edge.

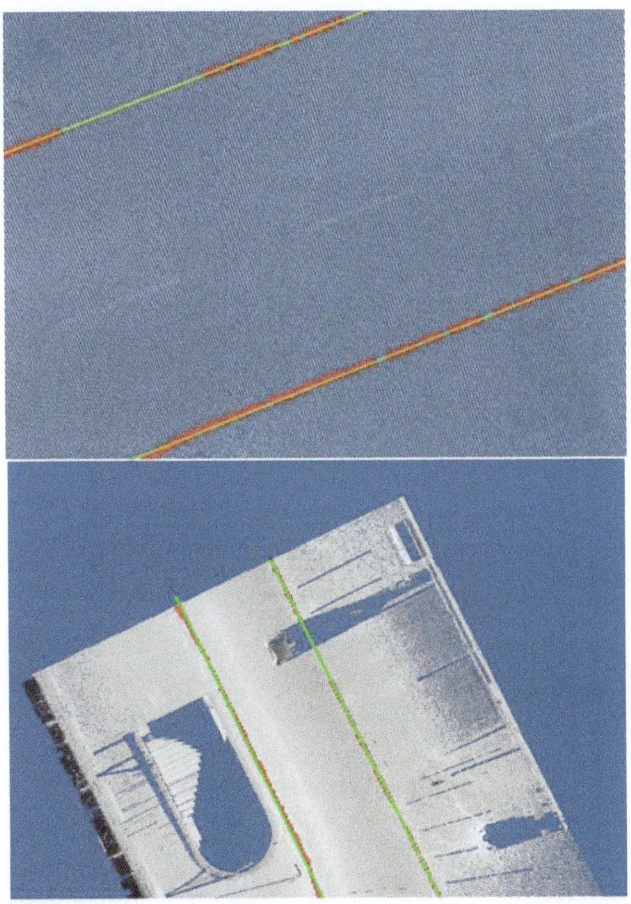

Figure 8. Result of curbstone edge detection and vectorisation (green straight-lines)

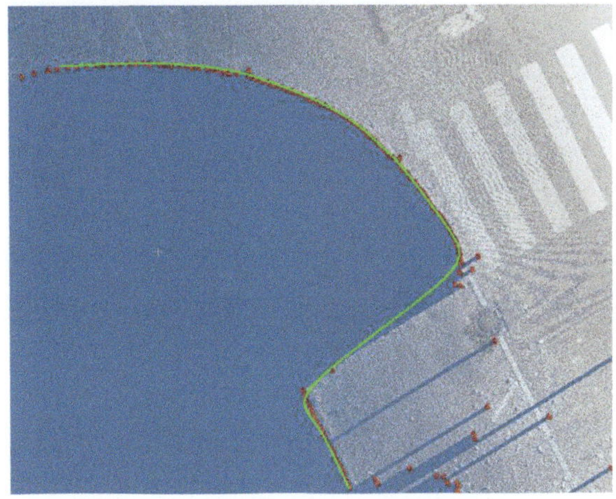

Figure 9. Result of curbstone edge detection and vectorisation using 4th order spline

### 3.4 Accuracy assessment

We have compared results of road surface extraction using HDB and HB variants with ground truth data. Ground truth measurements are digitized manually from the laser scanner data. Percentage of both completeness and correctness are calculated using following formulae:

$$Correctness = \frac{correctly\ detected\ road\ surface}{correctly\ +false\ detected\ road\ surface}$$

$$Completeness = \frac{correctly\ detected\ road\ surface}{correctly\ +not\ detecetd\ road\ surface}$$

Because of the existence of obstacles, some parts of the road surface are omitted. We have made use of dataset A (see Table 1) in such assessment. For the HD variant with two-neighbourhood region growing, the correctness value is 94.9% and completeness rate amounts to 93.8%. Further, regarding HDB variant with four-neighbourhood region growing the correctness value is 92.4% and completeness rate amounts to 95.5% (See Figure 10). As the dark blue area demonstrates the background; the area shown in red is the correctly-detected area. Bright blue regions show the wrongly detected and yellow areas present not-detected field. The omission error is more in HD because of the nature of two-neighbourhood region growing. Because of the existence of crack or extreme depression on the road surface, the process of two-neighbourhood region growing is terminated. Therefore in Figure 10 (down), some more yellow regions are obvious in the road surface. The commission error is more in HDB variant with four-neighbourhood region growing.

## 4. SUMMARY AND FURTHER WORK

In this study, two variants are introduced considering road surface and curbstone edge detection; HDB with four-neighbourhood and HB with two-neighbourhood. Boundaries of the detected road surfaces, as potential curbstone locations, are handled for further curb line detection. The results of the described procedures are used as input of the further research, i.e. road distress detection such as general unevenness, rut and pothole. The accuracy assessment of our results clarifies few manual refinement is necessary to remove wrongly extracted regions. The next step of our research in this area is to take the advantage of images which increases the possibility of automatic curbstone edge detection.

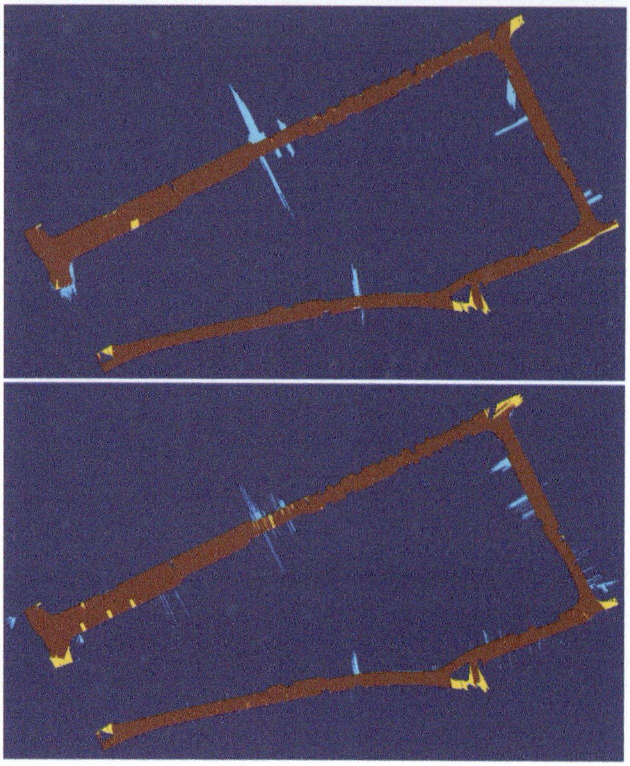

Figure 10. Accuracy assessment of automatic road surface extraction using HDB (top) and HB (down) variant


## 5. ACKNOWLEDGEMENTS

This work is funded by the German Federal Ministry of Education and Research within the project RoadInspect: Detection and assessment of road conditions with the help of the reflection spectroscopy - Development of a kinematic platform solution. Helpful discussions with Dr. Johannes Engels are highly acknowledged.


## 6. REFERENCES


El-Halawany, S., Moussa, A., Lichti, D. D., and El-Sheimy, N.: Detection of Road Curb from Mobile Terrestrial Laser Scanner Point Cloud, Int. Arch. Photogramm. Remote Sens. Spatial Inf. Sci., XXXVIII-5/W12, 109-114, doi:10.5194/isprsarchives-XXXVIII-5-W12-109-2011, 2011.

Gallo, O., Manduchi, R. and Rafii, A.: Robust curb and ramp detection for safe parking using the canesta ToF camera, Computer Vision and Pattern Recognition Workshops, CVPRW'08. Computer Society Conference. IEEE, 2008.

Michalke, T., Kastner, R., Fritsch, J. and Goerck, Ch.: A self-adaptive approach for curbstone/roadside detection based on human-like signal processing and multi-sensor fusion, Intelligent Vehicles Symposium (IV), IEEE, 2010.

Miraliakbari, A., Hahn, M., and Maas, H.-G.: Development of a Multi-Sensor System for Road Condition Mapping, Int. Arch. Photogramm. Remote Sens. Spatial Inf. Sci., XL-1, 265-272, doi:10.5194/isprsarchives-XL-1-265-2014, 2014.

Samadzadegan, F., Hahn, M. and Bigdeli, B.: Automatic Road Extraction from LIDAR Data based on Classifier Fusion, Urban Remote Sensing Joint Event, Joint. IEEE, 2009.

Siegemund, J., Pfeiffer, D., Franke, U. and Förstner, W.: Curb reconstruction using Conditional Random Fields, IEEE Intelligent Vehicles Symposium (IV), 2010.

Vosselman, G. and Liang, Z.: Detection of curbstones in airborne laser scanning data, Arch. Photogramm. Remote Sens. Laser scanning 2009, ISPRS, Vol. XXXVIII, Part 3/W8, 2009.

Wikipedia, Curb, http://en.wikipedia.org/wiki/Curb#cite_note-6 (Access 15.04.2014)

# PHENOLOGICAL TRACKING OF AGRICULTURAL FEILDS INVESTIGATED BY USING DUAL POLARIMETRY TANDEM-X IMAGES

S. Mirzaee [a], *, M. Motagh [b], H. Arefi [a], A. Nooryazdan [c]

[a] University of Tehran, Dept. of Surveying and Geomatics Eng., Tehran, Iran - (saramirzaee, hossein.arefi)@ut.ac.ir
[b] GFZ German Research Centre for Geosciences, Section of Remote Sensing, 14473, Potsdam, Germany - motagh@gfz-potsdam.de
[c] Forest, Range and Watershed Management of Iran, Lorestan province - reza_nooryazdan@yahoo.com

**KEY WORDS:** TanDEM-X, Interferometric Coherence, Phenological Change

**ABSTRACT:**

Remote sensing plays a key role in monitoring and assessing environmental changes. Because of its special imaging characteristics such as high-resolution, capabilities to obtain data in all weather conditions and sensitivity to geometrical and dielectric properties of the features, Synthetic Aperture Radar (SAR) technology has become a powerful technique to detect small scale changes related to earth surface.SAR images contain the information of both phase and intensity in different modes like single, dual and full polarimetric states which are important in order to extract information about various targets. In this study we investigate phenological changes in an agricultural region using high-resolution X-band SAR data. The case study is located in Doroud region of Lorestan province, west of Iran. The purpose is to investigate the ability of copolar and interferometric coherence extracted from TanDEM-X dual polarimetry (HH/VV) in bistatic StripMap mode for tracking the phenological changes of crops during growing season. The data include 11 images acquired between 12.06.2012 and 02.11.2012 and 6 images acquired between 30.05.2013 and 04.08.2013 in the CoSSC format. Results show that copolar coherence is almost able to follow phenological changes but interferometric coherence has a near constant behaviour with fluctuations mainly related to baseline variations.

## 1. INTRODUCTION

Agricultural monitoring is important as it is directly related to human food supply. In recent years much attention has been paid on monitoring phenological changes and classification of different crop types using remote sensing observations (Bouvet and Le Toan, 2009; Hoekman, 2003; Liu et al., 2013; Juan M. Lopez-Sanchez et al., 2011). Issues like recognizing growth behaviour, cultivation problems and crop yield estimation have been worked by different researchers (Dhar et al., 2009; Ferencz et al., 2004; Silleos et al., 2002). Among different sensor data, radar remote sensing with its reliable and frequent imaging capability all-weather functionality, sensitivity to target geometrical structure and orientation, has enhanced further our capabilities for agricultural monitoring.

Sabour et al.,( 2007) used dual polarization HH/VH time series of ENVISAT data for classification of crop types. Koppe et al., (2012) investigated the behaviour of X-band backscattering values over rice fields in different phenological states Zhao et al., (2014) discussed about the feasibility of C- and X-band polarimetric SAR images for crop harvest pattern recognition. Voormansik et al., (2013) investigated the capability of dual polarimetry TerraSAR-X (TSX) images for detecting grassland cutting practices.

In this research we evaluated polarimetric and interferometric coherence for monitoring phenological changes of different crop types. For this purpose we used TanDEM-X dual polarized (HH/VV) images in bistatic StripMap mode, provided by German aerospace centre (DLR). In order to study the agricultural fields, a classification was first done to distinguish different crop types, and then time series of coherence values were plotted for each crop. Finally a comparison was done between copolar and interferometric coherences.

## 2. PHENOLOGICAL TRACKING

### 2.1 Data

The dataset includes 17 TDX StripMap images in bistatic and coherent dual polarimetry (HH/VV) mode. 11 images cover the time interval from Jun $12^{th}$ to November $2^{nd}$ 2012 while the other 6 images cover from May $30^{th}$ to August $4^{th}$ 2013. All images were acquired in a descending mode at 07:25 local time (02:55 UTC) with a scene centre incidence angle of 22° and in coregistered single look slant range complex (CoSSC) format. Average range and azimuth spacing are 0.9m and 2.4 m respectively. Acquisition intervals are not equal, most of them are 11 days separated but some of them have an interval of 22 to 33 days. Bistatic baselines are listed in Table 1.

### 2.2 Study area

Our test site is located in the north east of Doroud city in Lorestan province, Iran. It is located in a region with centre coordinate of 33°31'55.00"N, 49° 6'8.00"E. The area includes agricultural fields with different crop types, which are mostly cereals and beans, rice and alfalfa. Figure 1 shows the study area with the TDX image of July 26th 2012 overlapped on Google earth.

* Corresponding author

| Acquisition date | Baseline (m) | HOA (m) |
|---|---|---|
| 06.12.2012 | 262.34 | 11.87 |
| 23.06.2012 | 260.98 | 11.94 |
| 15.07.2012 | 251.89 | 12.36 |
| 26.07.2012 | 243.09 | 12.82 |
| 06.08.2012 | 246.36 | 12.63 |
| 17.08.2012 | 234.48 | 13.28 |
| 28.08.2012 | 231.36 | 13.47 |
| 08.09.2012 | 226.51 | 13.75 |
| 11.10.2012 | 206.59 | 15.07 |
| 22.10.2012 | 191.92 | 16.23 |
| 02.11.2012 | 181.55 | 17.16 |
| 30.05.2013 | 104.61 | 29.74 |
| 21.06.2013 | 101.45 | 30.68 |
| 02.07.2013 | 102.79 | 30.30 |
| 13.07.2013 | 55.59 | 56.01 |
| 24.07.2013 | 50.39 | 61.75 |
| 04.08.2013 | 43.92 | 70.87 |

Table 1: Dataset acquisition with bistatic baselines and height of ambiguities

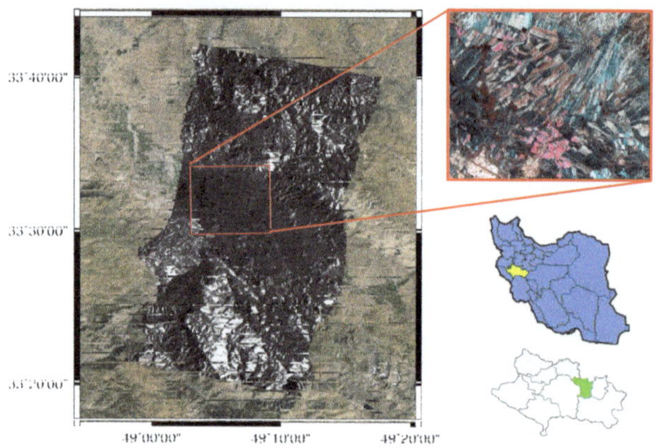

Figure 1: TSX amplitude image of July 26th 2012 over google earth image in Doroud and the subset area shown with RGB colour composite in Pauli basis over the agricultural fields.

### 2.3 Methodology

As different crop types have non-similar structure and phenological behaviour, a supervised classification with SVM classifier was first applied to separate crop classes. The classification was done for the two years by using polarimetric features of one satellite images (TSX). The procedure has been defined in the previous paper of Mirzaee et al., (2014). Then two types of coherence values were plotted for each class, which are copolar coherence with HH and VV channels of TSX, and interferometric coherence of HH and also VV of two satellites (TSX and TDX). The equation used for calculation of copolar coherence is as follows

$$\gamma_{HHVV} = \frac{\left\langle S_{HH} S_{VV}^{*} \right\rangle}{\sqrt{\left\langle |S_{HH}|^{2} \right\rangle \left\langle |S_{VV}|^{2} \right\rangle}} \quad (1)$$

Interferometric coherence is calculated with the same equation but instead of using HH and VV channels of one satellite, we use HH or VV channels of two satellites. Coherence values are averaged over each class and plotted through time.

## 3. RESULTS AND DISCUSSION

The overall accuracy obtained for classification was about 93% for 2012 and 90% for 2013. Classified map of each year is represented in Figure 2.

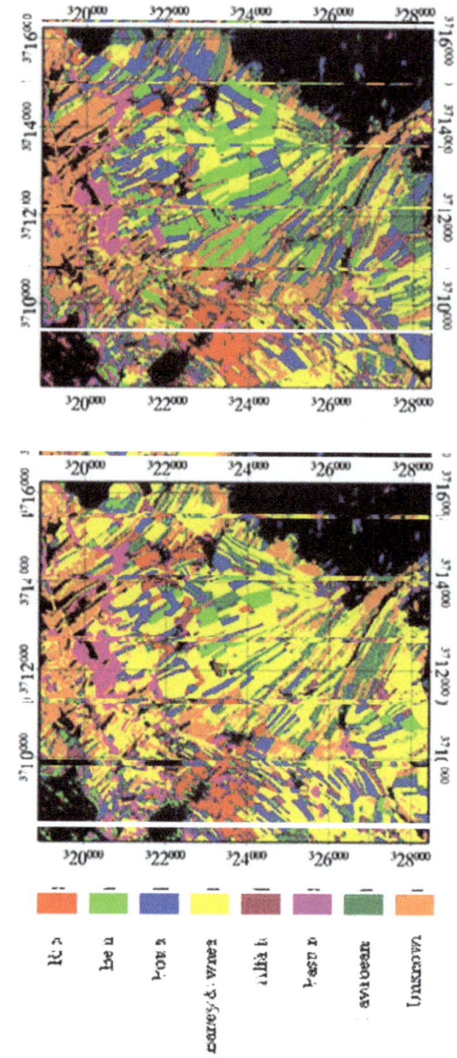

Figure 2: Classification map of the year 2012 (on the top) and 2013 (on the bottom)

As we can see from the results, the final classification map does not follow a specific pattern. The changes are due to alternating cultivation between crops. In this region farmers do not always cultivate a specific plant in order to save the quality of soil. So crop types like wheat, barley and beans are implanted alternatively in a field every 2 to 4 years. This is an important issue for the image preparation. Mapping crop types in an area through phenological changes requires acquiring dataset within one agricultural season in order to not have any changes, in particular for the areas with diversity of crop cultivation.

Time series of copolar coherence is shown in Figure 3 and interferometric coherence is shown in Figure 4.

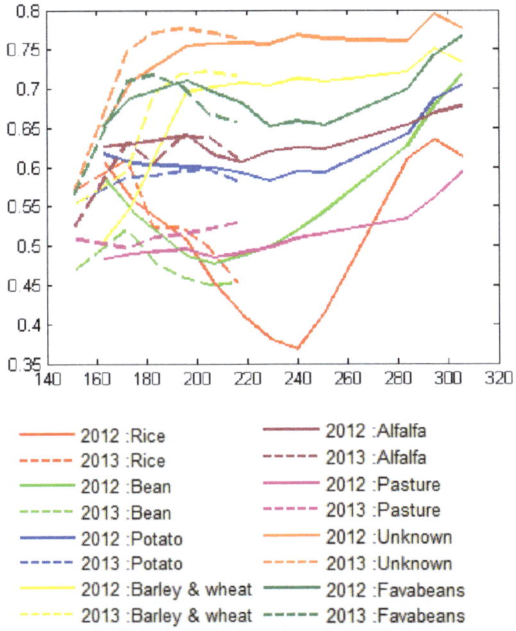


Figure 3: Copolar coherence time series for different crop types

Generally, the agricultural period divides into three main phenological stages: vegetative, reproductive and mature phase. At the vegetative phase crop height increases. The appearance of leaves and panicles occurs at the reproductive phase, and then at mature phase grains develop until harvesting.

The dataset includes almost all classes' phenological cycle except of the cereals (barley and wheat) which includes only the mature phase. The winter type of cereals, are implanted in November and December. Therefore their main growth period begins at the late march when the crop height increases. Plants are ready to harvest at the late June, so the harvest begins at day of year 160.

The rice class has distinguished properties compared to other classes due to presence of water and double bounce scattering mechanism. As it is stated by other researchers (Choudhury and Parihar, 2007; Lopez-Sanchez and Ballester-Berman, 2010), due to the single scattering mechanism of the rough water surface, the coherence between copular channels has a value of about 0.6 at the beginning of the vegetative phase. Once the plants are emerged, the copolar coherence decreases to less than 0.4 until the end of reproductive phase (doy 240). The decrement of coherence is partly due to high differential attenuation of HH and VV and partly to random volume scattering (Lopez-Sanchez et al.,2011). In the maturation phase, plant water content and also differential attenuation decrease, so backscattering values of two polarimetric channels become close to each other and copolar coherence begins to increase until the time of harvest.

Beans are implanted at the late spring. This crop type needs regular irrigation and lots of water at the vegetative phase.

At the beginning of the vegetative phase, leaves grow and crop height increases. This continues until 30 to 40 days after implantation (day of year 200). During this period the volume scattering increases so it leads to a decreasing trend in copolar coherence.

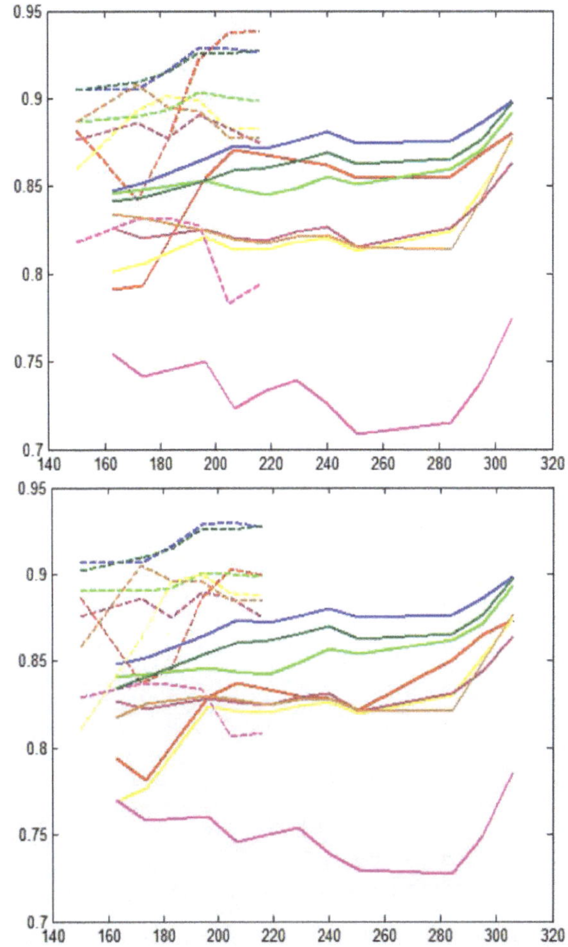


Figure 4: Interferometric coherence time series for different crop types, HH coherence in the top and VV in the bottom

As flower buds emerge and flowers open, after a month, petals start to fall and dry and also pods develop until the end of season and harvest time. During this period copolar coherence gradually increases due to dry volume of plant.

Potato and cereals also have similar trends but with different values. As far as the dataset only covers the end of growing season for cereals, we can see the increasing trend of copolar coherence as a result of dry and dense volume of plants until the harvest time.

Alfalfa is grown and harvested repeatedly during the vegetative season. So this plant's average curve does not have a clear trend and it is full of deviations.

The class which is named Fava beans actually consists of two classes. Fava beans are usually cultivated before wheat or barley to prepare and richen the soil. Its period of cultivation begins at late November and it is harvested until the end of May. After that the spring cereals cultivation begins. So an increasing trend related to the end of Fava beans class followed by a decrease and increase trend related to spring cereals are detected.

About the interferometric coherences which were calculated for both channels and are depicted in Figure 4, the assumption was that it should follow the height changes of crops, in a way that by increasing the plant height and due to creation of a volumetric layer of plants, the interferometric coherence should decrease. The most obvious behaviour in the results is that the interferometric coherence values highly depend on the baseline values, as the 2013 coherence values are all higher than 2012 values and their baselines are smaller.

Almost every class has a constant value over time except in the special occasions like the presence of water in the beginning of rice growth cycle, which results in lower values of interferometric coherence.

Among all the classes, pasture has the least coherence values that may be because of semi-moist properties of such lands.

The last issue is that because of differential attenuation between two polarimetric channels and also moisture conditions, the interferometric coherence values of HH and VV have a slight difference and it has a random behaviour as it is not similar for all the classes.

## 4. CONCLUSION

In this research we investigated the behaviour of different copolar and interferometric coherence values over agricultural fields. Results showed that the copolar coherence between polarimetric channels of one satellite (TSX) can almost track the changes and it is able to distinguish different crop types through their special characteristics resulting different behaviours in time.

The interferometric coherence on the other hand, has almost constant behaviour through time for each class except when there is a special occasion like the presence of water in the beginning of rice plant growing cycle. Also the interferometric coherences of HH and VV channels have almost similar trends with slight differences which may be the result of differential attenuation and also plant moisture content.

## ACKNOWLEDGEMENTS

The copyright of SAR data belongs to DLR (German Aerospace Center) and was provided under the research proposal XTI_LAND0505.

## REFERENCES

Bouvet, a., Le Toan, T., 2009. Monitoring of the Rice Cropping System in the Mekong Delta Using ENVISAT/ASAR Dual Polarization Data. IEEE Trans. Geosci. Remote Sens. 47, 517–526. doi:10.1109/TGRS.2008.2007963

Choudhury, I., Parihar, J.S., 2007. ESTIMATION OF RICE GROWTH PARAMETER AND CROP PHENOLOGY WITH conjucative use of radarsat and envisat, in: Envisat Symp.

Dhar, T., Gray, D., Menges, C., 2009. Agricultural performance monitoring with polarimetric SAR and optical imagery. Proc. 4th Int. Work. Sci. Appl. SAR Polarim. Polarim. Interferom. 26–30.

Ferencz, C., Bognár, P., Lichtenberger, J., Hamar, D., Tarcsai†, G., Timár, G., Molnár, G., Pásztor, S., Steinbach, P., Székely, B., Ferencz, O.E., Ferencz-Árkos, I., 2004. Crop yield estimation by satellite remote sensing. Int. J. Remote Sens. doi:10.1080/01431160410001698870

Hoekman, D.H., 2003. A new polarimetric classification approach evaluated for agricultural crops, in: European Space Agency, (Special Publication) ESA SP. pp. 71–79. doi:10.1109/TGRS.2003.817795

Koppe, W., Gnyp, M.L., Hütt, C., Yao, Y., Miao, Y., Chen, X., Bareth, G., 2012. Rice monitoring with multi-temporal and dual-polarimetric terrasar-X data. Int. J. Appl. Earth Obs. Geoinf. 21, 568–576. doi:10.1016/j.jag.2012.07.016

Liu, C., Shang, J., Vachon, P.W., McNairn, H., 2013. Multiyear Crop Monitoring Using Polarimetric RADARSAT-2 Data. IEEE Trans. Geosci. Remote Sens. 51, 2227–2240. doi:10.1109/TGRS.2012.2208649

Lopez-Sanchez, J., Ballester-Berman, J., 2010. An advanced electromagnetic model for rice fields at X-band: Development and interpretation of dual-pol TerraSAR-X images. EUSAR 2010 754–757.

Lopez-Sanchez, J.M., Ballester-Berman, J.D., Cloude, S.R., 2011. Monitoring and retrieving rice phenology by means of satellite SAR polarimetry at X-band. Int. Geosci. Remote Sens. Symp. 2741–2744. doi:10.1109/IGARSS.2011.6049781

Lopez-Sanchez, J.M., Ballester-Berman, J.D., Hajnsek, I., 2011. First results of rice monitoring practices in Spain by means of time series of TerraSAR-X dual-pol images, in: IEEE Journal of Selected Topics in Applied Earth Observations and Remote Sensing. pp. 412–422. doi:10.1109/JSTARS.2010.2047634

Mirzaee, S., Motagh, M., Arefi, H., Nooryazdan, M., 2014. Classification of agricultural fields using time series of dual polarimetry TerraSAR-X images. ISPRS - Int. Arch. Photogramm. Remote Sens. Spat. Inf. Sci. XL-2/W3, 191–196. doi:10.5194/isprsarchives-XL-2-W3-191-2014

Sabour, S.M.T., Lohmann, P., Soergel, U., 2007. Monitoring agricultural activities using ASAR ENVISAT data, in: European Space Agency, (Special Publication) ESA SP.

Silleos, N., Perakis, K., Petsanis, G., 2002. Assessment of crop damage using space remote sensing and GIS. Int. J. Remote Sens. doi:10.1080/01431160110040026

Voormansik, K., Jagdhuber, T., Olesk, A., Hajnsek, I., Papathanassiou, K.P., 2013. Towards a detection of grassland cutting practices with dual polarimetric TerraSAR-X data. Int. J. Remote Sens. 34, 8081–8103. doi:10.1080/01431161.2013.829593

Zhao, L., Yang, J., Li, P., Zhang, L., 2014. Characteristics Analysis and Classification of Crop Harvest Patterns by Exploiting High-Frequency MultiPolarization SAR Data. IEEE J. Sel. Top. Appl. Earth Obs. Remote Sens. 1–1. doi:10.1109/JSTARS.2014.2308273

# RESEARCH ON THE LAYERING A* ALGORITHM FOR REAL-TIME NAVIGATION

Xiao Tan[a,b]
a.State Key Laboratory for Information Engineering in Surveying, Mapping and Remote Sensing, Wuhan University, Wuhan 430079 , China
b.Department of Management Engineering, Naval University of Engineering, Wuhan 430033 , China
talklaugh@163.com

**KEY WORDS:** Embedded GIS, Route Programming, A* Algorithm, Road Network

**ABSTRACT:**

According to the existing problems in the applications of embedded navigation, this paper designs the hierarchical search A*algorithm, based on the transferring road network, to meet the need of real-time navigation. In the algorithm, a hierarchical search strategy is applied to route programming of large area, yet the duplicate searching A* algorithm, based on the transferring road network, is applied to the path computation, which is able to handle intersection turn restrictions and node weight, with little storage space but fast searching speed. Practically, the algorithm is proved to meet the technological need of real-time navigation both in computing speed and route rationality.

## 1. INTRODUCTION

Along with the current development of the embedded GIS and its related technologies to navigation, many embedded GIS products with navigation application function have appeared; these products can provide real-time path planning, voice guidance, and other functions, which can meet the basic demand of the user's navigation [1]. But in practical application, it also revealed some shortcomings, such as slow speed of data calls; low efficiency of path planning in large area; unadvisable path planning, and without considering traffic restriction factors, etc. [2]. Therefore, it is needed to study how to achieve the algorithm which can meet the application requirements of real-time navigation path planning in the limited resources embedded environment.

## 2. IMPLEMENTATION STRATEGY OF PATH PLANNING

In the actual driving process, in the case of easily finding the close distance, the drivers will generally choose to drive in the good high-grade road, rather than to go on the poor low-grade road which can cause traffic jam and with poor traffic capacity, unless the starting point or end point is in poor low-grade road. So in the process of path calculation, the hierarchical search strategy can be adopted, that is, search the low-grade road in the vicinity of the starting point and end point, and search the high-grade road in the midway. In addition, as for the path planning of a city or a small local area, due to the small amount of data, it is usually only needed to load the road network data once into memory, and then use the conventional path planning algorithm to calculate the optimal path. However, in most cases, the path planning calculation is used for vehicle navigation in nationwide large area and a huge amount of road network, so we must develop the reasonable path calculation strategy according to the regional scope of navigation application.

Based on the above considerations, in the path calculation, the study determines whether it is a path calculation of local small range or a large range first according to the starting point and end point, and then adopts the corresponding calculation strategies:

(1) The path calculation of local small range

The path calculation of local small range is to select all grades roads near the starting point and end point, and select the higher-grade road in the midway to constitute the network for the optimal path calculation.

(2)the path calculation of large range

The path calculation of large range mainly used the hierarchical search strategy [3][4]; first, search the path in the high-grade road network; then, conduct the local refining optimal path search near the starting point and end point by calculation method of local small range.

At present, 20 kinds of path planning algorithms can be used for optimal path analysis according to the classification statistics, of which, more classical algorithms are Dijkstra Algorithm[5], Bellman-Ford-Moore Algorithm, A* Algorithm [6] and others.

The A* algorithm makes an estimate on the distance of the node from the end point when selecting the next node, and to evaluate the possible measures of the optimal path of the road, in this way, it can optimally search the node with the optimal large possibility. So it can be seen from the search process, when it is near to the end point, the number of the searched nodes is fewer with small storage space and quick search speed.

Therefore, the study in this paper chose the A* algorithm which needs a small storage space and have quick search speed for the path planning calculation.

## 3. DISPOSAL OF TRAFFIC CONTROL INFORMATION AND INTERSECTION DELAY

### 3.1 Analysis of conventional network representation

Under normal circumstances, road network is abstracted by using a weighted directed graph, road intersection is represented as a node, a section between two intersections is expressed as arc, and some quantitative attributes of a section serve as a weight of segmental arc. As shown in Fig. 1, Fig. 1

**Corresponding Author**
Xiao Tan (1979- ), male, is from Huangshi City, Hubei Province, PhD, his main research fields include 3D GIS, navigation engineering.

(b) is a road network schematic diagram of Fig. 1 (a) by using a conventional weighted directed graph to represent road intersections.
In the process of driving, inevitable intersection delay directly results in weighted nodes. In addition, influenced by steering restriction on the intersection, one-way traffic, vehicle forbidding and other traffic control information, vehicles cannot reach roads originally connected each other in the actual driving process. Among the above factors affecting vehicle driving, one-way traffic, vehicle prohibiting, etc., can be solved through setting arc weights corresponding to road as ∞(or a sufficiently large positive number), but weighted nodes and steering restriction cannot be solved through the conventional road network representation and the shortest path algorithm. As shown in Fig. 1 (a), vehicles driving on Section A is prohibited to turn left into Section D, but Fig. 1 (b) is unable to express the fact that arc (v1,v2) from Node #1 cannot reach arc (v2, v5) via Node# 2. Furthermore, different steering behaviors (such as going straight, turning left, turning right, etc.) generally produce different intersection delay time. Fig. 1 (b) is unable to respectively express node weights caused by different steering behaviors after vehicles enter Intersection # 2.

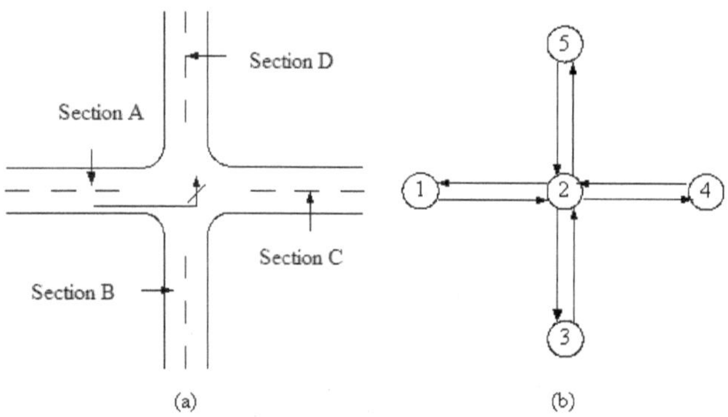


Fig.1 Schematic diagram for road intersection representation by using a conventional weighted directed graph

### 3.2 Transformation of road network construction based on segmental arc

In order to deal with the influence of intersection steering restriction and node weights on path planning, the traditional road network representation must be transformed. In the paper, the ideas related to the method of dual graph [7] [8] are introduced herein, and the method to transform road network construction is used to handle intersection steering restriction and node weights. The basic idea is as below: nodes-based weighted directed graph is transformed into arc (section)-based weighted directed graph, and then the optimal path is calculated on the transformed road network.
As shown in Fig. 2, segmental arcs (Sections A, B, C, D) in the original network are transformed into new nodes in the network. At this point, if the starting point and the end point are considered as a segmental arc with a length of 0, it is only required to define the weight from one segmental arc to the other segmental arc, then any shortest path algorithm is used to solve the optimal path from the starting point to the end point.
Based on the transformed road network, intersection steering restriction and node weights are handled through defining the weight from one segmental arc to the other segmental arc. Assume that the serial number of the ith segmental arc in the original network is Li, if Li exists in the subsequent arc Lj and associated node is P, the weight from Li to Lj can be defined as below:

$$W(L_i, L_j)=W(L_i)+W_p(L_i, L_j) \quad (5\text{-}1)$$

Where: $W(L_i)$ represents as the weight of arc Li in the original network, $Wp(L_i, L_j)$ represents the additional weight of node P, which is used to represent intersection delay. If it is prohibited to turn from Li to Lj, however, $Wp(L_i, L_j)$ represents ∞ (or a sufficiently large positive number).

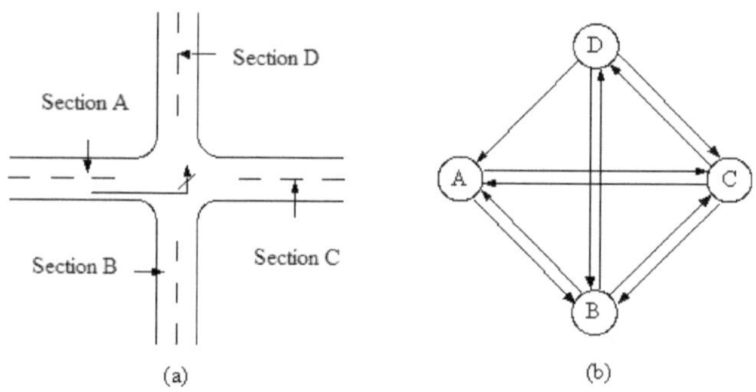


Fig.2 Schematic diagram for road intersection representation by using the method to transform road network

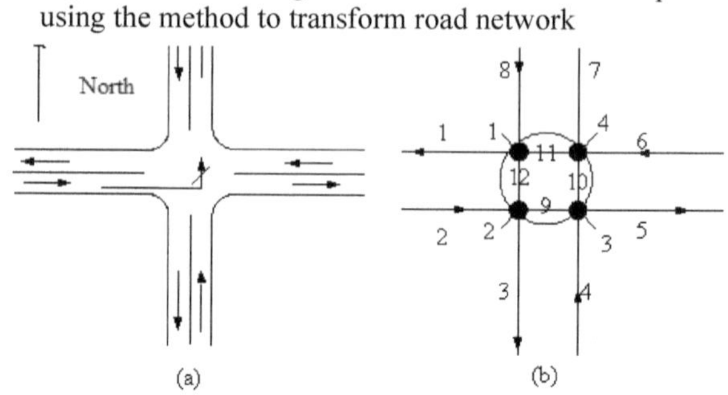


Fig. 3 Schematic diagram for navigable data representation of complex intersections

### 3.3 Binary search A* algorithm based on the transformation of road network

Intersection steering restriction and node weights can be handled by constructing the transformed road network based on segmental arc. Data of navigable electronic map produced by many manufacturers of navigable electronic map, however, do not describe a complex intersection as a node. Fig. 3 (a) shows an intersection, and Fig. 3 (b) shows collected navigable data. There is a median in front of intersection where four roads connected to the intersection pass, so every road is digitalized as two sections in the opposite direction when data are produced, in a total of eight sections, respectively marked as $L_1$, $L_2$,...$L_8$;These eight sections constitute four nodes in an intersection, respectively marked as $P_1$, $P_2$, $P_3$, $P_4$. Internal sections connected to four nodes are respectively marked as $L_9$, $L_{10}$, $L_{11}$ and $L_{12}$. It can be seen Fig. 3 (b) that $L_2$ must pass through $L_9$, $L_{10}$ before reaching $L_7$, so the method to transform road network cannot be directly used to describe the fact that $L_2$ cannot reach $L_7$.
Internal sections have no practical significance for path planning calculation, but after the completion of path planning, it is required to organize internal sections into a complete optimal path. Therefore, the paper designs the binary search A* algorithm based on the transformation of road network. The algorithm design idea is as below: When the first path calculation, internal sections are removed, an intersection is considered as the transformed road network of a node structure based on segmental arc, and the A*

algorithm is used to search the optimal path. When secondary path calculation, the optimal path first calculated and all internal sections where the path goes constitute the transformed road network based on segmental arc. Regardless of steering relationship, the $A^*$ algorithm is used to search the secondary optimal path. Thus, when the first path calculation, the method to transform road network is used to handle intersection steering restriction and node weights, to get the optimal path exclusive of internal sections; when second path calculation, the optimal path, in order to connect to the optimal path, road network used for path calculation only includes the optimal path first calculated and internal sections of the intersection where the path goes, so the calculating speed is fast.

The binary search $A^*$ algorithm based on the transformation of road network is implemented according to the following steps:

1) Based on the positions of the starting point $P_s$ and the end point $P_e$, as well as the current driving direction of vehicles, path matching is conducted, to get the positions corresponding to the sections where $P_s$ and $P_e$ are located, and to generate the segmental arc of the starting point ($P_s$, $P_s'$) and the segmental arc of the end segment ($P_e'$, $P_e$) accordingly;

2) According to the positions of $P_s'$ and $P_e'$, data of segmental arc and node data are extracted within a certain scope;

3) When organizing road network, internal sections are deleted, the method to the transformation of road network is used to transform road network data into an arc (section)-based weighted directed graph, intersection steering restriction and node weights are handled according to the segmental arc and its weight defined in Formula (5-1);

4) The $A^*$ algorithm is used to solve the optimal path from Arc ($P_s$, $P_s'$) to Arc ($P_e'$, $P_e$) on the transformation of road network (excluding internal sections);

5) Internal sections of the intersection where $L_1$ pass are extracted, and constitute road network data together with $L_1$;

6) The method to transform road network is used to transform the above road network data into an arc (section)-based weighted directed graph;

7) The $A^*$ algorithm is used to solve the optimal path from Arc ($P_s$, $P_s'$) to Arc ($P_e'$, $P_e$), namely, the optimal path from the starting point $P_s$ to the end point $P_e$.

## 4. IMPLEMENTATION OF THE HIERARCHICAL SEARCH A* ALGORITHM BASED ON THE TRANSFORMATION OF ROAD NETWORK

Based on the above analyses, the paper designs the hierarchical search $A^*$algorithm based on the transformation of road network to meet the demand for real-time navigation application. The basic idea of the algorithm is as below: The hierarchical search strategy is used for path planning of a large area. The binary search $A^*$ algorithm based on the transformation of road network, which has the characteristics of able to handle intersection steering restriction and node weights, small storage space and fast search, is used as the path planning algorithm.

The hierarchical search $A^*$ algorithm based on the transformation of road network is implemented according to the following steps:

1) Judgment whether to be the path planning for a large area according to the distance from the starting point $P_s$ to the end point $P_e$, if it is, it turns to (3); otherwise, it turns to (2);

2) Road network data within a certain scope are extracted from the road network containing all roads. The binary search $A^*$ algorithm based on the transformation of road network is used to directly calculate the optimal path L from $P_s$ to $P_e$, and then the path calculation is completed;

3) The points nearest to $P_s$ and $P_e$ is calculated in accordance with the principle of the closest distance in high grade road network, respectively marked as $P_s'$ and $P_e'$ ;

4) Road network data within a certain scope are extracted from the high grade road network. The binary search $A^*$ algorithm based on the transformation of road network is used to calculate the optimal path;

5) $P_i$ is cut from $L_1$. The principle of point selection is as below: On the premise of satisfying the path calculating speed, it is as far as possible away from $P_s'$ , and road network data within a certain scope are extracted from road network including all roads. The binary search $A^*$ algorithm based on the transformation of road network is used to calculate the optimal path;

6) The method of (5), $P_j$ is cut from $L_1$. The principle of point selection is as below: On the premise of satisfying the path calculating speed, it is as far as possible away from $P_e'$ , and the binary search $A^*$ algorithm based on the transformation of road network is used to calculate the optimal path $L_3$ from $P_j$ to $P_e$.

7) Given that a path in the middle of $P_i$ and $P_j$ of $L_1$ is expressed by $L_4$, after $L_2$, $L_4$ and $L_3$ are connected, to get the optimal path L, and the path calculation is completed.

## 5. EXPERIMENT AND CONCLUSIONS

The hierarchical search $A^*$ algorithm based on the transformation of road network designed in the paper has been applied in the embedded navigation application system developed by the author, which supports national road network data. The operating environment of the system is the operating system Windows CE 5.0, 500 MHZ, memory of 256 MB, 800×480 screen resolutions. The running interface is shown in Fig. 4.

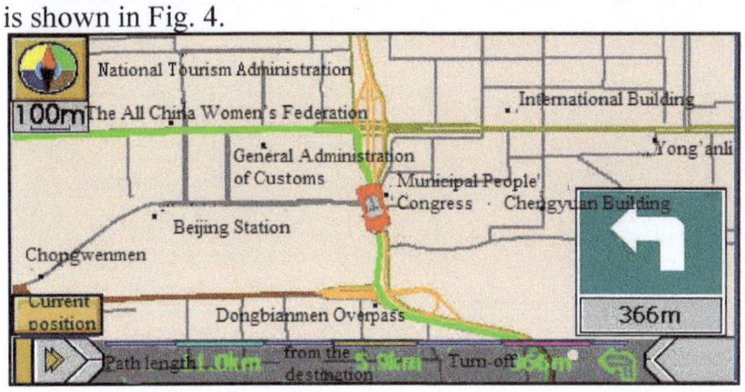


Fig. 4 Interface diagram for the embedded navigation application system

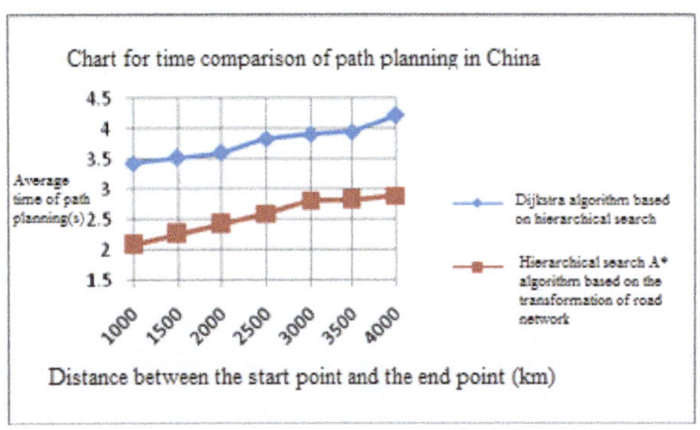


Fig. 5 Chart for time comparison of path planning in China by using two algorithms

There are massive data on the national road network, when

path planning of a large area, so it is impossible to completely load road network data in a computational domain into memory. In the paper, therefore, when verifying the algorithm's validity, the Dijkstra algorithm based on hierarchical search and the hierarchical search $A^*$ algorithm based the transformation of road network are respectively used to compare the efficiencies of path planning. Fig. 5 shows the chart for time comparison of path planning among different cities of China after using two algorithms are used respectively.

It can be seen from Fig. 5 that two algorithms use the hierarchical search strategy when path planning in China, so both of them have higher computational efficiency. In search of the next section, the hierarchical search $A^*$ algorithm based on the transformation of road network uses the distance away from the end point as the measurement to evaluate the possibility of this section locating in the optimal path, and can first search the section with a great probability, so, its search efficiency is superior to that of the Dijkstra algorithm based on hierarchical search. In addition, when path calculation, the hierarchical search $A^*$ algorithm based on the transformation of road network uses the method of transformation of road network to transform node weights into an arc (section)-based weighted directed graph, and then the optimal path is calculated in the transformation of road network, therefore, the algorithm can identify and deal with all kinds of traffic control information and the influence of intersection delay.

The practical navigation application shows that the hierarchical search $A^*$ algorithm based on the transformation of road network has many advantages in calculating speed, path rationality, etc., and is able to satisfy the technical requirements of real-time navigation application under the embedded environment.

**References:**

[1] Zhao Yilin. Vehicle Positioning and Navigation System [M] . Beijing: Publishing House of Electronics Industry, 1999.

[2] Zhang Dong, Liu Ailong, Zhang De. The Present Situation of Vehicle Navigation Techniques in China and the Multi-Topological Formation and Route Computing of National Navigation Map[C]. XXII International Cartographic Conference (ICC2005), ISBN: 0-958-46093-0, Produced by Global Congress. Spain, 2005.

[3] R.E.Korf. Real-Time Heuristic Search[J]. Artificial Intelligence, 1990, 42: 189-211.

[4] J.Shapiro, J.Waxman, D.Nir. Level Graphs and Approximate Shortest Path Algorithm[J]. Network,1992,22: 691-717.

[5] E.W.Dijkstra. A Note on Two Problems in Connexion with Graphs[J]. Nuberische Mathematik, 1959, 1: 269-271.

[6] P.E.Hart, N.J.Nilsson, B.Raphael. A Formal Basis for the Heuristic Determinations of Minimum Cost Paths. IEEE Trans.Syst.Sci.Cybernetics, 1968, 4(2): 100-107.

[7] Du Duanpu. Graph Theory of Operations Research[M]. Beijing: Beijing University of Aeronautics and Astronautics, 1990.

[8] J.Anez, T.Barra, B.Perez. Dual Graph Representation of Transport Networks[J]. Transportation Research-B, 1996, 30(3): 209-216.

**Acknowledgements**

The project is sponsored by the National Natural Science Fund (41471387)

# CAMERA CALIBRATION FOR UAV APPLICATION USING SENSOR OF MOBILE CAMERA

Y. Takahashi [a, b], H. Chikatsu [b]

[a] Research and Develoment Center, AERO ASAHI CORPORATION,
350-1165, 3-1-1 Minamidai, Kawagoe, Saitama, Japan - youji-takahashi@aeroasahi.co.jp
[b] Dept. of Civil and Environmental Engineering, Tokyo Denki University,
350-0394, Hatoyama, Saitama, Japan - chikatsu@g.dendai.ac.jp

**Commission V, WG V/4**

**KEY WORDS:** Camera Calibration, Performance Evaluation, Android, Motion Sensor, Consumer Grade Digital Camera

**ABSTRACT:**

Recently, 3D measurements using small unmanned aerial vehicles (UAVs) have increased in Japan, because small type UAVs is easily available at low cost and the analysis software can be created the easily 3D models. However, small type UAVs have a problem: they have very short flight times and a small payload. In particular, as the payload of a small type UAV increases, its flight time decreases. Therefore, it is advantageous to use lightweight sensors in small type UAVs.
A mobile camera is lightweight and has many sensors such as an accelerometer, a magnetic field, and a gyroscope. Moreover, these sensors can be used simultaneously. Therefore, the authors think that the problems of small UAVs can be solved using the mobile camera.
The authors executed camera calibration using a test target for evaluating sensor values measured using a mobile camera. Consequently, the authors confirmed the same accuracy with normal camera calibration.

## 1. INTRODUCTION

Recently, unmanned aerial vehicles (UAVs) have been noticed for periodic inspection to maintain and preserve Japanese civil engineering buildings. Previously, inspection work in structures such as bridges and under the elevated road, was performed by humans in dangerous environments. In such situations, it is believed that using UAVs will lead to safe inspections. Furthermore, multi-view images captured by UAVs during regular inspections and high-density point cloud data created using Structure from Motion (SfM) and Multi View Stereo (MVS) enables the creation of 3D models and quantitative evaluations of the displacement of a structure. However, low-cost UAVs that are currently sold have some issues: they are small payloads and have very short flight times. A low-cost UAV has an approximate flight time of 15 min. Therefore, there is a limit on the weight of a sensor that can be mounted on the UAV; it is necessary to devise such replace a sensor mounted on the UAV for each measurement. On the other hand, the use of high-performance and lightweight mobile devices (e.g., Android and iOS devices) has progressively and rapidly spread in Japan (Figure 1). These mobile devices have a digital camera as well as GPS and several sensors such as an accelerometer, a gyroscope, and a digital compass. By integrating some of the sensors built into a mobile device, it is possible to estimate the position and orientation of the device. In this case, noise contained in sensor values can be denoised using a Kalman filter (Fuse, 2014). Moreover, mobile devices are lightweight at approximately 0.15 kg, and it is presumed to be capable of solving the problems of UAVs.
In this study, the authors described and evaluated camera calibration techniques for UAVs using images and orientation parameters of sensor values from mobile devices.

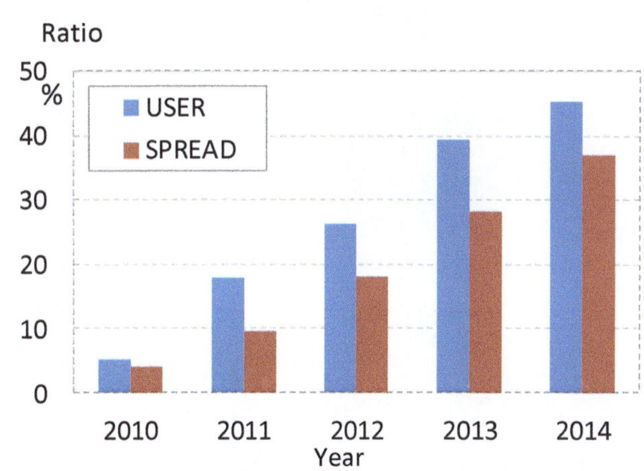


Figure 1. User and spread ratio

## 2. ANDROID DEVICE

### 2.1 Android

Android devices have digital cameras as well as various sensors. However, Android is included in various devices from different manufacturers, and some models do not have many mounted sensors. The left pane of figure 2 shows the various sensors of the Android device used in this paper, and the right pane shows the sensors in another device. Therefore, the Android device must be carefully selected.

Figure 2. Difference of mounted sensor

### 2.2 Android Coordinate System

Figure 3 shows the coordinate system of the accelerometer, magnetic field, and gyroscope measured from the Android device. The origin of the coordinate system is defined as the center of the Android device, and the coordinate axes are not changed even if the orientation of the device has changed.

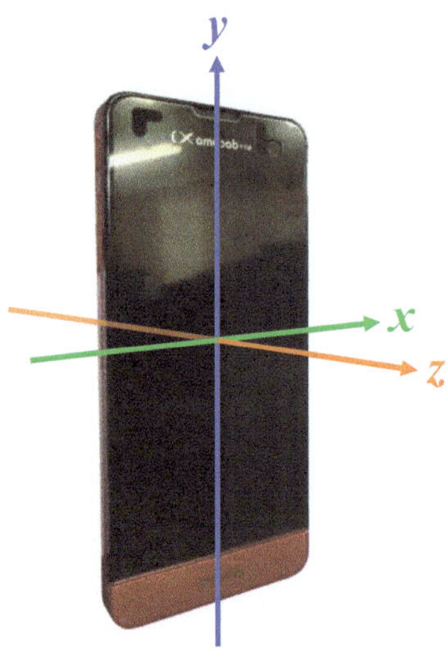


Figure 3. Coordinate system

In this case, X and Y axes must be exchanged by equation (1) when taken android device rotating 90 degrees. Therefore, the axes of coordinate system must be carefully oriented (Google Inc., 2015).

$$\begin{aligned} x &= -y \\ y &= x \\ z &= z \end{aligned} \tag{1}$$

where $x, y, z$ = Android sensor values.

## 3. EXPERIMENTAL PROCEDURE

Figure 4 shows the test target (H: 640 mm, W: 480 mm, D: 20 mm; 3 rows in the center) used in this paper. The white-circled points are control points for camera calibration and the other 86 black-squared points are check points. The digital cameras shown in Tables 1 and 2 took five triplet images of base-height ratio 0.34 with altitude varying between 0.6–0.8 m to keep a uniform photo scale. Furthermore, in the experiment using the device of Table 1, each optical axis was adjusted to be consistent with the image taken by a consumer grade digital camera. Moreover, image and sensor values were synchronize using Internet time because in this paper, the device is connected to a network of all devices.

○ : Control points □ : Check points

Figure 4. Test target

Camera calibration was performed through bundle adjustment using 13 control points. Therefore, the sensor values measured by the Android device of Table 2 is denoised using the Kalman filter, and rotation by a certain angle is performed. Furthermore, in order to compare this result performed to normal camera calibration.

After the camera calibration, the accuracy was estimated using average root-mean-square-error (RMSE) for check points. The XY and Z accuracies of the center coordinates of each black-circled point were ±0.05 mm, and these image coordinates were obtained as area centroid by image-processing procedures.

| | |
|---|---|
| Device Name | CyberShot DSC-HX60V |
| Pixel Number | 20MP |

Table 1. Consumer grade digital camera

| Device Name | Xperia Z(by SONY) |
|---|---|
| OS | Android 4.4 (Kitkat) |
| On Sensor | GPS<br>Accelerometer<br>Magnetic Field Sensor<br>GyroScope<br>et al. |
| Pixel Number | 13MP(out), 2MP(in) |

Table 2. Mobile camera

## 4. RESULTS

### 4.1 Consumer Grade Digital Camera and Android Device

Experimental results using the consumer grade digital camera and sensor value measured by the Android device are shown in figure 5, in which "Normal" is the result of normal camera calibration and "Sensor Angle" is the result of camera calibration using the sensor value as the initial value measured by the Android device. Furthermore, "Normalization" is divided by the standard error in equation (3) from each result. Consequently, a greater "Normalization" implies that a result has error lower than the standard error.

$$\sigma_{X0} = \sigma_{Y0} = \frac{H}{f}\sigma_p$$
$$\sigma_{Z0} = \sqrt{2}\frac{H}{f}\frac{H}{B}\sigma_p \qquad (3)$$

where $\sigma_{X0}$, $\sigma_{Y0}$, $\sigma_{Z0}$ = standard error
$H$ = altitude
$f$ = focal length
$B$ = base line
$\sigma_P$ = pointing accuracy (here is estimated 0.1 pixels)

Figure 5 shows almost the same result in both cases. Furthermore, Table 3 shows the relationship between the normal camera calibration result and camera calibration using the Android device measuring the orientation result of the triplet image. Table 3 shows that the measurement value of the Android device can be the same as that of the normal camera calibration. Therefore, the orientation obtained from the Android device is considered to be available as the initial value.

| Location | | Calibration Result | Sensor Values |
|---|---|---|---|
| Left | $\omega$ | -0° 32′ 44.59″ | 0° 10′ 58.95″ |
| | $\phi$ | 7° 28′ 10.16″ | 7° 29′ 38.04″ |
| | $\kappa$ | 0° 19′ 03.80″ | 2° 14′ 26.44″ |
| Right | $\omega$ | -0° 29′ 34.70″ | 0° 16′ 54.07″ |
| | $\phi$ | -8° 16′ 31.53″ | -8° 51′ 53.20″ |
| | $\kappa$ | 0° 23′ 26.00″ | 2° 08′ 49.62″ |

Table 3 Compare orientation

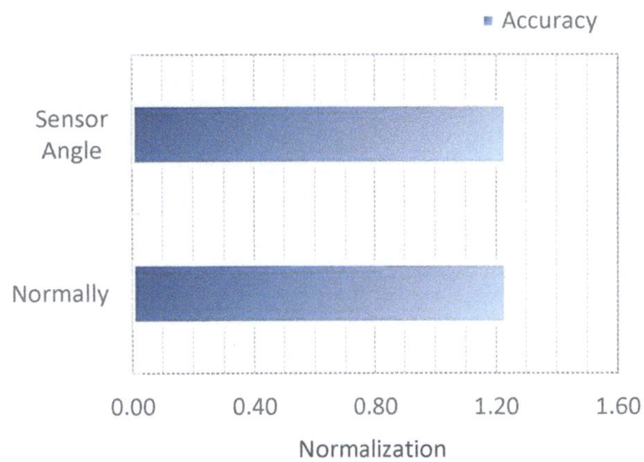


Figure 5. Consumer grade digital camera

### 4.2 Use of Only the Android Device

The equipment that can be mounted on a UAV is restricted as the flight time decreases with increasing payload. The Android device is lightweight and features any built-in sensors and a digital camera of high resolution. The Android device can measure simultaneously using the sensor and digital camera.

The experimental results using only the Android device is shown in figure 6, which shows a large RMSE compared with the standard error. However, the result of normal camera calibration and camera calibration using the sensor value is showed the same tendency. Therefore, the large RMSE is assumed to be caused by the calibration techniques used in this study. Accordingly, the authors think that the sensor values of Android devices can be used for photogrammetry.

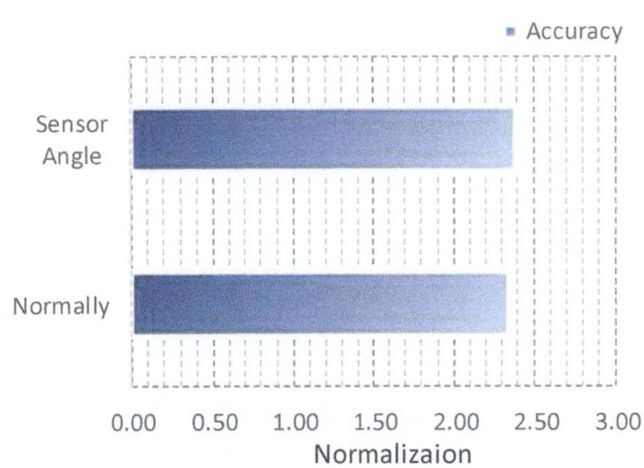


Figure 6. Mobile camera

## 5. CONCLUSION

This paper evaluated the camera calibration techniques using images and the orientation parameter of sensor values from mobiles device for UAVs. Normal camera calibration with a consumer grade digital camera yielded the same result as with calibration using the sensor value with an Android device.

In the experiment using the image and sensor values of the Android device, an RMSE greater than the standard error was obtained. This result has a different trend from the result obtained using a consumer grade digital camera. However, the authors believe that the large RMSE was caused by the camera calibration technique used because the same error was obtained in normal camera calibration. Therefore, the sensor value of an Android device can be used in a UAV, as a large difference is not observed in the measurement accuracy.

Future tasks include camera calibration using the sensor value and GPS of an Android device and mounting an Android device on a UAV.

## 6. REFERENCE

T., Fuse, and K., Matsumoto, 2014. development of a self-localization method using sensors on mobile devices. ISPRS Volume XL-5, pp. 237-242

A., Al-Hamad and N., El-Sheimy. 2014 ISPRS Volume XL-5, pp. 29-34

Y. I. Abdel-aziz, 1982. Accuracy of the Normal Case of Close-Range Photogrammetry, Photogrammetric Engineering and Remote Sensing, 48(2), pp.207-213

W., Tampubolon and W., Reinhardt., 2014. uav data processing for large scale topographical mapping ISPRS Volume XL-5, pp. 565-572

Google Inc., 2015. Android Developers API Reference, https://developer.android.com/reference/android/hardware/SensorEvent.html (17 Apr 2015).

# ASSESSING THE IMPACT OF URBANIZATION ON URBAN CLIMATE BY REMOTE SENSING PERSPECTIVE: A CASE STUDY IN DANANG CITY, VIETNAM

Nguyen Hoang Khanh Linh [a,*], Huynh Van Chuong [a]

[a] Faculty of Land Resources and Agricultural Environment, Hue University of Agriculture and Forestry, 102 Phung Hung, Hue City, Vietnam – (nguyenhoangkhanhlinh, huynhvanchuong)@huaf.edu.vn

**Commission ATMC – Atmosphere, weather and climate**

**KEY WORDS:** Danang, urban climate, urbanization, change detection,

**ABSTRACT:**

Urban climate, one of the challenges of human being in 21 century, is known as the results of land use/cover transformation. Its characteristics are distinguished by different varieties of climatic conditions in comparison with those of less built-up areas. The alterations lead to "Urban Heat Island", in which temperature in urban places is higher than surrounding environment. This happens not only in mega cities but also in less urbanized sites. The results determine the change of land use/cover and land surface temperature in Danang city by using multi-temporal Landsat and ASTER data for the period of 1990-2009. Based on the supervised classification method of maximum likelihood algorithm, satellite images in 1990, 2003, 2009 were classified into five classes: water, forest, shrub, agriculture, barren land and built-up area. For accuracy assessment, the error metric tabulations of mapped classes and reference classes were made. The Kappa statistics, derived from error matrices, were over 80% for all of land use maps. An comparison change detection algorithm was made in three intervals, 1990-2003, 2003-2009 and 1990-2009. The results showed that built-up area increased from 8.95% to 17.87% between 1990 and 2009, while agriculture, shrub and barren decreased from 12.98% to 7.53%, 15.72% to 9.89% and 3.88% to 1.77% due to urbanization that resulted from increasing of urban population and economic development, respectively. Land surface temperature (LST) maps were retrieved from thermal infrared bands of Landsat and ASTER data. The result indicated that the temperature in study area increased from 39°C to 41°C for the period of 1990-2009. Our analysis showed that built-up area had the highest LST values, whereas water bodies had the least LST. This study is expected to be useful for decision makers to make an appropriate land use planning which can mitigate the effect to urban climate.

## 1. INTRODUCTION

Urbanization is believed one of the most prevalent anthropogenic causes of the losing arable land, devastating habitats, and the decline in natural vegetation cover (Dewan and Yamaguchi, 2009). As a consequence, rural areas have been converted into urban areas with an unprecedented rate and making a noted effect on the natural functioning of ecosystems (Turner, 1994). Consequently, a profound understanding of land use change is very important to have a proper land use planning and sustainable development policies (Braimoh and Onishi, 2007).

According to Myint and Wang (2006), to fulfil such a sustainable urban development, urban and regional planners have to summarise from numerous decisions. However, these decisions must be extracted from huge data sources, which represent physical, biological and social parameters of urban areas in the continued spectrum of spatial and temporal domains. Hence, it is important to have an effective spatial dynamic tool, which is used to understand urban land-use and land-cover change (LULC) for predicting future change. Compared with various methods used to collect land use data, remote satellite technologies has proven its capacity in providing accurate and timely information on the geographic distribution of land use, especially for regional size areas (Sun *et al.*, 2009). With the support of Geographical Information Systems (GIS), satellite images can effectively estimate and analyse changes and LULC trends (Hathout, 2002).

Recently, various studies on urban heat island (UHI) have been based on remote satellite images instead of air temperature obtained from weather stations (Li *et al.*, 2009). It shows that data got from these *in situ* data can only give accurate local meteorological data, which are normally distributed as discrete points and hardly displayed as spatial variation (Owen *et al.*, 1998).To determine the increase of thermal environment in urban, the land surface temperature (LST) of study area is calculated from available infrared information of satellite images (X.,Zhou, Y.-C. Wang., 2011). Voogt and Oke (1998) and Weng (2009) indicated that LST parameter is able to modulate the air temperature above the earth surface and close related to surface radiation and energy exchange. In addition, previous authors gave the evidences that rapid urban sprawl and population growth in metropolitan change physical properties of urban land surface, in which, temperature in urban areas are a few degrees higher than in surrounding non-urbanized areas (Xian and Crane, 2005). In fact, UHI distributes not only in mega cities but also in medium and small cities. Therefore, this research had been investigated to illustrate the significant relationship between land use/cover (LULC) type and surface temperature even though in small scale.

.

## 2. STUDY AREA

Danang city is located in Middle Central of Viet Nam, between the 15°55'19" to 16°13'20"N and 107°49'11" to 108°20'20"E (Figure 1). It is a long-stretching narrow region and well known as a dynamic city of the Key Economic Zone in central Viet Nam. The area consists of hills and mountains in the northwest and the Eastern Sea in the east. The altitude varies from 400m to 1524m above sea level; next to is the upland with low mountains and the delta takes ¼ areas in the southeast; it covers an area of 1,283.42km$^2$, including Hoang Sa island district of 305 km$^2$.

* Corresponding author

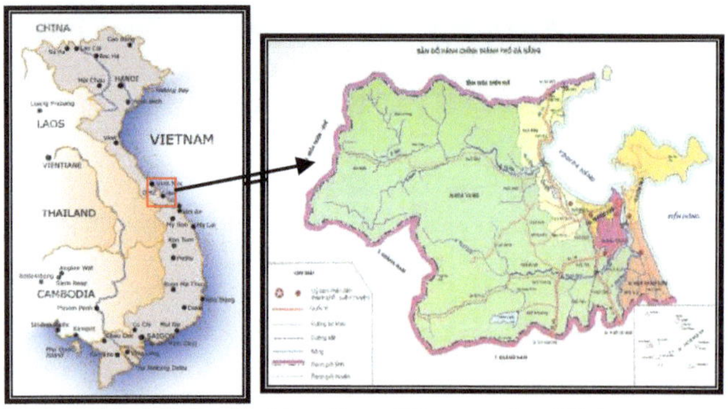


Figure 1. General location of Danang city, Vietnam (http://www.myvietnam.info/)(http://www.danang.gov.vn/)

Climate of Danang city is in the zone of typical tropical monsoon. The average annual temperature is about 26°C, average rainfall is about 2,505mm per year and average humidity is 83.4%. There are two main seasons annually: the wet (August-December) and the dry (January-July). In 2009, the total population is about 887,070 and the population density is 906.7 persons per km$^2$. Danang city is known as one of the most densely populated and urbanized area in Vietnam. With the economy development and population increasing, the local LULC in Danang city has changed seriously.

## 3. DATA AND METHOD

### 3.1 Data sources and Image preprocessing

Landsat imagery were used for this study. The following criteria were considered for choosing proper data: (1) the images should be long time enough for detecting the land use change; (2) study area should not have cloud cover. Unfortunately, the study area is located near coastal. Due to the influence of weather, there are not many data satisfied both conditions. The images always have some thick cloud cover or haze. In addition, the study area is not entirely contained within one scene of Landsat Therefore, having acquisition images near anniversary dates for changing detection as Jensen mentioned (2005) was unavailable. In this study, three periods of satellite images were selected to classify study area: Landsat-5 TM August 24, 1990; Landsat-7 ETM+ March 04 and April 14, 2003; Landsat-5 TM Jun 2009 (download free at http://earthexplorer.usgs.gov/ and http://glovis.usgs.gov/ ); and ASTER image (April 02, 2009) was purchased from Japan Space Systems/Earth Remote Sensing Division (ERSDAC) was used as reference image. The details of data were described in Table 1. For this study, the reference data were also used, included: (1) topographic map 2001, at scale of 1/50.000; (2) land use maps in 1997, 2003 and 2010, at scale of 1/25.000.

Because Landsat and ASTER imagery were collected at level 1T and 1B respectively, geometric correction do not require. However, images were acquired at different spatial resolution and projections. Therefore, all images were first rectified to Universal Transverse Mercator (UTM) coordinate system, Datum WGS 84, Zone 48 North for matching the geographic projection of the reference data. Images were also co-registered together within 25 well distributed GCPs (ground control points) and polynomial 1sd by means of OrthoEngine provided by PCI Geomatica 10.3 software. RMS <0.5 was received. In addition, Nearest Neighbour resampling was set for not changing heavily the radiometric characteristic of image. In this study, the iteratively re-weighted multivariate alteration detection (IR-MAD) transformation was used for automatic radiometric normalization for all images by means of ENVI 4.8 software; see (Canty, 2010, A. A. Nielsen *et al.*, 1998, Canty and Nielsen, 2008). ASTER 02/04/2009 was chosen as reference image. However, this image does not cover all the region of study area, therefore a subset of 1800x1100 pixels with 30m spatial resolution including 968.17km$^2$ was created for all images for further studying. This territory was chosen to ensure the specific study area was in the analysis image. Besides the requirement of the same dimension, images must have the same spectral resolution. Hence, the composite of standard false colours was used for this study: Landsat TM/ETM+ (432).

### 3.2 LULC classification and Change detection

Six land use/cover classes were defined for image classification based on the modified Anderson land use/cover scheme level I (Anderson *et al.*, 1976), included: (1) water, (2) forest, (3) shrub, (4) agriculture, (5) barren and (6) urban land. Supervised classification using maximum likelihood approach in ENVI 4.8 was individually applied for each image of study area to classify land use/cover. Maximum likelihood algorithm was preferred because this rule is considered to have accurate results because it has more accurate results than other algorithms (Mengistu D. A. and Salami A. T., 2007, Reis, 2008, Diallo Y. *et al.*, 2009).

Because of various image acquisition dates, training areas for the images of the years 1990, 2003 and 2009 were different during the classification. In addition, the training areas were verified by references data. As the next step, post-classification comparison change detection algorithm was selected to detect changes in LULC from 1990 to 2009 in study area. In order to minimize the problem in radiometric calibration of imagery of two different dates. For comparison of the classification results of two dates, a change detection matrix was created based on pixel-by-pixel (Jensen, 2005). Thereby, each type of from-to LULC change is identified.

### 3.3 Land Surface Temperature

#### *3.3.1. Retrieving brightness temperature*

Brightness temperature (BT) from thermal band of Landsat imagery was derived as following two steps: firstly, the satellite-based digital number is converted to at-sensor spectral radiance ($L_\lambda$) by using Eq (1) (Landsat Project Science Office, 2002):

$$L_\lambda = GAINS\ x\ DN + BIASES \tag{1}$$

Where DN is the digital number of thermal band, and the corresponding spectral radiance. The gain (*GAINS*) offset (*BIASES*) values can be obtained from the image header file. The second step is to convert spectral radiance values from thermal band to brightness temperature using the formula (2) (Schott and Volchok, 1985; Wukelic et al., 1989):

$$T_k = \frac{K_2}{ln(\frac{K_1}{L6}+1)} \tag{2}$$

Where $K_1$ and $K_2$ are calibration constants, $T_k$ represents temperature. For Landsat imagery, $K_1$= 607.76W*m$^{-2}$sr$^{-1}$, $K_2$= 1260.56K

#### *3.3.2. Transforming BT into LST*

The temperature calculated in Eq (2) is not the actual LST, but the at-satellite temperature. In this study, single-channel method by Jiménez-Muñoz, Juan C. Sobrino, José A (2003) and Sobrino et al., (2004):

$$LST = \gamma[\varepsilon^{-1}(\psi_1 L_\lambda + \psi_2) + \psi_3] \tag{3}$$

$$\gamma = \left\{ \frac{c_2 L_\lambda}{T_k^2} \left[ \frac{\lambda^4}{c_1} L_\lambda + \lambda^{-1} \right] \right\} \quad (4)$$

$$\delta = \gamma L_\lambda + T_k \quad (5)$$

Where $L_\lambda$ is sensor radiance, $T_k$ is brightness temperature, $\lambda$ is the effective wavelength (11.457 μm for Landsat TM6 band), $c_1 = 1.19104 \times 10^8 Wm^{-2}sr^{-1} \mu m^4$, $c_2 = 14387.7\ \mu m\ K$.$\psi_1$, $\psi_2$, and $\psi_3$ can be calculated as a function of the total atmospheric water vapour content ($\psi$)

$$\psi_1 = 0.1471\omega^2 - 0.15583\omega + 1.1234 \quad (5a)$$

$$\psi_2 = -1.1836\omega^2 - 0.37607\omega - 0.52894 \quad (5b)$$

$$\psi_3 = -0.0455\omega^2 + 1.8719\omega - 0.39071 \quad (5c)$$

### 3.4. Normalized Difference Vegetation Index

The Normalized Difference Vegetation index (NDVI) is a simple graphical indicator to measure the live green vegetation from remote satellite imagery. It can be calculated from these individual measurements as follows:

$$NDVI = \frac{(NIR - VIS)}{(NIR + VIS)} \quad (6)$$

Where VIS and NIR stand for the spectral reflectance measurement acquired in the visible (red) and near-infrared regions, respectively. NDVI values range from -1 to 1;

| *Type of sensor* | *Spatial resolution (m)* | *Band* | *Date* | *Path* | *Row* | *Average cloud coverage (%)* |
|---|---|---|---|---|---|---|
| Landsat-5 TM | 30 | 1-5, 7 | August 24, 1990 | 124 | 49 | 0 |
| Landsat-7 ETM+ | 30 | 1-5, 7 | March 04, 2003 | 125 | 49 | 34.7 * |
| | 30 | 1-5,7 | April 14, 2003 | 124 | 49 | 0.3 |
| Landsat-5 TM | 30 | 1-5, 7 | Jun 25, 2009 | 124 | 49 | 3.7 |
| ASTER | 15 | 1-3 | April 02, 2009 | - | - | 4.0 |

(* Although the average cloud coverage of Landsat-7 ETM+ is very high, there is almost no cloud in study area at that time).

Table 1. Characteristics of satellite data used in study area

## 4. RESULTS AND DISCUSSION

### 4.1 Land Use/ Cover Changes

Before doing any other interpretations, thematic LULC maps (1990, 2003 and 2009) were assessed their accuracy through four measurable means of error matrix: overall accuracy, producer's accuracy, user's accuracy and Kappa coefficient. A total of 300 stratified random pixels was taken for each LULC map and then checked with reference data. According to the accuracy assessment results of classified maps, the overall accuracy for Landsat TM 1990, Landsat ETM+ 2003 and Landsat TM 2009 was 84.44% and 89.00%, and 92.15% respectively; the Kappa Coefficient of those maps reached at 0.7534, 0.8005, and 0.9021 respectively. Collating with the minimum 85% accuracy stipulated by the Anderson classification scheme for satellite-derived LULC maps, these statistics were adequate for continuously studying (Anderson *et al.*, 1976, Kamusoko and AniYa, 2006).

The LULC maps of study area were generated for all three years (Figure 2) and classification area statistics were summarised in table 3. The classified areas were measured by multiplying the number of pixel with spatial resolution of remote data (i.e. 30m), in which the pixel number was determined after applying postclassification analysis. And then changes were defined based on the difference of pixel number between two dates. Based on Figure 2, forest and urban areas were the dominant LULC classes in spatial distribution pattern. Accordingly, forest area was counted for about 64%, 62.2% and 59.8% of the total area in 1990, 2003 and 2009 respectively; meanwhile urban area was occupied 6.5%, 11.3% and 17.9% of the total area in 1990, 2003 and 2009 respectively. The surface water body covers about 2.5%, 3.3% and 3.1% of the total region study in 1990, 2003 and 2009, respectively. The results also showed that from 1990 to 2009 LULC units under shrub, agriculture and barren decreased from 10.1% to 9.9%, 12.4% to 7.5% and 4.5% to 1.8%, respectively.

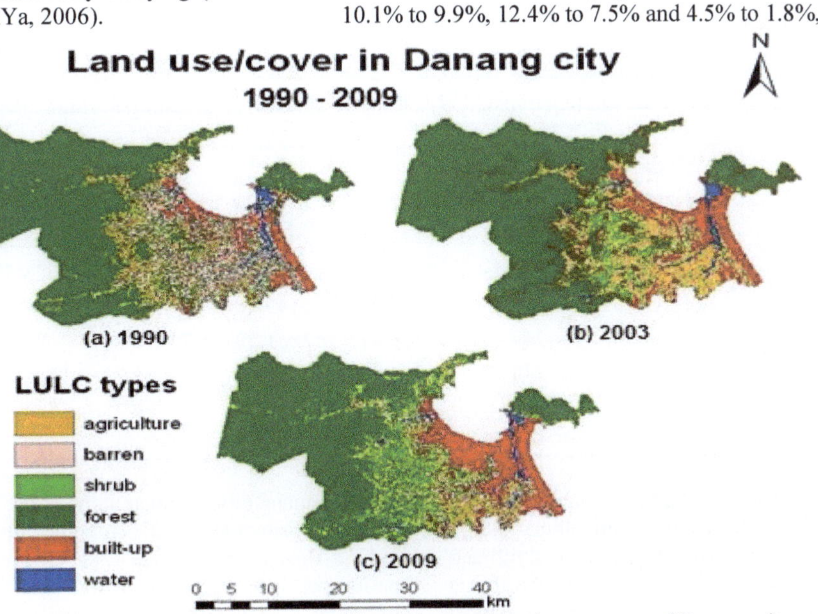


Figure 2. Land use/cover maps of Danang city area

To provide a further comprehensive calculation in losing and gaining among the six LULC classes, the from-to change matrix of land use/cover in Danang city were created in three intervals,1990-2003, 2003-2009 and 1990-2009 (Table 4).

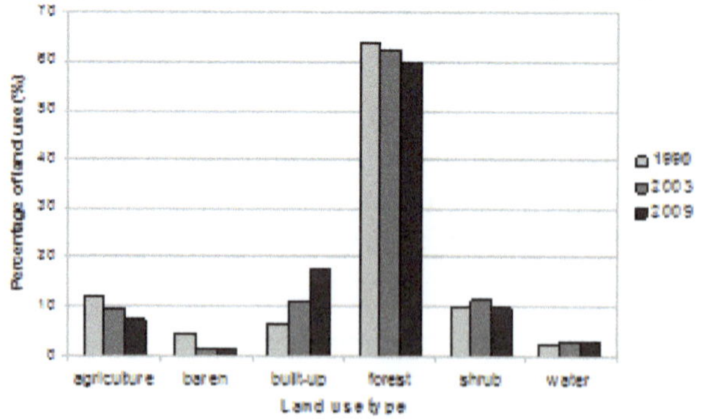


Figure 3. Areas land use/cover classes of Danang city

In cross tabulation, unchanged pixels were located along the major diagonal of the matrix while conversion values of classes were arranged in descending order.

| LULC class | 1990 | | 2003 | | 2009 | |
|---|---|---|---|---|---|---|
| | Area (ha) | (%) | Area (ha) | (%) | Area (ha) | (%) |
| Agri-culture | 12570.3 | 13.0 | 9512.0 | 9.8 | 7294.7 | 7.5 |
| Barren | 3759.0 | 3.9 | 1771.0 | 1.8 | 1708.9 | 1.8 |
| Urban | 8661.2 | 8.9 | 10900.7 | 11.3 | 17298.5 | 17.9 |
| Forest | 54264.1 | 56.0 | 60233.0 | 62.2 | 57936.2 | 59.8 |
| Shrub | 15216.5 | 15.7 | 11169.4 | 11.5 | 9575.8 | 9.9 |
| Water | 2346.1 | 2.4 | 3231.2 | 3.3 | 3003.6 | 3.1 |
| Total | 96817.2 | 100 | 96817.2 | 100 | 96817.7 | 100 |

Table 3. Results of and use/cover classification for 1990, 2003 and 2009 images

| | 1990 | | | | | | |
|---|---|---|---|---|---|---|---|
| 2003 | Agriculture | Barren | Urban | Forest | Shrub | Water | 2003 Total |
| Agriculture | 2651.22 | 1125.45 | 644.76 | 2303.10 | 2611.35 | 176.13 | 9512.01 |
| Barren | 240.12 | 314.01 | 305.28 | 412.47 | 461.25 | 37.89 | 1771.02 |
| Urban | 1053.00 | 1463.67 | 4690.44 | 2082.87 | 1338.84 | 271.89 | 10900.71 |
| Forest | 3710.16 | 453.42 | 130.95 | 54105.21 | 1761.21 | 72.00 | 60232.95 |
| Shrub | 4125.78 | 584.46 | 315.81 | 2740.41 | 3290.49 | 112.41 | 11169.36 |
| Water | 267.75 | 371.16 | 228.06 | 327.96 | 322.02 | 1714.23 | 3231.18 |
| 1990 Total | 12570.3 | 3759.0 | 8661.2 | 54264.1 | 15216.5 | 2346.1 | |
| Change 1990-2003 | - 2536.02 | -2541.15 | 4585.41 | -1739.07 | 1384.20 | 846.63 | |

(a) 1990-2003

| | 2003 | | | | | | |
|---|---|---|---|---|---|---|---|
| 2009 | Agriculture | Barren | Urban | Forest | Shrub | Water | 2009 Total |
| Agriculture | 2022.75 | 129.15 | 666.36 | 3016.53 | 1428.57 | 31.68 | 7294.7 |
| Barren | 103.5 | 101.34 | 145.26 | 896.31 | 427.59 | 32.94 | 1708.9 |
| Urban | 3990.69 | 750.06 | 8621.73 | 783.45 | 2417.58 | 697.77 | 17298.5 |
| Forest | 1166.76 | 201.42 | 381.69 | 52895.25 | 3126.06 | 125.28 | 57936.2 |
| Shrub | 2074.23 | 520.02 | 739.17 | 2525.13 | 3624.66 | 89.1 | 9575.8 |
| Water | 148.32 | 68.31 | 335.7 | 61.56 | 139.05 | 2221.56 | 3003.6 |
| 2003 Total | 9512.01 | 1771.02 | 10900.98 | 60232.95 | 11169.36 | 3231.18 | |
| Change 2003-2009 | -2210.22 | -59.22 | 6428.7 | -2226.69 | -1581.66 | -222.84 | |

(b) 2003-2009

| | 1990 | | | | | | |
|---|---|---|---|---|---|---|---|
| 2009 | Agriculture | Barren | Urban | Forest | Shrub | Water | 2009 Total |
| Agriculture | 1777.5 | 969.21 | 122.85 | 2407.23 | 1943.91 | 74.34 | 7294.7 |
| Barren | 361.44 | 72.18 | 91.71 | 937.17 | 236.07 | 8.37 | 1708.9 |
| Urban | 2915.55 | 1930.23 | 5126.58 | 3872.61 | 2789.73 | 626.58 | 17298.5 |
| Forest | 3817.62 | 236.88 | 295.92 | 51499.8 | 1945.35 | 100.89 | 57936.2 |
| Shrub | 2881.62 | 749.79 | 456.84 | 2851.29 | 2549.16 | 83.61 | 9575.8 |
| Water | 291.33 | 348.12 | 204.93 | 345.06 | 317.7 | 1467.36 | 3003.6 |
| 1990 Total | 12570.3 | 3759.0 | 8661.2 | 54264.1 | 15216.5 | 2346.1 | |
| Change 1990- 2009 | -4746.24 | -2600.37 | 11014.11 | -3965.76 | -197.46 | 623.79 | |

(c) 1990-2009

Table 4. Matrices of land use/cover changes in Danang city, Vietnam in hectares

During 1990-2003, results showed that agriculture, barren and forestry decreased strongly while urban area, shrub and water body increased, notably the raising of urban area. Table 4(a) indicated that the expansion of urban area was the most dramatic changes in the region, which was the result of promptly growing up of economic after applying "DOI MOI" policy. Based on statistic, 4585.41ha of urbanized area in this period was calculated, which was nearly as large as the coverage of urban area in 1990, thus representing an increase of 73%. Analysing the component of the conversion of growth in urban area, 42.6% was converted from forestry, 25.3% from barren and 22.3% from shrub. The coverage of water body between 1990 and 2003 was significant difference because of the construction of two new reservoirs in study area. In second period, from 2003 to 2009, urban area was the only one which continuously increased, whereas the other classes decreased. Statistic from Table 4(b) showed the amount of urbanized area as 6428.7ha. The increase of urban area could be represented as 59% of its 2003 area by 2009. Of the area that urban got from 2003 to 2009, 51.7% was converted from agriculture and 26.1% from shrub. In the span of 20 years, the coverage of urban area has incessantly increased from 8861.2ha in 1990 to 17298.5ha in 2009, thus representing an increase of 95.2%% (8437.3ha). Table 4(c) indicated that the increasing of urban area in Danang city mainly came from about 32.5%, 25.4%, 25.3% and 17.5% of forestry, agriculture, shrub and barren area, respectively. Based on Figure 3, the difference of spatial distribution of urban area could be clearly observed by the years. In 1990, the urban area dispersedly located along the costal line. By 2003, this area was expanded more concentrated along coastal zone and moved toward Sontra peninsula. From 2003 to 2009, the urban expansion changed the direction from costal toward in land.

### 4.2 Land surface temperature

Figure 4 shows the results of surface temperature distribution across the city of Danang. Observations map shows the highest surface temperature during study period ranges from 38°C - 41°C. In particular, the construction of high-density land surface temperature has the largest, with an average value is 37°C and are shown in red on the map; concentrated in the urban areas along the coastal region. In contrast, the forestland and the green area of agricultural land remaining lower temperatures. Particularly, the surface temperature of the water in the soil is low, with average values reaching 20°C and is shown in dark blue on the map. The results also show that the temperature in vegetated areas was lower in comparison with built-up areas. Because vegetation has the capability of evaporating, which help to accelerate the process of heat transfer between land surface and atmosphere. Meanwhile, temperature in urban area covered by impervious surfaces was higher due to its construction materials shown a tendency in absorbing and holding heat, and less evaporating.

Over the years, various studies have shown that the change of LULC overlay huge impact on the environment. To clarify this statement by remote sensing, through linear regression method we analyzed the relationship between the difference vegetation index (NDVI) and surface temperature (LST) of each land use/cover type. From the data in Table 5, it shows that NDVI of water surface gets the lowest value. Meanwhile, built-up area, covered by impervious surface has high temperature. Thus, the NDVI and LST of different land use/cover type tend inversely proportional to each other.

| No | LULC type | LST (°C) | NDVI |
|---|---|---|---|
| 1 | Water | 28.41 | 58.5 |
| 2 | Forestland | 31.13 | 228 |
| 3 | Agriculture | 32.42 | 203 |
| 4 | Bare soil | 33.67 | 184 |
| 5 | Built-up area low density | 37.01 | 135.5 |
| 6 | Built-up area high density | 34.96 | 164 |

Table 5. Values of LST and NDVI at different LCLV type in classified map 2009

Using random sampling method, 100 samples generated from the classified map 2009, in which pixels value NDVI of water had been removed. Next, the values NDVI and LST of different LULC at each sample point were determined. Results of linear regression analysis showed that NDVI and LST of land use/land cover has a negative correlation with a correlation coefficient R2 = 0.618. As can be seen from Figure 5, the points located at the right bottom corner have high value of NDVI, and conversely. Similarly, on the top left corner, the points got high value of LST, low value of NDVI. It means that, changing the type of LULC will affect directly the surface temperature within Danang.

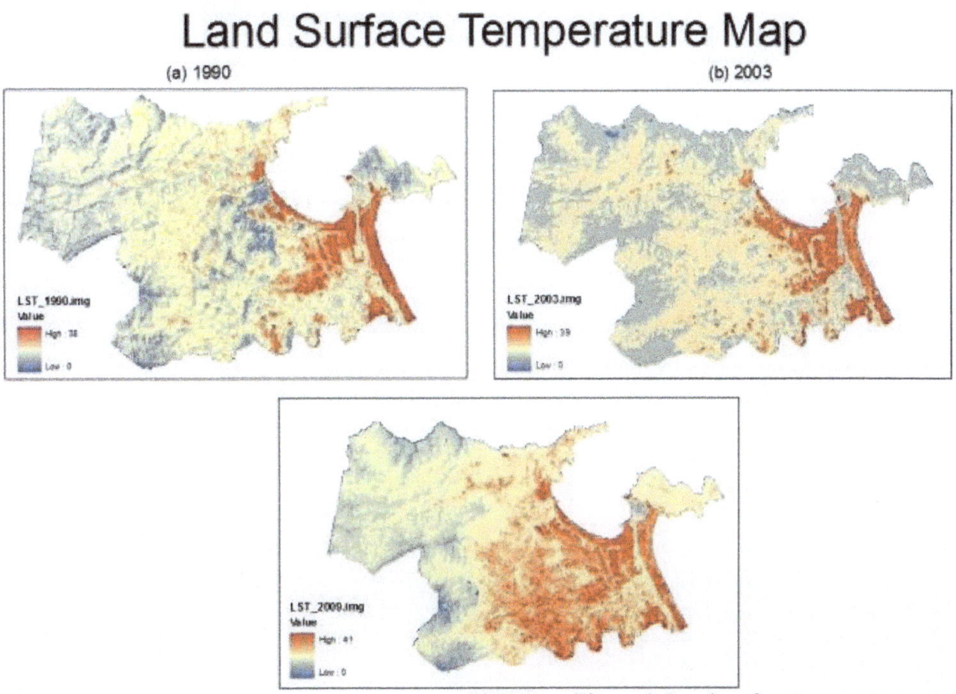


(c) 2009 Figure 4. Land surface temperate map of Danang city

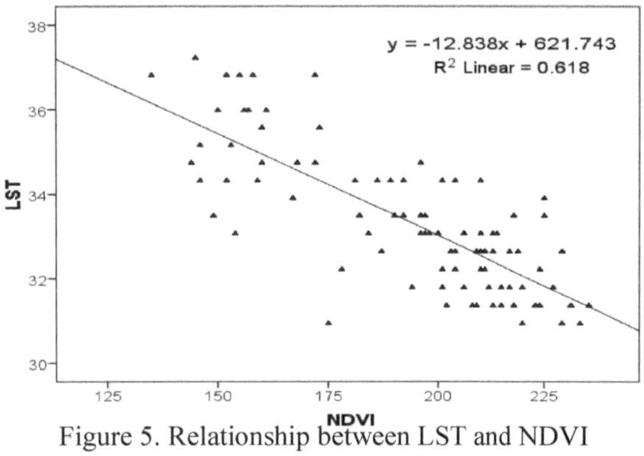


Figure 5. Relationship between LST and NDVI

## 5. CONCLUSIONS

By using the remote sensing and GIS techniques, this paper describes the analysis of LULC and landscape change in the Danang region, Vietnam during the period 1990-2009. The analysis carried out found that a notable decrease of agriculture, forest, barren and shrub because of conversion to urban land during the span of 20 years has taken place. Surface temperature in Danang city increases continuously from 38°C to 41°C. For further understanding the effect of changing LULC to environment, the relationship between LST and NDVI taken within classified map 2009 were perform. Accordingly, urban area showed high value of LST and low value of NDVI.

### References


A. A. Nielsen, Conradsen, K. & Simpson, J., 1998. Multivariate alteration dectection (MAD) and MAF postprocessing in multispectral, bitemporal image data: new approaches to change detection studies, *Remote Sensing of Environment,* 64, 1-19.

Anderson, J.R., Hardy, E.E., Roach, J. & Witmer, R.E., 1976. A land use/cover classification system for use with remote sensor data, *US Geological Survey Professional* 964: Sious Falls, SD.

Braimoh, A.K. & Onishi, T., 2007. Geostatistical techniques for incorporating spatial correlation into land use change models, *International Journal of Applied Earth Observation and Geoinformation,* 9, 438.

Canty, M.J., 2010. *Image analysis, classification and change detection in remote sensing with algorithms for ENVI/IDL,* Second edn, Vol., pp. Pages, Taylor and Francis.

Canty, M.J. & Nielsen, A.A., 2008. Automatic radiometric normalization of multitemporal satellite imagery with the iteratively re-weighted MAD transformation, *Remote Sensing of Environment,* 112, 1025-1036.

Dewan, A.M. & Yamaguchi, Y., 2009. Land use and land cover change in Greater Dhaka, Bangladesh: Using remote sensing to promote sustainable urbanization, *Applied Geography,* 29, 390-401.

Diallo Y., Hu G. & Wen X., 2009. Applications of Remote Sensing in Land Use/Land Cover Change Detection in Puer and Simao Counties, Yunnan Province, *Journal of American Science,* 5, 157-166.

Hathout, S., 2002. The use of GIS for monitoring and predicting urban growth in East and West St Paul, Winnipeg, Manitoba, Canada, *Journal of Environmental Management,* 66, 229-238.

http://www.danang.gov.vn/ Retrieved 10/04/2010, from http://www.danang.gov.vn/TabID/76/CID/711/ItemID/8743/default.aspx.

http://www.myvietnam.info/ Retrieved 20/12/2011, from http://www.myvietnam.info/

Li,J.-., Wang, X.-R., X.-J., Ma, W.-C. and Zhang, H., 2009, Remote sensing evaluation of urban heat island and its spatial pattern of the Shanghai metropolitan, China, Ecological Complexity 6, 413-420

Jensen, J.R., 2005. *Introductory digital image processing: a remote sensing perspective,* 3nd edn, Vol., pp. Pages, 3nd Edition, NJ: Prentice-Hall.

Jiménez-Muñoz, J. C., and Sobrino, J. A., 2003, A generalized single-channel method for retrieving land surface temperature from remote sensing data: Journal of Geophysical Research: Atmospheres, v. 108, no. D22.

Owen, T. W., Carlson, T.N. and Gillies, R.R.,, 1998, An assessment of satellite remotely-sensed land cover parameters in quantitatively describing the climate effect of urbanization.: international Journal of Remote Sensing, v. 19, p. 1663-1681.

Landsat Project Science Office, L. P. S., 2002, Landsat 7 Science Data User's Handbook. Available at. http://landsathandbook.gsfc.nasa.gov/.

Kamusoko, C. & AniYa, M., 2006. Land use/cover change and landscape fragmentation analysis in the Bindura district, Zimbabwe, *Land Degradation and Development,* 18, 221-233.

Mengistu D. A. & Salami A. T., 2007. Application of remote sensing and GIS inland use/land cover mapping and change detection in a part of south western Nigeria, *African Journal of Environmental Science and Technology,* 1, 99-109.

Myint, S.W. & Wang, L., 2006. Multicriteria decision approach for land use land cover change using Markov chain analysis and a cellular automata approach, *Canadian of Remote Sensing,* 32, 390-404.

Reis, S., 2008. Analyzing Land Use/Land Cover Changes Using Remote Sensing and GIS in Rize, North-East Turkey, *Sensors,* 8, 6188-6202.

Sorbrino, J. A., Jiménez-Muñoz, J. C., Paolini, L., 2004. Land surface temperature retrieval from Landsat TM5. Remote Sensing of Environment 90, 434-440

Schott, J.R., Volchok, W.T., 1985. Thematic mapper thermal infrared calibration. Photogrammetric Engineering and Remote Sensing 51 (9), 1351 -1357

Sun, Z., Ma, R. & Wang, Y., 2009. Using Landsat data to determine land use changes in Datong basin, China, *Environmental Geology,* 57, 1825-1837.

Turner, B.L., 1994. Local faces, global flows: The role of land use and land cover in global environmental change, *Land Degradation and Development,* 5, 71-78.

Voogt, J. A. a. o., T, R.,, 2003, Thermal remote sensing of urban areas: Remote Sensing of Environment, v. 86, no. 370-384.

Weng, Q., Lu, D, and Schrubring, J.,, 2004, Estimation of land surface temperature-vegetation abundance relationship for urban heat island studies.: Remote Sensing of Environment, v. 89, p. 467-483.

Wukelic, G, E., Gibbons, D, E., Martucci, L.M, Foote, H, P., 1989. Radiometric calibration of landsat thematic mappper thermal band. Remote Sensing of Environment 28, 339-347

Xiaolu Zhou, Y.-C. W., 2011, Dunamics of Land Surface Temperature in response to Land-use/cover Change: Geographical Research, v. 49, no. 1, p. 23-36.

# PRELIMINARY RESEARCH ON RADIANCE FOG DETECTION BASED ON TIME SERIES MTSAT DATA

Xiongfei Wen [a,*], Zhe Li [a], Sui Zhang [a], Shaohong Shen [a], Dunmei Hu [b], Xiao Xiao [a]

[a] Changjiang River Scientific Research Institute, Changjiang Water Resources Commission, Wuhan 430010, PR China-xfwen19@163.com

[b] Department of Pharmacy, Tongji Hospital Affiliated with Tongji Medical College, Huazhong University of Science and Technology, Wuhan, China 430032, PR China

**KEY WORDS:** Radiance fog, MTSAT, time series, Haar wavelet, frequency domain

**ABSTRACT:**

Fog is a kind of disastrous weather phenomenon. In this paper, the geostationary satellite MTSAT imagery is selected as the main data source to radiance fog detection. According to the unique feature of radiance fog from its generation to dissipation, especially considering the difference between clouds and fog during their lifecycle, the characteristics in frequency domain was constructed to discriminate fog from clouds, The time series MTSAT images were register with a modified Gauss Newton optimization method firstly, then, the Savitzky-Golay smoothing filter was applied to the time series remote sensing imageries to process the noises in the original signal, after that the non-orthogonal Haar wavelets was applied to convert the signal from time domain into frequency domain. The coefficient of high frequency component, including the properties: "max", "min", "the location of the min", "the interval length between the max and min", "the coefficient of linear fit for the high frequency", these properties are selected as the characteristic parameters to distinguish fog from clouds. The experiment shows that using the algorithm proposed in this paper, the radiance fog could be monitored effectively, and it is found that although it is difficult to calculate the thickness of the fog directly, while the duration of fog could be obtained by using the frequency feature.

## 1. INTRODUCTION

### 1.1 Background

As a relative frequently disastrous weather phenomenon, fog can reduce regional or local visibility, deteriorate air quality and cause problems for nautical operations, aviation and surface transportation. It also produces a great hazard to power equipment, crop growth and human health (Li, 2008). Therefore, it is very important to monitor fog in quasi real time. Conventional ground observation is a precise method for fog detection, but it requires much human, as well as economical, resources. Routine forecast methods, such as meteorological statistical approximations and numerical-prediction models, also cannot provide satisfactory forecast results because of their limitations, which include lacking of enough and accurate land surface/air observation data or relevant information, shortcomings inherent from these methods, etc (Ma, 2007).

Fog detection methods based on remote sensing technology use many kinds of satellite data, from geostationary to polar-orbiting satellites, through to observing the Earth's surface repeatedly on a large scale. It would be convenient in fog detection by using the high temporal resolution satellite. With the increasing development of satellite remote sensing technology, data sources will be more numerous, reliable and stable, which makes fast and dynamic fog detection possible.

### 1.2 Introduction of MTSAT

The Multi-functional Transport Satellite (MTSAT) series are a series of weather and aviation control satellites. They are geostationary satellites owned and operated by the Japanese Ministry of Land, Infrastructure and Transport and the Japan Meteorological Agency (JMA), and provide coverage for the hemisphere centered on 140° East. The MTSAT series succeeds the Geostationary Meteorological Satellite (GMS) series as the next generation of satellites covering East Asia and the Western Pacific. MTSAT-1R was successfully launched on February 26, 2005 and became operational on June 28, 2005, while MTSAT-2 successfully launched on February 18, 2006 and is positioned at 145° East(JMA, 2006).

The MTSAT series provides imagery for the Northern Hemisphere every 30 minutes in contrast to the previous hourly rate, enabling JMA to more closely monitor typhoon and cloud movement. The MTSAT series carries a new imager with a new infrared channel (IR4) in addition to the four channels (VIS, IR1, IR2 and IR3) of the GMS-5. MTSAT imagery is more effective than GMS-5 imagery in detecting low-level cloud/fog and estimating sea surface temperatures at night and has enhanced brightness levels, enabling a whole new level of image imagery (JMA, 2006). The technical specification of payloads of MTSAT is shown in Table 1.

| | Wavelength (μm) | Resolution | Brightness level | Frequency |
|---|---|---|---|---|
| VIS | 0.55 - 0.90 | 1 km (VIS) and 4 km (IR) at the sub-satellite | 10 bits for both VIS and IR channels (1,024 | S-band (reception: 2026 - 2035 MHz, transmission: |
| IR1 | 10.3 - 11.3 | | | |
| IR2 | 11.5 - 12.5 | | | |

* Corresponding author: Xiongfei Wen

| | | | | |
|---|---|---|---|---|
| IR3 | 6.5 - 7.0 | point | gradations) | 1677 - 1695 MHz) UHF (reception: 402 MHz, transmission: 468 MHz) |
| IR4 | 3.5 - 4.0 | | | |

Table 1. Technical specification of payloads of MTSAT

## 2. FOG DETECTION WITH TIME SERIES MTSAT IMAGERIES

### 2.1 Existing methods

Research on fog monitoring and forecasting based on remote-sensing techniques has been carried out for many years. In the early stage, fog and stratus were discriminated by their different reflectance characteristics from other clouds in the visible band. Hunt (Hunt, 1973) theoretically proved that the small droplets found in fog produce a lower emissivity at 3.7μm than at 10.8μm, while the emissivity at these two bands is roughly the same for larger droplets. The difference in emissivity will cause a differentiation between the brightness temperatures at the mid-infrared band, and the thermal infrared bands. Eyre et al. (Eyre 1984) and Turner et al. (Turner, 1986) use the brightness-temperature difference between the 10.8 and 3.7μm band of the National Oceanic and Atmospheric Administration (NOAA) Advanced Very High Resolution Radiometer (AVHRR) data, known as the 'dual-channel different temperature method', to detect fog and stratus cloud at night. This algorithm has been ported successfully to geostationary platforms, such as the Geostationary Operational Environmental Satellite Imager (GOES-8) and the Spinning Enhanced Visible and Infrared Imager (SEVIRI) on board the Meteosat Second Generation (MSG) (Cremak,2008). In recent years, with the extensive application of the Earth Observing System (EOS) Moderate Resolution Imaging Spectroradiometer (MODIS) data, fog detection during daytime using MODIS data becomes a new focus point. Bendix et al. (Bendix,2006) developed a radiative-transfer-based classification scheme for MODIS data and derived mini-mum/maximum albedo values for modeled fog layers in several spectral channels using the Streamer radiative-transfer model. These extreme values are then applied as thresholds in each of the channels, with all values in between classified as fog. Inspired by the difference of microphysical characteristics between fogs and clouds, Deng et al. (Deng, 2006)selected the spectral characteristics of MODIS channels 1, 6, 20 and 31 to simulate fog and cloud by utilizing the SBDART radiative-transfer model, and proposed a fog-monitoring algorithm based on the simulations by analyzing a fog case. Liu et al. (Liu,2011) selected MODIS channels 1, 6, 20 to simulate the reflectance of fog and high/middle/low cloud by utilizing the Streamer radiative-transfer model, and proposed NDFI (Normalized Difference Fog Index) as an new feature to discriminate fog from high/middle/low cloud based on the analyzing three fog case covering in Eastern China. Based on Liu's research, Wen et al. (Wen,2014) found the NDFI feature could discriminate fog from high /middle/low cloud in the perspective of spectrum, while fractal dimension feature could describe fog and high/middle/low cloud in the perspective of texture. The NDFI feature of fog is lower than clouds and the fractal dimension feature of fog is also lower than clouds, based on this, a new fog-monitoring algorithm based on combined the NDFI and fractal dimension is proposed.

Radiation fog generally appears at night or in the early morning, and then it will gradually dissipate as the sun rises. means there is a strong diurnal variation characteristics of radiation fog, many researchers have in depth research with MODIS in fog detection, while due to the limitations of its transit time (local time 10:30 AM), at this time radiation fog is likely already in its dissipating stage, using MODIS data alone is often difficult to be objective and accurate detection of radiation fog, and it is also difficult to obtain the fog distribution of time-spatial varying information. For this situation, the geostationary satellite such as MTSAT is a kind of suitable data sources.

### 2.2 Fog detection with time series imagery

Clouds and fog have unique micro-physical characteristics, and clouds and fog really exist certain similarities, especially low cloud and fog is difficult to discriminate. During the generation to dissipation of radiance fog, the change in the time series remote sensing images could show its unique feature in reflectance, brightness temperature, and difference value in brightness temperature during the lifecycle of fog, while clouds do not have the obvious characteristics during its generation to dissipation. The difference during the lifecycle between fog and clouds is more significant than only using one piece of remote sensing imagery.

The high temporal resolution stationary satellite imageries could be regarded as a kind of time series data, the time frequency conversion methods such as Fourier transform, and Wavelet analysis could be applied to process the time series data, the features including "brightness temperature", "reflectance", and "the difference in brightness temperature" for clouds, fog and clear land surface could be extracted and analyzed in time domain and frequency domain to construct the characteristic parameters from the time series stationary satellite imageries.

Wavelets are a recently developed mathematical tool for signal analysis. Wavelets are used as a kernel function in an integral transform, much in the same way that sines and cosines functions used in the Fourier analysis. Now the primary application of wavelets has been in the areas of signal processing, image compression, medical imaging, etc. Compared to Fourier analysis, a Fourier series requires that all time functions involved be periodic. In addition, Fourier analysis does not consider frequencies that evolve in time. Finally, Fourier techniques suffer from certain annoying anomalies such as Gibbs phenomenon. Wavelets, on the other hand, can be chosen with very desirable frequency and time characteristics as compared to Fourier techniques, the wavelet transform uses short windows at high frequencies and long windows at low frequencies. Thus the windowing of wavelet transforms is adjusted automatically for low or high frequencies or a low frequency, and each frequency component gets treated in the same manner without any reinterpretation of the results (Guan, 2007).

The time-series MTSAT images involves a set of MTSAT images taken from the same scene at different acquisition times. Considering the noise or the error during the data transmission or data producing procedure in the preprocess of the MTSAT time series images, it is necessary to register the time series images firstly. In this paper, we adopt a modified Gauss Newton optimization method to find out higher accurate registration of the time series images by calculating the minimum standard deviation between the registered images.

Then, considering the noise distributed in the time series images, the Savitzky-Golay smoothing filter should be applied to perform a local polynomial regression on a distribution to determine the smoothed value for each pixel at each time phase. Savitzky-Golay filter could preserve features in the original signal such as relative maxima, minima and width, which are usually flattened by other adjacent averaging techniques.

After the preprocess the input time series signal could be decomposed into approximate part and detail part by using discrete wavelet transform, the approximate part is obtained by the low-pass filter to reflect the approximate coefficients of the input signal, while the detail part is obtained by the high-pass filter to indicate detail coefficients of the input signal. The simplest discrete wavelet transform is Haar wavelets. Figure 1 shows the difference after the wavelet transform between low cloud and fog. In this section the non-orthogonal Haar wavelets is applied, the calculate formula is below:

$$d_k^{j-1} = \frac{c_k^j - c_{k+1}^j}{2} \quad (1)$$

$$a_k^{j-1} = \frac{c_k^j + c_{k+1}^j}{2} \quad (2)$$

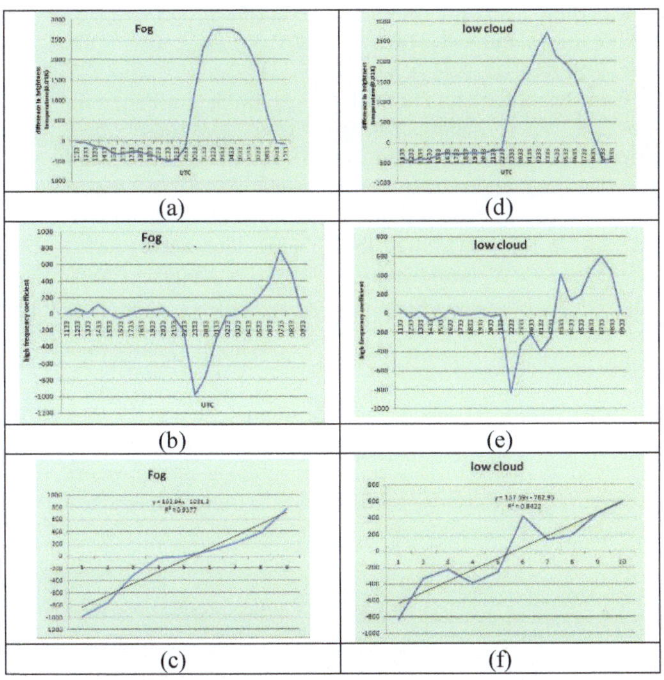


Figure 1. discriminate fog and low cloud by using non-orthogonal Haar wavelets: (a) original signal of fog, (b)fog's high frenquency part after Haar wavelet transform, (c) original signal of low cloud, (d)low cloud's high frenquency part after Haar wavelet transform,(e) the line fitting for fog signal, (f) the line fitting for low cloud signal

The coefficient of high frequency components, including the "max", "min", "the location of the min", "the interval length between the max and min", "the coefficient of linear fit for the high frequency", these properties are selected as the characteristic parameters to distinguish fog from clouds. It is difficult to calculate the thickness of the fog directly, while the duration of fog could be obtained by using "the interval between the max and min" described in this section.

## 3. EXPERIMENT AND VALIDATION

For validation of this new technique, a series of MTSAT images covering Eastern China were selected and processed in the manner described in section 2. The MTSAT during the 24 December 2006 to 25 December 2006 is selected, just shown in Figure 2, the fog distribution area includes Anhui, Jiangsu , Hebei , Henan and Hubei Province. As ground-truth data, Surface Weather Chart (SWC) datasets provided by the China Meteorological Administration (CMA) were used.

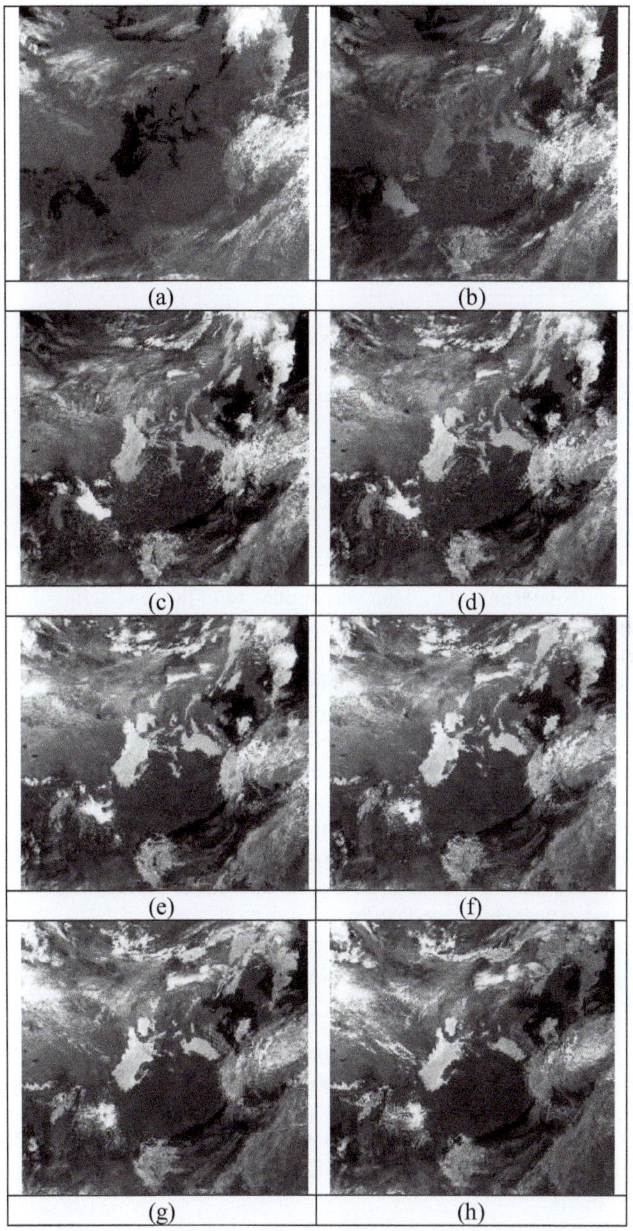

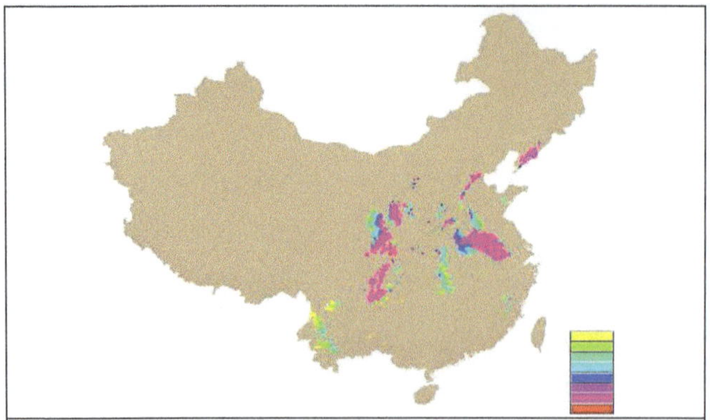

(i)

Figure 2. Case analysis with MTSAT from Dec 24-24,2006: (a)-(h) the different in band 4 and band1 from 2333UTC to 0633UTC, (i) the fog distribution map with different duration(from yellow to red, continuance from 1 hour to 8 hours)

| In-situ | Remote sensing | |
|---|---|---|
| | Fog | No fog |
| Fog | 36 | 65 |
| No fog | 25 | 412 |
| overall accuracy | 83.27% | |
| Kappa coefficient | 0.3529 | |

Table 2. Accuracy Assessment

## 4. CONCLUSION AND DISCUSSION

In this paper, according to the time series feature of radiance fog and cloud and the limitations in the existing fog detection method, the MTSAT data is selected as the main remote sensing data source. The time series MTSAT images were register with a modified Gauss Newton optimization method firstly, then, the Savitzky-Golay smoothing filter was applied to to the time series remote sensing imageries to process the noises in the original signal. After that, the non-orthogonal Haar wavelets was applied to make up the characteristics feature in frequency domain to discriminate radiance fog from clouds and land surface. The experiment shows that using the algorithm proposed in this paper, the radiance fog could be monitored effectively, and although it is difficult to calculate the thickness of the fog directly, the duration of fog could be extracted accurately with high temporal imageries. In the future research, the thickness of fog and the duration of fog should be combined to find the relationship between these feature.

## ACKNOWLEDGMENTS

This work was funded through a grant from the National Scientific Fund of China (No. 41201452, No. 41409021, No. 41401487) and the Basic Research Operating Expenses for the Central-level Non-profit Research Institutes of China (No. CKSF2015018/KJ, No. CKSF2015019/KJ). The authors would like to thank the staff of the Japan Meteorological Agency for their support.

## REFERENCES

[1] Li, Z., Yang, J. and Shi, C., 2008, The Physics of Regional Dense Fog, pp. 1–7, Beijing: China Meteorological Press.

[2] Ma, H., 2007, The research of remote sensing radiation fog detection in land based on EOS MODIS data. PhD thesis, Wuhan University, China.

[3] http://www.jma.go.jp/jma/jma-eng/satellite/about_mt/2.Meteorological_payload.html

[4] Hunt, G., 1973, Radiative properties of terrestrial clouds at visible and infrared thermal window wavelengths.Quarterly Journal of the Royal Meteorological Society, 99, pp.346–369.

[5] Eyre, J.R., Brownscombe, J.L. and Allam, R.J., 1984, Detection of fog at night using Advanced Very High Resolution Radiometer.Meteorological Magazine, 113, pp.266–271.

[6] Turner, J., Allam, R.J. and Maine, D.R., 1986, A case study of the detection of fog at night using channel 3 and 4 on the Advanced Very High Resolution Radiometer (AVHRR).Meteorological Magazine,115, pp. 285–290.

[7] Cremak, J. and Bendix, J., 2008, A novel approach to fog/low stratus detection using Meteosat 8 data.Atmospheric Research,87, pp. 279–292.

[8] Bendix, J., Thies, B., Nauss, T., et al, 2006, A feasibility study of daytime fog and low stratus detection.with TERRA/AQUA-MODIS over land. Meteorological Applications,13, pp. 111–125.

[9] DENG, J., BAI, J. and LIUJ., 2006, Detection of daytime fog using MODIS multispectral data.Meteorological Science and Technology,34, pp.188–193.

[10] Liu, Liangming; Wen, Xiongfei; Gonzalez, Albano; Tan, Debao; Du, Juan; Liang, Yitong; Li, Wei; Fan, Dengke; Sun, Kaimin; Dong, Pei; Xiang, Daxiang; Zhou, Zheng(2011). An object-oriented daytime land-fog-detection approach based on the mean-shift and full lambda-schedule algorithms using EOS/MODIS data. INTERNATIONAL JOURNAL OF REMOTE SENSING, 32(17), 4769-4785.

[11] Wen, Xiongfei; Hu, Dunmei; Dong, Xinyi; Yu, Fan; Tan, Debao; Li, Zhe; Liang, Yitong; Xiang, Daxiang; Shen, Shaohong; Hu, Chengfang; Cao, Bo (2014). An object-oriented daytime land fog detection approach based on NDFI and fractal dimension using EOS/MODIS data. INTERNATIONAL JOURNAL OF REMOTE SENSING, 35(13), 4865-4880.

[12] Guan, L., . Introduction to Wavelet analysis, 2007. Beijing : Higher Education Press.

# PRECISE HEAD TRACKING IN HEARING APPLICATIONS

A. M. Helle [a], J. Pilinski [a], T. Luhmann [a]

[a] IAPG, Institute for Applied Photogrammetry and Geoinformatics, Jade University of Applied Sciences, Oldenburg, Germany - (helle, iapg, luhmann)@jade-hs.de

**Commission V and III, WG V/5 and WG III/3**

**KEY WORDS:** Tracking, Medicine, Hearing, Development, Hardware, Infrared, Measurement techniques

**ABSTRACT:**

The paper gives an overview about two research projects, both dealing with optical head tracking in hearing applications. As part of the project "Development of a real-time low-cost tracking system for medical and audiological problems (ELCoT)" a cost-effective single camera 3D tracking system has been developed which enables the detection of arm and head movements of human patients. Amongst others, the measuring system is designed for a new hearing test (based on the "Mainzer Kindertisch"), which analyzes the directional hearing capabilities of children in cooperation with the research project ERKI (Evaluation of acoustic sound source localization for children). As part of the research project framework "Hearing in everyday life (HALLO)" a stereo tracking system is being used for analyzing the head movement of human patients during complex acoustic events. Together with the consideration of biosignals like skin conductance the speech comprehension and listening effort of persons with reduced hearing ability, especially in situations with background noise, is evaluated. For both projects the system design, accuracy aspects and results of practical tests are discussed.

## 1. INTRODUCTION

Video-based tracking methods have been applied in a lot of different fields for many years. The tracking and capture of body movements using cameras is often subsumed by the term motion capture. Applications of this technology can be found in medicine, robotics and in the entertainment industry for computer animation and motion capture for video games and movies. Popular examples are pedestrian detection for advanced driver assistance systems (Enzweiler et al. 2008) and the navigation of tools in surgery (Broers et al. 2007). Available tracking systems differ significantly in measuring accuracy, tracking frequency, measurement volume, costs and other specifications.

In order to ensure robust tracking of fast movements in these applications, frequencies of more than 20 Hz are required. The sensors usually have a resolution of about one megapixel, and synchronization between cameras can be utilized by a joint trigger unit.

The achievable measuring volume depends on various factors such as sensor resolution, focal length, image scale and also the numbers of used cameras. Generally it can be said that at constant conditions for the image scale, a larger volume can be achieved if the aperture angle is larger and the appropriate configuration uses more cameras.

## 2. OPTICAL TRACKING METHODS

Currently there is a wide range of optical tracking methods. In general, two- and three-dimensional tracking and the various methods of image analysis are to be differentiated. In marker-based approaches, for example, retro-reflective targets were frequently used in combination with infrared illumination. Other approaches work without signalization and use methods of pattern recognition or pattern matching (D'Apuzzo 2003). This paper deals with marker-based tracking.

Both presented techniques, the single and stereo camera tracking approaches, are implemented in the AXIOS software library Metrology that is used to process the images from the camera systems SingleCam and CamBar B2. The SingleCam used for first tests for the low-cost application has a resolution of 776 * 582 pixels with a focal length of 8 mm and a frame rate of up to 64 Hz. The camera is used to calculate the 6DOF parameters between two rigid bodies (locators) based on space resections (Luhmann 2009, see section 2.1). The specified measurement accuracy is 3 mm maximum length measurement deviation between two single points in the measurement volume. The CamBar B2 (Fig. 7, left) consists of two cameras with a sensor size of 736 * 582 pixels, a focal length of 8 mm and a frame rate of up to 64 Hz. The measurement accuracy is specified to 1 mm of maximum length measurement error between two single points in a measurement volume of ~1.9 $m^3$ and 0.25 mm for the 3D positioning accuracy of a point field. Both camera systems are pre-calibrated by the system supplier and equipped with an optical band-pass filter, which is permeable for the infrared range (810 nm). As retro-reflective targets reflect the light back into the direction where it comes from an infrared lighting is already integrated so that observed retro targets are perfectly visible for both of the camera systems.

### 2.1 Single camera tracking

By using just one camera, the orientation and position between two objects (locators) can be determined by a spatial resection if both are observed in the same image.
A reference probe is used to define a local coordinate system XYZ. The six parameters of the spatial position and orientation (6DOF) of a second probe with respect to the reference probe are determined. The second probe consists of at least three points given in a local coordinate system xyz. Local control point coordinates are determined through a photogrammetric measurement. For each set of given control points in the local systems, two resections are calculated. These describe the transformation of the camera coordinate system in the respective local system (Luhmann 2009).

$$\mathbf{X} = \mathbf{X}_0 + m \cdot \mathbf{R}_R \cdot \mathbf{x}' \text{: for the reference system} \tag{1}$$

$$\mathbf{x} = \mathbf{x}_0 + m \cdot \mathbf{R}_L \cdot \mathbf{x}' \text{: for the second system} \tag{2}$$

By solving equation (2) for $\mathbf{x}'$ we get:

$$\mathbf{x}' = \frac{1}{m} \cdot \mathbf{R}_L^{-1} \cdot (\mathbf{x} - \mathbf{x}_0) \tag{3}$$

If $\mathbf{x}'$ is inserted into (1), the scale factor $m$ is eliminated and we get:

$$\begin{aligned} \mathbf{X} &= \mathbf{X}_0 + \mathbf{R}_R \cdot \mathbf{R}_L^{-1} \cdot (\mathbf{x} - \mathbf{x}_0) \\ \mathbf{X} &= \mathbf{X}_0 + \mathbf{R} \cdot (\mathbf{x} - \mathbf{x}_0) \end{aligned} \tag{4}$$

Where $\mathbf{x}$ : position of a point P at the probe in xyz
$\mathbf{X}$: desired position of the same point P in XYZ
$\mathbf{x}_0$, $\mathbf{X}_0$: projection center in xyz and XYZ
$\mathbf{R}$: rotation matrix of the probe in XYZ.

A special problem for single camera 6DOF tracking is given by the need of robust point identification. Since the points appear in only one image, epipolar matching of corresponding points cannot be performed. The AXIOS system uses a sophisticated identification scheme based on individual point distribution of the locator design.

### 2.2 Stereo tracking

Tracking of a probe with two cameras instead of one offers a number of benefits, such as enhanced robustness (solving occlusions), simpler point identification (epipolar geometry) and improved 3D accuracy.
Stereo tracking systems allow to determine the 3D coordinates of a probe in an object coordinate system XYZ by calculating a spatial intersection from two synchronized, oriented images. The probe has a local coordinate system xyz that is determined through a photogrammetric calibration.

1. Transformation of image coordinates of left and right image

$$\begin{pmatrix} X' \\ Y' \\ Z' \end{pmatrix} = \begin{pmatrix} X_{01} \\ Y_{01} \\ Z_{01} \end{pmatrix} + \mathbf{R}_1 * \begin{pmatrix} x' \\ y' \\ z' \end{pmatrix} \qquad \begin{pmatrix} X'' \\ Y'' \\ Z'' \end{pmatrix} = \begin{pmatrix} X_{02} \\ Y_{02} \\ Z_{02} \end{pmatrix} + \mathbf{R}_2 * \begin{pmatrix} x'' \\ y'' \\ z'' \end{pmatrix} \tag{5}$$

2. Components of the basis

$$\begin{aligned} bx &= X_{02} - X_{01} \\ by &= Y_{02} - Y_{01} \\ bz &= Z_{02} - Z_{01} \end{aligned} \tag{6}$$

For the common spatial intersection the crooked rays intersect in XY-plane in the height of the calculated point of the probe.

$$\begin{aligned} X &= X_1 = X_2 \\ Z &= Z_1 = Z_2 \\ Y &= \frac{Y_1 + Y_2}{2} \end{aligned} \tag{7}$$

3. Scale factor

$$\lambda = \frac{bx(Z'' - Z_{02}) - bz(X'' - X_{02})}{(X' - X_{01})(Z'' - Z_{02}) - (X'' - X_{02})(Z' - Z_{01})} \tag{8}$$

$$\mu = \frac{bx(Z' - Z_{01}) - bz(X' - X_{01})}{(X' - X_{01})(Z'' - Z_{02}) - (X'' - X_{02})(Z' - Z_{01})}$$

4. Object coordinates, whereby $pY$ is a measure of quality and will be close to zero if rays intersect (Luhmann et al. 2014).

$$\begin{aligned} X &= X_{01} + \lambda(X' - X_{01}) \\ Z &= Z_{01} + \lambda(Z' - Z_{01}) \\ Y_1 &= Y_{01} + \lambda(Y' - Y_{01}) \\ Y_2 &= Y_{02} + \mu(Y'' - Y_{02}) \\ Y &= \frac{Y_1 + Y_2}{2} \\ pY &= Y_2 - Y_1 \end{aligned} \tag{9}$$

## 3. TRACKING IN CHILDREN HEARING TESTS

### 3.1 Audiological background

The directional perception of sound events is a process of binaural hearing in which the ear uses the time and intensity differences of an incoming sound signal to determine the horizontal direction. In general, high-frequency signals can be localized worse than low-frequency. The hearing system primarily uses volume level differences (Interaural Level Difference, ILD) at the higher frequencies and the time delay (Interaural Time Difference, ITD) at lower frequencies (Plotz et al. 2013, Geldermann 2012). Within the project "Evaluation of acoustic sound source localization in children (ERKI)" an upgradable examination unit is developed which allows, based on the "Mainzer Kindertisch", to analyze the capability for localizing sounds by (elementary) children using virtual sound sources. The task of the research project ELCoT was to develop a low-cost 3D measurement system for detection of head and arm movements in real time, which can be applied to new hearing tests in the cooperative project ERKI. Since high investment costs for the distribution of the examination unit with doctors and specialized clinics are not acceptable, the challenge lies in the development of a particularly cost-effective system, taking into account the required accuracy of 1-2° (Pilinski et al. 2014).

### 3.2 System setup

For detecting the body movements, markers can be used which are measured by a camera system mounted either at the room ceiling or on a tripod. The "Mainzer Kindertisch" is a device supplied with five fixed loudspeakers on a semicircular table. The loudspeakers are hidden by an opaque curtain; the test person is sitting in the center of the speaker semicircle (Fig. 1).

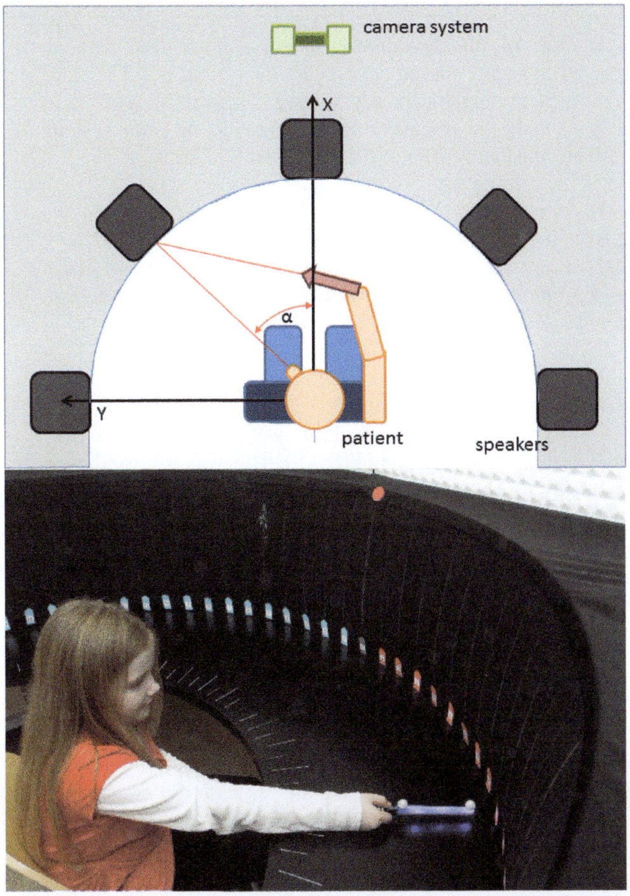


Figure 1. "Mainzer Kindertisch" with camera system, test person and loudspeaker semicircle

Compared to a conventional hearing test scenario with only five speaker positions, this setup provides improved capabilities. It is possible to generate sound events at each position in the semicircle by using phantom sound sources. This makes it necessary to develop an objective detection of the pointing motion to the acoustic event from the test person. The manual detection by the investigator, as it is currently common practice, would not be sufficient because of the required accuracy of 1-2° for the angle of the pointing gesture. As a pointing device a toy drill (Theo Klein / Bosch cordless drill) equipped with a probe is used. This is child-friendly and by the given ergonomic grip of the cordless drill it is ensured that the volunteer points in the targeted direction (Fig. 2, left). An infrared remote control has been integrated which enables the test persons to confirm the indicated direction by pressing a button. That means when the test person activates the button of the toy drill, the currently indicated direction (angle in degrees) is saved for the later analysis of the directional hearing.

In addition to the pointing gesture, the head movement of the test person will be tracked. On the one hand, a start signal of the hearing test is given, which means a new measurement can be started when the test person is looking straight ahead. On the other hand, additional information may be obtained with regard to how the person behaves during the test and if one turns his head, for example, to the sound source. For the head tracking the test person is wearing a light head belt. The head belt is also equipped with a probe that is tracked by the camera system continuously (Fig. 2, right). The mount is fixed to the head of the test person as good as possible and does not cover the ears, so as not to affect the incoming audio signal and the hearing.

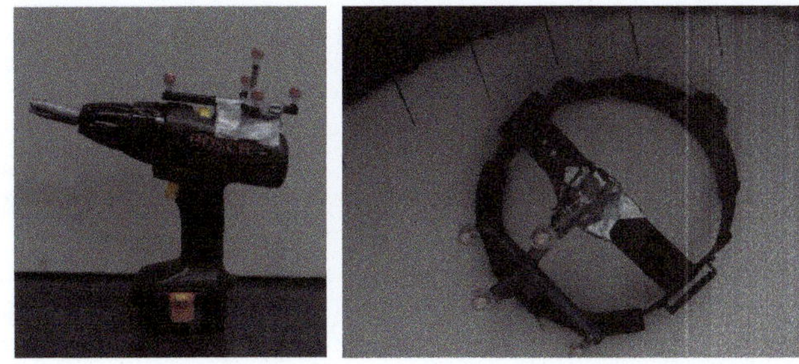

Figure 2. Left: Toy drill with probe, Right: Head mount with probe

Since the 6DOF movement of the two movable components (head and pointing tool) have to be registered within the loudspeaker coordinate system, a number of local reference points are attached to the table which are measured simultaneously with the other locators.

The recorded directions and calculated angles should be visible through an optical feedback to the test persons. So it is one goal to visualize the current pointing direction of the test person by using a LED strip. The light strip serves as a control for the test person. The real-time capability (20-30 fps) of the system is an important condition for this visualization, as is the suitable synchronization capabilities to the audiology system and LED strip.

The low-cost tracking system has been developed on the basis of an available professional AXIOS 3D SingleCam system connected with the AXIOS 3D library Metrology that is able to detect and track targeted probes. For the further development of the measuring system leading to a low-cost solution, a cost-effective USB industrial camera of the Visiosens GmbH was chosen. The main features are shown in Table 1:

| | |
|---|---|
| Image Sensor | 1/3" Aptina CMOS monochrome |
| Effective pixels | 1280 x 960 |
| Pixel size | 3.75µm |
| Mount | S-mount |
| Lens | 5mm focal length |

Table 1. Features of the USB industrial camera

In order to measure retro-reflective targets in an image, suitable lighting is necessary. Therefore eight LEDs are mounted around the lens, similar to a ring light (Fig. 3). The LEDs have an additional power supply. The industrial camera is a quite compact one; due to space restrictions an S-mount was chosen for the lens.

Figure 3. USB industrial camera with LED ring light

### 3.3 Results

In order to assess the level of accuracy of the developed measurement system, comparative measurements with the AXIOS 3D CamBar B2 were performed.

In each case an angle determination was performed using the low-cost camera system and the stereo camera system at 19 positions for the comparative measurements. The deviations of the low-cost camera system from the professional system are shown in Figure 4.

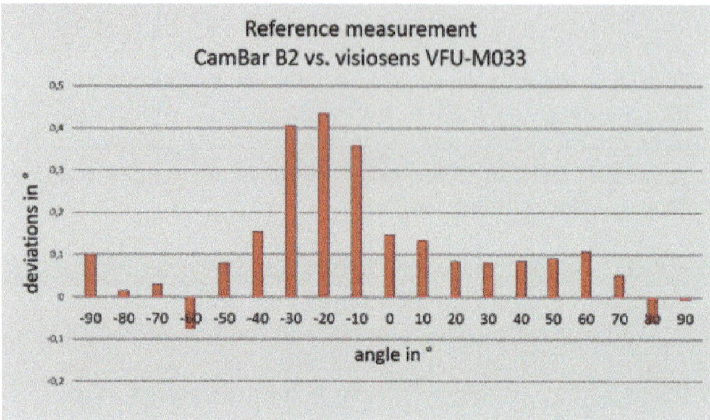


Figure 4. Reference measurement; deviations of the horizontal angle

The maximum deviation of the low-cost system is 0.43°. The average value of the deviation between the low-cost camera system and the professional stereo camera system is 0.11°. The required measurement accuracy of 1-2° is thus upheld. It is noticeable that the measurement accuracy of the system decreases to the centre (0°) of the "Mainzer Kindertisch".

In a laboratory test, a series of experiments with the developed low-cost camera system and, for comparison, with the AXIOS 3D SingleCam was performed. Both camera systems were used at the same time and in each case the angle indicated by the test person was determined.

The study involved ten normal hearing test persons, which were divided into two groups. Group A first got sound sources with "Pink Noise" and Group B first got sequences of the International Speech Test Signal (ISTS). This was followed by a change of the noise.

The boxplots (Fig. 5) show similar results in the localization performance for both camera systems. The dispersion of the indicated locations with both cameras is greater at the sides, which corresponds to the human hearing.

The two systems produced some outliers, where either the test person might have been inattentive or the camera detected a wrong angle. Besides, the low-cost camera was not able to capture some angles, which has to be improved. But, altogether, the low-cost system can be used like the professional system.

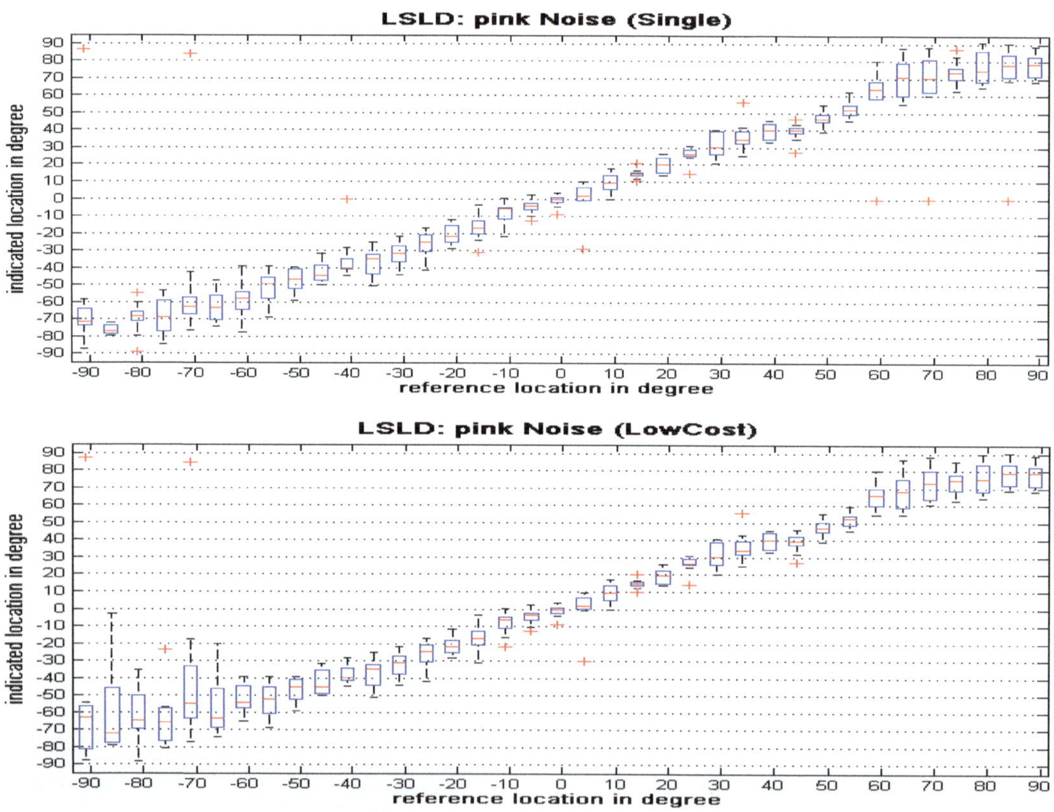


Figure 5. Representation of the localization performance; stimulus used was "Pink Noise". The results are shown as boxplots (median value, the 75%- and the 25% -quartile, as well as the inner limitation reference to the boxplots). The outliers are shown with a "+". Top: Results with SingleCam; Bottom: Results with low-cost camera

## 4. TRACKING IN WAVE FIELD SYNTHESIS SYSTEM

### 4.1 Audiological background

For the normal-hearing population the human sense of hearing allows to filter background noise in a conversation to concentrate on the significant sounds. This is known as "cocktail party effect". Hearing impaired people often have problems concentrating in a conversation as they are distracted by other speakers or noise sources e. g. in a cafeteria. They have a higher listening effort. The term 'listening effort' refers to the attention and cognitive resources required to understand speech (Gosselin et at. 2011). In recent years, hearing aid manufacturers have made it a priority to find ways to improve the performance of hearing aids in noisy listening situations. One way is through the development of signal-processing algorithms such as noise-reduction algorithms (Desjardins et al. 2014).

As head movements affect the spatial hearing but are not needed for directional hearing, one aim of the research project framework "Hearing in everyday life (HALLO)" is to analyze how far hearing impaired people use head movements to better locate sound sources and thus better understand complex stimuli. Several experiments contribute to the development and evaluation of measuring parameters and techniques to quantify the listening effort. The observation of biosignals like skin conductance and heart rate during hearing tests should allow conclusions on the speech comprehension as there might be a correlation (Mackersie et al. 2015). The rotation of the head is required to be calculated with a relative accuracy of 1°.

### 4.2 System setup

The experimental setup consists of a wave field synthesis system in the center of which a patient listens to complex acoustic events (Fig. 6). The wave field synthesis system is a set of 104 speakers arranged in a rectangle that can reproduce complex acoustic stimuli. After hearing a stimulus the patient rates its level of difficulty. During the experiment several biosignals and the head movement are recorded. For observing the biosignals a NeXus 10-MKI is used. For tracking the head movement an AXIOS stereo camera CamBar B2 (Fig. 7, left) and the proprietary AXIOS software library Metrology are used. The camera is attached to a wooden beam above the patient to have a good view on the markers used. Head movements during listening can be classified into three groups: tilting, inclining and rotating (Blauert 1997). Thus the test person wears a light head belt with a targeted probe that represents the head movement in 6DOF and is tracked by the camera system continuously (Fig. 7, right). The head belt is fixed to the head of the test person as firmly as possible and does not cover the ears, so as not to affect the incoming audio signal and the hearing.

To compensate slight camera movements during the test a reference probe on the floor is observed continuously. All movements of the head probe are related to the reference probe. To associate the head movement with the direction of the stimuli which is generated by the wave field synthesis system, all data have to be transformed into the same coordinate system. For this purpose a simultaneous camera calibration is performed to measure the wave field synthesis system together with the reference probe that represents the origin of the coordinate system.

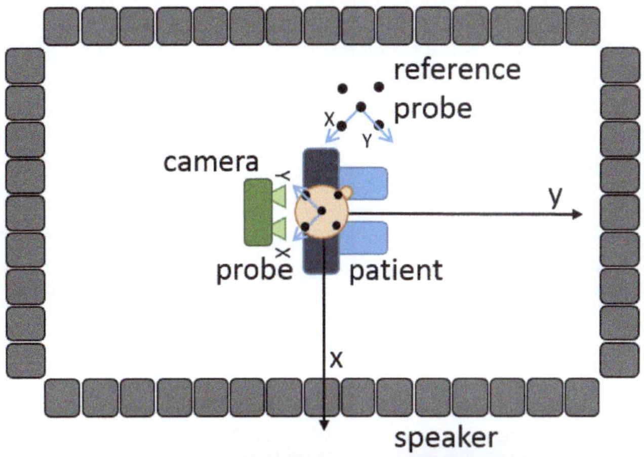

Figure 6. Patient is sitting in the center of the wave field synthesis system. The stereo camera above observes the points on the patient's head and the reference system on the floor simultaneously

Figure 7. Left: AXIOS stereo camera CamBar B2, Right: light head belt with a targeted probe

As the Metrology library offers the 6DOF parameters (x, y, z, ω, φ, κ) for each probe in the camera coordinate system, a transformation of the probe at the head belt into the coordinate system of the wave field synthesis system is calculated. In order to get the rotation and translation of the head, the relative movement to the first captured frame is calculated.

To synchronize all generated and recorded data during the practical tests the Linux-based real time database KogMo-RTDB is used (Goebl et al. 2007). Several systems that record (CamBar, NeXus) or generate (wave field synthesis system) data during a test series can send their information to the database independently from the frequency the other "senders" deliver their data. All received data are saved in a container format. For each object that is saved in the database a timestamp is saved, too. Thus the whole data set can be replayed for analysis. Apart from the same coordinate system the synchronized data is an important resource to associate the head movements with the stimuli and to distinguish relevant from arbitrary movements.

### 4.3 Results

Since the stereo camera is attached to a wooden beam that is connected to the floor, slight camera movements may occur during the hearing tests. In order to check the impact of vibration of the wooden beam, a measurement with the vibration sensor ZEB/GS3T (Fig. 8) was performed. The vibration in x-, y- and z-direction was sent to an oscilloscope. Fig. 9 displays the vibration while a person walks through the room. The xy-plane lies parallel to the aluminum plate. The diagram maps a measurement over 120 ms. The sensitivity of the vibration sensor is specified to 10 $\frac{\text{mV}}{\text{mm}/_\text{s}}$: if the oscilloscope displays an amplitude of 10 mV, this corresponds to a vibration speed of 1 mm/s. The deviation of the wooden beam in z-direction is calculated as follows:

$A = 4$ mV (amplitude) $\triangleq v(t) = 0.4$ mm/s (vibration speed)
$\tau = 0.06$ s (cycle duration)
$f = 1 / \tau = 16.67$ Hz (frequency)
$\omega = 2 \cdot \pi \cdot f = 104.72 \, \frac{1}{\text{s}}$ (angular velocity)
$x_0 = v_0 / \omega = 0.00382$ mm (deviation of the wooden beam)

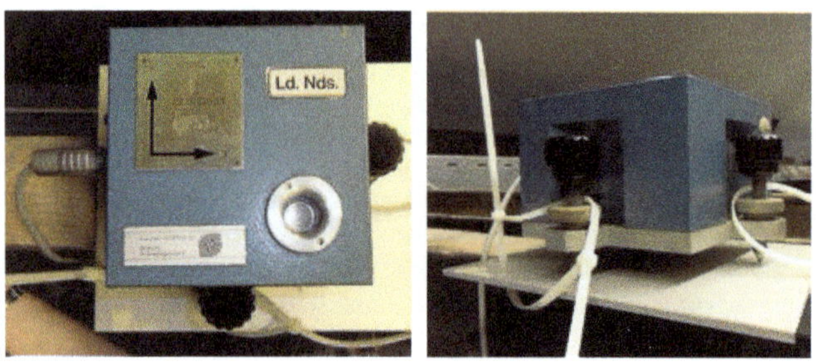


Figure 8. Top view (left) and side view (right) of the vibration sensor. The cable ties only are for security reasons.

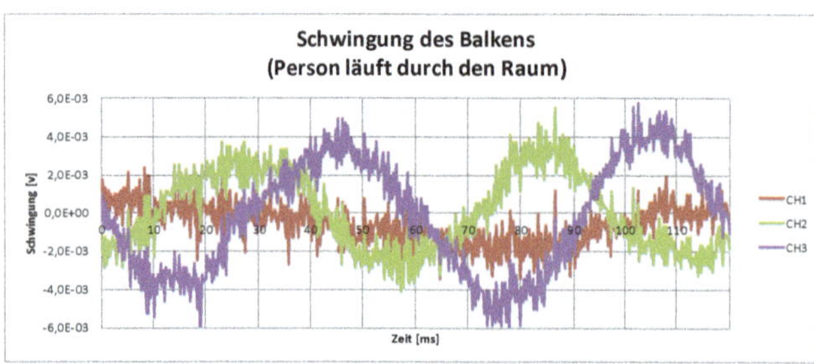


Figure 9. Measured vibration speed in x- (red), y- (green) and z-direction (violet) while a person walks through the room

As shown in Fig. 9, significant vibrations occur. Therefore a stable reference coordinate system provided by fixed targets (Fig. 6) is used to compensate relative movements between camera and loudspeaker system, as described in section 4.2.

Two experiments are conducted in order to verify the measurement accuracy of rotation and position of the head: a targeted probe is successively fixed to a compound slide and a tachymeter Leica TC2002 is used to observe the shifting in x- and y-direction and horizontal torsion with the AXIOS CamBar B2 (Fig. 10). The accuracy of the compound slide and tachymeter is shown in Table 2. The position of the targeted probe and thus the head of a patient can be determined with a standard deviation of 0.01 mm in x- and y-direction, the rotation with a standard deviation of 0.06°.

Figure 10. Experimental setups for verifying the accuracy of a rotation (left) and position (right).

| | | |
|---|---|---|
| Compound slide | x-direction | 0.1 mm |
| | y-direction | 0.1 mm |
| Tachymeter | angle | 0.15 mgon |

Table 2. Accuracy of compound slide and tachymeter used for verifying the measurement accuracy

For the three dimensional reconstruction of the wave field synthesis system a Nikon D4 camera with a wide-angle lens is used. A photogrammetric multi-image configuration has been acquired that allows for the 3D measurement of relevant points of the loudspeaker system and of the reference system points as well. In addition, the camera can be calibrated simultaneously.

The simultaneous camera calibration has been tested in a laboratory before it will be applied to the wave field synthesis system. Coded targets were distributed among the whole measuring volume and scales were laid out in all axial directions (Fig. 11). The reference probe has been calibrated before to represent the origin and provide approximate values for the bundle adjustment. The relation between both coordinate systems is established through the origin of the wave field synthesis system. It is defined by the intersection of both diagonal lines of the rectangular wave field synthesis system. To measure the diagonal lines during the calibration the edges are marked with targets. If the head movement and the origin of the wave field synthesis system are known in one coordinate system the head movement can be transformed into the coordinate system of the wave field synthesis system and compared to the stimuli generated by it.

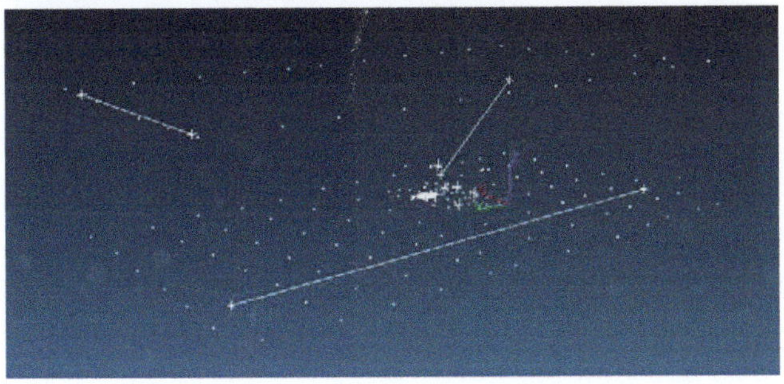

Figure 11. Coded targets in a 50x50 cm grid and scales for the simultaneous camera calibration. The origin of the measured system is defined in one corner of the reference probe.

## 5. CONCLUSION AND OUTLOOK

In this paper the system design, accuracy aspects and (first) results of two research projects were presented. For the project ELCoT a low-cost camera tracking solution was presented using an USB industrial camera with LED ring light and a filter to block visible light. The developed solution is practicable and can be used to measure directional hearing. Only the robustness of the entire system should be increased in the future.

To make the system as cost-effective as possible more test series with cost-effective, adhesive markers (Fig. 12) should be performed. First tests have already shown that low-cost probes (material costs of a few euros) are also suitable for the hearing tests with respect to the required accuracy. Furthermore, the robustness of the system should be increased and outliers should be eliminated reliably. In general, the low-cost measurement system can be transferred to other applications.

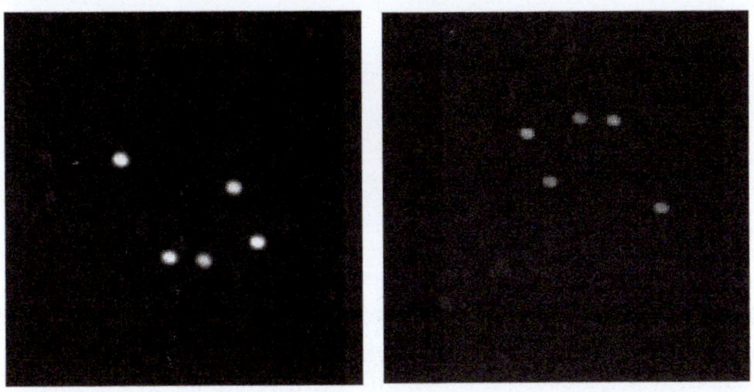

Figure 12. Left: Measuring image of a low-cost probe, Right: Measuring image of the toy drill probe with professional markers

Within the project HALLO a proprietary stereo system is used for tracking a patient's head during hearing tests, performed in a wave field synthesis system. Apart from the head movements other biosignals are analyzed in order to study the speech comprehension and listening effort of hearing impaired people in cocktail party situations.

Extensive test series are planned for the next project stages that will show the real behavior of patients and the relevance of head measurements during hearing events. Currently real test scenarios like a speech conversation in a car or on the street are recorded. These scenarios will be reproduced through the wave field synthesis system to observe the test persons' behavior and learn how difficult such scenarios appear to them.

Furthermore, another position for the camera system is imaginable. A wide-angle lens could be used to see the targeted wave field synthesis system as well as the test person simultaneously. The advantage of this setup could be that the reference probe is not needed anymore and thus the measuring effort is reduced.

## REFERENCES


Blauert, J. (1997): *Räumliches Hören.* Stuttgart, Hirzel-Verlag

Broers, H. & Jansing, N. (2007): *How precise is navigation for minimally invasive surgery?* In: International Orthopedics, Springer Verlag Berlin Heidelberg 2007.

D'Apuzzo, N. (2003): *Surface measurement and tracking of human body parts from multi station video sequences.* Ph.D. Thesis, Institut für Geodäsie und Photogrammetrie, Mitteilungen No.81, ETH Zürich, Schweiz, 2003.

Desjardins, J. L. & Doherty, K. A. (2014): *The effect of hearing aid noise reduction on listening effort in hearing-impaired adults.* Ear & Hearing, Vol. 35, No. 6, p. 600-610.

Enzweiler, M., Kanter, P. & Gavrila, D.M. (2008): *Monocular Pedestrian Recognition Using Motion Parallax.* IEEE Intelligent Vehicles Symposium, Eindhoven, The Netherlands, 2008.

Geldermann, C. (2012): *Lokalisation von Phantomschallquellen am MAINZER Kindertisch zur Entwicklung eines Diagnose-verfahrens von Auditiven Verarbeitungs- und Wahrnehmungs-störungen.* Bachelorarbeit am Institut für Hörtechnik und Audiologie der Jade Hochschule Oldenburg, 2012.

Goebl, M. & Färber, G. (2007): *A real-time-capable hard- and software architecture for joint image and knowledge processing in cognitive automobiles.* In: Proceedings of the 2007 IEEE Intelligent Vehicles Symposium, pages 734-740, IEEE Press, Istanbul, Turkey, June 13-15, 2007

Gosselin, P. A. & Gagne, J.-P. (2011): *Older adults expend more listening effort than young adults recognizing speech in noise.* Journal of Speech, Language, and Hearing Research, Vol. 54, p. 944-958, June 2011

Luhmann, T. (2009): *Precision potential of photogrammetric 6 DOF pose estimation with single images.* ISPRS Journal of Photogrammetry and Remote Sensing, Vol. 65, pp 275-284.

Luhmann, T., Robson, S., Kyle, S., Boehm, J. (2014): *Close-Range Photogrammetry and 3D Imaging.* 2nd ed., Walter de Gruyter, Berlin, 684 p.

Mackersie, C. L., MacPhee, I. X., Heldt, E. W. (2015): *Effect of hearing loss on heart rate variability and skin conductance measured during sentence recognition in noise.* Ear & Hearing, Vol. 36, No. 1, p. 145-154.

Pilinski, J., Luhmann, T., Schmidt, K. (2014): *Development of a real time low-cost tracking system for medical and audiological problems.* LowCost3D – Sensors, Algorithms, Applications, Berlin, 2.+3.12.2014.

Plotz, K., Schmidt, K., Kissner, S., Geldermann, C., Bitzer, J. & Schönweiler, R. (2013): *ERKI – Erfassung des Richtungshörens bei Kindern - Entwicklung eines verbesserten Verfahrens durch*

*Nutzung virtueller Quellen zur Erfassung des Richtungshörens bei Kindern am Mainzer-Kindertisch.* Deutsche Gesellschaft für Phoniatrie und Pädaudiologie. 30. Wissenschaftliche Jahrestagung der Deutschen Gesellschaft für Phoniatrie und Pädaudiologie (DGPP). Bochum, 20.-22.09.2013. Düsseldorf: German Medical Science GMS Publishing House; 2013.

Visiosens (2014): *Visiosens VF Camera Platform – Customised for High-Tech Industries.* Visiosens GmbH. http://visiosens.de/uploads/media/Datasheet_Visiosens_VFxx-xxxx-xH_v07_2012-04-19_03.pdf (24.01.2014)

# WAY-FINDING ASSISTANCE SYSTEM FOR UNDERGROUND FACILITIES USING AUGMENTED REALITY

K. Yokoi [a,*], N. Yabuki [a], T. Fukuda [a], T. Michikawa [b], A. Motamedi [a]

[a] Division of Sustainable Energy and Environmental Engineering, Osaka University, Suita, Osaka, Japan
yokoi@it.see.eng.osaka-u.ac.jp, (yabuki, fukuda, motamedi)@see.eng.osaka-u.ac.jp
[b] Center for Environmental Innovation Design for Sustainability, Osaka University, Suita, Osaka, Japan
michikawa@ceids.eng.osaka-u.ac.jp

**WG IV/7 and WG V/4**

**KEY WORDS:** Augmented Reality, Structure from Motion, Way-finding, Navigation Assistance, Underground Facilities

**ABSTRACT:**

Way-finding is one of main challenges for pedestrians in large subterranean spaces with complex network of connected labyrinths. This problem is caused by the loss of their sense of directions and orientation due to the lack of landmarks that are occluded by ceilings, walls, and skyscraper. This paper introduces an assistance system for way-finding problem in large subterranean spaces using Augmented Reality (AR). It suggests displaying known landmarks that are invisible in indoor environments on tablet/handheld devices to assist users with relative positioning and indoor way-finding. The location and orientation of the users can be estimated by the indoor positioning systems and sensors available in the common tablet or smartphones devices. The constructed 3D model of a chosen landmark that is in the field of view of the handheld's camera is augmented on the camera's video feed. A prototype system has been implemented to demonstrate the efficiency of the proposed system for way-finding.

## 1. INTRODUCTION

Underground spaces play an important role in big cities. They have been mainly developed as commercial spaces in association with the development of the subways. These spaces are used not only for shopping sites, but also for commuting route and for avoiding extreme weather conditions. Indeed, many big cities have developed large underground spaces (e.g. Tokyo and Osaka in Japan).

Underground spaces pose one major problem for pedestrians that is the difficulty with finding the route to the destination, called way-finding problem. The underground spaces are usually developed incrementally and the underground spaces become complex network on connected labyrinths. Way-finding in these connected underground spaces is not only inconvenient for pedestrians but also critical in the case of emergency (e.g. flooding and fires), as the pedestrians should be evacuated from the underground spaces quickly. Regarding the evacuation behavior, Taniguchi et al. (2014) researched about the evacuating plan in case of Tsunami in Umeda underground spaces, Osaka, and simulated the situation by creating a Virtual Reality (VR) world of the underground.

The way-finding problem is mainly caused by the loss of the sense of direction and orientation due to lack of landmarks visible from a distance (such as tall buildings or city monuments), or inability to visually identify recognizable natural features (such as sun, mountains and hills). The same problem exists outdoors in urban areas where the visibility is obstructed by dense high-rise buildings. The Global Positioning System (GPS) is not accurate indoors and in dense urban areas. Hence, the current method for way-finding and navigation relies only on the placement of map and signage. However, these methods have limitations and are less intuitive compared to visual way-finding methods using known landmarks.

Regarding signs, one of the issues is that the number of the destinations which are written on the signs are limited. Furubayashi et al. (2013) researched about the discontinuity of the directional sign in subway stations. They developed a mathematical data model and a system for examining whether the sign can efficiently guide pedestrians to all exits. According to their case study, there are intermittent signs as well as destinations within the station that cannot be reached by using only installed signage.

Moreover, reading maps and floor plans can be difficult for many pedestrians as they need to locate their current location on the map, figure out their orientation, locate their destination, and decide on the route to a desired the destination. Additionally, they need to constantly check the map to make sure the right path is being taken. As a result, many pedestrians may choose to go aboveground to be able to see landmarks.

This paper presents a method to assist solving the way-finding problem using augmented reality (AR). AR is a technology to combine virtual objects (CG models) and the real-time feed of video images. This enables us to enhance real world image by displaying invisible objects. Our proposed system uses this technology to display invisible landmarks to pedestrians in underground spaces. When the users direct tablet devices to desired directions, the system shows the landmarks overlaid on the camera images in real time. The landmarks are arranged based on positioning systems and sensors in the device. As the landmarks are overlaid to the real-time image, the system can facilitate reaching a desired destination by providing a positional relationship between underground spaces and above ground spaces.

## 2. RELATED WORK

Regarding the application of AR for guidance, Yamauchi (2012) developed an AR application that shows the 3D models of each floor in a smart phone by using markers on the ceiling, which are identified by the camera. This application enables the user to understand the current location and the route to the ground level more easily and intuitively. However, in this research, users have to compare a 3D map in their devices with the real world scene. Therefore, the problems related to difficulties with map-reading are remained.

TagCast. Inc (2014) developed a navigational aid application for underground facilities. The application uses AR to show direction signs on the smartphone. It helps pedestrians navigate to their chosen destinations. However, the application only provides directions to the users and does not provide them with an overall view of the path and the location relationships between the current position and the destination.

Regarding other guidance methods, Tardieu et al. (2009) proposed an auditory way-finding system. In their research, three sound signals were used: Orientation, Confirmation, and Timeline. The first sound signal directs the users to the destination. The second sound signal confirms at each step of the walk that the direction is right. The third signal informs the users, at each step, about the progression of the walk. In the way-finding experiment with the auditory system, compared to the experiment without the system, more participants completed the steps without any difficulty. However, sound signal does not provide pedestrians with an overall view of the path and the location relationships between the current position and the destination.

Regarding research related to showing the invisible objects occluded by other objects using AR, Kameda et al. (2004) introduced a method for visualizing invisible areas that are monitored by surveillance cameras to the user. Given a 3D model of the occluded objects and video of the invisible area, the method displays the blind spots by texture mapping of the video and the 3D objects using semi-transparent rendering. However, this system requires the presence of video cameras in the area and the large usage of resources. Schall et al. (2009) introduced a visualization system for underground infrastructure (such as pipes) using AR. The system supports recognizing the relationship between the facilities in construction in order to avoid unexpected accidents. These research projects demonstrate the application of AR technology in efficiently visualizing invisible objects.

## 3. PROPOSED SYSTEM

### 3.1 Overview of the System

As discussed in Section 2, AR-based navigational aid systems often provide users with direction signs for guiding them within the facility. They usually do not give the user an overall view of their orientation or the relative distance to known city landmarks. However, the common walking experience includes observing natural/man-made landmarks and subconsciously tracking the relative distances to these known points. Although a pedestrian does not exactly know where he is geolocated while walking, the relative distance to these landmarks gives him/her an approximation of his location, the direction where the destination is located, and the time it takes to reach it.

As explained in Section 1, our proposed system is an AR-application for hand-held devices. When the users point the device to a desired direction in the underground floor, the system shows live stream of images from the camera with 3D objects of the landmarks overlaid on them. The user can then recognize the direction to the landmarks, accordingly. Figure **1** shows an application scenario for the proposed system.

Our proposed system does not aim to provide users with direction guidance to reach specific destinations. Instead, it provides user with visual information about invisible, known landmarks. By using the proposed system, the user can enhance his way-finding experience in underground spaces or dense urban areas.

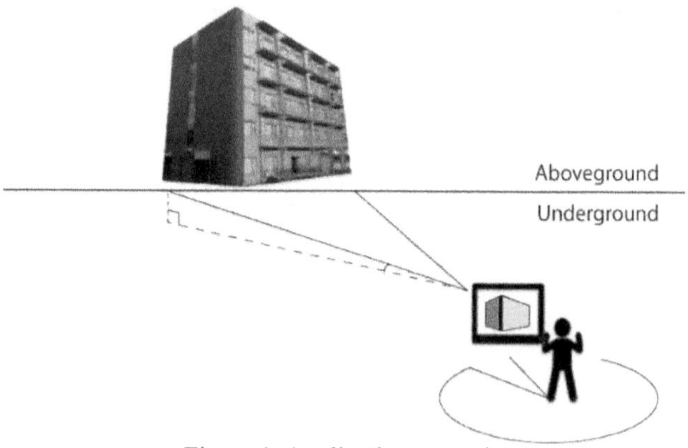


Figure 1. Application scenario

### 3.2 Application Scenarios:

**Scenario 1**: *User aims for a specific destination*: The pedestrian uses the system to get information about how far he is from the destination or if he is heading to the right direction.

**Steps:** (1) the 3D model of the destination is chosen from a list; (2) the 3D model is shown on the screen when the tablet is held towards its direction; (3) The size/view angle of the model is changed based on the distance and the relative location of the user to the destination; (4) The view can be accompanied by Plan Position Indicator (PPI) and navigational aid such as voice and arrow signs.

**Scenario 2:** *User is strolling or walking to an undecided destination (for example for the purpose of shopping, jogging, or sightseeing)*: The pedestrian uses the system to approximate how far he is from a set of known landmarks and which direction he is heading relative to known landmarks.

**Steps:** (1) user selects a set of landmarks; (2) the 3D model of selected landmarks are shown on the screen when the tablet is held towards their direction; (3) The size/view angle of the models are changed based on the relative distance and the angel.

In order to realize the system two major steps should be taken: (1) *Preprocessing*: to construct and arrange 3D models of landmarks in a virtual spaces before using the system; (2) *Augmentation*: to display the overlaid images on the hand-held device. These two steps are described in details in the following subsections.

### 3.3 Pre-processing

Our system requires 3D models of the landmarks to be used in the application. These objects should be modelled prior to the use of the system as a pre-processing step. Several methods for creating 3D models of the landmarks exist. 3D objects can be manually created by CAD systems (e.g. Autodesk 3ds Max and

Revit), however, this method is labour intensive and time-consuming. Additionally, the use of laser scanners provides us with "as-built" models automatically from point cloud data, however, the scanning devices are not usually used due to high cost.

This research uses Structure from Motion (SfM) (Dellaert et al., 2000) techniques to create 3D point cloud data of the landmarks. Using this approach, 3D objects can be easily reconstructed from a set of photographs using software applications such as Bundler (Snavely et al., 2006), OpenMVG (Moulon, 2012), and PhotoScan (Agisoft LLC., 2006). This approach has a potential to create the 3D models using available photos. These photos can be obtained from the image hosting services, such as flickr.com or any other source of images. For example, Agarwal et al. (2011) used photos in the cloud image hosting services to reconstruct historical landmarks. In this research, 150,000 photographs, which is returned when entering the search term "Rome" on flickr.com, were used. The system uses a collection of parallel distributed matching and reconstruction algorithms, designed to maximize parallelism at each stage and minimize serialization bottlenecks. Therefore, reconstructing 3D points from images took only 21 hours. The SfM approach is used in our system as it provides the desired accuracy and speed fitted to our objectives. Since the SfM does not consider scaling, the scaling factor is manually adjusted. Figure 2 shows the constructed 3D model of the building in Figure1 created using SfM technique.

Figure 2. The model created by SfM

In our method, the landmarks need to be manually geolocated. However, there are techniques to calculate the location coordinates of the landmarks by using the GPS information embed in its photographs (Junejo and Foroosh. 2008).

Although SfM is used in our system (as it is fast, relatively accurate and high performance), other techniques to construct the 3D object for landmarks can be also used. Additionally, available 3D models in virtual map applications (such as Google Earth) or CAD applications can be used in our proposed system.

### 3.4 Augmentation

The main function of the system is to combine the 3D objects of the landmarks and the live feed from the video camera of the tablet. In order to combine them, the current position of the user is acquired from some sensors in the tablet device. For example, the position is acquired from real-time positioning systems (such as Global Positioning System (GPS), UltraWide Band (UWB), Radio Frequency Identification (RFID) and location markers) and, the orientation and its changes are also acquired from the digital compass (magnometer) and gyroscopes. As most of the current smartphones and tablet devices are equipped with these sensors, the system can be easily utilized on the common devices. By having these information, combined images are created. For that, the information of the current position and orientation is acquired. The tablet's camera is setup. If the landmarks are in the field of view of the camera, the 3D models are overlaid on the video scene.

## 4. CASE STUDY

### 4.1 Description of Prototype System

In our prototype systems, Microsoft Windows 8.1 was used as the development environment, and Metaio SDK (Metaio Inc. 2013) was used for the AR development. The software development is done using Java programming language. For the field experiment, Sony Xperia acro So-02C android smartphone was used. In the field test, GPS technology was utilized for positioning and PhotoScan SfM software was employed to construct 3D models.

The SfM technique shows high efficiency for creating complex 3D models (such as the one on this research) due to its ability to construct the models automatically. The steps for creating the model using SfM are visualized in Figure 3 and are as follow: (1) Preparation of photos: the photos of the landmarks took from a various angle of view should be prepared, and the size of photos should be unified using photo editing software. It is also desirable to prepare aerial photograph for better results. (2) Reconstructing 3D point data from photographs; (Snavely et.al. 2006). (3) Estimating high density point data from Step 2 (Furukawa et.al. 2010). (4) Creating mesh data from Step 3. (5) Sticking the texture to the step 4's model.

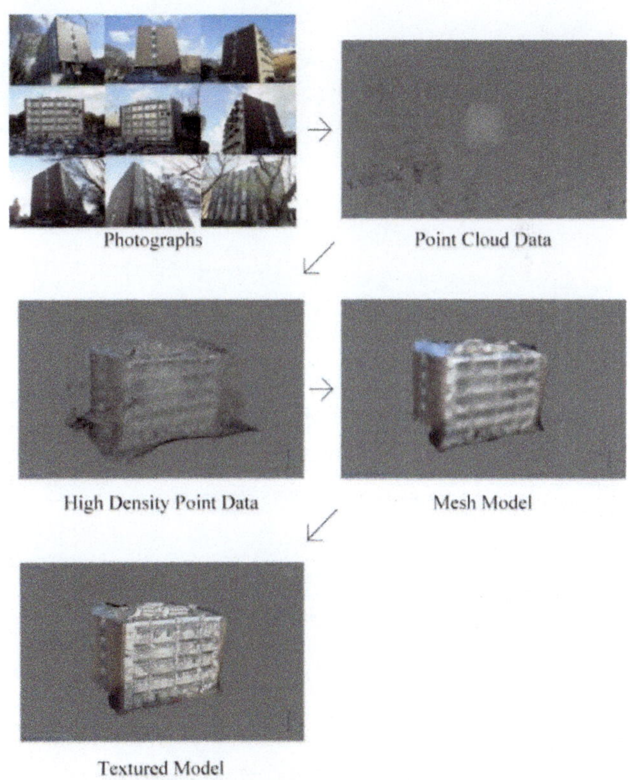


Figure 3. SfM Steps

### 4.2 Experiment

The experiment was done in the basement first floor of the U1W building of Osaka University (Figure 4). The augmented landmark was set to be the model of US1 Building in the same campus (Figure 5). Figure 6 shows the relative location of the two buildings in the experiment. US1 building is situated at lat. $34^{\circ}$ 821' N, and long. $135^{\circ}$ 523' E, and elevation. 60.034 m. U1W is situated at lat. $34^{\circ}$ 823' N, and long. $135^{\circ}$ 520' E, and elevation. 58.020 m. Regarding location relationship, US1 building is located in southeast of U1W building with an approximate distance of 350 meters, as shown in Figure 6. 162 pieces of photographs were prepared as input for SfM. The pictures were then resized to 640 pixel × 640 pixel. It took 5 minutes 54 seconds to finish creating the 3D point cloud data using our SfM software.

In our prototype application, Plan Position Indicator (PPI) were showed on the screen in order to help users understand the relative location and distance of the landmark and the direction of the device. Figure 7 and Figure 8 shows snapshots of the developed application. In these figures, the blue icon shows the location of the user, and yellow icon shows the location of the landmark (represented by the AR model). Additionally, the field of view and the visibility range is also shown.

Figure 4. U1W Building Indoor Environment

Figure 5. Osaka University US1 Building

Figure 6. Location of U1W and US1

### 4.3 Results

Figure 7 shows the prototype application in action. In the camera screen, PPI view is shown at the upper left and the augmented landmark is shown in the middle. Our field test confirmed that the location of the model and the shown direction of the handheld on the PPI view are correct (Figure 8). Additionally, the test confirmed that the augmented model of US1 building is shown while directing the phone towards it.

Figure 7. Snapshot of Application

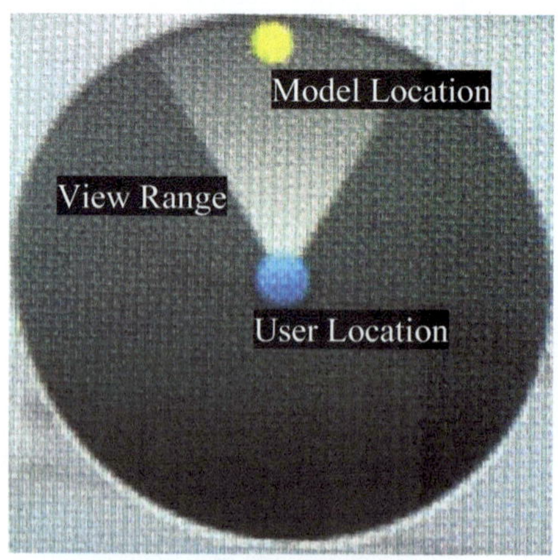


Figure 8. Plan Position Indicator

### 4.4 Discussion

The created model for our case study, as shown in Figure 7, does not include all the details of the building and suffers from low quality. Additionally, the model is not complete and misses parts of the building. For example, the opposite side of the model was not completely restored by SFM. This is because adequate photos from the opposite side could not be taken due to the problem with the photographing location. The photos that are taken from a distance and include the main features of the building are more efficient for the SfM. However, in the dense urban areas or if buildings are occluded by trees, such photos cannot be taken which influence the results. If using SfM technique is not feasible due to limitations in providing photos, other modelling method can be utilized to include the 3D object in the system.

Additionally, the system shows some inaccuracies in terms of locating the landmark when the smartphone is directed to its direction, and displaying the augmented model. We suspect this error is due to the inaccuracy of the GPS positioning in indoor environments as well as the errors related to the internal digital compass and the gyroscope. The problem of positioning accuracy can be tackled by utilizing more accurate indoor Real-Time Location Tracking Systems (RTLS) instead of GPS.

## 5. CONCLUSION AND FUTURE WORK

The paper presented a method to facilitate way-finding for pedestrians in underground facilities and dense urban environments. The AR-based application overlays the 3D model of a chosen landmark on the live camera feed of the user's handheld. The augmentation provides the user with the ability to see invisible landmarks, which helps him/her with way-finding intuitively. The SfM technique is utilized to create the model of landmarks from available photos. The case study shows that the 3D models can be added to the video stream when the handheld is directed towards the landmark. The projected model can also be accompanied by navigation information.

The accuracy of the system can be improved by utilizing more accurate RTLS methods (such as, UWB, RFID, and Wireless LAN). Additionally, the detailed investigation on the sources of error and the improvement of the accuracy is considered a future work for this research. Regarding the quality of the 3D model created by SfM, it is suggested to provide aerial photographs. However, other methods can be used to create accurate 3D models. The performance of various software applications to create high density models should also be evaluated. Moreover, the ability to visualize multiple landmarks, add relative distance between the user and the landmarks to the handheld screen and embedding the perspective to the views are considered future steps to complete the prototype system. The case study is planned to be implemented in a larger underground facility in order to validate the benefits of the proposed system.

## REFERENCE

Agarwal, S., Snavely, N., Simon I., Seitz, S., Szeliski, R., 2011. Building Rome in a day, Communications of the ACM, Vol.54 Issue 10, October 2011, pp.105-112.

Agisoft LLC. 2006. Report on Agisoft "Agisoft PhotoScan" http://www.agisoft.com/, (7 April. 2015).

Dellaert, F., Seitz, S., Thorpe, C., and Thrun, S., 2000. Structure from Motion without Correspondence, *IEEE Computer Society Conference on Computer Vision and Pattern Recognition,* Vol.2, pp.557-564.

Furubayashi, S., Yabuki, N., Fukuda,T., 2013. A data model for Checking Direvtion Singnage at Railway Stations, Proceedings of the First International Conference on Civil and Building Engineering Informatics, *ICCBEI 2013, November 7-8, Tokyo, Japan,* pp.167-174.

Furukawa, Y., Ponce, J., Team, W., and Inria, E.N.S. 2010. Accurate , Dense , and Robust Multi-View Stereopsis, *IEEE Trans. Pattern Anal. Mach. Intell.* Vol. 32, Issue 8, pp.1362-1376.

Junejo, I.N., Foroosh, H., 2008. GPS coordinate estimation from calibrated cameras, *Pattern Recognition, ICPR 2008. 19th International Conference on,* pp.1-4.

Kameda, Y., Takemasa, T., Ohta, Y., 2004. Outdoor See-Through Vision Utlizing Surveilillance Cameras, *Proceedings of the Third IEEE and ACM international Symposium on Mixed and Augmented Reality* , ISMAR 2004, pp.213-218.

Metaio Inc. 2013. Report on the metaio "metaio Developer Portal", http://dev.metaio.com/sdk/tutorials/hello-world/ (14 Feb. 2015).

Moulon, P., 2012. OpenMVG: "open Multiple View Geometry", http://imagine.enpc.fr/~moulonp/openMVG/index.html, (12April. 2015).

Schall, G., Mendez, E., Kruijff, E., Veas, E., Junghanns, S., Reitinger, B., Schmalstieg, D. 2009. Handheld Augmented Reality for underground infrastructure visualization, *Pres Ubiquit Comput,* 13: pp.281-291.

Snavely, N., Seitz, S.M., and Szeliski, R. 2006. Photo Tourism : Exploring Photo Collections in 3D, *SIGGRAPH '06 ACM SIGGRAPH 2006 Pap.*, vol. 1, no. 212, pp. 835–846.

TagCast Inc. 2014. Report on TagCast "Case study - Provide TAGCAST for a large experiment lead by Land, infrastructure and Transportation Ministry of Japan -", https://play.google.com/store/apps/details?id=jp.tagcast.shibuya.navi&hl=ja (18 Apr. 2015).

Tanighuchi, Y., Aida. S., Yoshinaka, S., Takizawa, A., 2014. Study on Tsunami Evacuation Plan at the Underground Mall in front of Osaka Station, *Summaries of technical papers of annual meeting 2014*, pp.147-148.

Tardieu, J., Susini, P., Poisson, F., Kawakami, H., and McAdams, S., 2009. The design and evaluation of an auditory way-finding system in a train station, *Appl. Acoust.*, vol. 70, no. 9, pp. 1183-1193.

Yamauchi, S. 2012. Research on the spatial perception and the guidance method in underground space, *Hosei University graduate school design engineering graduate course bulletin,* Vol.1.

# ADAPTIVE PARAMETER ESTIMATION OF PERSON RECOGNITION MODEL IN A STOCHASTIC HUMAN TRACKING PROCESS

W. Nakanishi [a, *] , T. Fuse [a] and T. Ishikawa [a]

[a] Dept. of Civil Engineering, The University of Tokyo, 7-3-1 Hongo, Bunkyo-ku, Tokyo 1138656 Japan
- (nakanishi@civil, fuse@civil, ishikawa@trip).t.u-tokyo.ac.jp

**KEY WORDS:** State space modelling, Human tracking, Sequential image, Adaptive parameter estimation, Close Range

**ABSTRACT:**

This paper aims at an estimation of parameters of person recognition models using a sequential Bayesian filtering method. In many human tracking method, any parameters of models used for recognize the same person in successive frames are usually set in advance of human tracking process. In real situation these parameters may change according to situation of observation and difficulty level of human position prediction. Thus in this paper we formulate an adaptive parameter estimation using general state space model. Firstly we explain the way to formulate human tracking in general state space model with their components. Then referring to previous researches, we use Bhattacharyya coefficient to formulate observation model of general state space model, which is corresponding to person recognition model. The observation model in this paper is a function of Bhattacharyya coefficient with one unknown parameter. At last we sequentially estimate this parameter in real dataset with some settings. Results showed that sequential parameter estimation was succeeded and were consistent with observation situations such as occlusions.

## 1. INTRODUCTION

Human tracking method is classified into two manners. One is that in each frame human candidates are detected and then the same person is identified in adjacent frames. The other is that, in each frame, each human position is predicted from that of previous frame and then accurate position is calculated by a stochastic model using observed information. As we can consider pedestrian behaviour model for successive frames, this method has an advantage to tracking many persons simultaneously in complex situations in which occlusions and proximities are occurred (e.g. Ali and Dailey, 2009; Nakanishi and Fuse, 2012). Mathematically this method should be solved as a stochastic process. Typical formulation for this method is based on Bayesian statistics; predicted position is described by prior distribution, observed information corresponds to data and accurate position to posterior. When we use this method, we should define a likelihood function that describes likelihood of observed information occurred by given predicted position. We call this likelihood function "observation model" in human tracking method (explained later in detail). This paper aims at an estimation of parameters of observation model, that is, person recognition model, using a sequential Bayesian filtering method. We especially deal with a stochastic process for human tracking mentioned above.

A main problem on human tracking from sequential images is to identify the same person in successive frames. When we acquire colour information from an image, we use a human detection model to detect and then identify a person in adjacent frames. Any models used here have some parameters that correspond to sensitivity and thresholds. Usually this parameters are set empirically or adjusted to data in advance of tracking process. This means that we have to set models and parameters according to observations every time. We cannot directly apply a setting on one observation site to another. Moreover, we might have to change model settings on the same observation site on different time or day.

Thus in this paper we apply sequential Bayesian filtering to an estimation of this parameters and update parameters according to observations at each frame. We call this updating as an adaptive estimation. In addition to an advantage that parameters are automatically adjusted to change in observation situations as mentioned above, this adaptive estimation can deal with differences in prediction accuracy in a tracking process. When we use stochastic process like general state space model for tracking, an estimated human position is determined by both prediction by previous human position and detection by present observation. Therefore if prediction is quite accurate, then sensitivity of detection can be high; Sensitivity parameters can be set large, or thresholds parameters can be set high. In contrast, if prediction is uncertain, then detection sensitivity should be low.

In previous research (Nakanishi and Fuse, 2012) we assume a tracking process as a data assimilation process and describe it in a form of general state space model. Human position and size are defined as unobservable variables: state vectors, and stochastic pedestrian behaviour model is introduced into a system model. We assume that state vectors generate colour images according to a person recognition model: an observation model. We applied Bhattacharyya coefficient (Bhattacharyya, 1943) to formulate an observation model. While even we use only Bhattacharyya coefficient to formulate an observation model, we have slight knowledge about function forms and parameter settings that bring good observation models. Although experimental tests have been conducted in some function forms and parameters (Nakanishi and Fuse, 2014), parameters are decided in advance there.

In this paper we sequentially estimate a parameter of some observation models using general state space model. From this estimation we consider good parameters correspond to prediction and observation situations; how parameters should be according to prediction and observation accuracy.

## 2. MODELLING FRAMEWORK

### 2.1 General State Space Model

Firstly we explain a general state space modelling (Figure 1) and its application to human tracking (Nakanishi and Fuse, 2012). In this modelling, we can estimate a dynamics of unobservable variables. A state vector $\mathbf{x}_t$ is a set of unobservable variables at time $t$ that we will estimate a true value. An observation vector $\mathbf{z}_t$ is a set of observable variables at time $t$. A system model $p(\mathbf{x}_t|\mathbf{x}_{t-1})$ is a conditional density function that represents a dynamics of a state vector. An observation model $p(\mathbf{z}_t|\mathbf{x}_t)$ is also a conditional density function that represents a likelihood of realisation of $\mathbf{z}_t$ given $\mathbf{x}_t$. After we obtain $\mathbf{z}_{1:t} = \{\mathbf{z}_1, \mathbf{z}_2, \ldots, \mathbf{z}_t\}$, a series of observations from time 1 to $t$, the posterior distribution of $\mathbf{x}_t$ is calculated by Bayes' theorem as follows:

$$\begin{aligned} p(\mathbf{x}_t \mid \mathbf{z}_{1:t}) &\propto p(\mathbf{z}_t \mid \mathbf{x}_t)p(\mathbf{x}_t \mid \mathbf{z}_{1:t-1}) \\ &= p(\mathbf{z}_t \mid \mathbf{x}_t)\int p(\mathbf{x}_t \mid \mathbf{x}_{t-1})p(\mathbf{x}_{t-1} \mid \mathbf{z}_{1:t-1})d\mathbf{x}_{t-1} \end{aligned} \quad (1)$$

In this equation, $p(\mathbf{z}_t|\mathbf{x}_t)$ is an observation model: human recognition model, $p(\mathbf{x}_t|\mathbf{x}_{t-1})$ is a system model and $p(\mathbf{x}_{t-1}|\mathbf{z}_{1:t-1})$ is the estimation result at time $t$-1.

If we need the estimated value for $\mathbf{x}_t$, following values are usually used (Jaynes, 2003).

(a) Posterior mean: An expected value of $p(\mathbf{z}_t|\mathbf{x}_t)$.

$$\bar{\mathbf{x}}_t = \int \mathbf{x}_t p(\mathbf{x}_t \mid \mathbf{z}_{1:t})d\mathbf{x}_t \quad (2)$$

(b) Posterior mode: Maximum a posteriori

$$\hat{\mathbf{x}}_t = \arg\max_{\mathbf{x}_t} p(\mathbf{x}_t \mid \mathbf{z}_{1:t}) \quad (3)$$

(c) $n$ percentile of the posterior: A value that cumulative frequency of posterior equals to $n$ percent

$$\hat{\mathbf{x}}_t \; such that \int_{-\infty}^{\hat{\mathbf{x}}_t} p(\mathbf{x}_t \mid \mathbf{z}_{1:t})d\mathbf{x}_t = \frac{n}{100} \quad (4)$$

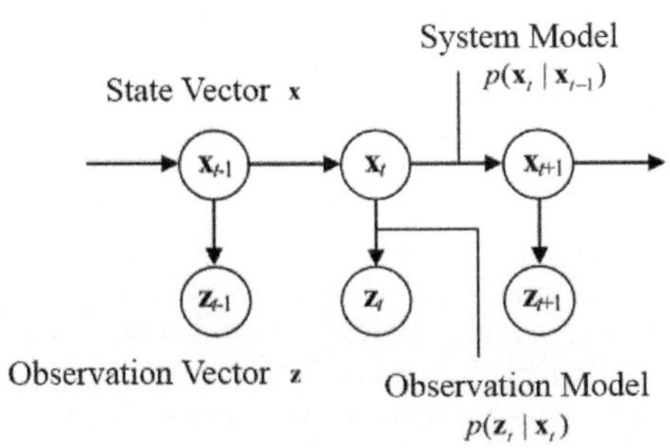


Figure 1. General state space model

We use particle filter (Gordon, *et al.*, 1993; Kitagawa, 1996) for calculating equation (1), for some models are assumed to be non-linear. In this work we use 65536 particles to approximate posteriors. We also use Halton sequences (Halton, 1964) to generate prior (predictive) distributions. They are typical quasi-random number sequences. Thus good approximation of posteriors with less number of particles is expected.

### 2.2 Self-Organizing Modelling

Now we consider that a set of parameters $\boldsymbol{\theta}$ of observation model is also unobservable. This means that parameters of a person recognition model can take different value from time to time. We introduce $\boldsymbol{\theta}$ into state vector as follows (Kitagawa, 1998):

$$\tilde{\mathbf{x}}_t = \begin{pmatrix} \mathbf{x}_t \\ \boldsymbol{\theta}_t \end{pmatrix} \quad (5)$$

Then we can describe system and observation models as follows.

$$p(\tilde{\mathbf{x}}_t \mid \tilde{\mathbf{x}}_{t-1}) \equiv \begin{pmatrix} p(\mathbf{x}_t \mid \mathbf{x}_{t-1}) \\ p(\boldsymbol{\theta}_t \mid \boldsymbol{\theta}_{t-1}) \end{pmatrix} \quad (6)$$

$$p(\mathbf{z}_t \mid \tilde{\mathbf{x}}_t) \equiv p(\mathbf{z}_t \mid \mathbf{x}_t, \boldsymbol{\theta}_t) \quad (7)$$

Here we assume that a system model of parameters $\boldsymbol{\theta}$: $p(\boldsymbol{\theta}_t|\boldsymbol{\theta}_{t-1})$ is random walk model. Then we can calculate a posterior in the same way as equation (1),

$$\begin{aligned} p(\tilde{\mathbf{x}}_t \mid \mathbf{z}_{1:t}) &\propto p(\mathbf{z}_t \mid \tilde{\mathbf{x}}_t)p(\tilde{\mathbf{x}}_t \mid \mathbf{z}_{1:t-1}) \\ &= p(\mathbf{z}_t \mid \tilde{\mathbf{x}}_t)\int p(\tilde{\mathbf{x}}_t \mid \tilde{\mathbf{x}}_{t-1})p(\tilde{\mathbf{x}}_{t-1} \mid \mathbf{z}_{1:t-1})d\tilde{\mathbf{x}}_{t-1} \\ &= p(\mathbf{z}_t \mid \mathbf{x}_t, \boldsymbol{\theta}_t)\begin{pmatrix} \int p(\mathbf{x}_t \mid \mathbf{x}_{t-1})p(\mathbf{x}_{t-1} \mid \mathbf{z}_{1:t-1})d\mathbf{x}_{t-1} \\ \int p(\boldsymbol{\theta}_t \mid \boldsymbol{\theta}_{t-1})p(\boldsymbol{\theta}_{t-1} \mid \mathbf{z}_{1:t-1})d\boldsymbol{\theta}_{t-1} \end{pmatrix} \end{aligned} \quad (8)$$

Marginal distributions are calculated as

$$p(\mathbf{x}_t \mid \mathbf{z}_{1:t}) = \int p(\tilde{\mathbf{x}}_t \mid \mathbf{z}_{1:t})d\boldsymbol{\theta}_t \quad (9)$$

$$p(\boldsymbol{\theta}_t \mid \mathbf{z}_{1:t}) = \int p(\tilde{\mathbf{x}}_t \mid \mathbf{z}_{1:t})d\mathbf{x}_t \quad (10)$$

We obtain this posteriors at each time instant $t$; Equation (9) is estimated positions and (10) is estimated parameters at each time. These are the estimated results in this paper.

## 3. MODEL SETTINGS

### 3.1 State Vector and Observation Vector

We use "BEHAVE" dataset (Blunsden and Fisher, 2010; Figure 2) for analyses. This dataset provides sequential images and ground truth of human positions and sizes of rectangles on these images. We define four-dimensional $\mathbf{x}_t$, representing human position and its size, considering an inscribed ellipse in the rectangle (equation (11)). We also define $\mathbf{z}_t$ as colour values at each pixel inside the ellipsoid (equation (12)). We define state and observation vectors at one second intervals. This means that we predict each second human position of one second after. And we also use colour information from images every one second.

$$\mathbf{x} \equiv \left(x, y, a_x, a_y\right)^T \tag{11}$$

$$\mathbf{z}_{xy} \equiv \left(r_{xy}, g_{xy}, b_{xy}\right)^T \tag{12}$$

where $x, y$ = coordinates of each pixel on images
$a_x$, $a_y$ = ellipse's semidiameter of $x$ and $y$ respectively
$r_{xy}$, $g_{xy}$, $b_{xy}$ = corresponding colour value at pixel $(x, y)$

Figure 2. Example of used dataset (Blunsden and Fisher, 2010)

### 3.2 System Model

System model in this paper describe a difference of positions and sizes of the same person in successive frames. We consider only changes in positions and assume that sizes are predicted precisely at each frame. Then we set system model in two manners. One is that we set it as difference of true values. This means that relative value of position change is precisely predicted. The other is that we set it with error terms in addition to relative position changes. Details are explained later.

### 3.3 Observation Model

We build a model of $p(\mathbf{z}_t|\mathbf{x}_t)$ as follows, using Bhattacharyya coefficient $B$ and a parameter $c$ (Figure 3):

$$B = \prod_{r,g,b} \sum_i \sqrt{p_i q_i} \tag{13}$$

$$p(\mathbf{z} \mid \mathbf{x}) = f(B) = f\left(\sum_m \sqrt{p_m q_m}\right) \tag{14}$$

$$f(B, \mathrm{c}) = \frac{ce^{cB}}{e^c - 1} \quad (0<c) \tag{15}$$

where $B$ = Bhattacharyya coefficient
$i$ = pixel value
$r, g, b$ = corresponding colour
$p, q$ = normalized histogram of previous and predictive position respectively
$p_m$ = relative frequency of pixel value $m$ in histogram $p$

As shown in equation (13), the domain of Bhattacharyya coefficient is [0, 1]. Here an integration of equation (15) with interval [0, 1] is one for any $B$ and $c$. This fact theoretically guarantee that adaptive estimation can be done. This parameter $c$ corresponds to $\boldsymbol{\theta}$ in Chapter 2 in following part of this paper.

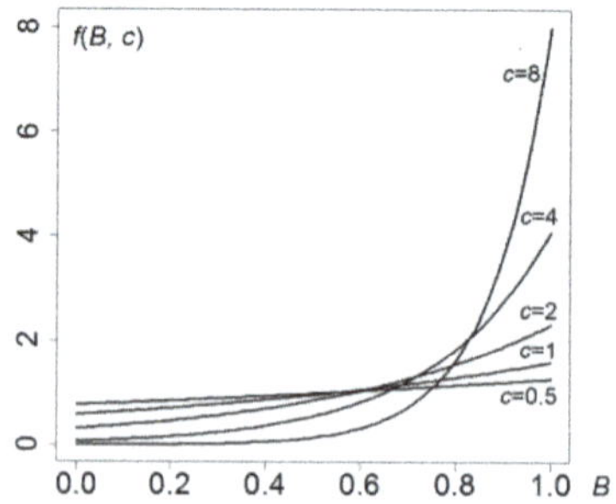


Figure 3. Observation model with parameter $B$ and $c$

## 4. EXPERIMENTAL RESULTS

### 4.1 Settings

We conduct experiments on dataset mentioned above in five settings shown in Table 4.

Firstly we set initial state vectors as a true value from the dataset and system models as a true model: the difference between true values of successive frames. We set several patterns for the parameter's initial value and system models of that parameter. These are setting 1-3. Initial value of $c$ in setting 1 and 2 are decided as the maximum likelihood estimator that is calculated using dataset itself. In contrast, that of setting 3 is smaller value. The difference between setting 1 and 2 is variance of system model of $c$; in setting 2 system model of c has large variance than that of setting 1.

Secondly we change system models and initial values of human position. These are setting 4 and 5. This is natural assumptions; the former setting is corresponding to uncertainty in prediction models, and the latter in detection models. This time we get not only the posterior of the parameter but also the posteriors of human position.

### 4.2 Results and Discussions

It is difficult to evaluate an accuracy of estimation of state vectors, for state vectors themselves cannot be observed. So in this paper we discuss results qualitatively. We take a result of one person as an example to explain and discuss about results. We confirmed that similar tendencies are common among other persons. Results are shown in Figure 5. Solid lines show posterior mean and dashed lines show 25, 50 and 75 percentiles respectively.

Table 4. Setting for analyses

| Setting | Initial value of $c$ | Initial error of position | System model of $c$ | Noise of system model of position |
|---|---|---|---|---|
| 1 | 8.52 | 0 | $c_t \sim N(c_{t-1}, 0.1)$ | 0 |
| 2 | 8.52 | 0 | $c_t \sim N(c_{t-1}, 0.5)$ | 0 |
| 3 | 2.00 | 0 | $c_t \sim N(c_{t-1}, 0.1)$ | 0 |
| 4 | 2.00 | 0 | $c_t \sim N(c_{t-1}, 0.1)$ | $N(\mathbf{0}, \text{diag}(25))$ |
| 5 | 2.00 | 5 pix for $x$-$y$ each | $c_t \sim N(c_{t-1}, 0.1)$ | $N(\mathbf{0}, \text{diag}(25))$ |

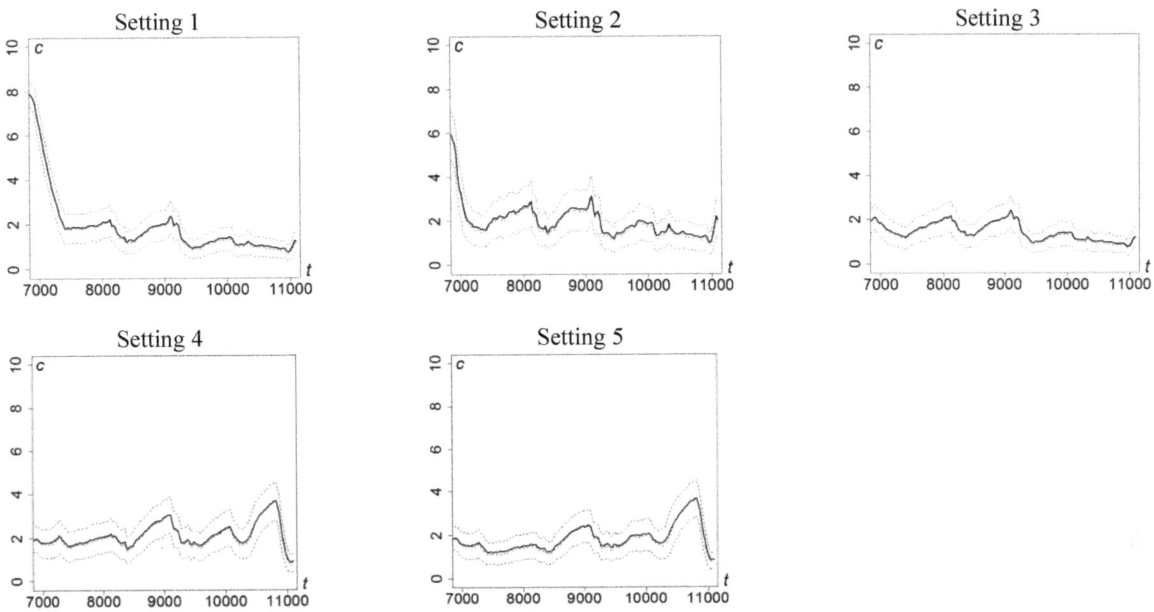


Solid lines show expected values of posteriors; Dashed lines show 25, 50 and 75 percentiles respectively; $t$ is frame number and framerate is 25 [fps].

Figure 5. Posterior distributions of parameter $c$ in each setting

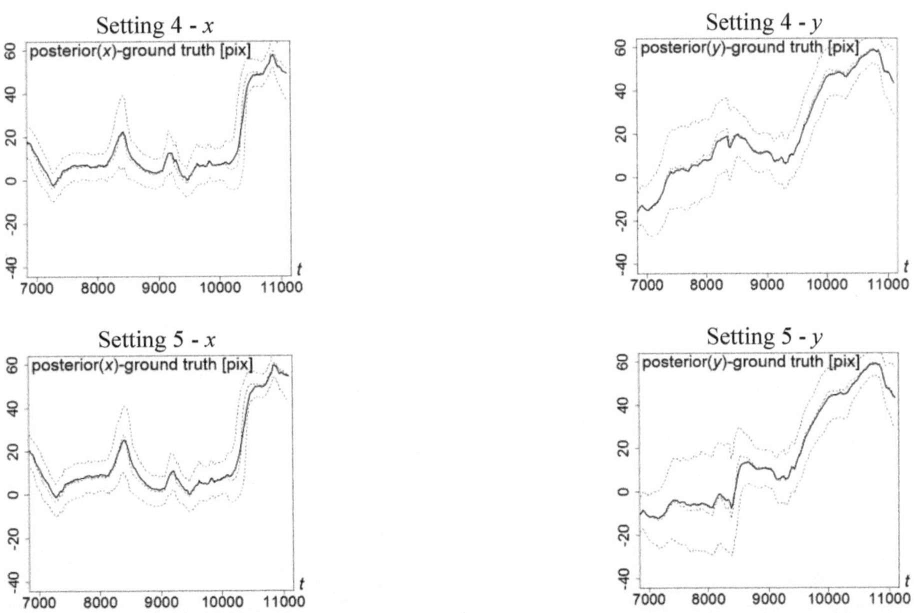


Solid lines show expected values of posteriors; Dashed lines show 25, 50 and 75 percentiles respectively; $t$ is frame number and framerate is 25 [fps].

Figure 6. Posterior distributions of positions in each setting

In setting 1-3, sequential Bayesian estimation are succeeded and posterior distributions are similar to each other. We also found that the value of parameter is low when a person is occluded by other person: around frame #8200 and #9200. This is the expected tendency, for Bhattacharyya coefficient is low when occlusions occur. This showed that adaptive parameter estimation was useful in that values of parameter changed according to observation situations.

In setting 4, when noise of system model is large, the accuracy of position estimation became low when there were occlusions. This is consistent with the results of setting 1-3 above. And after frame #10000 we found that the parameter value became slightly high. This is because once we lost an accurate position of a person, then we compared two histograms composed of background pixels. This fact is also shown from position's marginal posteriors. Figure 6 shows posteriors of residual (relative value to ground truth). If shape of this distribution is sharp around zero, estimations are good. Actually from around frame #10000 we lost the accurate position in each coordinates $x$ and $y$. Especially $y$ coordinates was difficult to predict precisely, for a little difference in $y$ value cannot be affected to shape of colour histogram; clothes tended to be the same colour in upper and lower pixels of a certain pixel, and also background pixels be the same colour in upper and lower of a certain person. In setting 5, very similar tendency to setting 4 was shown.

In summary, the observation model in this paper can deal with prediction errors that are less than 10 pixels, otherwise it is difficult to correct errors and estimate precise positions of persons. In order to achieve successful tracking in such a case, parameter value should be kept low during low accuracy of prediction. Also it may bring about good result to introduce foreground detection model into proposed model and to use only foreground pixels to make histograms.

## 5. CONCLUSION

In this paper we estimated the parameter of person recognition model adaptive to observation data by using sequential Bayesian filtering. After formulate the estimation, we applied the proposed method to dataset. From results we showed that appropriate parameter changes from time to time according to the situation of observation and prediction. Also we discussed about the tendency of parameter change by comparing parameter values and some settings for parameter estimation. Basically the results were consistent with settings and situations. Thus we could confirm that sequential Bayesian filtering is a useful way to deal with time and space varying parameter estimation.

Future works are as follows. Firstly we conduct such analyses on other models. For example, we combine the models with foreground detection models to deal with larger prediction errors. Also we employ models with more than one parameter to deal with more complex situations. At last we try to find the way to formulate observation models themselves, not only their parameters, according to the dataset. To achieve such estimations with larger calculation amount, we have to consider how to approximate posteriors by smaller particles. An application of merging particle filter (Nakano *et al.*, 2007) or Rao-Blackwellisation (Casella and Robert, 1996) is one possible way.

**References:**

Ali, I. and Dailey, M., 2009. Multiple human tracking in high-density crowds, *Advanced Concepts for Intelligent Vision Systems*, LNCS 5807, pp.540-549.

Bhattacharyya, A., 1943. On a Measure of Divergence between Two Statistical Populations Defined by Their Probability Distributions, *Bulletin of the Calcutta Mathematical Society*, 35(14), pp.99-109.

Blunsden, S. J. and Fisher, R. B., 2010. The BEHAVE Video Dataset: Ground Truthed Video for Multi-person Behavior Classification, *Annals of the BMVA*, 2010(4), pp.1-11.

Casella, B. Y. G. and Robert, C. P., 1996. Rao-Blackwellisation of Sampling Schemes, *Biometrika*, 83(1), pp.81-94.

Gordon, N. J., Salmond, D. J. and Smith, A. F. M., 1993. Novel Approach to Nonlinear / Non-Gaussian Bayesian State Estimation. *Radar and Signal Processing, IEE Proceedings F*, 140(2), pp.107-113.

Halton, J. H., 1964. Algorithm 247: Radical-inverse quasi-random point sequence, *Communications of the ACM*, 7(12), pp.701-702,

Jaynes, E. T., 2003. *Probability Theory: The Logic of Science*, Cambridge University Press, pp.172-175.

Kitagawa, G., 1996. Monte Carlo Filter and Smoother for Non-Gaussian Nonlinear State Space Models. *Journal of Computational and Graphical Statistics*, 5(1), pp.1-25.

Kitagawa, G., 1998. A Self-Organizing State-Space Model, *Journal of the American Statistical Association*, 93(443), pp.1203-1215.

Nakanishi, W. and Fuse, T., 2012. Multiple Human Tracking in Complex Situation by Data Assimilation with Pedestrian Behavior Model. *ISPRS Archives*, 39-B3, pp.409-414.

Nakanishi, W. and Fuse, T., 2014. Sensitive Analysis of Observation Model for Human Tracking Using a Stochastic Process, *ISPRS Archives*, 40-5, pp.445-450.

Nakano, S., Ueno, G. and Higuchi, T., 2007. Merging Particle Filter for Sequential Data Assimilation, *Nonlinear Processes in Geophysics*, 14(4), pp.395-408.

# DACTYL ALPHABET GESTURE RECOGNITION IN A VIDEO SEQUENCE USING MICROSOFT KINECT

S. G. Artyukhin[a] , L. M. Mestetskiy [a]

[a] MSU, Faculty of Computational Mathematics and Cybernetics, Russia, 119991, Moscow, GSP-1, 1-52, Leninskiye Gory

**KEY WORDS:** Gesture recongniton, Kinect, Morphological skeleton, Dactyl sign, Segmentation

**ABSTRACT:**

This paper presents an efficient framework for solving the problem of static gesture recognition based on data obtained from the web cameras and depth sensor Kinect (RGB-D - data). Each gesture given by a pair of images: color image and depth map. The database store gestures by it features description, genereated by frame for each gesture of the alphabet. Recognition algorithm takes as input a video sequence (a sequence of frames) for marking, put in correspondence with each frame sequence gesture from the database, or decide that there is no suitable gesture in the database. First, classification of the frame of the video sequence is done separately without interframe information. Then, a sequence of successful marked frames in equal gesture is grouped into a single static gesture. We propose a method combined segmentation of frame by depth map and RGB-image. The primary segmentation is based on the depth map. It gives information about the position and allows to get hands rough border. Then, based on the color image border is specified and performed analysis of the shape of the hand. Method of continuous skeleton is used to generate features. We propose a method of skeleton terminal branches, which gives the opportunity to determine the position of the fingers and wrist. Classification features for gesture is description of the position of the fingers relative to the wrist. The experiments were carried out with the developed algorithm on the example of the American Sign Language. American Sign Language gesture has several components, including the shape of the hand, its orientation in space and the type of movement. The accuracy of the proposed method is evaluated on the base of collected gestures consisting of 2700 frames.

## 1. INTRODUCTION

A gesture is a form of non-verbal communication or non-vocal communication in which visible bodily actions communicate particular messages, either in place of, or in conjunction with, speech. Gestures include movement of the hands, face, or other parts of the body. In the last decade, more and more attention is paid to the automatic recognition of gestures. This is because the gestures is a convenient way to enter information into a computer. The appearance of depth sensors such as Kinect, and the computing power of personal systems provide an opportunity to solve the problem of gesture recognition in real time. Important in the field of gesture recognition is the problem of recognition of sign language. Gestures of such languages are divided into two types: dynamic, in which the important movement and change hands posture, and static, which are determined only by the hand posture. In sign language most gestures are dynamic. Their diversity makes it difficult to recognize. More simple language recognition is dactyl letters and numbers. Dactyl sign language - is the language in which each letter and number corresponds to a gesture usually static. There are more and more devices in the market which help to solve the problem of gesures recognition. One such device is the Kinect, developed by Microsoft. Kinect consists of two cameras and an infrared projector and allows to obtain both color image and depth map. Also, Microsoft has developed a library that allows to recognize human posture, but there is no standard solutions for hand gestures. We solve the problem of static gesture recognition based on data obtained from the web cameras and Kinect depth sensor. Static gesture given by a pair of images: color image and depth map. The database gestures stored in one or more frames for each gesture. At the stage of the recognition algorithm is applied to the input video sequence (a sequence of frames) which we want to mark, that is put in correspondence each frame of the sequence with bases gesture, or decide that there is no appropriate gesture in the database. Classification of frames of a video sequence is independent, without interframe information. Then, a sequence of successful marked frames in equal gesture is grouped into a single static gesture.

## 2. RELATED WORK

Many approaches for gesture recognition are described in the literature. Most articles describe the methods that work with a small set of highly different gestures. Some methods are not applicable to real-time. To Some techniques required plain background or special gloves.

The first group of methods are defined by geometrical model for each hand gesture, then it can be reconstructed from the image. For example, in (Wang2009) multi-colored glove is used. Glove arranged so that in most cases it is possible to restore the hand model. Glove helps to accurately segment the image for almost all type of background, but recovery is performed by storing a large database and finding a suitable gesture. It lead to a great deal of time and memory. In (Stenger2006) also the three-dimensional model of the hand is restored, the method is based on storage the large number of examples for each pose hands. The method proposed in (Malik2003), based on skin color segmentation and searching for fingers. Selecting only ends of the fingers narrows the set of gestures for which method is applicable.

The second group includes methods based on direct feature generation from the image. For example, in (Suryanarayan2010) the image is normalized with the palm width and height, a uniform grid is introduced, and the degree of filling of each cell is used as feature. For qualitative classification requires stored for each rotation gesture, which leads to increase the storage.

The third group of methods is based on a comparison of the hand shape with the standards of the base. In (Beristain2010) method for comparing the shape bassed on discrete skeletonization and the method of comparing the skeletons are described. Skeletonization is powerful method for comparing shapes. It reduces the amount of information with minimum loss and let to use effective methods to extract special points. Methods of Discrete skeletization have higher complexity, in contrast by continuous methods. Discrete skeletons can be discontinuous that complicates the solution of the problem of classification.

Method of gesture classification based on continious skeletons proposed in (Kurakin2012). Authors condsider dynamic gestures in this work. Each frame is described by the position of several special points, that is not enough to detect a large number of static gestures.

In this paper we solve the problem of image and depth maps segmentation in order to select the hand on the frame. Segmentation is the process of dividing image into several segments. The result of image segmentation a set of segments that together cover the whole image. Several algorithms and universal methods is developed for image segmentation. We list the main types of segmentation methods:

- based on clustering
- based on the analysis of the histogram
- based on edge detection
- based on the calculating minimum graph cut
- method of watershed and others.

Most of them are described in (Shapiro, 2001). Since the general solution for the problem of segmentationtation image does not exist, these methods often have to combine with knowledge of the subject area to effectively solve this problem.

## 3. PROPOSED METHOD

In this paper we propose a method combined segmentation frame for depth map and images. The primary segmentation is based on the depth map and provides information about the position and fuzzy hand border. Then color image is analyzed to clarify the boundaries. Feature generation for hand shape is based on continuous skeleton. Introduceing of clear interpretational features allows to create classifications based on simple rules.

### 3.1 Segmentation

The image of segmentation is an unsolved problem in the general case. Usually the problem narrowed to a specific area. In this paper we want to select a hand for further features generation. By itself, the hand has a clear spatial structure and color, but because of the different lighting, color and complex background necessary to use additional tools for segmentation. In this paper, the problem is solved by a combination of segmentation by the depth map and the color image.

The target object of segmentation is hand. Let with the image of $I$ (Figure 1) algorithm is applied to the input depth map $D$ (Figure 2) — matrix of the same size as the $I$, whose elements are the distance from the camera to the object corresponding pixel in the image $I$. Assume that the hand is the foreground. This makes it possible to localize the position of the hand using threshold binarization. Because of the large error of the depth map, we use image $I$ to find the exact boundaries.

Combined segmentation algorithm is follows. The first step is the sellection of $S_{front}$ — foreground depth map (Figure 3). For this we introduce the parameter $t_f$ and select those pixels whose depth is different from the minimum no more than $t_f$:

$$S_{front} = \{s_{ij} = [d_{ij} < mind_{ij} + t_f]\}. \qquad (1)$$

Apply the morphological erosion operation (Figure 4). Set the resulting region of $S_{erosion}$

$$S_{erosion} = S_{front} \ominus S_0^r, \qquad (2)$$

where $S_0^r$ circle with radius $r$. This operation is necessary because the boundary of the foreground could get background pixels. In the region of $S_{erosion}$ contains only the pixels of the target, so they can be used to calculate the average color of the target $avg\_color$. Then we search the edge of the image by algorithm Canny. The final step is to run the search algorithm BFS on image pixels starting with the field $S_{erosion}$. In BFS used 4-connected system of neighborhood pixels. Stop criterion of the wide search is that pixel belonging to the edge or condition $[\rho(current\_color, avg\_color) > t_c]$, where $\rho$ — is the sum of RGB component distance. All pixels that we visited mark 1 in the matrix $S$ (Figure 5).

Figure 1: Source RGB image

Figure 2: Source depth map

Figure 3: Foreground

Figure 4: Foureground core

Figure 5: Segmentation result

### 3.2 Feature space

This paper proposes a method of generating features based on the shape obtained after segmentation image and depth map. Continious skeleton (Mestetskiy2008) is build by the two-dimensional projection of the hand. We consider skeleton as graph with width function on it's edges.

If there is no hole in the figure, the skeletal graph is a tree, otherwise it has cycle. The presence of cycles in a connected graph can be determined by running the search algorithm in width from each vertex.

With a special terminal branches classification algorithm described in the next section, we can determine the number of fingers and end points of terminal branches that correspond to the fingers. We also search for a branch forming the wrist and its end point, searching a point of the skeleton, in which the width function is maximum. Call it base point. Then we can calculate the angle between the end point of the branch of the wrist and each endpoint branches of fingers relative to a bas point. On the Figure 6 highlighted circle by red color correspond to the end points of the fingers, yellow — end points of the wrist, green — base point.

Using binary search method will find the minimum param of regularization in which the skeletal graph is transformed into a chain. This chain is called a main. If none of the branches of the skeleton was not classified as a finger, a good feature that describe the shape is the width of the main chain.

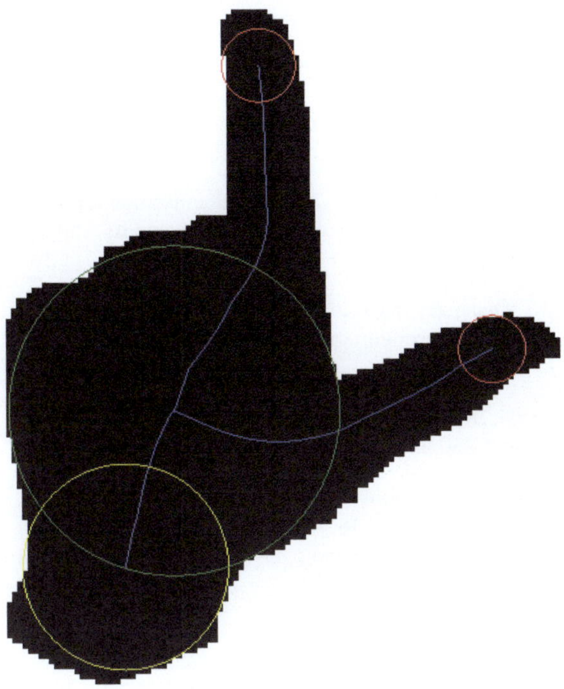

Figure 6: Terminal branches

Thus, the feature space will be as follows

- $N$ — number of fingers,
- $M$ — number of loops in skeletal graph (number of holes in hand shape),
- $A = (\alpha_1, \dots, \alpha_N)$ — angles between the end point of the wrist and each endpoint of fingers
- $W = (w_1, \dots, w_{50})$ — value of width function of main chain in uniform 50 points.

### 3.3 Terminal branch classification

In this section we describe an algorithm classification of terminal branches. Each branch of the skeleton is described by its position and width function. Normalize the length of each branch so that it is equal to 1 and normalize width function relative to the maximum width. If we consider plot or resulting width functions of terminal branches that match fingers, we will notice that they all behave the same way. The functions of the remaining terminal branches are very different from them. Given a training set consisting of a set of functions corresponding to fingers $\{W_i(x)\}_{i=1}^{N}$, we construct two functions

$$W_{max}(x) = \max_i W_i(x) + \epsilon, \tag{3}$$

$$W_{min}(x) = \min_i W_i(x) - \epsilon. \tag{4}$$

If the width function falls between $W_{min}$ and $W_{max}$ we assign a terminal branch to fingers, Figure 7 is an example of the resulting functions, built for 10 points.

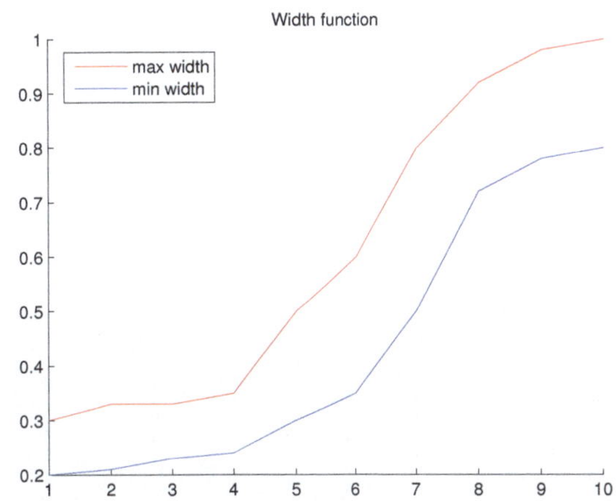


Figure 7: Fingers brunch width function

### 3.4 Gesture classification

Given a database of gestures $G_b$, the set of labels $Y$ and the target function $f^* : G \longrightarrow Y \cup \{\lambda\}$, where $\lambda$ is empty gesture. We need to construct an algorithm for approximation $f^*$.

In the problem of gesture recognition hypothesis of compact is true, so we can use the classification method of nearest neighbor. For each gesture from the database will hold segmentation and generation features will continue to keep the base in the form of features. In the same type of gesture should be the same number of holes and fingers. If the first two features do not match, then this is exactly different gestures, otherwise you need to compare the function of the width of the main chain and the angles between the fingers and the wrist. Find closest gesture from the database, if the proximity is less than a predetermined threshold, then the gesture is associated with the gesture from the database, otherwise $\lambda$ (empty gesture).

Introduce $\rho_W(g_1, g_2)$ and $\rho_A(g_1, g_2)$:

$$\rho_W(g_1, g_2) = \sum_{i=1}^{|g_1.W|} (g_1.W_i - g_2.W_i)^2 \tag{5}$$

$$\rho_A(g_1, g_2) = \frac{\sum_{i=1}^{g_1.N} (g_1.A_i - g_2.A_i)^2}{g_1.N} \tag{6}$$

Than the proposed classification method is a method of nearest neighbor with the two measures $\rho_W(g_1, g_2)$ and $\rho_A(g_1, g_2)$, which are used defferent cases. $\rho_W(g_1, g_2)$ is used when number of fingers is equal to 0 ot 1, $\rho_A(g_1, g_2)$ otherwise.

## 4. EXPERIMENT

### 4.1 Gesture base

To check the quality of the resulting methods we formed the gesture base. There are involved 10 persons. Each was asked to show 27 sign of the ASL dactyl alphabet, each gesture has 10 frame. Thus, the set of gestures consists of a 2700 gestures, which are divided into 27 classes. All gestures were recorded on a webcam and Kinect depth sensor. To check the quality of the segmentation problem solution 25 gestures was manually processed and hand was selected. To assess the classification quality of terminal branches was marked 30 frames, which was attended by 100 terminal branches. Examples of the gestures from base are shown in Figure 8.

Figure 8: Gesture base, sign D

### 4.2 Results

Quality of classifying gesures preblem solution is the number of correctly recognized gestures, divided by the sample size. Quality solution to the basic problem of gesture recognition for the training set is 0.94, 0.93 on control (Table 1). The results show a good generalization performance of the method. Average time recognition of gesture is 0.2 seconds, which allow to solve the problem in real time. As a rule, the process of displaying a single gesture extended in time, so you can use several consecutive frames and classify gesture by voted majority. For a group of 10 gestures, the quality increases to 0.97. In this experiment, a sign was based on one representative per class. Increasing the base will improve the quality of recognition.

| Dataset | Quality | Empty class | Wrong class |
|---|---|---|---|
| train | 0.941 | 0.023 | 0.036 |
| test | 0.929 | 0.024 | 0.047 |

Table 1: Quality of gestures classification

## 5. CONCLUSIONS

In this paper we propose a method for gesture recognition of sign language. For gesture recognition is necessary to set the base. For each gesture is sufficient presence of the one example in the database. Described method is very efficient in terms of execution time and resources requirment. It can recognize up to 5 gestures per second. Base stored as attributes and it saves memory. Thus, the system can be used in real-time. The method consists of several independent algorithms that can be optimized separately. This allows us to find the optimal set of parameters which give us maximum of proposed functional quality. Method proved it's quality on collected gesture base. The method can be used in practice for any of hand gestures for which the condition of compact is true.

## 6. ACKNOWLEDGEMENTS

This work was supported by Russian Foundation for Basic Research, research projects 14-01-00716, 14-07-00965 and 12-07-92695-IND_a.

## REFERENCES

Beristain A., Grana M., 2010. A stable skeletonization for tabletop gesture recognition. Computational Science and Its Applications ICCSA.

Kurakin A., Zhang Z., Liu Z., 2012. A real-time system for dynamic hand gesture recognition with a depth sensor, Proc. of EUSIPCO.

Malik S.,2003. Real-time hand tracking and finger tracking for interaction. CSC2503F Project Report.

Mestetskiy L., and Semenov A., 2008. Binary image skeleton - continuous approach. VISAPP (1), pages 251258. INSTICC - Institute for Systems and Technologies of Information, Control and Communication.

Shapiro L. G., Stockman G. C.,2001. Computer Vision. New Jersey, Prentice-Hall.

Stenger B., Thayananthan A., 2006. Model-based hand tracking using a hierarchical. IEEE Transactions on pattern analysis and machine intlelligence, vol. 28, no. 9.

Suryanarayan P., Subramanian A., Mandalapu D., 2010. Dynamic hand pose recognition using depth data. 20th International Conf. on Pattern Recognition (ICPR).

Wang R. Y., Popovic J., 2009. Real-time hand-tracking with a color glove. ACM Transactions on Graphics. Vol. 28, no. 3.

# 36

# GEOMETRY AND TEXTURE MEASURES FOR INTERACTIVE VIRTUALIZED REALITY INDOOR MODELER

Kalaivani Thangamani[a,*], Ryosuke Ichikari[a], Takashi Okuma[a], Tomoya Ishikawa[b], Takeshi Kurata [a]

[a] SErvice SEnsing, Assimilation, and Modeling Research Group, Human Informatics Research Institute,
National Institute of Advanced Industrial Science and Technology (AIST), 1-1-1 Umezono, Tsukuba 305-8568, Japan -
(thangamani.kalaivani, r.ichikari, takashi-okuma, t.kurata)@aist.go.jp
[b] Site Sensing Inc., Japan - t.ishikawa@site-sensing.com

**KEY WORDS:** Virtualized reality indoor models, Texture distortion, Inpainting, Shape from Texture

**ABSTRACT:**

This paper discusses the algorithm to detect the distorted textures in the virtualized reality indoor models and automatically generate the necessary 3D planes to hold the undistorted textures. Virtualized reality (VR) interactive indoor modeler, our previous contribution enables the user to interactively create their desired indoor VR model from a single 2D image. The interactive modeler uses the projective texture mapping for mapping the textures over the manually created 3D planes. If the user has not created the necessary 3D planes, then the texture that belong to various objects are projected to the available 3D planes, which leads to the presence of distorted textures. In this paper, those distorted textures are detected automatically by the suitable principles from the shape from texture research. The texture distortion features such as the slant, tilt and the curvature parameters are calculated from the 2D image by means of affine transformation measured between the neighboring texture patches within the single image. This kind of affine transform calculation from a single image is useful in the case of deficient multiple view images. The usage of superpixels in clustering the textures corresponding to different objects, reduces the modeling labor cost. A standby database also stores the repeated basic textures that are found in the indoor model, and provides texture choices for the distorted floor, wall and other regions. Finally, this paper documents the prototype implementation and experiments with the automatic 3D plane creation and distortion detection with the above mentioned principles in the virtualized reality indoor environment.

## 1. INTRODUCTION

### 1.1 Virtualized reality indoor modeler

Virtualized Reality (VR) modeling generates photo-realistic 3D models from real-world clues such as photographs from the real scenes. These models enable the viewers to feel the virtual environment from any viewpoint. While virtual reality is simply the artificially created CAD models, the virtualized reality starts with the real scene and virtualizes it (Kanade1995). Generating photo-realistic indoor models from 2D digital photos and overcoming the textural constraints such as the texture distortion are discussed in this paper.

Interactive virtualized reality indoor modeler which was our previous contribution (Ishikawa2011), allows the user to create the VR model from a single 2D photo from the ordinary digital camera.The interactive VR indoor modeling tool starts the process by finding the vanishing point in the 2D image. The vanishing point in the 2D image is used to find the camera parameters and hence fixing the world coordinates for the VR model. The height of the ground, in the world space is interactively set by the user. The planes are added interactively in the world space by the user; Projective texture mapping is used to map the textures over the created 3D planes. The indoor models created from the single photos are integrated to generate the complete indoor environment. Figure 2 shows the local indoor model and the complete VR model for the Ganko restaurant chain, Tokyo.

The VR model faces the occlusion and the distortion which are explained in detail in the coming sections. There are often untextured regions / occluded regions on some of the 3D planes since it is not easy to take a set of photos so that every region of the

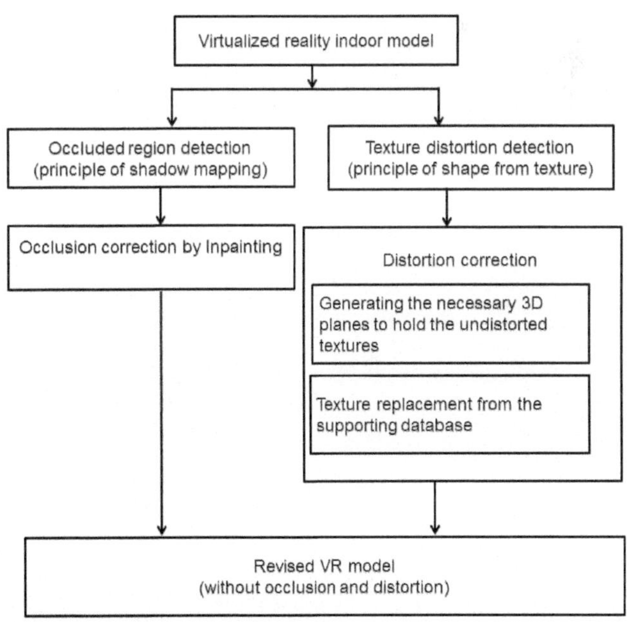


Figure 1: Virtualized reality indoor model and occlusion and distortion handling

*Corresponding author

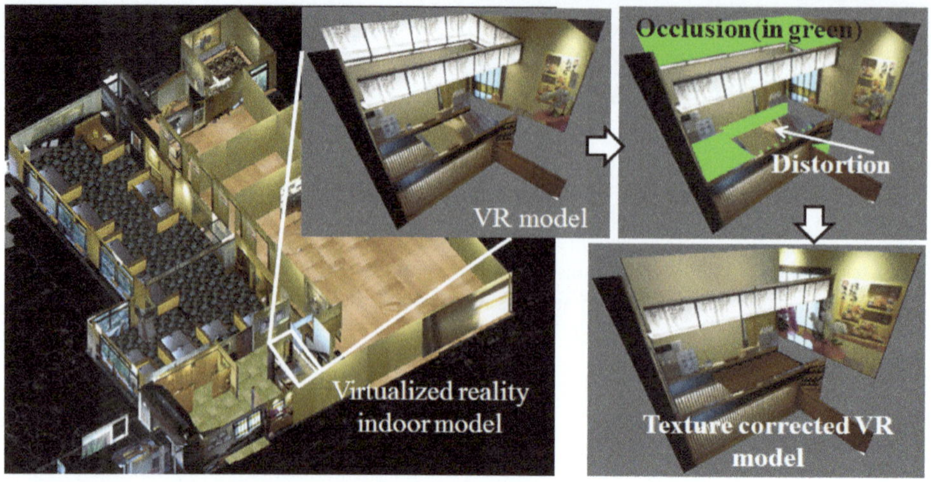


Figure 2: Virtualized reality indoor model and occlusion and distortion handling

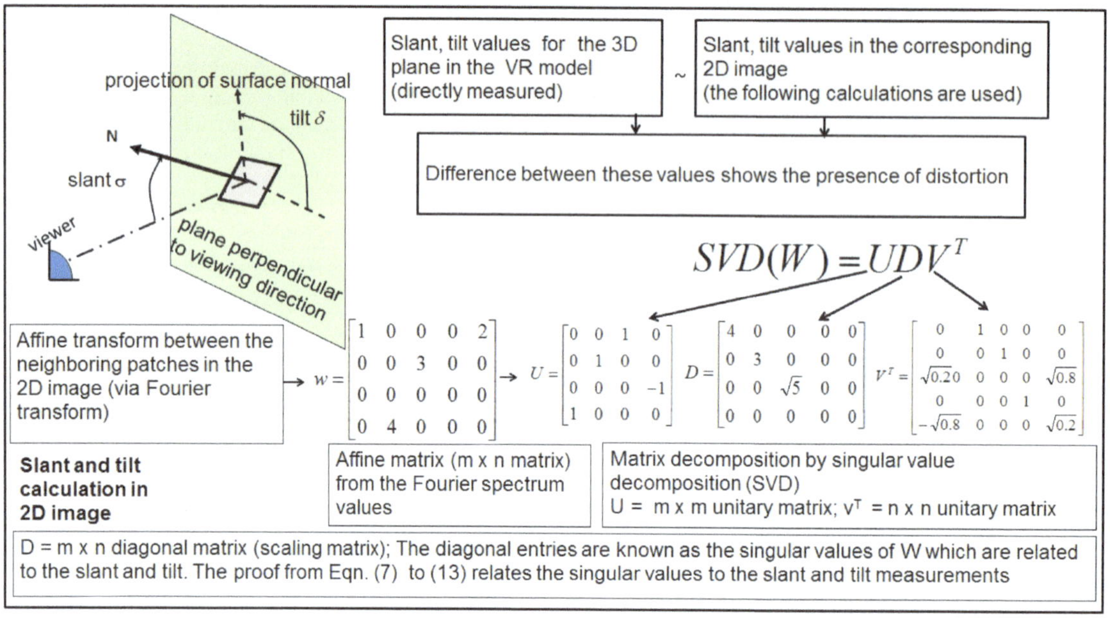


Figure 3: Distortion detection algorithm

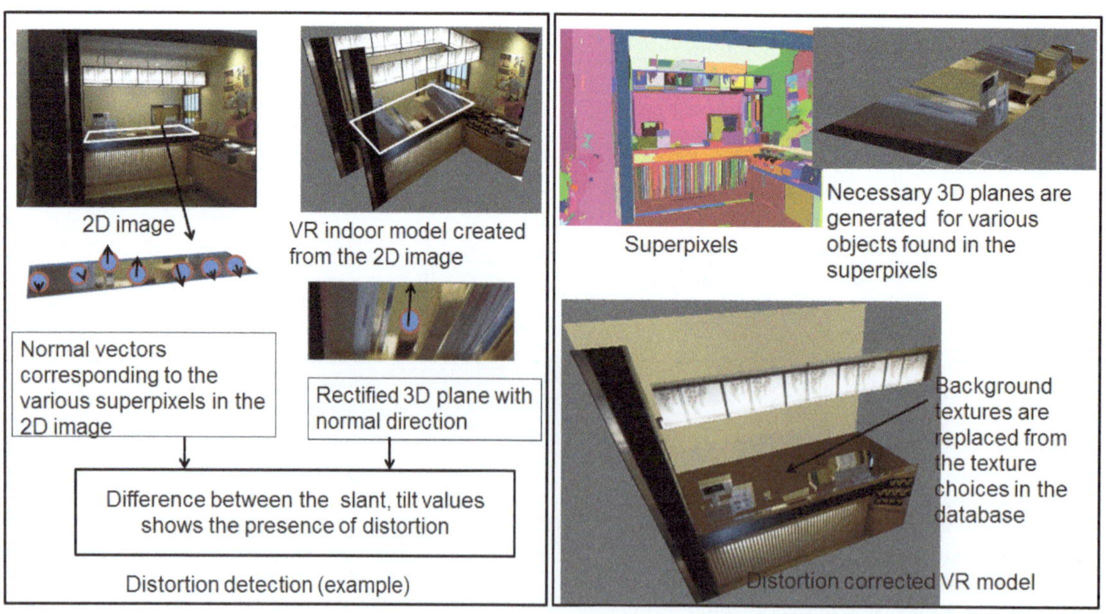


Figure 4: Virtualized reality indoor model and distortion handling

3D model is included at least in one of those photos. The occluded region is shaded in green in the Fig. 2 and these regions are textured with the Inpainting techniques (Thangamani2012). We have successfully tested the Inpainting algorithms to handle the occlusion in the VR indoor models.

The other problem in the VR model is the distorted textures. The users enjoy creating their indoor models but it is tedious for them to create every single object in the 2D photo. The lack of the necessary 3D planes causes the projected textures to get mapped on the available 3D planes. This action leads to the presence of distored textures. Texture distortion are also caused by the image resolution used in the input photo. This paper discusses about the distortion caused by the lack of the necessary planes in the VR indoor model.

### 1.2 Literature review

The problem of texture distortion was referred in some of the 3D modeling works. Texture distortion was found in the works of Hoeim et al.(Hoeim2005), in their developed 3D model, from a single image. They separated the segmented pixel clusters in the single image, in to three categories namely ground, wall and the sky and map the textures to the horizontal, vertical and the sky planes. They maintained a learning system and presented the automatic 3D modeling. Due to the limited 3D planes, the mapped textures faced the problem of distortion.

While photographing the scene, the textures in the real world are rotated, translated, scaled and mapped as the 2D image. To back project the 2D textures once again to the 3D space, needs oneself to travel back the rotation, translation and scaling processes. One should able to decompose the mapped 2D texture matrix to the rotation, translation and scaling matrices and find the distortion parameters. This is the fundamental concept in the shape from texture research. The term "distortion parameters" used by the shape from texture research, refer the basic parameters that hold the relation between the 3D to 2D transformation. The works by Gibson(Gibson1950) is the starting point for this shape from texture, and structure from motion research. Later works by Garding (Garding1992),(Garding1993) shows the theoretical and experimental ways in realizing the structures from the texture. They measure the shape parameters by measuring the texture gradients in the image. Then the work by Malik (Malik1997) travels much closer to the shape from texture and calculate the distortion parameters by measuring the affine transform from the single image. This paper considers the work by Malik (Malik1997) as the basis for distortion detection in our modeling scenario.

Fourier transform was used by (Malik1997) for finding the affine transform between the neighboring patches within the single image. Fourier transform maps the pixel values in the frequency domain. Any raw data observed against time is called as the data in time domain, and for any analysis, the time domain has to be converted into the frequency domain. Frequency domain, states the number of times any particular action occurred; the frequency count. Fourier transform is one such time domain to frequency domain converter. The method of applying Fourier transform to find the affine matrix by (Lucchese1996) between the image pairs holds the good start for the Fourier transform usage. The similar textured region in the neighboring patches in the image, are considered for fixing the neighboring patches in the affine calculation. The idea of superpixels are useful for segmenting the similar pixel clusters. Superpixels (Ren2003) correspond to small, nearly-uniform segmented regions in the image. They are defined to be local, coherent and preserve most of the structure necessary for segmentation at the scale of interest. Automatic pop-up by (Hoiem2005) used superpixels as one of their criterion in categorizing the pixel clusters in their 3D modeling works. The efficient graph based segmentation (Felzenswalb2004) is used for segmenting the image regions into superpixels. This graph based method segments the region based on the intensity, color, motion, location and other local attributes.

## 2. CONTRIBUTION AND OVERVIEW

This paper proposes the texture distortion algorithm derived from the shape from texture principles, for detecting the distorted textures, in our previous VR modeling (Ishikawa2011) system. The flowchart in Fig.1 shows the overall process flow for checking the geometry and texture in the VR indoor model generated from the modeling tool (Ishikawa2011). Once the user designs the VR model, it is analysed for the presence of occlusion and distortion. Occluded textures are found out by the concept of shadow mapping and handled by the Inpainting techniques. Inpainting process is not discussed in this paper since our successful results with Inpainting are discussed in detail in our previous works (Thangamani2012). Distortion is detected by the principles from shape from texture. The distortion features such as the slant, tilt and the curvature parameters define the 3D to 2D geometry and texture transformation. The slant and tilt are measured for the 3D planes in the VR model and the same are compared with their 2D superpixels. The degree of variation in the values, gives the degree of distorted textures. We are not considering the curvature parameters since our modeling system uses only the planar region for the VR modeling. The detected distortion parameters, together with the superpixel, are used for generating the necessary 3D planes for the various objects found in the 2D image. These newly generated planes hold the undistorted textures which are meant for them. The distorted background textures are replaced by the textures in the local database.

Figure 3 shows the definition for the slant and tilt, and explains the distortion detection algorithm in detailed steps. The slant angle $\sigma$ is defined to be the angle between the surface normal $N$ and the viewing direction $p$. Tilt $\delta$ is referred as an angle in positive x axis, during the parallel projection of the surface normal. These parameters are directly measured in the VR model, but certain techniques are to be followed in calculating for the 2D image. The superpixels corresponding to the 3D model plane are subjected to Fourier tranform and the response are used to create the affine matrix. This affine matrix is decomposed by singular value decomposition. The singular values of the affine matrix are related to the slant and tilt parameters. The proofs in the next section explains the relation in detail. Once the distortion parameters are found out, they are used for generating the new 3D planes to hold the undistorted textures. The superpixels are used to define the various objects in the scene and helps in the automatic 3D planar region generation. There is also a supporting texture data base, formed by categorizing the available textures in the VR indoor model, based on their GLCM (Haralick1973) energy levels. This database suggests the possible texture choices for replacing the distorted background textures.

The example in fig. 4, shows the distortion detection and distortion handling in one such VR model. The 3D planes in the VR model is taken one after the other and their slant and tilt are measured (the sketch for explaining the slant and tilt in Fig. 3 shows the clue for the direct measurement). The superpixels corresponding to the rectified plane textures are subjected to the affine matrix calculation. The normal vectors are shown for the 2D region and the rectified 3D plane. The various superpixels in the 2D region hold the normals in different directions. But these textures are mapped to the single plane whose normal is also shown for comparison. The difference between the distortion parameters shows the presence of distortion. The detected slant and tilt values are used for generating the necessary 3D planes to hold the undistorted textures. The distorted background textures are replaced by the textures from the local database.

## 3. SHAPE FROM TEXTURE- REVIEW

There are totally 5 parameters which are said to be the distortion parameters; namely, the slant $\sigma$, the tilt $\delta$ and the 3 curvature parameters (normal curvature along the slant direction, normal curvature along the tilt direction, geodesic torsion). These are the parameters that hold the relation between the 3D and 2D mapping. The slant angle $\sigma$ is defined to be the angle between the surface normal $N$ and the viewing direction $p$. Tilt $\delta$ is referred as an angle in positive x axis, during the parallel projection of the surface normal.

### 3.1 Relationship between the texture distortion map and 3D shape

The following principles and formulae are derived from the works by (Malik1994). The reader is referred to access (Malik1997), (Garding1992), (Garding1993), for the complete proof. The texture distortion in an particular direction on the image plane is measured as an affine transformation between a pair of image patches. The following derivation are written for the spherical projection but the VR modeling uses the planar projection.

The relation between the slant and tilt are discussed as follows. When a smooth surface S is mapped by central projection to a unit sphere $\sum$ centered at a focal point, the back projection map $F$ from $\sum$ to $S$ is defined as $F(p) = r(p) = r(p)p$ where $p$ is a unit vector from the focal point to a point on the image sphere, and $r(p)$ is the distance along the visual ray from the focal point through $p$ to the corresponding point $r = F(p)$ on the surface $S$.

The slant angle $\sigma$ is defined to be the angle between the surface normal $N$ and the viewing direction $p$, so that $cos\sigma = N.p$.

$F * (p)$ (differential of $F(p)$) can be expressed as

$$F * (p) = \begin{bmatrix} r/cos\sigma & 0 \\ 0 & r \end{bmatrix} = \begin{bmatrix} 1/m_p & 0 \\ 0 & 1/M_p \end{bmatrix} \quad (1)$$

where $r$ is the distance to the object from the center of the viewing sphere, and $\sigma$ is the slant angle. We see that $m_p = cos\sigma/r$ is the scaling of the texture pattern in the minor axis i.e., in the tilt direction and $M_p = 1/r$ is the scaling in the major axis i.e., in the orthogonal direction. The shape of the surface is captured in the shape operator, which measures how the surface normal $N$ changes as one moves in various directions in the tangent space of the surface.

The shape operator, in the $(T, B)$ basis as

$$\begin{bmatrix} -\nabla_T N \\ -\nabla_B N \end{bmatrix} (p) = \begin{bmatrix} k_t & \tau \\ \tau & k_b \end{bmatrix} \begin{bmatrix} T \\ B \end{bmatrix} \quad (2)$$

where $k_t$ is the normal curvature in the $T(tilt)$ direction, $k_b$ the normal curvature in the $B(slant)$ direction and $\tau$ is the geodesic torsion.

Our aim is to find the matrix $A$, which represents the affine transformation between the spherically projected texture between the two nearby pixel points. This matrix will be a function of the local parameters and shape parameters. The orientation parameters are $\sigma$, the slant of the surface and $t$, the direction of tilt of the surface. The shape parameters are $rk_t, rk_b$, and $r_\tau$. The variable $r$ is the distance from the center of the viewing sphere to the given point on the surface.

Malik et al., derives the $A$ as follows,

$$A = Rot(\delta_t). \begin{bmatrix} k_m cos\delta_T & k_m sin\delta_T cos\sigma \\ k_M sin\delta_T / cos\sigma & k_M cos\delta_T \end{bmatrix} \quad (3)$$

where

$$k_m = 1 + \nabla m/m \circ \begin{bmatrix} \nabla t \\ \nabla b \end{bmatrix} \quad (4)$$

$$k_M = 1 + \nabla M/M \circ \begin{bmatrix} \nabla t \\ \nabla b \end{bmatrix} \quad (5)$$

We find the affine transformation between the neighboring patches in the superpixels. The affine transformation will depend on the direction and magnitude of the vector displacement between the two patches in the superpixel. Affine transforms are estimated in a number of directions to recover the surface orientation and shape parameters. Shape recovery via affine transformation is the useful suggestion by Malik et al., (Malik1997) as it successfully overcomes the traditional formulation of applying texture gradients to find the shape parameters. The actual affine transformation will not be in terms of the $(t, b)$ basis. Instead, the affine matrix is related as, $\hat{A} = UAU^{-1}$. Here comes the necessity of the Singular Value Decomposition (SVD) to realize the individual matrices. The rotation matrix, $U$, is given by

$$Rot(\theta_t) = \begin{bmatrix} cos\theta_t & -sin\theta_t \\ sin\theta_t & cos\theta_t \end{bmatrix} \quad (6)$$

where $\theta_t$ is the tilt angle at the point of interest.

### 3.2 Shape recovery by SVD

There are five unknowns, the slant $\sigma$, the tilt direction $\theta_t$ and the three shape parameters $(rk_t, rk_b, r_T)$ . Each estimation of an affine transform in an image direction yields four non-linear equations. Two directions are sufficient for solving the equations and finding the distortion parameters. Malik et al.,(Malik1997) proposed two shape recovery algorithms. The first is the linear algorithm based on singular value decomposition (SVD) of the affine matrices. The second algorithm is based on the least squares formulation of an error criterion.

Singular value decomposition (SVD) is adapted for this work. The simple logic of the SVD shows that the 2D image is the combination of the translation, rotation and scaling. If the resultant matrix is decomposed, then the diagonal matrix represents the scaling in the vertical and the horizontal axis (minor and major axes). These singular values can be processed further to find the slant and the tilt angle for the particular patch of interest. If the patches in the superpixels are subjected for their slant and tilt test, the orientation of the corresponding objects are found out.

The following derivation part relates the singular values to the minor and major scaling axes. Malik et al., justify this by noting that the singular values $S_1$, $S_2$ of matrix $\hat{A}_i$ are related to the eigen values $\lambda_1$, $\lambda_2$ of the matrix $\hat{A}_i^T \hat{A}_i$ by the relationship $\lambda_1 = S_1^2$ and $\lambda_2 = S_2^2$. Then expressions are computed for trace($\hat{A}_i^T \hat{A}_i$) = trace($A_i^T A_i$) and $det(\hat{A}_i^T \hat{A}_i) = det(A_i^T A_i)$. Dropping the subscripts and the superscripts $i$ for a better view,

$$S_1^2 + S_2^2 = trace(A^T A) \tag{7}$$

$$= k_m^2 cos^2\delta_T + k_M^2 sin^2\delta_T/cos^2\sigma + k_m^2 sin^2\delta_T cos^2\sigma + k_M^2 cos^2\delta_T \tag{8}$$

*and*

$$s_1^2 s_2^2 = det(A_i^T A_i) = k_m^2 k_M^2 \tag{9}$$

$$(s_1 + s_2)^2 = trace(A^T A) + 2\sqrt{det(A^T A)} \tag{10}$$

$$= k_m^2 cos^2\delta_T + k_M^2 sin^2\delta_T/cos^2\sigma + k_m^2 sin^2\delta_T cos^2\sigma + k_M^2 cos^2\delta_T + 2k_m k_M(cos^2\delta_T + sin^2\delta_T) \tag{11}$$

$$= (k_m + k_M)^2 - k_m^2 sin^2\delta_T(1 - cos^2\sigma) + k_M^2 sin^2\delta_T((1/cos^2\sigma) - 1) \tag{12}$$

$$= (k_m + k_M)^2 - (k_m sin\delta_T sin\sigma)^2 + (k_M sin\delta_T tan\sigma)^2 \tag{13}$$

This suggests that we can approximate $s_1 + s_2 \approx k_m + k_M$. So we can estimate $k_m$ and $k_M$ as the singular values of $\bar{A}$. Then the slant and tilt direction are found from the $k_m$ and $k_M$ values. This assumption is true for any planar surface and for the curved surfaces.

## 4. ADAPTING THE SHAPE FROM TEXTURE PRINCIPLE IN THE VR MODELING

The proofs stated in the previous sections are aimed at the spherical projection but our VR modeler uses the planar projection. The conversion between these two are necessary for realizing the application.

The 2D image is segmented into superpixels and they are subjected to the affine transformation. The size of the superpixel should be reasonable so that it could afford for the least patch size, set up for the Fourier transform process. The VR modeling system uses its own technique (Ishikawa2011) to fix the orientation for the 3D plane generation. These techniques are collaborated with the distortion calculation measures, so that the calculated slant and the tilt values after the SVD process can be used for deciding the orientation of the 3D plane. The modeling tool suggests the user to click over the object boundary to create a new plane, but these steps are made automatic by upgrading the modeling system to use the superpixel boundary for the creation of the new plane.

## 5. TEXTURE DATABASE

In the indoor VR model, there are many repeated textures which are meant for the floor, wall and other objects. The undistorted texture samples can be stored and used for the places of need. Designing a texture database to maintain the textures in the entire model is really useful. Kholgade et al. (Kholgade2014) uses the textures from the internet database to fill the occluded textures in their 3D model. We have already introduced the GLCM (Haralick1970) based texture categorization for fast Inpainting process (Thangamani2012) in the VR model. The same can be extended to categorize and sort the repeated textures in the VR model. The texture patches are classified and stored in the hash table for quick search during the distortion handling processes. The hash function relates the incoming texture patch to the address of the particular hash table bin. During the query step, the necessary patch is retrieved by a single/ minimum search. The texture information is adequately specified by a set of gray tone spatial dependence matrices otherwise known as Gray level co-occurrence matrix (Haralick1970) which are computed for various angular relationships ( 0°, 45°, 90° or 135°) and distances between neighbouring resolution pairs on the image. More details can be found in (Thangamani2012) for categorizing the texture patches and storing in the hash table.

## 6. RESULTS AND DISCUSSION

This paper detects the texture distortion and removes the distorted textures by generating the necessary planar regions so that every plane holds the undistorted texture. The principles from shape from texture, usage of superpixels together with the features of the modeling tool enabled the automatic 3D plane generation. The database formed from the available textures are also referred for good choices for the floor and the wall textures. The proposed prototype system is tested in the virtualized reality model and the preliminary results are stated in the table 1. Qualitative analysis in table 1, compares the VR local model before and after texture treatment (occlusion and distortion detection). The individual 3D planes in the VR model are shown separately. The planes that are occluded are inpainted and the distorted ones are replaced by the similar textures in the groups. In some cases, the occluded textures are also replaced by the similar textures in the groups. The textures which do not need any change are preserved as such for the final model. There are also shortcomings, found in our way of calculating the affine transformation from a single image, due to the deficient patch size in some of the superpixels. Our algorithm should be extended and optimized to handle the variety of input targets and modeling conditions.

## 7. CONCLUSION AND FUTURE WORKS

This paper proposed the algorithm for detecting the distorted textures in the 3D planes of the virtualized reality indoor model. Out of the five distortion parameters, only the slant and tilt parameters are experimented for automatic creation of the planar regions in our VR tool. Extending the trials to use all the five distortion parameters to design the non-planar regions and mesh surfaces would be of our future work. In that case, we need to adapt the suitable texture mapping method; the simple projective texture mapping is not enough to handle the regions that are out of the view zone. The techniques such as the PPTM (Parallel Projective Texture Mapping) are of good choice for projecting the textures over the non-planar regions.

Currently, our database is designed to hold the textures from the indoor model and these textures are used as the choices for replacing the distorted ones. This database can be extended to use the texture reference from the internet such as the SNS (Social Networking Services) textures. The textures from the internet can also be categorized by the GLCM filters and stored in the hash bins for our texture reference.

Table 1: Qualitative analysis

| 2D image used | VR local model (before texture treatment) | Depth map | VR local model with occluded (in green) and distorted region | Rectified planes in the local model | | | VR local model (after texture treatment) |
|---|---|---|---|---|---|---|---|
| | | | | Before and after inpainting | Before and after distortion handling | That did not undergo any texture correction | |
|  |  |  |  |  |  |  |  |
|  |  |  |  |  |  | All the textures underwent texture correction |  |
|  |  | 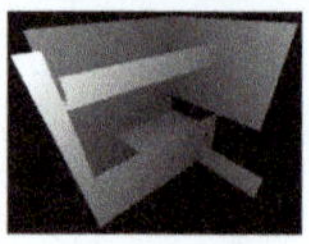 | 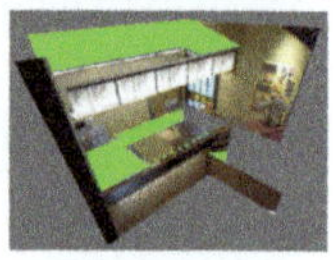 | 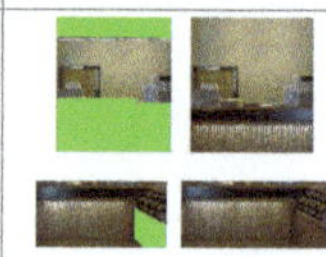 |  |  |  |
|  |  |  |  | 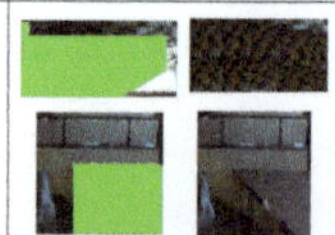 | 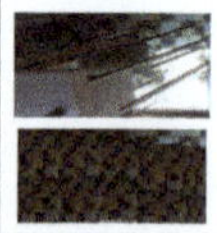 |  |  |
|  |  | 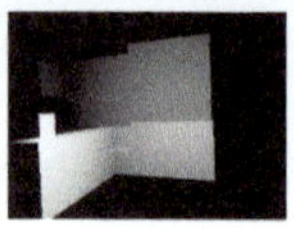 |  | 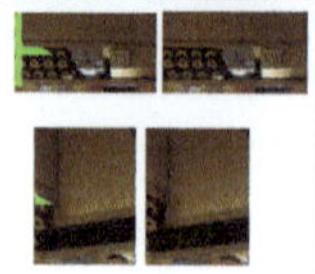 |  | All the textures underwent texture correction |  |


## ACKNOWLEDGEMENTS

This research is partly supported by Grant-in-Aid for Young Scientists (B) No.26870905.


## REFERENCES


Kanade, T., 1995. Virtualized Reality: Concepts and Early Results. *IEEE Workshop on the Representatives of Visual Scenes.*

Ishikawa, T., Thangamani, K., Kourogi, M., Gee, A.P., Mayol-Cuevas, W., Hyun, J., Kurata, T., 2011. Interactive 3-D indoor modeler for virtualizing service fields. *Int. J. Virtual Reality, SI: Mixed and Augmented Reality*, 19, ISSN. 1359-4338, pp. 1-21.

Thangamani, K., Ishikawa, T., Makita, K., Kurata, T., 2012. Fast and texture-structure preservable inpainting and its application for creating virtualized-reality indoor models. *Int. J. Computer science and informatics*, 1(3), pp. 85-101.

Hoiem, D., Fros, A.A., Hebert, M., 2005. Automatic photo pop-up. *ACM Transactions on Graphics*, 24(3), pp. 577-584.

Gibson, J., 1950. The perception of the visual world. *Boston: Houghton Mifflin*

Garding, J., 1992. Shape from texture for smooth curved surfaces in perspective projection. *J. of mathematical imaging and vision*, 2(4), pp. 327-350.

Garding, J., 1993. Shape from texture and contour by weak isotrophy. *J. of Artificial intelligence*, 64(2), pp. 243-297.

Malik, J., and Rosenholtz, R., 1997. Computing local surface orientation and shape from texture for curved surfaces. *International journal of computer vision*, 23(2), pp. 149-168.

Lucchese, L., 1996. Estimation of affine transformations between images pairs via Fourier transform. *Proceedings of International conference on image processing*, 3, pp. 715-718.

Ren, X., and Malik, J., 2003. Learning a classification model for segmentation. *Proceedings of the 9th International conference on computer vision*, 1, pp. 10-17.

Felzenswalb, P.F., and Huttenlocher, D.P., 2004. Efficient graph-based image segmentation. *International Journal of computer vision*, 59(2), pp. 167-181.

Haralick, R.M., 1973. Textural features for image classification. *IEEE T. Systems, Man and cybernetics*, 3(6), pp. 610-621.

Kholgade, N., 2014. 3D object manipulation in a single photograph using stock 3D models. *SIGGRAPH 2014.*

# Practical implementation of semi-automated as-built BIM creation for complex indoor environments

Sanghyun Yoon[a], Jaehoon Jung[a], Joon Heo[a]

[a] School of Civil and Environmental Engineering, Yonsei University, 134 Shinchon-dong, Seodaemun-gu, Seoul 120-749, Republic of Korea

**Commission/WG**

**KEY WORDS:** As-built BIM, Laser scanner, Point clouds, RANSAC

**ABSTRACT:**

In recent days, for efficient management and operation of existing buildings, the importance of as-built BIM is emphasized in AEC/FM domain. However, fully automated as-built BIM creation is a tough issue since newly-constructed buildings are becoming more complex. To manage this problem, our research group has developed a semi-automated approach, focusing on productive 3D as-built BIM creation for complex indoor environments. In order to test its feasibility for a variety of complex indoor environments, we applied the developed approach to model the 'Charlotte stairs' in Lotte World Mall, Korea. The approach includes 4 main phases: data acquisition, data pre-processing, geometric drawing, and as-built BIM creation. In the data acquisition phase, due to its complex structure, we moved the scanner location several times to obtain the entire point clouds of the test site. After which, data pre-processing phase entailing point-cloud registration, noise removal, and coordinate transformation was followed. The 3D geometric drawing was created using the RANSAC-based plane detection and boundary tracing methods. Finally, in order to create a semantically-rich BIM, the geometric drawing was imported into the commercial BIM software. The final as-built BIM confirmed that the feasibility of the proposed approach in the complex indoor environment.

## 1. INTRODUCTION

Digital representation of physical and functional characteristics of a facility, known as Building Information Models (BIMs), is now widely used in the Architecture, Construction, Engineering, and Facility Management (AEC/FM) domain due to its potential strength of supporting the planning stage and modelling of building life cycle (Gu and London, 2010; Hichri et al., 2013). Along with the proliferation of BIM, there is an increasing demand of constructing a BIM for actually built or currently existing buildings. Having an accurate as-built model of the existing structure can significantly improve the facility management process (Woo et al., 2010).

For as-built BIM creation, laser scanners play a crucial role: they can rapidly and accurately reconstruct the 3D scene of the surrounding environments represented by so-called 'point cloud' data (Canciani et al., 2013). However, the process of as-built BIM creation is labor-intensive and error-prone operations since it is largely a manual process (Anil et al., 2011). Automatic extraction of as-built BIM from point clouds has been studied by many researchers, but fully-automated method has not yet been developed due to complex indoor structure (Huber et al., 2011; Jung et al., 2014; Valero et al., 2012a; Valero et al., 2012b; Xiong et al., 2013).

Alternatively, our research group has focused on improving the performance and productivity of the as-built BIM creation by semi-automatic approach (Jung et al., 2014). In this paper, we aimed to verify the feasibility of the developed approach by implementing it to the complex indoor environment, Charlotte stairs in Lotte World Mall, Korea. The following sections explain detailed steps of methodology in consecutive order. General overview of the process is depicted in Fig.1. The process is divided into four main steps: (1) data acquisition; (2) data pre-processing; (3) data processing; (4) BIM creation.

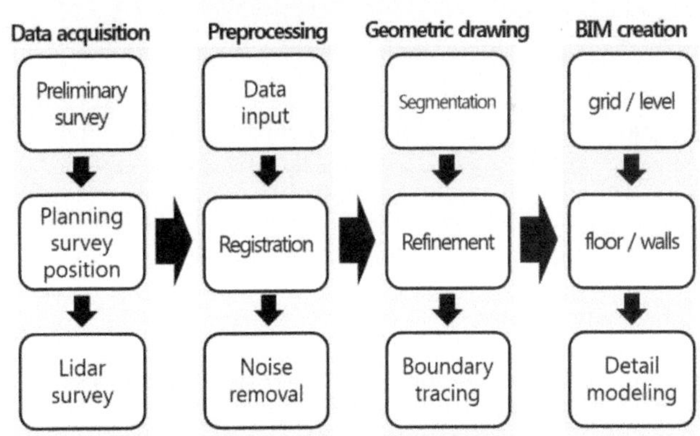


Figure 1. General overview of the process

## 2. PRACTICAL IMPLEMENTATION

### 2.1 Data acquisition

The test site, Avenuel Atrium includes a huge void space, from basement level to 8th floor, and two complex spiral structures, Charlotte stairs (Fig. 2). The stairs are connecting the 1st and 2nd floors, crossing each other. For the modelling of Charlotte stairs, we collected the point-cloud data from basement level to 2nd floor of the void space. Due to the occlusions by clutter such as fences, columns, and decorations, we needed to move the scanner locations several times. In this study, FARO Focus 3D, which is compact and simple to use for indoor laser scanning, was utilized. To scan the entire Charlotte stairs, the total number of 11 scanning was conducted in different locations: five times for the 1st floor; four times for the 2nd floor where the stairs are connected; twice for the basement. Fig. 3 shows the example of locations on the floor plan (1st floor) where the data was acquired in.

Figure 2. Charlotte stairs

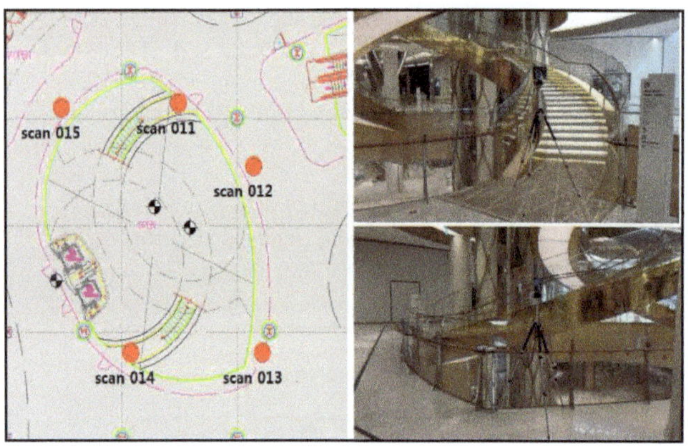


Figure 3. Scan position in 1st floor and picture of position number 11, 13

## 2.2 Data pre-processing

3D point clouds acquired from the previous step are required to go through a pre-processing phase before being used for the generation of geometric drawing. In this phase, the sets of scanned points are combined into a single point cloud in a common coordinate system. Also, the point cloud needs to be filtered for the removal of unwanted data (Tang et al., 2010).

**2.2.1 Registration**: The point clouds acquired from different locations are respectively represented by local coordinate system. Therefore, each local coordinate system should be unified as a common coordinate system through a process known as registration (Tang et al., 2010). Since the test site suffers from many occlusions and specialized target positioning because of the void space, the manual registration was conducted instead of fully automated methods (Gressin et al., 2013; Huber and Hebert, 2003; Makadia et al., 2006; Xie et al., 2010). For registration commercial software, Autodesk Recap Pro which provides target-less registration, was used. The scanned point-cloud data was imported to the software and the locations of the particular points that are clearly recognizable in each scan data, such as a corner or distinguishable spot of the environment was identified. After that, using target-less registration function, we could merge 11 scanned point-cloud data into a common coordinate system.

**2.2.2 Noise removal**: During the scanning process, noisy data caused by moving objects, decorations and laser reflections diffused onto glass surface could be included in the point clouds. Therefore, the noise removal is prerequisite to provide the proper input for as-built BIM creation. In this study, the registered point-cloud data was imported to Leica Cyclone 8.0, and the noisy data was manually removed.

**2.2.3 Coordinate transformation**: The filtered point-cloud data needs to be transformed to the World Coordinate System (WCS) for the purpose of its future application with GIS or display on the map. We surveyed 18 control points in the test site using a total station. The surveying was based on the well-distributed and clearly identifiable points such as corners of walls and stairs. After which, Helmert 3D transformation was conducted to transform the coordinate system from local to world.

## 2.3 Generation of 3D geometric drawings

The registered and filtered point clouds still includes a huge number of point data, which later lead to slow-down or system failure during the modelling process. To facilitate the use of point-cloud data in creating as-built BIM, such simplification or reduction of point data is prerequisite. This goal can be achieved by generation of 3D geometric drawing, proposed in our preceding studies (Heo et al., 2013; Jung et al., 2014). Geometric drawing is composed of three sub-procedures: (1) segmentation; (2) refinement; and (3) boundary tracing. The following sections briefly explain each process .

**2.3.1 Segmentation:** The segmentation process is based on RANSAC algorithm (Okorn et al., 2010) which can extract inliers from a set of observations containing abundant outliers. Throughout RANSAC algorithm point-cloud data which have the identical plane model parameter are divide to each plane.

**2.3.2 Refinement**: This process begins with defining the size of Refinement Grid Cell (RGC) as the cell size of binary image. After the size of RGC is defined, the points which are included in the segmented plane are projected onto a 2D binary image. From this binary image the labelling process is performed to extract the plane of interest and remove the others that are considered to be noise.

**2.3.3 Boundary tracing**: Boundary tracing process requires a parameter which is called Tracing Grid Cell (TGC). From the binary image that only includes the plane of interest, the boundary pixels are searched respectively and sequentially. For the hollow parts such as windows and open doors, the inverse binary image is created to search the boundary pixels with the same procedure. Finally, traced boundaries are inversely returned to their original coordinates.

**2.3.4 3D geometric drawing**: Fig. 4-a shows the Charlotte stairs in the test site, and Fig. 4-b shows the created 3D geometric drawing of the structure. The boundaries that traced out from the segmented planes, are shown as white lines and un-segmented points by the RANSAC segmentation are represented as red points. Using both the traced lines and the remaining points, the structure can be intuitively accessible. In addition, the data sizes of the original point clouds were significantly reduced from 881 MB to 8 MB. This reduction allows to efficiently manipulate the 3D model in the BIM software without slow-down or system failure.

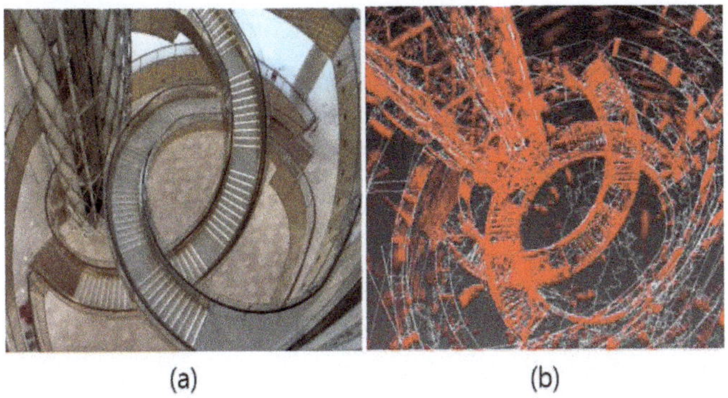


Figure 4. Detail components of 3D geometric drawings

## 2.4 As-built BIM creation

Typically, as-built BIM creation involves three aspects: (1) modelling geometry of the components; (2) assigning an object category and material properties to a component; and (3) establishing relationships between components (Tang et al., 2010). However, in this work, the material properties of each structures were assumed since geometric modelling is the main focus of the present work. The geometric drawing and the remaining points of the test site was exported as CAD-oriented DXF format and used as an input for a commercial BIM software Autodesk Revit 2014. As-built BIM production in Revit can be broadly divided into three steps: 1) grid and level creation; 2) walls and floor creation; and 3) detail modelling. For the grid-level creation step, the height levels of the structure should be defined. For the test site, 6 height levels were defined according to the specific structures such as the floors, ceilings, and stairs. The building boundaries were created according to the geometric drawing, and the gridlines were then traced. Fig. 5-a shows the defined height levels and gridline developed in the modelling. Following the grid and level creation, the walls and floor creation proceeded as shown in Fig. 5-b. Subsequently, as shown in Fig. 5-c, the modelling of the details, such as stairs and fences were conducted. The final result of the as-built BIM is depicted in Fig. 6.

# 3. CONCLUSIONS

In this study, the developed semi-automated as-built BIM approach has shown to be useful in creating as-built BIM model for complex indoor environment. In order to obtain the entire point-cloud data from a highly cluttered and occluded test area, we moved the laser scanner 11 times, and merged them using the target-less registration method. In addition, by using the total station measurements, we transformed the point-cloud data set into the world coordinate system for its future application with GIS or display on the map. The proposed approach also provided a geometric drawing, including detail-rich components while reducing the data size, which allows efficient modelling in BIM software.

However, the current RANSAC-based segmentation process showed some limitations in segmenting highly non-linear structures such as curved fences and circular columns. Given the growing trend of irregularly-shaped buildings, techniques for segmentation of such non-linear structures should be investigated in ongoing work. In addition, prior to BIM creation, noise removal for scanning data was performed manually by the modeler, which is very time-consuming and tedious work. This process therefore will require automation or some other techniques of improving productivity.

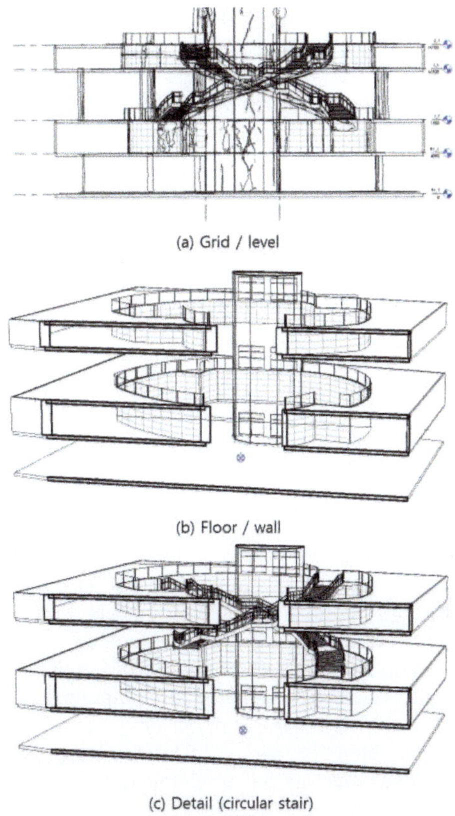


Figure 5. As-built BIM creation steps

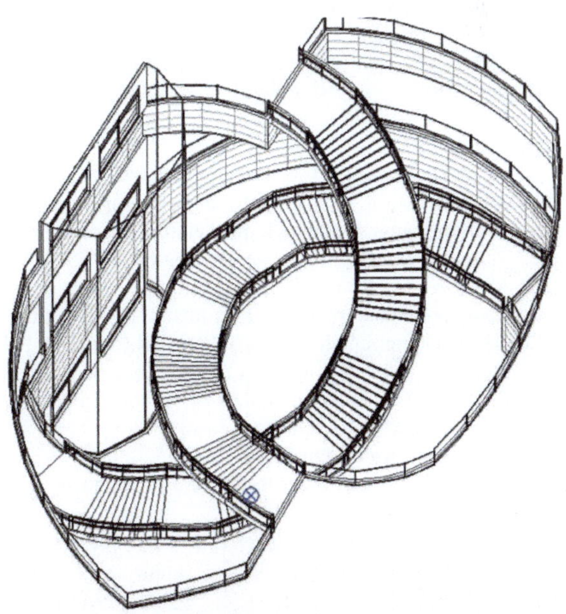

Figure 6. Completed model of as-built BIM

## REFERENCE

**References from Journals**:

Anil, E.B., Tang, P., Akinci, B., Huber, D., 2011. Assessment of the quality of as-is building information models generated from point clouds using deviation analysis, IS&T/SPIE Electronic Imaging. International Society for Optics and Photonics, pp. 78640F-78640F-78613.

Canciani, M., Falcolini, C., Saccone, M., Spadafora, G., 2013. From point clouds to architectural models: algorithms for shape reconstruction. International Archives of the Photogrammetry, Remote Sensing and Spatial Information Sciences 40, 25-26.

Gressin, A., Mallet, C., Demantké, J., David, N., 2013. Towards 3D lidar point cloud registration improvement using optimal neighborhood knowledge. ISPRS Journal of Photogrammetry and Remote Sensing 79, 240-251.

Gu, N., London, K., 2010. Understanding and facilitating BIM adoption in the AEC industry. Automation in construction 19, 988-999.

Heo, J., Jeong, S., Park, H.-K., Jung, J., Han, S., Hong, S., Sohn, H.-G., 2013. Productive high-complexity 3D city modeling with point clouds collected from terrestrial LiDAR. Computers, Environment and Urban Systems 41, 26-38.

Hichri, N., Stefani, C., De Luca, L., Veron, P., 2013. Review of the «as-built BIM» approaches. Int. Arch. Photogramm. Remote Sens. Spat. Inf. Sci. 40, 107-112.

Huber, D., Akinci, B., Oliver, A.A., Anil, E., Okorn, B.E., Xiong, X., 2011. Methods for automatically modeling and representing as-built building information models, Proceedings of the NSF CMMI Research Innovation Conference.

Huber, D.F., Hebert, M., 2003. Fully automatic registration of multiple 3D data sets. Image and Vision Computing 21, 637-650.

Jung, J., Hong, S., Jeong, S., Kim, S., Cho, H., Hong, S., Heo, J., 2014. Productive modeling for development of as-built BIM of existing indoor structures. Automation in Construction 42, 68-77.

Makadia, A., Patterson, A., Daniilidis, K., 2006. Fully automatic registration of 3D point clouds, Computer Vision and Pattern Recognition, 2006 IEEE Computer Society Conference on. IEEE, pp. 1297-1304.

Okorn, B., Xiong, X., Akinci, B., Huber, D., 2010. Toward automated modeling of floor plans, Proceedings of the Symposium on 3D Data Processing, Visualization and Transmission.

Tang, P., Huber, D., Akinci, B., Lipman, R., Lytle, A., 2010. Automatic reconstruction of as-built building information models from laser-scanned point clouds: A review of related techniques. Automation in construction 19, 829-843.

Valero, E., Adán, A., Cerrada, C., 2012a. Automatic method for building indoor boundary models from dense point clouds collected by laser scanners. Sensors 12, 16099-16115.

Valero, E., Adan, A., Cerrada, C., 2012b. Automatic construction of 3D basic-semantic models of inhabited interiors using laser scanners and RFID sensors. Sensors 12, 5705-5724.

Woo, J., Wilsmann, J., Kang, D., 2010. Use of as-built building information modeling, Construction Research Congress, pp. 538-547.

Xie, Z., Xu, S., Li, X., 2010. A high-accuracy method for fine registration of overlapping point clouds. Image and Vision Computing 28, 563-570.

Xiong, X., Adan, A., Akinci, B., Huber, D., 2013. Automatic creation of semantically rich 3D building models from laser scanner data. Automation in Construction 31, 325-337.

**Acknowledgements**

This research was supported by a grant (11 High-tech G11) from the Architecture & Urban Development Research Program funded by the Korean Ministry of Land, Infrastructure and Transport.

# RESEARCH ON RICE ACREAGE ESTIMATION IN FRAGMENTED AREA BASED ON DECOMPOSITION OF MIXED PIXELS

Huanxue Zhang[a], Qiangzi Li[a,*], Fan Lei[b], Xin Du[a], Jide Wei[b]

[a] Institute of Remote Sensing and Digital Earth, Chinese Academy of Sciences, Datun Road, Chaoyang District, Beijing 100101, China –zhanghx@radi.ac.cn; liqz@radi.ac.cn duxin@radi.ac.cn

[b] Hunan Province Land and Resources Planning Institute, Furong Road, Yuhua District, Changsha 410007, China – weijide2006@163.com

**KEY WORDS:** Rice, Acreage estimation, Mixed-pixels, Pure-pixels, Accuracy, Remote sensing

**ABSTRACT:**

Rice acreage estimation is a key aspect to guarantee food security and also important to support government agricultural subsidy system. In this paper, we explored a sophisticated method to improve rice estimation accuracy at county scale and we developed our approach with China Environment Satellite HJ-1A/B data in Hunan Province, a fragmented area with complex rice cropping patterns. Our approach improved the estimation accuracy by combing supervised and unsupervised classification upon decomposition of mixed pixels model, and the rice estimation results, validated by ground survey data, showed a close relationship (RMSE≈3.40) with survey figures, the estimated accuracy (EA) reached 83.74% at county level according to the sub-pixel method, and the accuracy can be increased about 12% compared to the pure-pixel method. The results suggest that decomposition of mixed pixels method has great significance to the improvement of rice acreage estimation accuracy, and can be used in mountainous and broken planting area.

## 1. INTRODUCTION

Paddy rice is one of the most important primary crops globally as the foundation of the economy of many countries, accounting for more than 11% of the global cropland area (Maclean, 2002). China is one of the biggest rice cultivation countries. Paddy rice accounts for more than 42% of the crop yield in China and its cultivation is strongly related to food security and economic sustainable development for China (Chen, 2011). So the Chinese government establishes the agricultural subsidy system, in this way, farmers can apply for an additional income subsidy, whose number is based on the type and acreage of the crops they grow. In the past, legal planting area data is obtained by the report way of the local Bureau of statistics (Han, 2007). It is estimated that about 5 billion is paid wrongly because the report applications were filled out incorrectly. For example, with the decrease of rice growing price, the farmers' planting aspiration is lower, more and more double-season rice was planted to single-season rice in southern China, but to get an income subsidy, the farmers may list the crop type of a field as double-season rice, which is heavily subsidized, while in fact single-season rice, which is not subsidized, is grown. To reduce a huge amount of money spent on incorrect applications, a high accuracy estimation of paddy rice acreage becomes a key link for China.

More recently, remote sensing has been widely used in large scale crop acreage estimation because it has wide swath and can provide effective and timely spatial and temporal information on crop growth conditions (Yang, 2007, Chen, 2007, Okamoto, 1999, Koppe, 2013). Multi-temporal low resolution images and NDVI Time series dataset have been used to identification various crop types by extracting crop phenological information (Peng, 2011, Xiao, 2006&2005&2002, Wang, 2010, Fang, 1998). Methods based on these images could cover large area and can be got at no cost but cannot provide accurate estimation due to mixture pixels (Toan, 1997). Moreover, multi-temporal and multi-polarize SAR data has been used to reflect the changes of rice crop height or roughness from others vegetation (Koppe, 2013). Most of these satellite-based rice maps could provide reliable estimation results, however these methods and images were constrained by narrow swath and low temporal resolution (low repeat cycles), limited spatial coverage at regional scale and are often costly.

The objective of this research is to explore a new method to improve the area estimation accuracy at county level and provide a technological support for the county grain subsidy. Previous research got the rice identification accuracy of 69% using multi-temporal HJ-1A/B data in the western part of Guangdong province (Chen, 2011). Although the result can be adopted at regional level, it can't meet the accuracy requirement to support the agricultural subsidy system, which needs at least 80% at county level. In the study, we compared the rice mapping using traditional hard classification method and decomposition of mixed pixels method based on HJ-1A/B data, and then contrasted the rice acreage estimation results obtained by calculating non-arable coefficient and cloud proportion. The result indicates that the acreage estimation accuracy can achieve 83.74% and it can be improved 12% according to sub-pixel decomposition method, and this method can meet the accuracy requirement at county level.

## 2. STUDY AREA

Hunan Province was selected as the case study area, as rice has been planted for more than thousands of years here and it is the major rice-producing area in China as well as one of the key bases of rice monitoring by the Agriculture Ministry of China. Hunan province is situated in the southern-central part of China (Lat. 24$^0$38'-30$^0$08' N, Lon.108$^0$47'-114$^0$15' E) and is made up of 14 cities and 122 counties. The province belongs to eastern Asia monsoon region with an annual average temperature of 17$^0$C and annual precipitation of 1450 mm. The climate of this

* Corresponding author

region is a continental subtropical monsoon humid climate. The province has a total area of 211,800 $km^2$, and cropland area of 87,949 $km^2$, among which rice fields account for 43,589 $km^2$ according to 2011 statistical data of local agriculture department. According to the paddy rice growth phenology collected from China Meteorological Bureau, we found total three rice seasons in Hunan Province. The agriculture statistical data indicated that in this area major rice cropping system is rotation system of early-season and late-season rice per year, secondly the single-season rice cropping per year. Double paddy rice includes an early and late rice crops with sowing for early rice starting in March, first harvest in early July, followed by transplanting of late rice and the second harvest in late October. Single-season paddy rice, only involves one rice crop per year for a given field, which always has a relative longer growth period compared to double rice, starting in May, harvest in early September. This research mainly estimates the late-season rice acreage in the study area. There are mainly five growth periods in rice life cycles: transplanting period; seedling developing period; ear differentiation period; heading period; mature period. Temporally, these five periods for late-season rice are usually June 15-20, June 25-30, August 5-15, August 25-September 10 and October 10-30 per year.

## 3. DATA AND PRE-PROCESSING

### 3.1 Remote sensing data

The China Environment Satellite constellation is designed mainly for environment and disaster monitoring by Chinese government. As a part of the satellite constellation, two optical satellites (HJ-1-A and HJ-1-B) are launched on September 6, 2008 (Guo, 2012). HJ satellite data is available to the public freely and operated by China Centre for Resources Satellite Data and Application, with the interface and relevant information available at http://www.cresda.com/n16/n1115/n1432/index.html in Chinese (Wang, 2010). The CCD data has four spectral bands with a spectrum ranging from 430 to 900 nm, a better spatial resolution of 30 m than MODIS and NOAA AVHRR, a higher temporal resolution of 4 days and bigger imaging swath of 700 km than TM and SPOT (Chen, 2011). The multi-spectral information, moderate spatial and temporal resolution, and the large coverage capability of the HJ data make it to be a highly suitable data source for improving agricultural mapping accuracy on regional scale and reduce the monitor cost.

Based on late rice calendar, 17 HJ-1A/B images with relatively less cloud or cloud shadow from July 30 to October 15 over the Hunan province were collected during the growth periods of late-season rice in 2012 (Fig.1). This period covers the whole growth stage of late rice from turning green to harvest.

Pro-processing of these images mainly included radiance calibration and geo-correction. The radiance calibration of the HJ data was to convert the DN value of the raw image to surface spectral reflectance. Atmospheric correction was done using a 6S model (Vermote, 1997). The geo-correction was completed referring to already geo-rectified TM images.

### 3.2 Paddy fields data

The 2nd National Resource Survey Project, under the support of the Chinese Ministry of Land and Resources and the Chinese Academy of Sciences, completed the analysis of System Probatoired' Observation dela Tarre (SPOT) images acquired from 2007 to 2009 for all of China. The images were geo-referenced and ortho-rectified, using field collected ground control points and fine-resolution aerial remote sensing images. Visual interpretation of SPOT images was conducted to generate a thematic map of land cover in village at county scale of 1:10,000 and a thematic map of land cover in city at a scale of 1:500. The result is vector and county boundary dataset, so it was converted into a gridded database at 5m spatial resolution. The 5m resolution gridded paddy rice layer dataset over the Hunan Province was used to mask non-arable area (Fig.1).

### 3.3 Ground survey data

In the province, 122 samples at 1Km*1Km scale were obtained by spatial random sampling, which were used as ground training and validation data for the estimation of paddy rice (Fig.1). The ground survey was carried out on 1-25 September 2012. Firstly, the paddy field borders were lined out on images by GPS. And then the linear features, odd features and other crops in each paddy field were recorded. Half of the ground survey data was randomly selected as training samples, which were used for supervised classification, and the other half was selected as validation samples, which were used to validate the paddy rice mapping and acreage results.

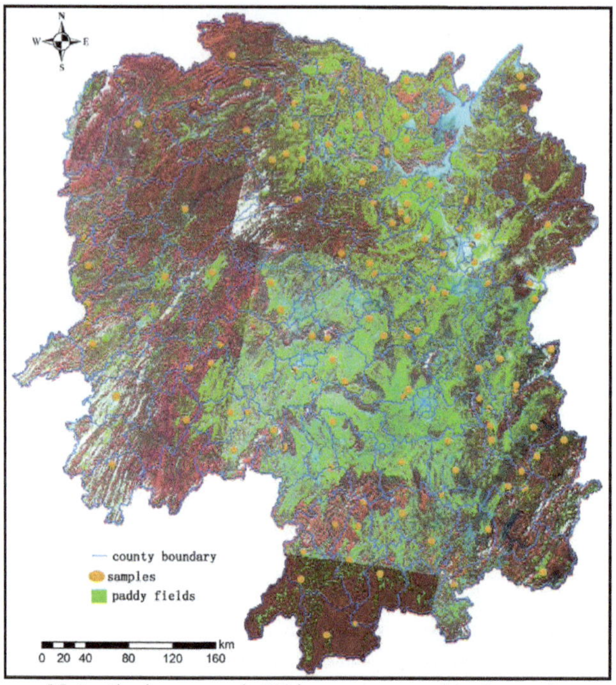


Note: the base map is HJ images we used in the study

Figure 1. The distribution of HJ images, paddy fields and ground survey data in Hunan Province

## 4. METHODOLOGY

This study explored the potential of mixed-pixels decomposition method to improve the rice acreage estimation using multi-temporal HJ-1A/B images. Firstly, 17 HJ-1A/B images were supervised classified to get the pure pixels, and then a rice map was derived from the mosaic of pure pixels. Secondly, mixed pixels were obtained by combining supervised and unsupervised classification, and then another rice map was got by the mosaic of pure pixels and rice fraction from mixed pixels. At last, the rice acreage estimation result was obtained by calculating the cloud proportion and arable coefficient, and then we compared and analysed the estimation results by the two methods. Fig.2 outlined the analysis algorithms and implementation.

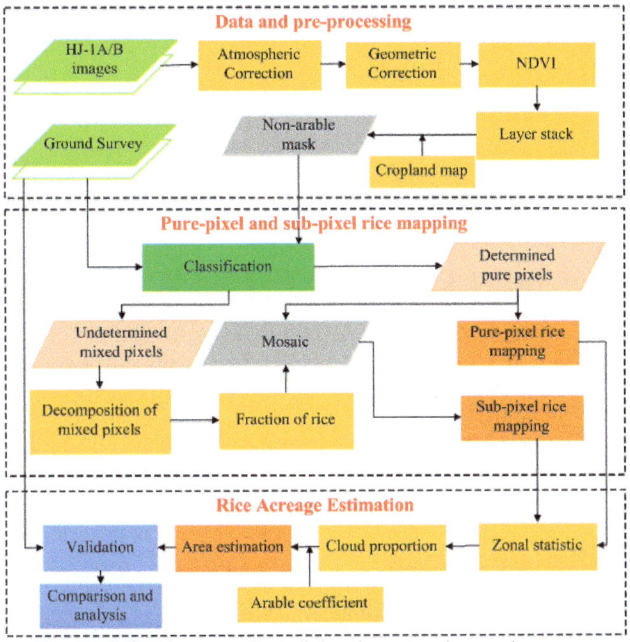


Figure 2. Implementation of the rice mapping and acreage estimation

### 4.1 Extraction of non-arable mask images

The crops are highly reflective in the near infrared and highly absorptive in the visible red, the contrast between these channels can be used as an indicator of the status of the vegetation and crops (Sakamoto, 2011).There are many studies have shown that vegetation indices (VI), such as the Normalized Difference Vegetation Index (NDVI), enhanced vegetation index (EVI) and Soil-Adjusted Vegetation Index (SAVI), which have been used to discriminate crops and monitor crop growth conditions (Kamthonkiat, 2005, Friedl, 2002, Goward, 2012). Among these VI, the NDVI is the most widely used and the other indices are its refined form. In this study, we calculate the NDVI (Eq.1) of every image as a 5th band to be layered stack to the image.

$$NDVI = \frac{R_{nir} - R_{red}}{R_{nir} + R_{red}} \tag{1}$$

Where $R_{nir}$ and $R_{red}$ represent the reflectance of near infrared and red bands, respectively, which correspond to fourth and third bands for HJ-1A/B images.

After obtained the images with 5 bands, masks are made to exclude no-paddy fields by using paddy fields map provided by the 2nd National Resource Survey Project.

### 4.2 Pure-pixel rice mapping

Over the last decades, numerous algorithms to extract land cover information from remotely sensed images have been designed (Guo, 2012, Friedl, 2002, Liu, 2012, Zhang, 2008, Pan, 2012). Being a well-known statistical technique in other scientific disciplines, pixel-based classification was the first method to be applied to the multispectral digital image data. There are many classification methods such as maximum likelihood classifier (MLC), Supported Vector Machine (SVM) and decision tree which have been used in mapping crop lands. In the study, the most favourable classifier, MLC was selected as classification, which was made using VITO's GLIMPSE software (Eerens, 2004).

The training and validation samples were conducted by expert knowledge based on field survey. Based on the phenology information, late-season and single-season rice were coinciding with the rainy season which typically starts at the end of March, and lasts until September. So, when the HJ imaging time is 15 September, four classes including late season rice, middle season rice, bare land and cloud, were trained for classification. And the else were trained for three classes, including late season rice, bare land and cloud. Then an adaptive maximum likelihood classification method was used to classify all the pixels into a certain class according to the ML-RULE. At last, the late rice identified result were obtained.

### 4.3 Sub-pixel rice mapping

#### 4.3.1 Mixed pixels extraction

A mixed pixel is a picture element representing an area occupied by more than one ground cover type. In the area, the mixed pixels are mainly located at the edges of large objects like agricultural fields, and the mountainous area.

Instead of considering every pixel as a mixed pixel, in the study we only thought the edges of paddy rice fields and the mountainous area as the mixed-pixel region. The mixed pixels were extracted as following: a) The 17 HJ images were classified to 100 classes by K-means clustering, which is one of the most common unsupervised methods. b) We identified the late-season rice class among the 100 classes c) The unsupervised rice identification result was masked by the mosaic result of late rice pure pixels, and then the mixed pixels were obtained.

#### 4.3.2 Decomposition of mixed pixels

As for the extraction result of mixed pixels, we selected the Linear mixture modelling, which is the most widely used method for extracting compositional information from remotely sensed images containing many mixed pixels. The fraction of rice can be calculated according to Eq. (2) and (3).

$$VI = \sum f_i VI_i = f_r VI_r + f_{bs} VI_{bs} \tag{2}$$

$$\sum f_i = f_r + f_{bs} = 1 \tag{3}$$

Where $VI$ represents the NDVI of mixed pixels, $VI_r$, $f_r$ represent the NDVI and fraction of rice pure pixels, $VI_{bs}$, $f_{bs}$ represent the NDVI and fraction of bare land pure pixels.

### 4.4 Rice acreage estimation

Because the precipitation from August to October accounts for over 30% of annual precipitation in the Hunan province, and each HJ-1 A/B image has a large imaging swath, so many images collected were covered by clouds. Although we select several images to cover the same region, the province also has some regions that are covered by clouds. To minimize the potential impact of those areas with cloud cover in rice growing period on our estimation, we calculated the cloud proportion of every county to reason rice acreage under cloud cover.

There is also a need to reduce the influence of thin and small non-arable objects, such as ridges of fields, canals etc., we also calculate the non-arable coefficient by dividing the net cropland area using gross cropland area which was provided by local agriculture departments. The non-arable coefficient (r) of every county can be obtained using Eq. (4).

$$\mathrm{r} = \frac{non\text{-}truth\ paddy\ filed}{paddy\ field} \tag{4}$$

Considering the rice fraction of every pixel and the analysis above, the final rice acreage can be calculated as following:

$$S = \frac{\sum_{p=1}^{n} A_p * f_r * (1-r)}{1-r_c} \quad (5)$$

where $S$ represent regional crop planting acreage, $A_p$ represent crop pixel acreage, $f_r$ represent the fraction of rice, $r$ represent non-arable coefficient, $r_c$ represent cloud proportion.

### 4.5 Survey data for validation of rice acreage results

Accuracy assessment of rice acreage estimation results at county scale is a challenging task, as the rice mapping result can overestimate or underestimate areas of individual land cover types due to the mosaic errors of 17 HJ images and the offset of geometrical rectification. Because of the human forces limitations and budget constraints, it is not able to conduct extensive labour intensive field surveys to validate the acreage result of every county in such a large area. In the research, we adopt the validation part of field survey acreage data to evaluate the estimated rice acreage results.

The root mean square error (RMSE) which is an alternative statistical measure and the EA were used to compare the estimated and survey figures. The two indices were defined by:

$$\mathrm{RMSE} = \sqrt{\frac{\sum_{i=1}^{N}(A_g - A_s)^2}{N}} \quad (6)$$

$$\mathrm{EA} = \left[1 - \sqrt{\frac{\sum_{i=1}^{N}(A_g - A_s)^2}{N}} / \left(\frac{\sum_{i=1}^{N} A_s^{\,2}}{N}\right)\right] * 100\% \quad (7)$$

Where$A_g$represent estimated crop planting acreage, $A_s$ represent ground survey crop acreage, $N$ represent the number of total validation samples.

## 5. RESULTS

### 5.1 Mixed pixels identification and the results

Mixed pixels can be effectively identified through the method above mentioned, which are nearly half of the total. Most of mixed pixels are distributed at the fringe of large clusters. And Fig.3 gave an example of mixed pixels rice identification results in Dingcheng District.

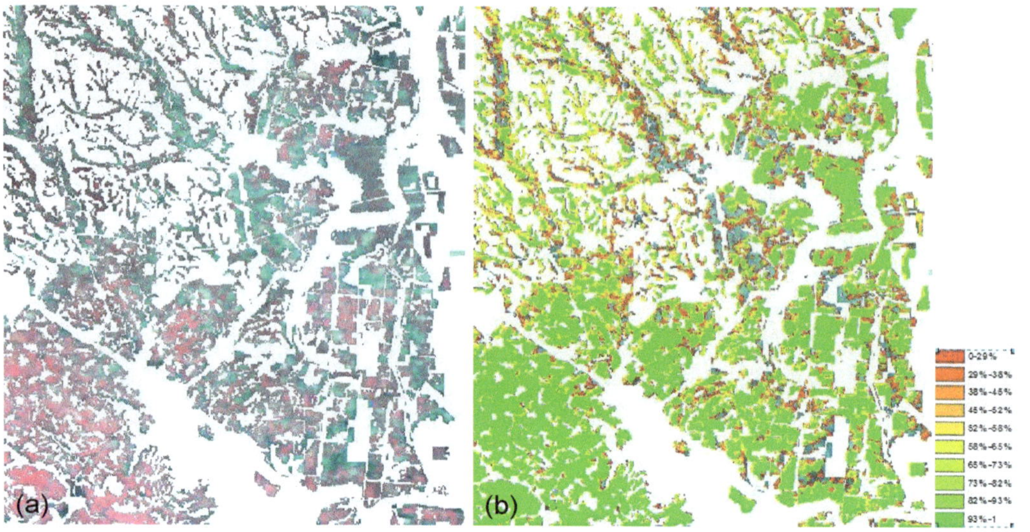


Figure 3. One example of decomposition of mixed pixels in Dingcheng District (a. HJ CCD images; b. the fraction of late-season rice)

### 5.2 Spatial distribution of late-season rice fields in Hunan Province

The distribution of late-season paddy rice fields in Hunan province were mapped from pure-pixel method using only maximum likelihood classifier (hereafter referred to as Purerice) and sub-pixel method combing Maximum likelihood classification and K-means clustering classifier (hereafter referred to as Mixrice) from 17 HJ-1A/B images. In this study, validation samples obtained from ground surveys data are employed to assess the classification accuracy of every image. Table 1 indicates the overall accuracy and Kappa coefficient of the 17 images using Maximum likelihood classification. The accuracy is all above 90% except for the images 2-84-20120820 and 4-88-20120829. This result indicates that we have accurately identified the rice pure pixels.

| imaging orbit and time | Overall accuracy (%) | Kappa |
|---|---|---|
| **1-88-20121004** | 99.74 | 0.9945 |
| **2-80-20120818** | 96.20 | 0.9347 |
| **2-80-20120919** | 96.96 | 0.9494 |
| **2-84-20120820** | 87.66 | 0.8306 |
| **2-84-20120919** | 92.86 | 0.8938 |
| **455-84-20120916** | 99.33 | 0.9875 |
| **456-88-20121005** | 99.56 | 0.9931 |
| **457-54-20120915** | 96.41 | 0.9387 |
| **4-84-20120906** | 93.77 | 0.9139 |
| **457-84-20120915** | 94.48 | 0.9122 |
| **4-80-20120829** | 95.5 | 0.9285 |
| **4-84-20120906** | 93.77 | 0.9139 |
| **4-88-20120829** | 89.21 | 0.8554 |
| **6-84-20120828** | 98.08 | 0.9720 |
| **6-85-20121002** | 99.8 | 0.9970 |
| **7-81-20121001** | 99.07 | 0.9870 |
| **5-85-20120801** | 95.39 | 0.9305 |

Table 1. The overall accuracy and Kappa coefficient of HJ 1A/B images using MLC method

Figure 4b illustrates the final spatial distribution pattern of Mixrice. Paddy rice fields were found to be mostly concentrated in the central-eastern region and northeastern major lake regions (Dongting Lake regions of Hunan Province), with the exception of southern and western regions where altitude and topography prohibit the rice growth.

The spatial pattern of late-season rice from Mixrice (Fig. 4b) has some notable differences compared to that of the Purerice (Fig. 4a). In the middle of the province, we can easily find the Mixrice mapping identified many rice fields with low fractional coverage, such as the Dongkou county, Shaoyang county, Hengyang county, Leiyang district where rice fraction is lower than 30%, while the Purerice mapping greatly underestimated rice planting area in these regions. This is because these areas are mountainous and increased complexity of topography restricts the size of rice fields that can occur, so the heterogeneous land cover in these areas is generally a mix of cropland and bare land, with most of the rice agriculture occurring in smaller river valleys or on terraced slopes.

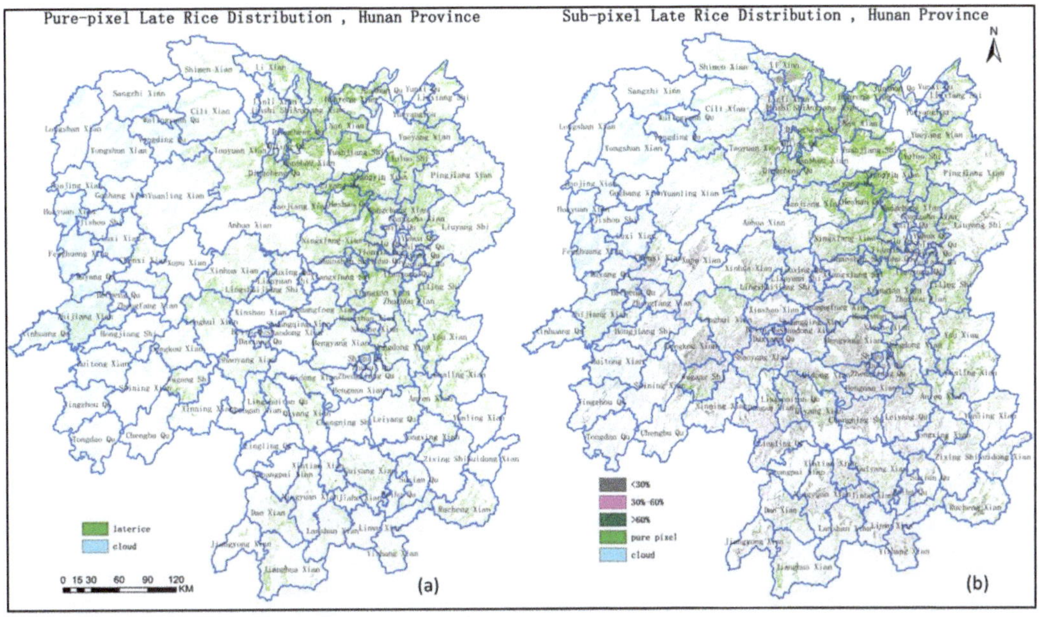


Figure 4. Pure-pixel, sub-pixel comparisons of late-season rice distribution

### 5.3 Comparison of late rice acreage between pure-pixel and sub-pixel derived results

After obtained the late-season rice spatial distribution, we calculated the rice acreage of every county using Eq.5. Figures 5 (a) and (b) respectively showed the acreage result of late rice estimated from Purerice and Mixrice mapping at county level. Rice acreage at provincial and municipal level is just the sum of every county. The late season rice estimated acreage of Hunan province is 1415,580 hectares and 1776,160 hectares separately using pure-pixel method and mix-pixel rice estimation method. The figures showed that the rice acreage was underestimated using only hard classification method, even if we calculated the non-arable coefficient and cloud proportion.

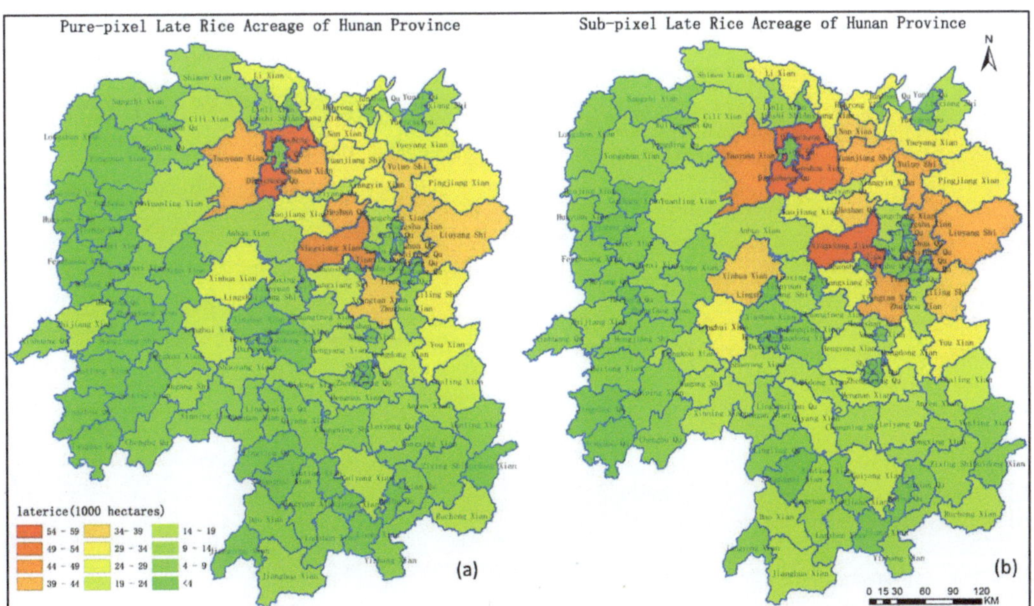


Figure 5. Pure-pixel, sub-pixel comparisons of late-season rice acreage

Compared Fig.5a and 5b, we can more intuitively find acreage obtained from the Purerice mapping is lower than the acreage of Mixrice mapping in some mountainous and rice sporadically planting counties especially in the middle and southern regions of the province, such as Qiyang County, Dao County, Longhui County, Dongkou County and so on. This is consistent with the analysis in front. In addition, from the figures we can also find even in the rice mainly producing areas (Dongting Lake in Hunan Province), the rice acreage derived from Purerice mapping is also underestimated, such as Ningxiang County, Yuanjiang County, Taoyuan County and so on, this is because even though these regions have relatively large rice planting fields, the mixed pixels still exist at the edge of the fields.

### 5.4 Quantitative evaluation of rice acreage estimation results

Certain challenges arise when we want to validate the rice acreage estimation result at county level. It is impossible to investigate the acreage of every county, because it will need large amounts of manpower, timing and material resources. In the study, rice acreage results validation is based on the ground survey figures. We calculated the RMSE, which is an alternative statistical measure that can be used to compare the datasets, and the EA to better verify acreage results. Fig.6(a) and (b) showed the relationship between survey figures and estimated late rice acreages respectively using pure-pixel and sub-pixel methods. Comparing the figures, the sub-pixel method obviously showed a better relationship between estimated and survey figures. The RMSE was 3.40 hectares for Mixrice (Fig.6b), while the RMSE was 5.91 hectares for Purerice (Fig.6a). The EA reached 83.74% according to Mixrice method (Fig.6b), while the acreage EA is only 71.73% derived from Purerice (Fig.6a). Compared to pure-pixel method, the accuracy has a 12% improvement using sub-pixel method. The strong relationship between the Mixrice and survey figures and the good estimation accuracy using Mixrice mapping suggest that sub-pixel method may be a useful tool to provide accuracy rice acreage estimation at county scale, and these results are very encouraging for government subsidy system.

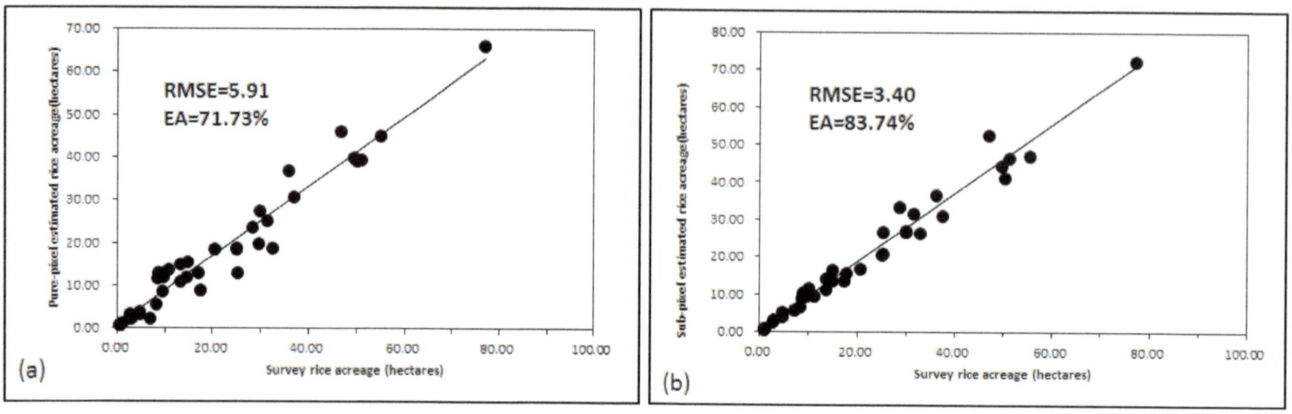


Figure 6. Pure-pixel, sub-pixel comparisons of late-season rice acreage estimated accuracy between estimated and survey figures

## 6. DISCUSSIONS AND CONCLUSIONS

In this study, we separately used the hard classification method and decomposition of mixed pixels method to identify the late-season rice of Hunan province, and estimated the rice acreage at county scale based on rice fraction, non-arable coefficient and cloud proportion. The feasibility of operational rice acreage estimation by sub-pixel method using multi-temporal HJ-1A/B data presents much better compared to using pure-pixel method. The estimated accuracy reached about 83.74% at county level according to the sub-pixel method, while it was only 71.73% using pure-pixel method and the accuracy can be increased about 12% by decomposition of mixed pixels.

With the phenomenon of double-season rice changed to single-season rice becomes more and more serious in southern China, highly accurate estimation of rice acreage is becoming more and more important. Decomposition of mixed pixels method based on Maximum likelihood classification and K-means clustering in this study provided reliable rice acreage estimation after compared with the ground survey data and can be applied to other rice-producing provinces in China or other countries in Asia to generate an updated continental database of paddy rice agriculture.

This research results also demonstrate the China Environment Satellite HJ-1A/B has great potential in the development of an operational system for monitoring rice crop growth and estimating rice crop acreage in southern China. If microwave sensor data on board HJ-1-C satellite is available later, the HJ constellation will have large scale and all-weather imaging capability, which will be a powerful data source for earth observation.

## ACKNOWLEDGEMENTS

The work in this paper was supported by the National Natural Science foundation of China (41071277), the Key Research Program of the Chinese Academy of Sciences (KZZD-EW-08-05) and the project (00-Y30A02-9001-12/13-2-2). The authors also thank Hunan Province Land and Resources Planning Institute for providing the land-use mapping data.

## REFERENCES

Chen, J., Huang J., and Hu, J., 2011. Mapping rice planting areas in southern China using the China Environment Satellite data. *Mathematical and Computer Modelling*, 54(3), pp.1037-1043.

Chen, L., 2007. Uncertainties in rice yield estimation using remote sensing data. Zhejiang University.

Eerens, H., 2004. GLIMPSE: Global Image Processing Software user manual. VITO in house document, Belgium.

Fang, H. L., 1998. Rice crop area estimation of an administrative division in China using remote sensing data. *International Journal of Remote Sensing*, 19(17), pp.3411-3419.

Fang, H. L., Wu, B. F., Liu H. Y., and Huang, X., 1998. Using NOAA AVHRR and landsat TM to estimate rice area year-by-year. *International Journal of Remote Sensing*, 19(3), pp. 521-525.

Friedl, M. A., McIver, D. K., Hodges, J. C. F., et al, 2002. Global land cover mapping from MODIS: algorithms and early results. *Remote Sensing of Environment*, 83(1), pp.287-302.

Goward, S. N., Markham, B., Dye, D. G., Dulaney W. and Yang, J. L., 1991. Normalized difference vegetation index measurements from the advances very high resolution radiometer. *Remote Sensing of Environment*, 35(2), pp.257-277.

Guo, H., 2012. China's earth observing satellites for building a digital earth. *International Journal of Digital Earth*, 5(3), pp.185-188.

Han, L., Pan, J., Jia, B., Zhu, X., Liu, L., Wang S., and Zhang J., 2007. Acquisition of paddy rice coverage based on multi-temporal IRS-P6 satellite AWIFS RS-data. *Transactions of the Chinese Society of Agricultural Engineering*, 23(5), pp.137-143.

Kamthonkiat, D., Honda, K., Turral, H., Tripathi N. K. and Wuwongse, V., 2005. Discrimination of irrigated and rainfed rice in a tropical agricultural system using SPOT VEGETATION NDVI and rainfall data. *International Journal of Remote Sensing*, 26(12), pp.2527-2547.

Koppe, W., Gnyp, M. L., Hütt, C., Yao, Y., Miao, Y., Chen X. and Bareth, G., 2013. Rice monitoring with multi-temporal and dual-polarimetric TerraSAR-X data. *International Journal of Applied Earth Observation and Geoinformation*, 21, pp.568-576.

Liu, J., Tian, Q., Huang Y. and Du, L., 2012. Crop extraction based on multi-temporal HJ satellite CCD Data in Jiaxiang County. *Remote Sensing information*, (2), pp.67-70.

Lobell D. B.andAsner, G. P., 2004. Cropland distributions from temporal unmixing of MODIS data. *Remote Sensing of Environment*, 93(3), pp.412-422.

Maclean, J. L., Dawe, D. C., Hardy B. and Hettel, G. P., 2002. Rice almanac: Source Book for the Most Important Economic Activity on Earth. Int. Rice Res. Inst..

Okamoto K. and Kawashima, H., 1999. Estimating of rice-planted area in the tropical zone using a combination of optical and microwave satellite sensor data. *International Journal of Remote Sensing*, 20(5), pp.1045-1048.

Pan, Y., Li, L., Zhang, J., Liang, S., Zhu X. and Sulla-Menashe, D., 2012. Winter wheat area estimation from MODIS-EVI time series data using the Crop Proportion Phenology Index. *Remote Sensing of Environment*, 119, pp.232-242.

Peng, D., Huete, A. R., Huang, J., Wang F. and Sun, H., 2011. Detection and estimation of mixed paddy rice cropping patterns with MODIS data. *International Journal of Applied Earth Observation and Geoinformation*, 13(1), pp.13-23.

Sakamoto, T., Shibayama, M., Kimura A. and Takada, E., 2011. Assessment of digital camera-derived vegetation indices in quantitative monitoring of seasonal rice growth. *ISPRS Journal of Photogrammetry and Remote Sensing*, 66(6), pp.872-882.

Toan, T. Le, Ribbes, F., Wang, L., Floury, N., Ding K. and Kong, J., 1997. Rice crop mapping and monitoring using ERS-1 data based on experiment and modeling results. *Geoscience and Remote Sensing, IEEE Transactions on*, 35(1), pp.41-56.

Vermote, E. F., Tanré, D., Deuzé, J. L., Herman M. and Morcrette, J. J., 1997. Second Simulation of the Satellite Signal in the Solar Spectrum, 6S: an overview. *Geoscience and Remote Sensing, IEEE Transactions on*, 35(3), pp.675-686.

Wang, Q., Wu, C., Li Q., and Li, J., 2010. Chinese HJ-1A/B satellites and data characteristics. *Science China Earth Sciences*, 53(1), pp.51-57.

Xiao, X., Boles, S., Frolking, S., Li, C., Babu, J. Y., Salas W. and Moore, B., 2006. Mapping paddy rice agriculture in South and Southeast Asia using multi-temporal MODIS images. *Remote Sensing of Environment*, 100(1), pp.95-113.

Xiao, X., Boles, S., Liu, J., Zhuang D. and Liu, M., 2002. Characterization of forest types in Northeastern China, using multi-temporal SPOT-4 VEGETATION sensor data. *Remote Sensing of Environment*, 82(2), pp.335-348.

Yang, X. H. 2007. Study on the remote sensing information extraction of rice based on neural network and support vector machine. Zhejiang University.

Mingwei, Z., Qingbo, Z., Zhongxin, C., Jia, L., Yong, Z., and Chongfa, C., 2008. Crop discrimination in Northern China with double cropping systems using Fourier analysis of time-series MODIS data. *International Journal of Applied Earth Observation and Geoinformation*, 10(4), pp.476-485.

# EXTRACTION OF EXTENDED SMALL-SCALE OBJECTS IN DIGITAL IMAGES

V. Yu. Volkov [a, b]

[a] Dept. of Radioengineering, The Bonch-Bruevich State Telecommunications University, Saint-Petersburg, Russia – vl_volk@mail.ru
[b] Dept. of Radioengineering, State University of Aerospace Instrumentation, Saint-Petersburg, Russia

**Commission III, WG III/3**

**KEY WORDS:** Filtering, Segmentation, Extraction, Automatic threshold control

**ABSTRACT:**

Detection and localization problem of extended small-scale objects with different shapes appears in radio observation systems which use SAR, infra-red, lidar and television camera. Intensive non-stationary background is the main difficulty for processing. Other challenge is low quality of images, blobs, blurred boundaries; in addition SAR images suffer from a serious intrinsic speckle noise. Statistics of background is not normal, it has evident skewness and heavy tails in probability density, so it is hard to identify it.
The problem of extraction small-scale objects is solved here on the basis of directional filtering, adaptive thresholding and morthological analysis. New kind of masks is used which are open-ended at one side so it is possible to extract ends of line segments with unknown length. An advanced method of dynamical adaptive threshold setting is investigated which is based on isolated fragments extraction after thresholding. Hierarchy of isolated fragments on binary image is proposed for the analysis of segmentation results. It includes small-scale objects with different shape, size and orientation. The method uses extraction of isolated fragments in binary image and counting points in these fragments. Number of points in extracted fragments is normalized to the total number of points for given threshold and is used as effectiveness of extraction for these fragments. New method for adaptive threshold setting and control maximises effectiveness of extraction. It has optimality properties for objects extraction in normal noise field and shows effective results for real SAR images.

## 1. INTRODUCTION

The task of detection and localization of small extended objects in noisy images occur in the electronic surveillance systems using radars with SAR, infrared and laser systems, as well as television cameras (Gao, 2010; Gonzalez, Woods, Eddins, 2004). This task is relevant, because these facilities typically have an artificial origin and are of Prime interest.
Upon detection, extraction and localization of such objects had substantial difficulties in building effective algorithms and structures processing, as in taken images there is intense and non-stationary background, also contains elements that are structurally similar to the signals, the signal/background is usually small, and the registered digital image has a low quality, small number of quantization levels, patency character and fuzzy borders of natural and artificial structures (rivers, roads, bridges, buildings).
Statistics background is very different from a Gaussian, the distribution is clearly asymmetric, and the tails of the distributions like lognormal density normal or mixed (contaminated-normal), and when small volumes of samples are identified with difficulty. Such a character background virtually eliminates the use of the known methods of thresholding, since improper formation thresholds can cause loss of useful objects at a very early stage of processing. You cannot use traditional methods of detecting contours in images in order to highlight natural features (rivers, roads, borders, windbreaks, etc.), which are based on the formation of the spatial derivatives (gradients and Laplacians), because the result will be a significant growth impulse noise without visible effect for selection of quality circuit.
The basic principles that allow solving this difficult problem, are the location-based filtering, adaptive thresholding and selection of useful sites on the connectivity of neighboring pixels given the length of the useful structures (Gonzalez, Woods., Eddins, 2004).

## 2. PROBLEM STATEMENT AND METHOD OF OBJECT DETECTION

Suppose there is a classification problem, so that objects have to be extracted belong to certain classes. But these classes have no precise description and this is the object of investigation. One of the main features for object segmentation is its extension. Useful objects are usually extensive, after binarization they consist of connected points, and look like lines or blobs.
Solving of segmentation problem includes thresholding and selection of objects with different extensions. Threshold processing gives binary image and the following segmentation is realized with special method of extraction. Dynamical threshold setting is the problem which should be solved with taking into account results of segmentation.
The aim of this paper is to develop method for segmentation and extraction of extensive objects with unknown sizes and orientations. Method proposed includes extraction of isolated fragments after binarization of pre-filtered image. Method also includes logical filtering with corresponding masks and allows extracting isolated fragments with different sizes and orientations.
Different isolated fragments differ in their sizes and orientation. Hierarchy of isolated fragments is proposed which relates to

their extensions and orientations and *characterizing mask* of fragment is inserted.

Setting of threshold is the main problem for qualitative segmentation. Low threshold level gives much noise, and the following processing becomes inefficient, too high level results in destroying of useful objects which may be split up to small fragments. The best threshold should be set after analysis of segmented fragments with the use of some quality indicator for extraction and segmentation. It is desirable to get some attributes which characterize the quality of segmentation.

The simplest attribute is proposed here as the number of points at each step of extraction. It should be normalized to the initial points in binary image and represents the *effectiveness of extraction* at corresponding step.

It is worth noting that threshold setting is dependent on pre-filtering processing. The general idea of threshold setting and control via results of segmentation is admissible for different pre-filtering algorithms. It also may be used for local thresholds in sliding windows.

Consider there is an image in digital form, containing useful objects, which have a relatively small length in relation to the size of the entire image and an arbitrary orientation. Shape of objects of interest can be linear or speckle, and their length is specified by specifying a maximum size or length of the object in pixels, and set the minimum and maximum bounds on the length of objects. The problem feature is that the emergences of small-scale objects of interest practically no effect on the integral characteristics of the image.

There are two examples of extended small-scaled objects in SAR images which are shown in Fig. 1 and Fig. 2. These objects are artificially highlighted with a white oval. First image contains two oriented linear objects and the second image contains two blob-like objects inside corresponding white ovals.

The general structure of digital image processing includes a pre-filter, binary quantization (threshold processing), and subsequent morphological processing (Fig. 3). The input image after registration is submitted in digital form (two-dimensional array on a rectangular grid of points).

Figure 1. Small-scale oriented objects to be extracted

The problem of automatic setting of the threshold in autonomous information and control systems is very important for segmentation (Akcay, Aksoy, 2007, Sezgin, Sankur., 2004, Weszka, Rosenfeld, 1978).

Figure 2. Small-scale blob-like objects to be extracted

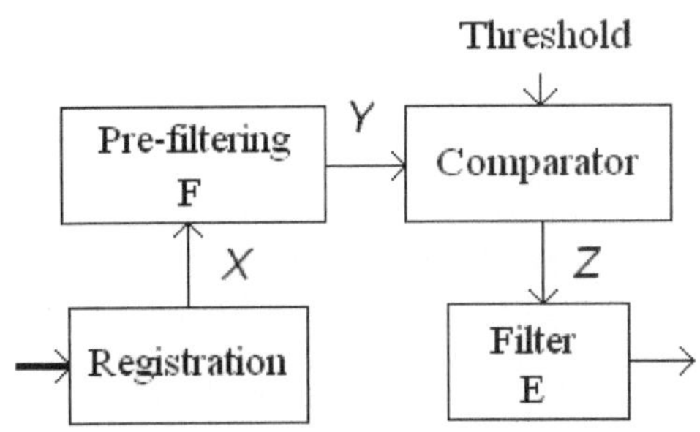


Figure 3. Image processing structure for objects extraction

Well-known installation methods of global and local thresholds typically use histograms or local properties of the pixels in the image (Gonzalez, Woods., Eddins, 2004, , Sezgin, Sankur., 2004). In our case, the threshold processing should depend on the results of binarization (Volkov, 2009a, 2009b).

The purpose of this paper is to study adaptive method threshold segmentation for the detection and selection of objects based on structural decomposition of a binary image into elementary, isolated objects, analysis of the impact of the threshold on the results of the decomposition, and algorithm development for installation and changes the threshold in accordance with the results of the decomposition.

## 3. PRELIMINARY FILTERING

Pre-filtering aims to improve the image and highlight the differences and boundaries. It is assumed that the useful objects always have a higher intensity relative to the background, otherwise it is necessary to invert the image. In this case we applied the differentiating filters (Laplacian type), which permit to use a global threshold for binary quantization exceeding the intensity threshold quantization.

When filtering oriented linear objects we used space-oriented mask filter of the following form (Fig. 4), which would have effectively allocate endpoints of the segments of unknown length. In this particular case the coefficients are $a = 1$, $b = -1$.

These masks should be used for searching fragments with different orientations for extraction all small-scale extended objects in the image.

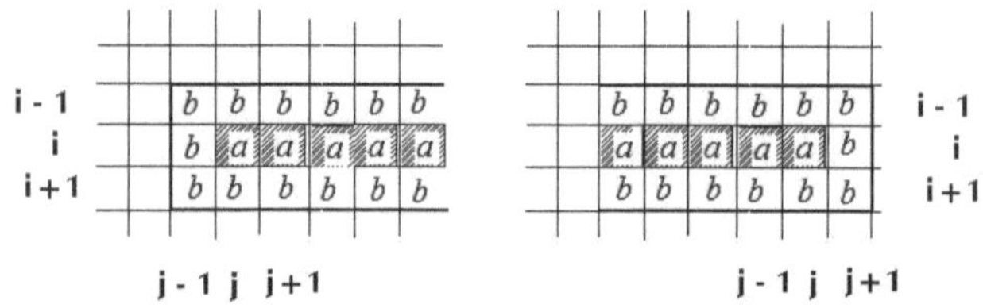


Figure 4. Space-oriented filter masks

For blob-like small-scaled objects we used non-oriented filter mask which is presented in Fig. 5.

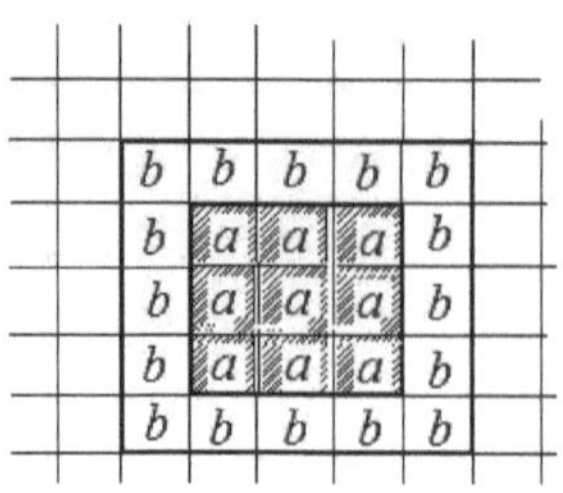


Figure 5. Non-oriented filter mask

Along with averaging in differentiating filters used other operators such as the sample median and selection the maximum value (in cells with coefficients *b*).
The results of preliminary filtering are shown in Fig. 6 and Fig.7 below.

Figure 6. The output of pre-filtering of image in Fig. 1

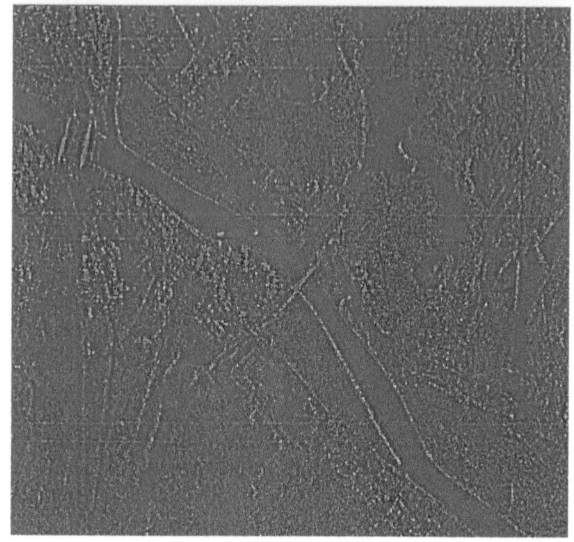

Figure 7. The output of pre-filtering of image in Fig. 2

## 4. THRESHOLD PROCESSING

This stage is very important. Incorrect threshold often results in irreversible losses of information. Suppose we are interested in objects with high intensity, and processing results in high level for pixels if threshold level is exceeded, and zero level otherwise.

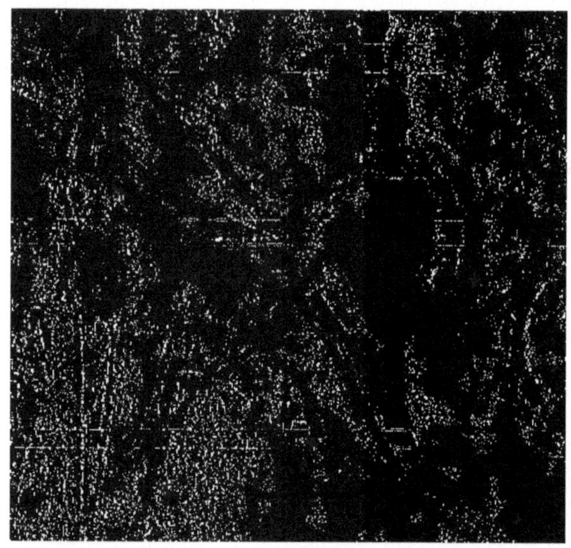

Figure 8. Binary image after low threshold level

Fig. 8 - Fig.10 illustrate changes of binary images upon threshold variations from low to high levels for image shown in Fig. 7. As it may be easily observed from Fig. 8, low threshold (two pictures on top) is not appropriate because of segmentation problems: useful and noisy objects do not differ in extensions. As far as threshold level rises, differences between extensions of objects start to appear, but at very high levels (bottom pictures) we can see destroying of useful extensive objects. The best threshold level gives maximal differences in extensions between useful and noise objects.

Figure 9. Binary image after intermediate threshold level

Figure 10. Binary image after high threshold level

We can also see that for high thresholds bottom pictures has weak dependence on type of pre-filtering processing.

The main idea is to set threshold level according to segmentation results. For this purpose hierarchy of isolated fragments is proposed, and *effectiveness of extraction* is inserted as indicator of segmentation degree. It may be used for threshold setting and control.

## 5. HIERARCHY OF ISOLATED FRAGMENTS IN BINARY IMAGE

Our aim is to find out attributes of image which characterize extension properties of objects and allow us to control threshold level for qualitative segmentation. Suppose we have binary image after threshold processing. The concept of *characterizing mask* for isolated fragments is introduced for analysis of segmentation results. It relates to definition of continuity and adjacency of pixels, here usual definitions are used (Gonzalez, Woods., Eddins, 2004, Volkov, 2009a, Volkov 2009b).

Fragments are isolated if they have no mutual pixels. Suppose isolated fragment consists of several adjacent pixels on the binary image.

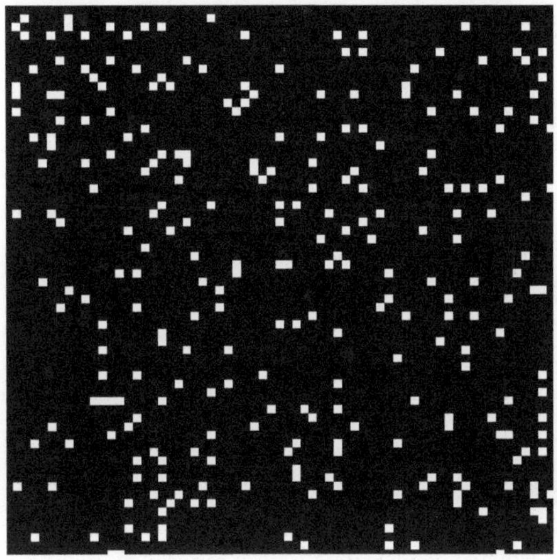

Figure 11. Gaussian noise field (top picture) and the result of binarization (bottom picture) with threshold level equals to 1.5

Extension properties of isolated fragment may be characterized by size of minimal rectangular mask which entirely covers this fragment. Objects may include several isolated fragments. Extensive objects usually contain extensive fragments.

Fig. 11 in the bottom contains results of threshold binarization for Gaussian noise field with zero mean and standard deviation equals to unity (top picture) with threshold level equals to 1.5. Top picture in Fig.12 represents isolated points. After deleting isolated points there are lots of connected fragments with different sizes and shapes (bottom picture in Fig. 12).

It is possible to represent hierarchy of small isolated fragments which is shown in Fig. 13 up to characterizing mask 3x3. Isolated point has characterizing mask 1x1, it is shown at high left corner of figure. Isolated pairs may have characterizing masks 1x2, 2x1 or 2x2 corresponding to their orientations.

Triple of points may form horizontal line (at high right corner of figure), vertical line (at low left corner), or diagonal lines (slash and inverted slash). Other fragments with three points may have different characterizing masks. Triple of points may also look like "corner", and have characterizing mask 2x2.

Extensive fragment with horizontal orientation has characterizing mask 2x3 if it is more in sizes than 2x2 but entirely covers by mask 2x3. Similar fragment with vertical "Gate" or "scoop" with three points has 2x3 or 3x2 masks.

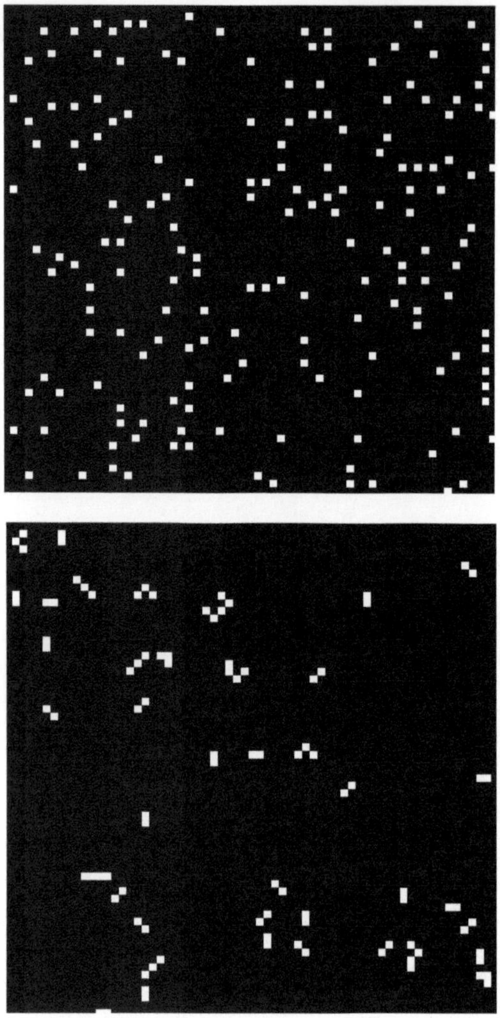

Figure 12. Isolated points extracted (top picture) and remaining small-scale objects (bottom picture) after thresholding of Gaussian noise field

Four points give large variety of isolated fragments. Only square 2x2 is shown on Fig. 13. Square masks do not consider fragment orientations.

Hierarchy of fragments is their ordering and may be obtained by the choice of characterizing masks to be considered. Masks 2x3, 3x5, 3x7, 5x7 and other similar masks characterize horizontal fragments; masks 3x2, 5x3, 7x3, 7x5 describe vertical fragments.

The choice of set of characterizing mask allows us to obtain different attributes for describing fragments with different extensions and orientations.

The simplest hierarchy uses only square characterizing masks. Then we have 1x1, 2x2, 3x3, 5x5, 7x7 and so on, as characterizing masks of extensive fragments with increasing extensions. This hierarchy does not take into account orientations of fragments.

## 6. THRESHOLD SETTING AND CONTROL BY THE USE OF CONNECTED FRAGMENTS IN NOISE IMAGE

It is easy to verify that the number of extracted small-scale objects of the given type in binary image is small at both very low and high thresholds. Thus, with some intermediate value of the threshold number of such objects is maximal.

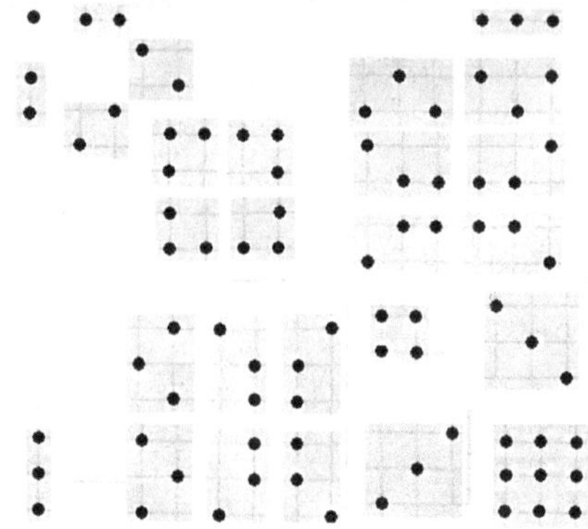

Figure 13. Hierarchy of small isolated fragments

However, along with the decrease in the number of the objects by increasing the threshold also decreases the total number of points in which intensities exceed the threshold.

Let after binary quantization with a predetermined threshold value $T$ the image contains only $N(T)$ pixels exceeding the threshold.

At each step of extraction we use corresponding characterizing masks 1x1, 2x2 and so on. Extracted fragments are deleted from the image. Then at each step of the extraction binary image loses $N_1(T)$, $N_2(T)$ ...etc. points. Since all these numbers depend on the quantization threshold, it is necessary to perform the normalization, and to consider the relative values $N_1(T)/N(T)$, $N_2(T)/N(T)$ and so on.

These values can be considered as the estimates of *effectiveness of extraction* of fragments at the appropriate step with given quantization threshold. For homogeneous Gaussian noise field we can calculate probability that intensity in each pixel exceeds the threshold

$$P(T) = 1 - \Phi(T) \tag{1}$$

where $\Phi(T)$ is a cumulative distribution function.

Probability to occur an isolated point will be

$$P_1(T) = \Phi^8(T)[1 - \Phi(T)] \tag{2}$$

and probability for any fragment with characterizing mask 2x2 is

$$\begin{aligned} P_2(T) &= 2\Phi^{10}(T)[1-\Phi(T)]^2[1+\Phi^2(T)] \\ &+ \Phi^{12}(T)[1-\Phi(T)]^3[1+3\Phi] \end{aligned} \tag{3}$$

Normalized values $E_1 = P_1(T)/P(T)$, $E_2 = P_2(T)/P(T)$ and so on were calculated and represented on the top of Fig. 14. Numbers of lines correspond to characterizing masks 1x1, 2x2, 3x3, etc.

Extraction efficiency for isolated points (line 1), fragments with characterizing mask 2x2 (line 2), and mask 3x3 (line 3).

Lines for connected fragments have evident maxima at the threshold value 1.3. This is the best threshold for extracting small connected fragments from Gaussian noise field. These results are confirmed by simulation curves represented on the bottom of Fig. 14.

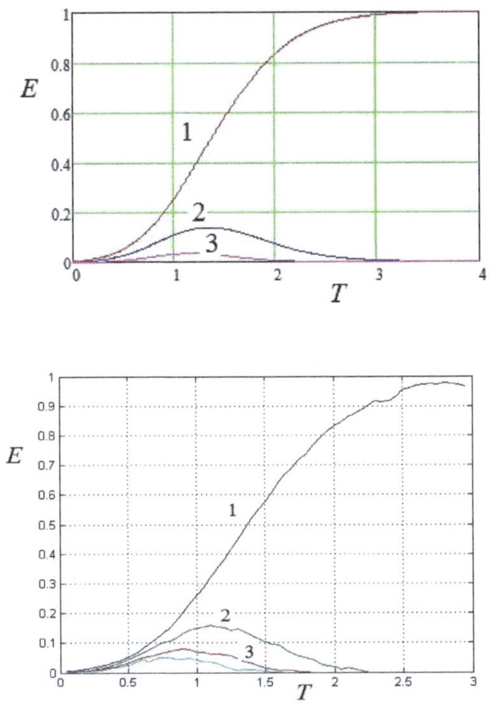


Figure 14. Extraction efficiency for isolated points (line 1), fragments with characterizing mask 2x2 (line 2), and mask 3x3 (line 3)

## 7. ADAPTIVE THRESHOLDING STRUCTURE

Processing structure containing adaptive threshold setting and control is shown in Fig. 15. Filter F produces pre-filtering, filter E realizes morphological operation for extraction isolated fragments with given characterizing masks. Binary quantizer obtains binary image with zero output level if input intensity is below a threshold and maximum level if input intensity exceeds a threshold. The threshold value should be varying and statistics are obtained which result in extraction efficiency.
Maximum selector obtains the threshold value which maximizes extraction efficiency for prescribed characterizing mask, length selector extracts small-scale objects with prescribed lengths.

## 8. EXTRACTION OF SMALL-SCALE OBJECTS FROM REAL SAR IMAGES

The results of the extraction and allocation of small-scale objects on real radar images presented in Fig. 16 and Fig 17 for the two images from Fig. 1 and Fig. 2 accordingly.
Top pictures represent outputs of binary quantizers which threshold were set for maximum extraction efficiency for small connected fragments. Objects of interest are highlighted by the ovals. Dependency plots of extraction efficiency upon threshold values are presented in the middle row of pictures.
Numbers 1, 2 and 3 correspond to connectivity of pixels in fragments. Pictures on the bottom show objects of interest extracted by the use of length selection.

## 9. CONCLUSIONS

New method of extraction and allocation small-scale objects is described which is based on pre-filtering, adaptive thresholding and morphological selection. It allows extracting small-scale objects with different extension and orientation.

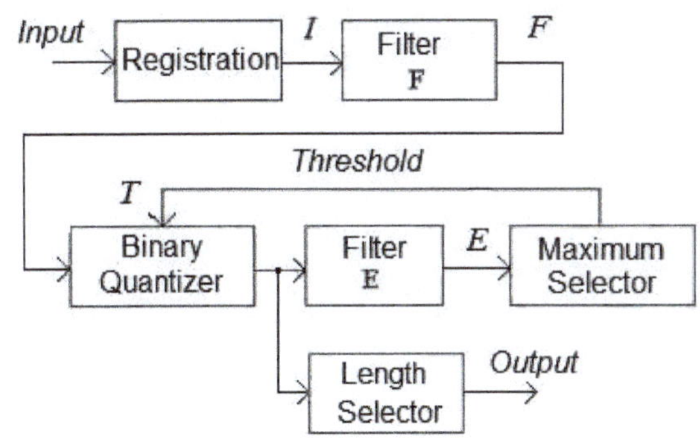


Figure 15. Processing structure with adaptive threshold setting and control

Hierarchy of isolated fragments is proposed for analysis of connected fragments. Analysis of small connected fragments is useful for obtaining indicators for threshold setting and control. Extraction efficiency is introduced as the relative number of points in fragments with respect to all points exceeding threshold. The best threshold level should give maximal extraction efficiency for a given size of characterizing mask.
Adaptive method for threshold setting and control has optimal property which was checked by modeling Gaussian field with extensive object region. Thresholds obtained are settled near the value of optimal maximal likelihood threshold for detection of shift on Gaussian field.

## REFERENCES

Akcay H., Aksoy S., 2007. Morphological segmentation of urban structures. *Urban Remote Sensing Joint Event*, 11-13 April, Paris, pp. 1-6.

Gao G., 2010. Statistical modeling of SAR images: A Survey. *Sensors*, pp. 775-795.

Gonzalez R.C., Woods R.E., Eddins S.L., 2004. *Digital Image Processing using MATLAB*. Englewood Cliffs, NJ: Prentice-Hall.

Sezgin M., Sankur B., 2004. Survey over image thresholding techniques and quantitative performance evaluation. *Journal of Electronic Imaging*, 13(1), pp. 146–165.

Volkov V., 2009a. Segmentation and extraction of extensive objects on digital Images. In: *Proceedings of the 2009 International Conference on Image Processing, Computer Vision and Pattern Recognition*. IPCV2009, II. Las Vegas, Nevada, USA, CSREA Press, pp. 656-662.

Volkov V., 2009b. Thresholding for segmentation and extraction of extensive objects on digital images. In: *Proceedings 32 Annual German Conference on Artificial Intelligence*. KI 2009. Paderborn, Germany, Springer, pp. 623-630; Lecture Notes in Computer Science (including subseries Lecture Notes in Artificial Intelligence and Lecture Notes in Bioinformatics) 5803 LNAI , pp. 623. http://www.springer.com/computer/ai/book/978-3-642-04616-2

Weszka J., Rosenfeld A., 1978. Threshold evaluation techniques. *IEEE Trans.*, SMC-8, pp. 622-629.

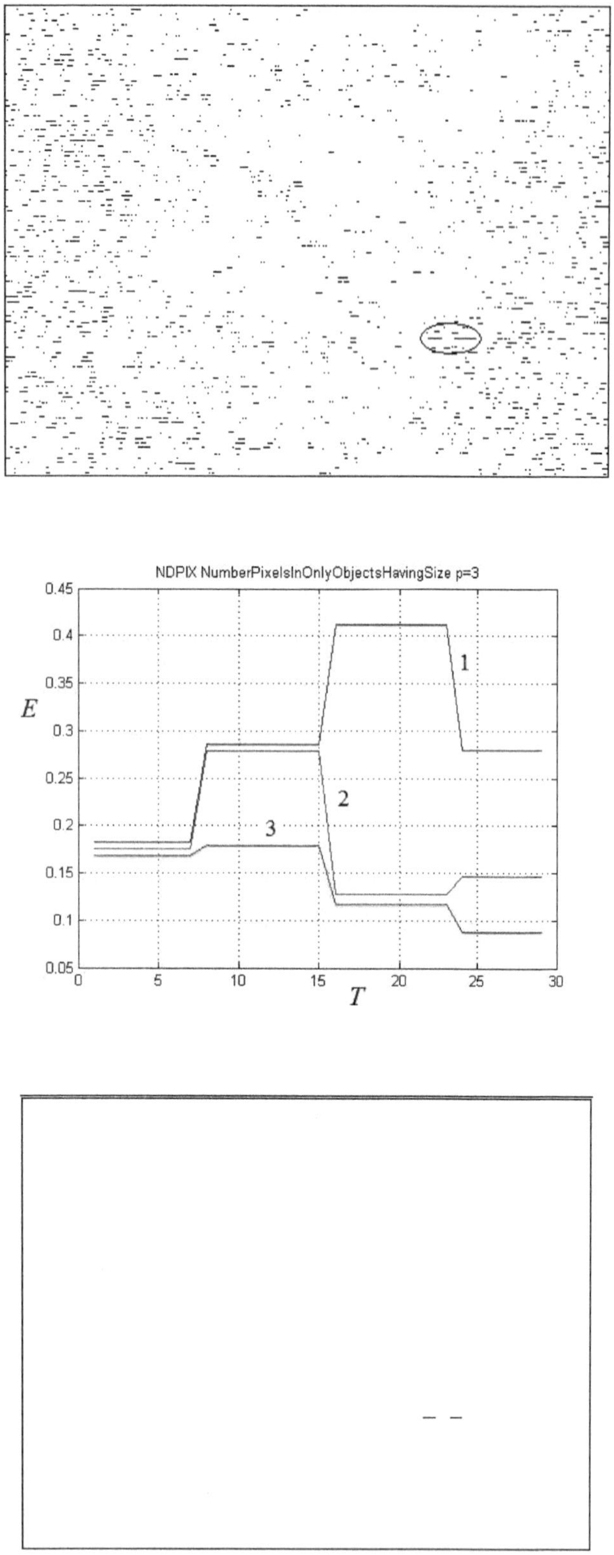


Figure 16. Extraction of small-scale objects from the image in Fig. 1

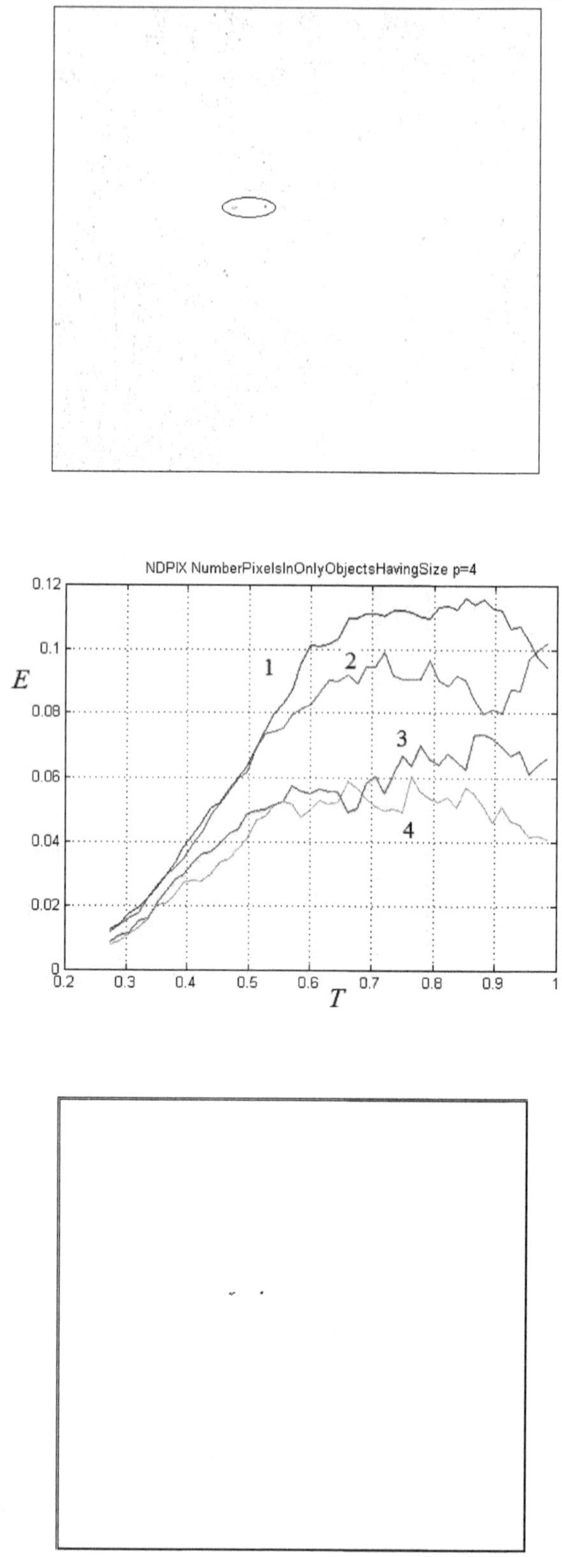


Figure 17. Extraction of small-scale objects from the image in Fig. 2

# SELF-LOCALIZATION METHOD BY INTEGRATING SENSORS

T. Fuse [a, *], K. Matsumoto [b]

[a] Dept. of Civil Engineering, University of Tokyo, Hongo 7-3-1, Bunkyo-ku, Tokyo, 113-8656, Japan - fuse@civil.t.u-tokyo.ac.jp
[b] Rakuten, Inc., Higashi Shinagawa 4-12-3, Shinagawa-ku, Tokyo, 140-0002, Japan

**Commission V, WG V/4**

**KEY WORDS:** Sensor Fusion, Self-Localization, Mobile Devices, Augmented Reality, Visualization, Navigation

**ABSTRACT:**

Recently, development of high performance CPU, cameras and other sensors on mobile devices have been used for wide variety of applications. Most of the applications require self-localization of the mobile device. Since the self-localization is based on GPS, gyro sensor, acceleration meter and magnetic field sensor (called as POS) of low accuracy, the applications are limited. On the other hand, self-localization method using images have been developed, and the accuracy of the method is increasing. This paper develops the self-localization method by integrating sensors, such as POS and cameras, on mobile devices simultaneously. The proposed method mainly consists of two parts: one is the accuracy improvement of POS data filtering, and another is development of self-localization method by integrating POS and camera. The POS data filtering combines all POS data by using Kalman filter in order to improve the accuracy of exterior orientation factors. The exterior orientation factors with POS filtering are used as initial value of ones in image-based self-localization method. The image-based self-localization method consists of feature points extraction and tracking, relative orientation, coordinates estimation of the feature points, and orientation factors updates of the mobile device. The proposed method is applied to POS data and images taken in urban area. Through experiments with real data, the accuracy improvement by POS data filtering is confirmed. The proposed self-localization method with POS and camera make the accuracy more sophisticated by comparing with only POS data filtering.

## 1. INTRODUCTION

Recently, development of high performance CPU, cameras and other sensors on mobile devices have been used for wide variety of applications. One of the most popular applications is augmented reality (AR). The AR technique superimposes various data on the scene which the users see at the time instead of comprehensive and detailed three dimensional modelling such as virtual reality. The AR uses sequential images taken from same view points of users as environmental scene, and then reality of visualization increase compared with the virtual reality. AR requires self-localization, namely orientation of the mobile device.

Since the self-localization of the mobile device is based on GPS, gyro sensor, acceleration meter and magnetic field sensor (called as position and orientation system (POS) in this paper) of low accuracy, the applications are limited to tags superimposition on the sequential images.

On the other hand, self-localization method using images have been developed, and the accuracy of the method is increasing. The method, however, requires initial values of the orientation factors and distance of baseline. GCPs can be used to specify the absolute coordinates. But it takes a lot of time and effort, and the applicability is restrictive.

This paper develops the self-localization method by integrating sensors, such as POS and cameras, on mobile devices simultaneously. The proposed method mainly consists of two parts: one is the accuracy improvement of POS data filtering, and another is development of self-localization method by integrating POS data and images.

## 2. POS DATA FILTERING

### 2.1 Overview

In this study, an iPhone4s is used as a mobile device. The centre of the mobile device is defined as the origin of the coordinate system for the mobile device. In the coordinate system, $z$ axis is set as the vertical direction of device monitor, and $y$ axis is norm direction of the monitor plane, and $x$ axis is defined according to right-handed coordinate system. The rotations with regard to the axis $x$, $y$, $z$ (roll, pitch, yaw) are expressed as $\omega_g$, $\varphi_g$, $\kappa_g$, respectively. For absolute coordinate system, origin is same as the device coordinate system. $Z$ axis is set as the direction of gravitational force, and $X$ axis is north-seeking. When the both coordinate systems coincide, the rotation angles are define as 0, namely $\omega_g=0$, $\varphi_g=0$, $\kappa_g=0$.

As mentioned before, POS includes GPS, gyro sensor, acceleration meter and magnetic field sensor. The corresponding data are three dimensional coordinates of the sensor (longitude: $E$ (degree), latitude: $N$ (degree), ellipsoidal height: $h$ (m)), angular velocity ( $\dot{\omega}_g$ , $\dot{\varphi}_g$ , $\dot{\kappa}_g$ (rad/s)), acceleration ( $\ddot{x}_a$ , $\ddot{y}_a$ , $\ddot{z}_a$ (G)) and azimuth angle ( $\kappa_{ti}$ (rad)), respectively.

Normally, each sensor is used separately (Daehne and Karigiannis, 2002), or some sensors are selected among them (Agrawal and Konolige, 2006; Kourogi and Kurata, 2005; Wagner *et al.*, 2009). The proposed method combines all sensors data in order to improve the accuracy of exterior orientation factors. Figure 1 shows the pipeline of the POS data filtering.

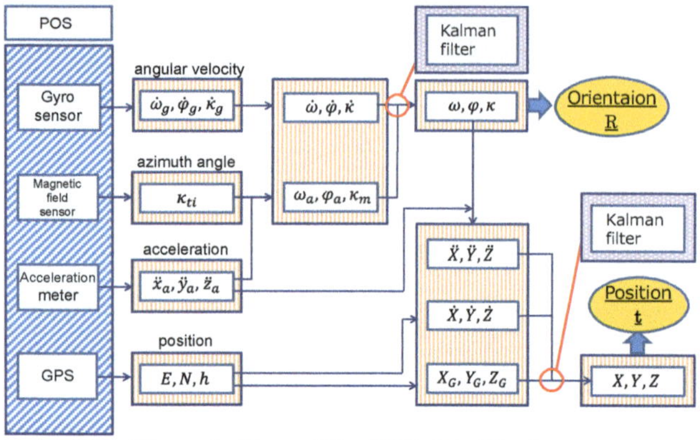


Figure 1. Flow of the POS data filtering

## 2.2 Filtering method of rotation

Initial rotation of the mobile device is calculated from acceleration meter and magnetic field sensor, and rotation during movement is calculated from the gyro sensor. The data of gyro sensor have drift error. In order to correct the drift error, filtering method with the data of the acceleration meter and the magnetic field sensor is applied. In this step, the data of such sensors are combined by using Kalman filter (Schall *et al.*, 2009). The assumption, that the mobile sensor moves linearly, is accepted here.

**2.2.1 Calculation of rotation:** When the mobile device remains stationary, the acceleration meter detect only acceleration of gravity *g*. The relationship between the acceleration and initial rotation ( $\varphi_a$, $\theta_a$, $\kappa_m$ ) can be expressed as follows.

$$\begin{bmatrix} \ddot{x}_a \\ \ddot{y}_a \\ \ddot{z}_a \end{bmatrix} = \begin{bmatrix} -g\sin\varphi_a \\ g\cos\varphi_a\sin\omega_a \\ g\cos\varphi_a\cos\omega_a \end{bmatrix} \quad (1)$$

From the equation (1), roll and pitch can be calculated.

$$\omega_a = \tan^{-1}\frac{\ddot{y}_a}{\ddot{z}_a}, \quad \varphi_a = \tan^{-1}\frac{-\ddot{x}_a}{\sqrt{\ddot{y}_a^2+\ddot{z}_a^2}} \quad (2)$$

Yaw should be transformed according to the above defined coordinate system:

$$\kappa_m = \pi - \kappa_{ti} \quad (3)$$

For calculation of rotation during movement, data of the gyro sensor are added to the initial rotation. Derivatives of the rotation with respect to time ( $\dot{\omega}$, $\dot{\varphi}$, $\dot{\kappa}$ ) are represented by using angular velocity ( $\dot{\omega}_g$, $\dot{\varphi}_g$, $\dot{\kappa}_g$ ).

$$\begin{bmatrix} \dot{\omega} \\ \dot{\varphi} \\ \dot{\kappa} \end{bmatrix} = \frac{1}{\cos\varphi}\begin{bmatrix} \cos\varphi & \sin\omega\sin\varphi & \cos\omega\sin\varphi \\ 0 & \cos\omega\cos\varphi & -\sin\omega\cos\varphi \\ 0 & \sin\omega & \cos\omega \end{bmatrix}\begin{bmatrix} \dot{\kappa}_g \\ \dot{\varphi}_g \\ \dot{\omega}_g \end{bmatrix} = B\begin{bmatrix} \dot{\kappa}_g \\ \dot{\varphi}_g \\ \dot{\omega}_g \end{bmatrix} \quad (4)$$

By integrating equation (4), the rotation during movement can be acquired.

$$\begin{bmatrix} \omega \\ \varphi \\ \kappa \end{bmatrix} = \int\begin{bmatrix} \dot{\omega} \\ \dot{\varphi} \\ \dot{\kappa} \end{bmatrix}dt + \begin{bmatrix} \omega_a \\ \varphi_a \\ \kappa_m \end{bmatrix} \quad (5)$$

**2.2.2 Filtering of rotation:** According to integration, the error is accumulated. It is important to point out here that the acceleration meter and the magnetic field sensor are free from such cumulative error. With the acceleration meter and the magnetic field sensor, the rotation during movement is corrected. Filtering theory is applied in order to correct the cumulative error.

For the efficient computation, Kalman filter is adopted as filtering method (Ristic *et al.*, 2004). The Kalman filter consists of dynamic equation and observation equation as follows:

$$\mathbf{x}_t = \mathrm{F}_t\,\mathbf{x}_{t-1} + \mathrm{G}_t\,\mathbf{v}_t \quad (6)$$
$$\mathbf{y}_t = \mathrm{H}_t\,\mathbf{x}_t + \mathbf{w}_t \quad (7)$$

where $\mathbf{x}_t$ = state vector at time *t*
$\mathbf{y}_t$ = observation vector at time *t*
$\mathrm{F}_t$ , $\mathrm{H}_t$, $\mathrm{G}_t$ = matrices of transition, system noise, observation at time *t*, respectively
$\mathbf{v}_t$ = system noise vector at time *t* ~N(0, $\mathrm{Q}_t$)
$\mathbf{w}_t$ = observation noise vector at time *t* ~N(0, $\mathrm{R}_t$).

In this study, these vectors and matrices are constructed as follows:

$$\mathbf{x}_t = \begin{bmatrix} \omega \\ \varphi \\ \kappa \\ \dot{\kappa}_g \\ \dot{\varphi}_g \\ \dot{\omega}_g \end{bmatrix}, \quad \mathbf{y}_t = \begin{bmatrix} \omega_a \\ \varphi_a \\ \kappa_m \\ \dot{\kappa}_g \\ \dot{\varphi}_g \\ \dot{\omega}_g \end{bmatrix}, \quad \mathrm{F}_t = \begin{bmatrix} 1 & 0 & 0 & & & \\ 0 & 1 & 0 & & \mathrm{B}_t & \\ 0 & 0 & 1 & & & \\ 0 & 0 & 0 & 1 & 0 & 0 \\ 0 & 0 & 0 & 0 & 1 & 0 \\ 0 & 0 & 0 & 0 & 0 & 1 \end{bmatrix} \quad (8)$$

The $\mathrm{B}_t$ is same as equation (4). Matrices $\mathrm{H}_t$ and $\mathrm{G}_t$ are set as identity matrices. In equation (8), suffix *a*, *m*, *g* express data of the acceleration meter, the magnetic field sensor, gyro sensor, respectively.

Since the Kalman filter is based on linear equation with normal distribution, the equations can be solved analytically. The equations (9) and (10) are related to forecast, and (11) –(13) filtering.

$$\mathbf{x}_{t|t-1} = \mathrm{F}_t\,\mathbf{x}_{t-1|t-1} \quad (9)$$

$$\mathrm{V}_{t|t-1} = \mathrm{F}_t\,\mathrm{V}_{t-1|t-1}\mathrm{F}^T{}_t + \mathrm{G}_t\,\mathrm{Q}_t\,\mathrm{G}_t^T \quad (10)$$

$$\mathrm{K}_t = \mathrm{V}_{t-1|t-1}\,\mathrm{H}_t^T(\mathrm{H}_t\,\mathrm{V}_{t-1|t-1}\mathrm{H}_t^T + \mathrm{R}_t)^{-1} \quad (11)$$

$$\mathbf{x}_{t|t} = \mathbf{x}_{t|t-1} + \mathrm{K}_t\,(\mathbf{y}_t - \mathrm{H}_t\,\mathbf{x}_{t|t-1}) \quad (12)$$

$$\mathrm{V}_{t|t-1} = \mathrm{V}_{t|t-1} - \mathrm{K}_t\,\mathrm{H}_t\,\mathrm{V}_{t|t-1} \quad (13)$$

State vector $\mathbf{x}_{t|t}$ is a result of filtering method for rotation.

### 2.3 Filtering method of position

Although the position of the mobile device is measured with GPS directly, the accuracy of the position can be improved by relating to the other sensors used for rotation calculation. Here the Kalman filter is also applied to all POS data to estimate the position.

**2.3.1 Calculation of position:** Since GPS data are originally longitude $E$, latitude $N$, ellipsoidal height $h$, the coordinates are transformed to the Plane Rectangular Coordinate System ($X_G$, $Y_G$, $Z_G$). The coordinates are used for initial position of the mobile device. For measurement of position during movement, data of GPS and the acceleration meter are considered.

The relationship between the measured acceleration and second derivatives of position ( $\ddot{X}$ , $\ddot{Y}$ , $\ddot{Z}$ ) can be expressed as follows.

$$\begin{bmatrix} \ddot{X}_t \\ \ddot{Y}_t \\ \ddot{Z}_t \end{bmatrix} = -\left( R_\kappa R_\varphi R_\omega \begin{bmatrix} \ddot{x}_a \\ \ddot{y}_a \\ \ddot{z}_a \end{bmatrix} - \begin{bmatrix} 0 \\ 0 \\ g \end{bmatrix} \right) \tag{14}$$

Here, rotation matrices are represented as R. Additionally, velocity ( $\dot{X}_t$ , $\dot{Y}_t$ , $\dot{Z}_t$ ) can be calculated from GPS data.

$$\begin{bmatrix} \dot{X}_t \\ \dot{Y}_t \\ \dot{Z}_t \end{bmatrix} = \frac{1}{\Delta t} \begin{bmatrix} X_{t-1} - X_{t-2} \\ Y_{t-1} - Y_{t-2} \\ Z_{t-1} - Z_{t-2} \end{bmatrix} \tag{15}$$

Where, $\Delta t$ is GPS data acquisition interval. By using both of velocity and acceleration, position at time $t$ is calculated.

$$\begin{bmatrix} X_t \\ Y_t \\ Z_t \end{bmatrix} = \begin{bmatrix} X_{t-1} \\ Y_{t-1} \\ Z_{t-1} \end{bmatrix} + \Delta t \begin{bmatrix} \dot{X}_{t-1} \\ \dot{Y}_{t-1} \\ \dot{Z}_{t-1} \end{bmatrix} + \frac{1}{2}\Delta t^2 \begin{bmatrix} \ddot{X}_{t-1} \\ \ddot{Y}_{t-1} \\ \ddot{Z}_{t-1} \end{bmatrix} \tag{16}$$

**2.3.2 Filtering of position:** In the similar fashion of filtering method of rotation, Kalman filter is applied for improving position. Dynamic and observation equations are same as equations (6), (7). State vector, observation vector and matrix $F_t$ are set for position filtering.

$$\mathbf{x}_t = \begin{bmatrix} X_t & Y_t & Z_t & \dot{X}_t & \dot{Y}_t & \dot{Z}_t & \ddot{X}_t & \ddot{Y}_t & \ddot{Z}_t \end{bmatrix}^T$$

$$\mathbf{y}_t = \begin{bmatrix} X_{Gt} & Y_{Gt} & Z_{Gt} & \dot{X}_t & \dot{Y}_t & \dot{Z}_t & \ddot{X}_t & \ddot{Y}_t & \ddot{Z}_t \end{bmatrix}^T$$

$$F_t = \begin{bmatrix} 1 & 0 & 0 & \Delta t & 0 & 0 & \Delta t^2/2 & 0 & 0 \\ 0 & 1 & 0 & 0 & \Delta t & 0 & 0 & \Delta t^2/2 & 0 \\ 0 & 0 & 1 & 0 & 0 & \Delta t & 0 & 0 & \Delta t^2/2 \\ 0 & 0 & 0 & 1 & 0 & 0 & \Delta t & 0 & 0 \\ 0 & 0 & 0 & 0 & 1 & 0 & 0 & \Delta t & 0 \\ 0 & 0 & 0 & 0 & 0 & 1 & 0 & 0 & \Delta t \\ 0 & 0 & 0 & 0 & 0 & 0 & 1 & 0 & 0 \\ 0 & 0 & 0 & 0 & 0 & 0 & 0 & 1 & 0 \\ 0 & 0 & 0 & 0 & 0 & 0 & 0 & 0 & 1 \end{bmatrix} \tag{17}$$

Analytical solution of the dynamic and observation equations can be conducted in the same manner of equations (9) – (13).

## 3. IMAGE INTEGRATING METHOD

The position and rotation based on POS data filtering are used as initial value of the orientation factors in image-based self-localization method. The position is already represented in the geodetic coordinate, and then the distance of baseline is also supplied in the real scale. The image-based self-localization method consists of feature points extraction and tracking, relative orientation, three dimensional coordinates estimation of the feature points, and orientation factors updates of the mobile device. Figure 2 shows an overview of the proposed method.

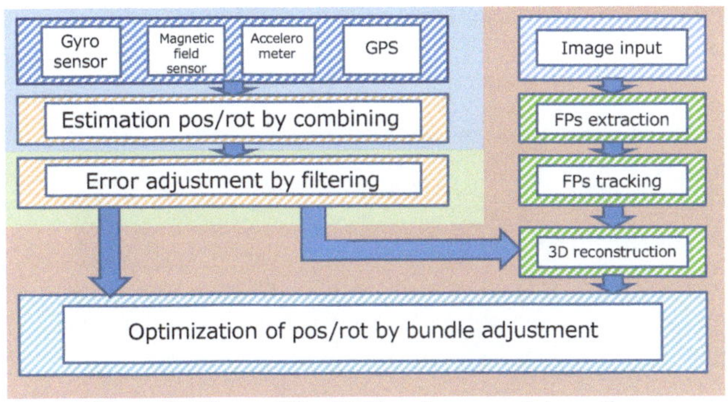


Figure 2. Flow of the POS and camera integration

### 3.1 Feature points extraction and tracking

Recently, sophisticated feature points extraction and tracking algorithms have been developed. One of the most reliable algorithms is SURF (Speeded-UP Robust Features) (Bay *et al.*, 2008). Figure 3 shows an example of feature points extraction by the SURF in our experiments.

Figure 3. Feature points extraction by SURF

Even if the SURF is applied to feature points extraction and tracking, incorrect matching points are still exist. The feature points tracking are refined by using not only adjacent frames also sequential frames. Firstly, extracted feature points are searched in sequence between adjacent frames. After the tracking process within a certain number of frames, position of feature points are re-projected into first frames (back matching). If the displacement between first and last position of the points is larger than a threshold, the feature points are discarded. With the result of the matching, three dimensional coordinates of the feature points can be calculated. When the depth of the points is larger than a threshold, the feature points are also discarded.

After the above mentioned thresholding process, incorrect matching points still remain (Figure 4). Especially in the case of application in urban area, similar textures make such incorrect matching. In this study, RANSAC (Random Sample Consensus) (Fischler and Bolles, 1981) is also applied (Figure 5). RANSAC algorithm is a method of outlier removal. Finally, the remaining points are accepted as feature points.

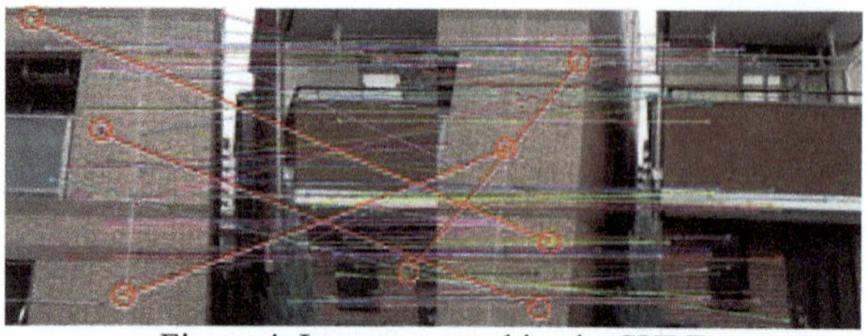

Figure 4. Incorrect matching by SURF

Figure 5. Correction of matching results by RANSAC

### 3.2 Three dimensional coordinates estimation

The position and rotation of a mobile sensor are already acquired through POS data filtering described in the previous chapter. With the position and rotation, and feature points tracking results, three dimensional coordinates of the feature points can be calculated. Camera calibration is conducted in advance. For the calibration, checkerboard is used.

First of all, an initial tree dimensional coordinates is built based on intersection (Stewenius *et al.*, 2006). For the optimization of intersection, RANSAC algorithm (Fischler and Bolles, 1981) is also applied.

### 3.3 Orientation factors updates

Once initial value of the orientation factors and three dimensional coordinates of the feature points are acquired, the orientation factors updates are conducted by bundle adjustment (Triggs *et al.*, 2000; Luhmann *et al.*, 2014).

The feature points have coordinates $\mathbf{p}_i = (X_i, Y_i, Z_i)$. The each sequential frame has a three dimensional coordinates $\mathbf{q}_j = (X_j, Y_j, Z_j)$ as the camera position. At the frames $j$, the feature point $i$ has camera coordinate system $(u_{ij}, v_{ij})$. A transformation between the camera coordinate and the world coordinate systems represents collinearity equation.

$$\bar{u}_{ij} = u_{ij} - \Delta u_{ij} = -c\frac{a_{11}(X_i - X_j) + a_{12}(Y_i - Y_j) + a_{13}(Z_i - Z_j)}{a_{31}(X_i - X_j) + a_{32}(Y_i - Y_j) + a_{33}(Z_i - Z_j)}$$
$$\bar{v}_{ij} = v_{ij} - \Delta v_{ij} = -c\frac{a_{21}(X_i - X_j) + a_{22}(Y_i - Y_j) + a_{23}(Z_i - Z_j)}{a_{31}(X_i - X_j) + a_{32}(Y_i - Y_j) + a_{33}(Z_i - Z_j)} \tag{18}$$

where $c$ = focal length

$\Delta u_{ij}, \Delta v_{ij}$ = factors of interior orientation

$a_{kl}$ = factors of rotation matrix

The position and rotation updates are computed iteratively by minimizing a robust objective function of the re-projection error.

$$E(\mathbf{p}_1, \cdots, \mathbf{p}_M, \mathbf{q}_1, \cdots \mathbf{q}_N) = \frac{1}{2}\|e\|^2 = \frac{1}{2}\sum_{i=1}^{N}\sum_{j=1}^{M}\left\{(u_{ij} - \bar{u}(\mathbf{p}_i, \mathbf{q}_j))^2 + (v_{ij} - \bar{v}(\mathbf{p}_i, \mathbf{q}_j))^2\right\} \tag{19}$$

$$\mathbf{x} = \arg\min \sum Obj\left(\frac{|e_k|}{\sigma_k}, \quad \sigma_T\right) \tag{20}$$

Tukey bi-weight objective function $Obj(,\sigma_T)$ is applied as a robust objective function and $\mathbf{x}$ is a set of parameters. Iteration of reweighted least squares method is used to allow the robust estimator to converge.

There are two types of the bundle adjustment: full bundle adjustment and local bundle adjustment. The local bundle adjustment uses only some recent frames. The full bundle adjustment is more accurate than the local bundle adjustment, but computational load is more expensive. In the sense of computation, the local bundle adjustment is more preferable at the expense of accuracy (Arth *et al.*, 2009). The local bundle adjustment can be applied recursively (Mclauchlan, 2000).

$$E_{1:j+1} = E_{1:j} + E_{j+1} \tag{21}$$

$E_{1:j}$ expresses the objective function by using from 1st frame to *j*th frame. According to the recursive form, bundle adjustment can be conducted effectively. It is important to point out here that the accuracy depends on the number of frames with the recursive form. We examined the relationships between number of frames and computation time / sum of squared error. In this study, based on the length of baseline, the local bundle adjustment is applied for improving the accuracy.

In order to solve the bundle adjustment problem, Levenberg-Marquardt method (Hartley and Zisserman, 2004) is applied. The objective function $E$ is approximated by the following formula:

$$E(\mathbf{x} + \delta\mathbf{x}) \approx E(\mathbf{x}) + \mathbf{g}^T\delta\mathbf{x} + \frac{1}{2}\delta\mathbf{x}^T H \delta\mathbf{x} \tag{22}$$

$$(H - \lambda I)\delta\mathbf{x} = \mathbf{g} \tag{23}$$

where $\mathbf{g} = \left.\frac{dE}{d\mathbf{x}}\right|_{\mathbf{x}}$ (gradient)

$H = \left.\frac{d^2E}{d\mathbf{x}^2}\right|_{\mathbf{x}}$ (Hessian)

$\lambda$ = dumping factor

## 4. EXPERIMENTS

The proposed method is applied to POS data and images taken in urban area. An iPhone4s is used as a mobile device, which is equipped with GPS, gyro sensor, acceleration meter, magnetic field sensor (POS) and a camera. Camera calibration is conducted in advance. In order to evaluate the accuracy, orientation data of Mobile Mapping System (MMS) of Trimble are compared. The MMS is also equipped with the iPhone4s.

The experimental site is in urban area (Figure 6), at which condition of GPS receiver is appropriate.

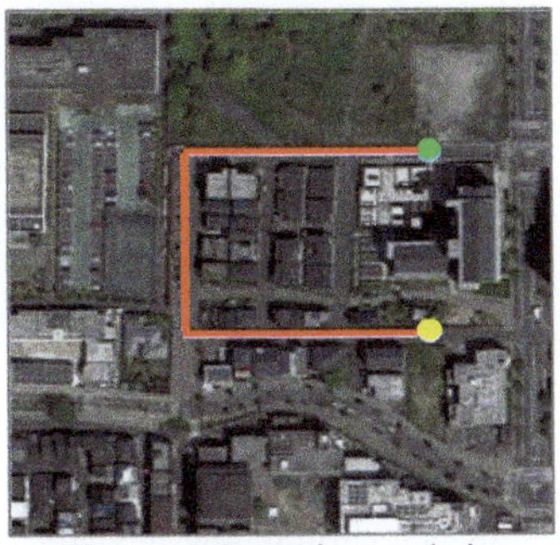

Figure 6. Experimental site

### 4.1 Experiment of POS data filtering

First experiment is application of POS data filtering. Figure 7 compares estimation results of roll, pitch and yaw with MMS data.

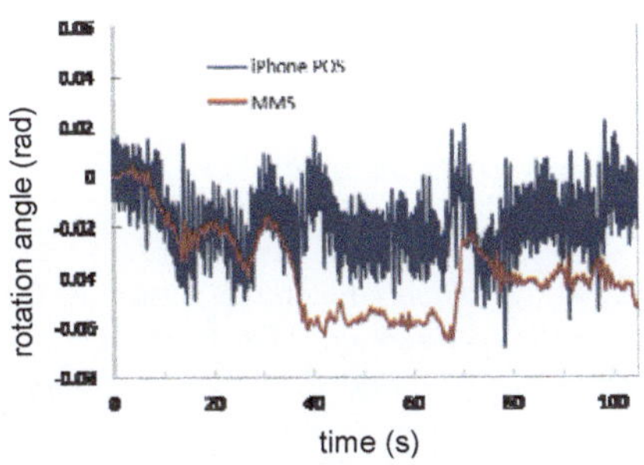


(a) roll

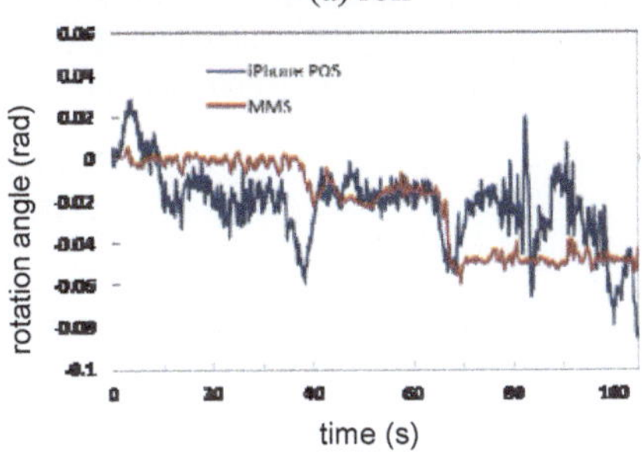


(b) pitch

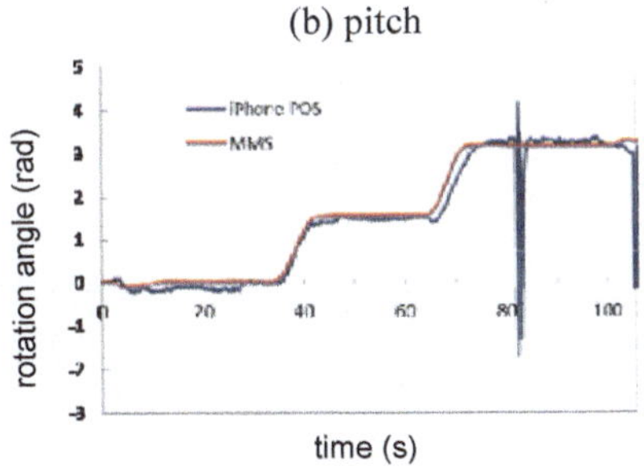


(c) yaw

Figure 7. Results of rotation angle with POS sensor fusion

The displacement between estimated value and MMS data with respect to roll, pitch, and yaw is 0.024 (rad), 0.031 (rad), and 0.17 (rad) on average, respectively. In the result of yaw, the gap was large, because of limitation of function in iPhone, that is negative value of yaw angle is converted to zero automatically.

Position accuracy is also examined (Figure 8). The root mean square of the positions displacement is 5.67 (m)

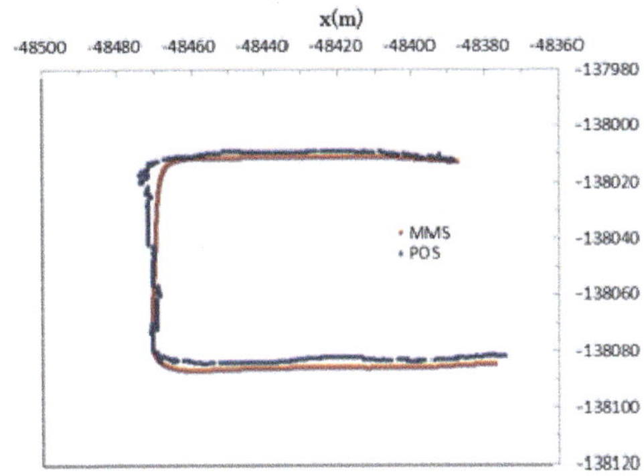


Figure 8. Result of position with POS sensor fusion

### 4.2 Experiment of POS and camera integration

As shown in Figure 8, accuracy of POS data filtering became worse at the area in a curve. Focusing on such area, result of POS and camera integration was examined.

Figure 9 shows the accuracies of rotation angle by all sensors (POS and camera) integration. From the comparison between the accuracies of the proposed method and ones of POS data filtering, clear improvement was not confirmed.

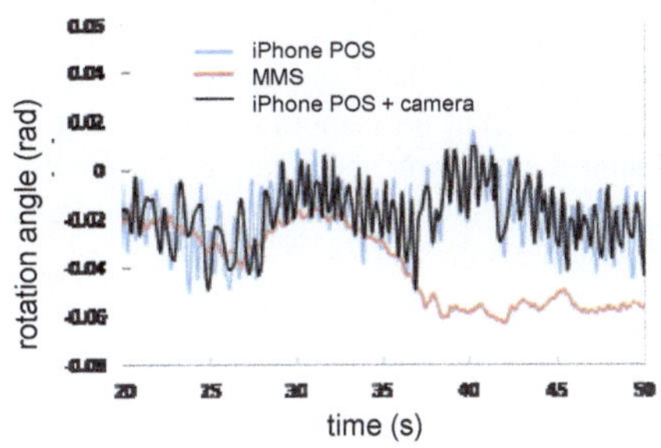


(a) roll

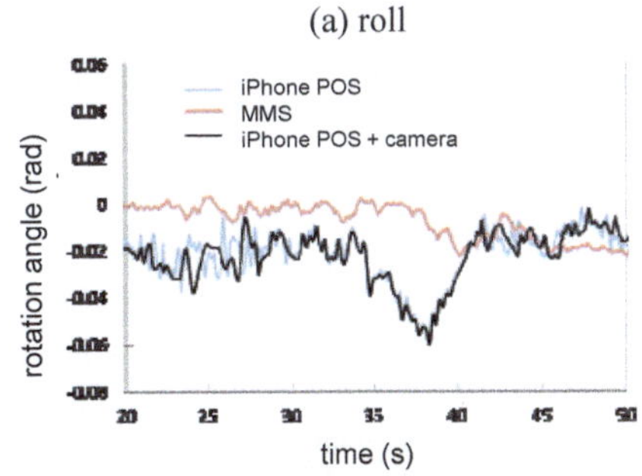


(b) pitch

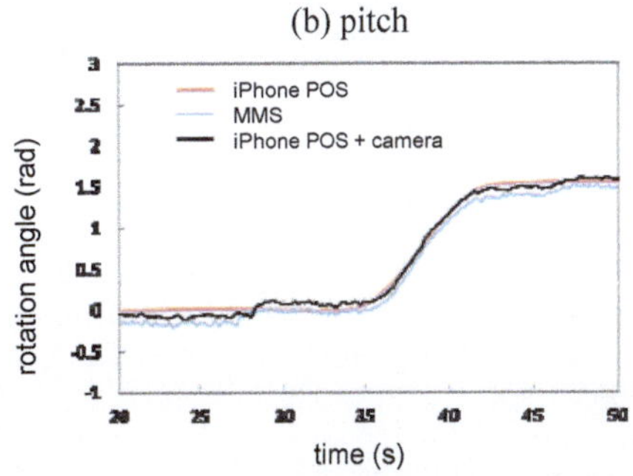


(c) yaw

Figure 9. Results of rotation angle with all sensors integration

Figure 10 shows the result of all sensors integration. Compared with the previous result, the accuracy improvement can be confirmed. During the application area, the root mean square of the positions displacement with only POS data filtering is 3.59 (m). On the other hand, one with POS and camera integration improved to 3.26 (m).

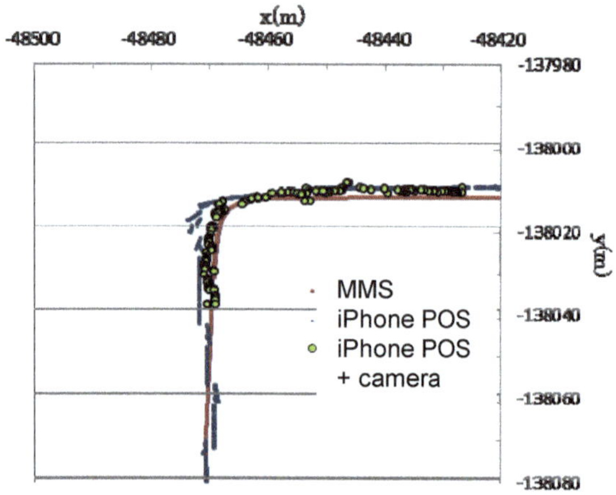


Figure 10. Result of position with all sensors integration

## 5. CONCLUSIONS

This paper develops the self-localization method using sensors, such as POS and cameras, on mobile devices simultaneously. The proposed method applies the Kalman filter in order to combine all sensors data except for camera (POS data filtering). Additionally, by using the position and rotation from POS data filtering as initial value of bundle adjustment, POS and camera integration method is achieved.

Through experiments with real data, the accuracy improvements of position and rotation by POS data filtering were confirmed. The results of final integration method improved the accuracy. It means that proposed self-localization method with POS and camera make the accuracy more sophisticated compared with only POS data filtering. Especially, the improvement at the area in a curve is noticeable. According to the experiments, the significance of the proposed method is confirmed.

As a further work, accuracy of three dimensional coordinates estimation of feature points will be evaluated by comparing with laser scanner data on MMS. Additionally, integrated filtering method between POS data filtering and bundle adjustment will become challenging investigation. As a result, promising method can be constructed, and then more impressive visualization will be accomplished.

## REFERENCES

Agrawal, M. and Konolige, K., 2006. Real-time localization in outdoor environments using stereo vision and inexpensive GPS. *Proceedings of the 18th International Conference on Pattern Recognition*, pp.1063-1068.

Arth, C., Wagner, D., Klopschitz, M., Irschara, A., Schmalstieg, D., 2009. Wide area localization on mobile phones. *Proceedings of 8th IEEE International Symposium on Mixed and Augmented Reality*, pp.73-82.

Bay, H., Ess, A., Tuytelaars, T. and Vangool, L., 2008. Speeded-up robust features (SURF). *Computer Vision and Image Understanding*, 110(3), pp.346-359.

Daehne, P. and Karigiannis, J., 2002. Archeoguide: system architecture of a mobile out door augmented reality system. *Proceedings of 1st IEEE International Symposium on Mixed and Augmented Reality*, pp. 506-514.

Fischler, M. A. and Bolles, R. C., 1981. Random sample consensus: a paradigm for model fitting with applications to image analysis and automated cartography. *Communications of the ACM*, 24(6), pp.381-395.

Hartley, R. and Zisserman, A., 2004. *Multiple View Geometry in Computer Vision.* Cambridge University Press, Cambridge.

Kourogi, M. and Kurata, T., 2005. Personal positioning based on walking locomotion analysis with self-contained sensors and wearable camera. *Proceedings of IEEE/ACM International Symposium on Virtual Reality Conference*, pp. 71-78.

Luhmann, T., Robson, S., Kyle, S. and Boehm, J., 2014. *Close Range Photogrammetry and 3D Imaging, 2nd Edition.* Walter de Gruyter GmbH, Berlin.

Mclauchlan, P.F., 2000. A batch / recursive algorithm for 3D scene reconstruction. *Proceedings of IEEE Computer Society Conference on Computer Vision and Pattern Recognition.* pp. 2738-2743.

Reitmayr, G. and Drummond, T.W., 2005. Going out: robust model-based tracking for outdoor augmented reality. *Proceedings of IEEE on Virtual Reality Conference*, pp. 71-78.

Ristic, B., Arulampalam, S., Gordon, N., 2004. *Beyond the Kalman Filter: Particle Filter for Tracking Applications.* Artech House Publishers, Boston.

Schall, G., Wagner, D., Reitmayr, G., Taichmann, E., Wieser, M., Schmalstieg, D., Hofmann, W.B., 2009. Global pose estimation using multi-sensor fusion for outdoor augmented reality. *Proceedings of 8th IEEE International Symposium on Mixed and Augmented Reality*, pp.153-162.

Stewenius, H., Engels, C. and Nister, D., 2006. Recent developments on direct relative orientation. *ISPRS Journal of Photogrammetry and Remote Sensing*, 60, pp284-294.

Triggs, B., McLauchlan, P., Hartley, R. and Fitzgibbon, A., 2000. Bundle adjustment – a modern synthesis. In Triggs, B., Zisserman, A. and Szeliski, R. eds., *Vision Algorithm: Theory and Practice*, Springer-Verlag, Berlin.

Wagner, D., Schmalstieg, D., Bischof, H., 2009. Multiple target detection and tracking with guaranteed framerates on mobile phones. *Proceedings of 8th IEEE International Symposium on Mixed and Augmented Reality*, pp.57-64.

# A RESEARCH ON THE HIERARCHY AND COMPLETENESS OF ROOF TOPOLOGY FOR ROBUST BUILDING RECONSTRUCTION FROM AIRBORNE POINT CLOUD

B. Xu [a,*] , W.S. Jiang [a], Q.S. Zhu [a]

[a] State Key Laboratory of Information Engineering in Surveying, Mapping and Remote Sensing, Wuhan University, Wuhan, Hubei, China – (lmars_xubo, jws, zhuqs)@whu.edu.cn;

**Commission IV, WG IV-7 & Commission V, WG V-4**

**KEY WORDS:** 3D Building Reconstruction, Point Cloud, Level of Details, Roof Topology Graph, Topology Completeness

**ABSTRACT:**

In this work, we concentrate on the hierarchy and completeness of roof topology, and the aim is to avoid or correct the errors in roof topology. The hierarchy of topology is expressed by the hierarchical roof topology graph (HRTG) in accord with the definition of CityGML LOD (level of details). We decompose the roof topology graph (RTG) with a progressive approach while maintain the integrality and consistency of the data set simultaneously. Common feathers like collinear ridges or boundaries are calculated integrally to maintain their completeness. The roof items will only detected locally to decrease the error caused by data spare or mutual interference. Finally, a topology completeness test is adopted to detect and correct errors in roof topology, which results in a complete and hierarchical building model. Experiments shows that our methods have obvious improvements to the RTG based reconstruction method, especially for sparse data or roof topology with ambiguous.

## 1. INTRODUCTION

### 1.1 Motivation

Automatic reconstruction of urban 3D models from point clouds has attracted numerous studies during the last two decades (Haala & Kada, 2010; Musialski et al., 2013). Owing to the developments of data acquisition technology, increasingly dense and reliable point clouds can be provided. This provides the possibility to recognize subtle details and complex structure, while also brings huge challenges to current reconstruction methods (Elberink, 2008). Reference (Xiong, Elberink, & Vosselman, 2014) claims that even when only considering planar roof surfaces, the building type complexity can still be a major problem.

A basic step for the reconstruction of complex building roof is to describe the topology relationship between roof planes, where a roof topology graph (RTG) is widely used. However, several challenges have to be confronted for them. Firstly, the RTGs are not always right in the presence of noisy, low point density and resulting errors in pre-processing steps. The reconstruction methods will fail under false RTG (Xiong et al., 2014). Secondly, the stored topological relationships by RTG is rather limited and often not exploited. The scale or hierarchy of roof planes are not considered and the step structures cannot be well expressed. Thirdly, the completeness and correctness of roof topology are not guaranteed. Holes or consistency boundaries might appear in the constructed models. As such, strategies to improve the roof topology and ways to find and correct the errors in roof topology become important.

In this work, we improve the definition of roof topology mainly in two aspects. A hierarchical roof topology graph (HRTG) is proposed instead of the traditional RTG in accord with the definition of building LOD(level of details, see (Gröger, 2008). The main goal is to avoid the errors in roof topology by generating overall solution for common feathers and partial solution for roof items. A major roof surface will be less interrupted by its accessory structure and the search scope when detecting building primitives can also be reduced. Meanwhile, a topology completeness test is adopted to detect errors in the model. Arcs at the roof top are forced to form the half-edge structure (McGuire, 2000), and areas where misrepresentations of step structures or holes in the models exist can be easily detected and then corrected by a hypothesize-and-verify process.

### 1.2 Relate Works

The roof topology graph (RTG) has been widely used for building roof reconstruction from point clouds. The concept of RTG is simple: a vertex in the graph stands for a roof plane and an edge represents the adjacency between two planes. Reference (Elberink & Vosselman, 2009) develops it by adding labels to the edge in graph that considers features like normal orientation, convex/concave, and tilted/horizontal. The local structure in RTG has been used to detect the intersection of multiply planes (minimal close circle analysis)(S. N. Perera, Nalani, & Maas, 2012), simple building primitives, i.e., in (Verma, Kumar, & Hsu, 2006), or outer boundary (outmost circle) in (G. S. N. Perera & Maas, 2014). A graph matching approach is adopted to search more complex primitives in (Elberink & Vosselman, 2009). The RTG are also used to interpret the building structures by some other works, i.e. (Sampath & Shan, 2010; Zhou & Neumann, 2011). However, reference (Xiong et al., 2014) points out that the RTG are not always right and the above reconstruction methods may fail under false RTG. Under such consideration, a graph-to-graph correcting approach based on graph edit dictionary is proposed and about one quarter of erroneous graph is claimed to be corrected. As such, the error in the RTG still remains a major problem in roof reconstruction.

As reference (Boguslawski, Gold, & Ledoux, 2011) argued, the stored topological relationships among model elements is rather limited and often not exploited. Generally, an overall solution that considers relations among roof elements turns to produce better results than totally data-driven methods. Model primitives are widely used for that goal and some highly model-driven methods (Karantzalos & Paragios, 2010; Lafarge, Descombes, Zerubia, & Pierrot-Deseilligny, 2008; Lafarge & Mallet, 2011; Vanegas, Aliaga, & Benes, 2012) impliedly added in topology constraints thus can directly giant topology-correct models. Considering the limit of pre-specified building primitives, reference (Sohn, Huang, & Tao, 2008) reconstruct geometric topology between adjacent linear features by adopting a BSP (Binary Space Partitioning) tree. Reference (Boguslawski et al., 2011) designs a dual data structure in which a vertex stands for a primal cell and an edge is the connection between cells. Then, the Euler-type operators can be adopted on the primitives and maintain their connectivity. However, hierarchy of RTG is rarely considered in those works, thus the connotative messages insider model (i.e. collinear, parallel ridges or inside items) are not exploited. When the building structure is complex or several buildings are adjacent, the RTG will be rather complex. Errors in RTG may be produced in the presence of resulting errors in pre-processing steps or mutual interference between roof elements. Moreover, some step structures (i.e. two planes have both intersected ridge and step edge) are lack of constraints in RTG and the completeness of constructed models are not guaranteed. Our HRTG is constructed to decompose the complex models step by step and make full use the implicit knowledge in the models simultaneously. The topology completeness test is also adopted to detect and correct errors in roof topology.

The remainder of this paper is organized as follows: Section 2 firstly gives the overview of our reconstruction procedure. Then details of the hierarchical roof topology graph (HRTG) and the topology completeness test are introduced. Experimental results are analysed in Section 3, followed by conclusions in Section 4.

## 2. METHODOLOGY

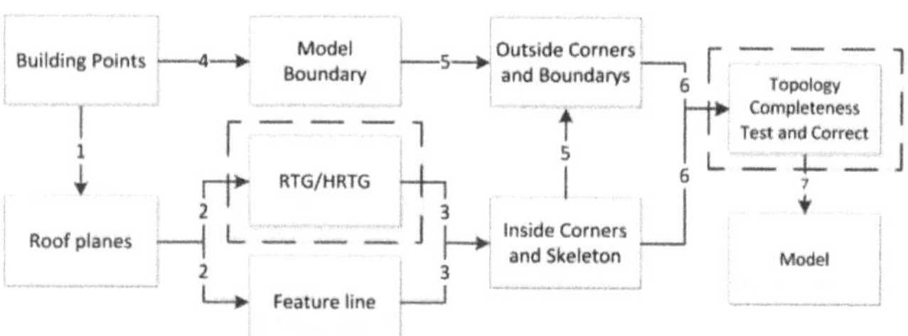


Fig 1. Reconstruction workflow

Similar to some existing work (Elberink & Vosselman, 2009; G. S. N. Perera & Maas, 2014; Verma et al., 2006; Xiong et al., 2014), our reconstruction strategy follows the procedure of segmentation, roof topology construct and feature extraction with the assistance some building primitives. The whole process are summarized in Fig .1 and our main contribution are marked by dotted boxes, where the HRTG and topology completeness test are proposed. As we mainly focus on the hierarchy and completeness of building topology, no more details about the work flow will be given here.

### 2.1 Hierarchical roof topology graph

The hierarchical or level of details (LOD) of building is an important factor that has been widely discussed by current researches. The concept is important because the following reasons:

1. The roof plans always have large range of scales and different reconstruction strategy needs to be adopted respectively. For major structures, data-driven methods are suggested to produce more accurate results, while model-driven methods need to be used for roof items or sub-structures because of date sparse or incompleteness.
2. To be man-made objects, the buildings have many connotative semantic features, such as main direction, symmetry, and parallel/vertical edges. Those features need to be calculated from the main structures or as a whole, and then delivered to sub-structures.
3. Decomposing the buildings into several child models or building primaries is necessary for the reconstruction of buildings with complex structure. Some methods need to match parts of the building with pre-constructed primaries library also.

However, the hierarchy of buildings are not easy or convenient enough to be expressed by the roof topology graph (RTG). The RTG only stores the plane-to-plane adjacency between neighbour planes and take all planes the same. What's more, the construction of RTG is mainly based on neighbour analysis of edge points, with few integral constraints. This makes it error-prone under spare data, in the presence of resulting errors in pre-processing steps, or when mutual interferences between roof elements exist. In this work, we attempt to describe the hierarchy of buildings via the level of details.

A hierarchical roof topology graph (HRTG) is progressively constructed that decompose the buildings step by step. The levels in HRTG are in accord with definition of building LOD. A vertex in an upper level will be a child graph in the downer level. The LOD we concerned are from LOD 0-LOD2. Due to limit space, definition about building LOD are not given here, more message can refer to work of (Verdie, Lafarge, & Alliez, 2015). Here, we mainly discuss how the HRTG in each level are constructed and how structure message in building hierarchy are used for more robust reconstruction process.

#### 2.1.1 LOD 0

LOD 0 care not the height of point clouds and only need to calculate the overall boundary of the interconnected buildings.

#### 2.1.2 LOD 1

Same as the CityGML LOD1, the rooftops are considered as horizontal in this level, and the major work is to separate the model by step structures. The step edges in RTG are disconnected and then the RTG can be decomposed by searching the connected subgraphs in graphs. Each subgraph is a vertex in the HRTG in this level. Two issues need to be considered in this step: the excluding of ambiguous edges and deciding a proper step wall between two child models.

False or ambiguous edges may exist between two adjacent buildings, especially when two roofs planes have similar height, orientations or in the present of falsely classified roof points. To avoid the problems, an extra test is adopted upon the detected ridges with a simple idea that points from two sides of the ridge should come from different roof surface. The point-plane distance deviation for points in conflict areas is calculated. A intersect ridge will be refused if the new deviation is bigger than both the origin value and a pre-given threshold.

The step structures need to be calculated to avoided holes in model if the data are original connected in 2D. If obvious height different can be detected, step ridge can be directly generated by method in (Oude Elberink, 2010). Sometimes however, step structure may have similar height, different main orientations or irregular plane boundary. Vertical cutting planes (cutting lines in 2D) need to be specially calculated under those situations. The cutting lines are determined by two elements: orientation and position. The orientation is decided by the nearest robust horizontal roof ridge from both sides (two orientations are tested if they are different), and the optimal position is decided by minimization the fitting error between nearby boundary points and cutting lines. The constructed child models are then extended to the vertical cutting planes to keep the connectivity of the whole building.

#### 2.1.3 LOD 2

This level discusses the intersection of multiple roof planes and the calculation of the inside corners or ridge skeletons. Instead of calculate all ridges as a whole in the whole RTG, some extra steps that further decompose the RTG is adopted. The main goal is use the structure features to constraint the identification of roof ridges. The decomposition of RTG can reduce the interrupt from neighbour planes and the influence of inaccurate segmentation results. Three decomposition rules (marked by A, B and C) are defined and used to further divide the models if identified:

A) Main horizontal or convex ridges, inspired by work of (Fan, Yao, & Fu, 2014). The main horizontal ridges are always local maxima points which are easily detected and useful for division. The horizontal/convex ridges are guided or verified by initial intersect edges and extended to the boundary of child model if possible.

B) Skeletons that divide the whole data into several parts. Based on the close cycle analysis, the intersect ridges can be calculated and then neighbour ridges can be connected to construct the inside skeletons. Some ridges can be extended to the model boundary thus divide the model into several parts.

C) Planes-in-plane or planes only connect to a single plane. These plans are always roof items or sub-structures, and are constructed locally. Some simple model primaries are used in this step for planes with small scales. Meanwhile, their points are projected into the major structure to avoid a major roof become broken pieces.

If the above three processes can successfully divide the point clouds into pieces that only contain single plane, proper model can be generated. However, the goal may not be easily under some difficult situation, i.e. improper segmentation results, sparse point clouds or incomplete data. Most current RTG based methods will fail then. An advantage of our method is that we decompose the point clouds when we construct the HRTG. Those problematic areas will be only limited in a small area divided by the skeleton and only contain a small number of planes. For those areas, several additional processes are adopted by turn until the precision of constructed model is accepted.

1) If the area is at one side of the main horizontal ridges, use the symmetry and side projection (Rau & Lin, 2011) to produce the major roof and construct the rest objects.
2) Another iteration of the segmentation and reconstruction process is applied locally.
3) Match the data with pre-defined model primaries.

#### 2.1.4 Conclusion

The construction of HRTG is not only a topology combination of roof planes, but also a process of data disaggregation. When decomposing the RTG into different parts, we also divide the dates and maintain their connectivity in 2D simultaneously. The HRTG in each step is in accord with the CityGML thus we can directly general the building model under different levels.

### 2.2 Topology completeness test

As argued in (Xiong et al., 2014), current RTG methods suppose that the RTG is right which may not always true. It frequently occurs that a small plane or roof edges are lost or false distinguished, which final results error in RTG or model. Even if the RTG is right, errors will still occur in constructed models, as step structures are often lack of constraints in RTG and some boundaries are not regularly parallel and vertical. What's more, the dates can also be sparse, occluded or even become broken pieces. In this section, we attempt to improve the definition of roof topology and find ways to detect and even correct the errors in roof topology. The topology completeness test is defined for such goal.

#### 2.2.1 Definition

The test is adopted based on the assumption below:
1) The constructed model should be connected in space if their point clouds are originally connected in 2D.
2) For each edge in the constructed model, two sides need to be found to form the half-edge data structure (Fig.2). Each side can be one of the following three types: (a) A roof plane; (b) A inside step wall; (c) An outside wall.

#### 2.2.2 Error detection

According to the above definition, only three types of arcs are accepted in the constructed model, whose two sides can be defined: (1) Roof ridge: (a) & (a); (2) Model boundary: (a) & (c); (3) Inside step edge: (a) & (b).

The detection process firstly mixes the roof plane boundaries with the intersected ridges or step edges which construct the mixed boundaries, and then compare them with the whole model boundary to find the undecided arcs. When errors in the roof topology graphs occurs, only one side can be detected for the arcs (Fig.3). These undetermined arcs will gather around the error region and compose a close circle (red dotted line) together with the integral model outside boundary. Due to the limit space, only two types of errors are provided and others are the same (four types are given in work of (Xiong et al., 2014)). Our topology correction is adopted upon those close circles.

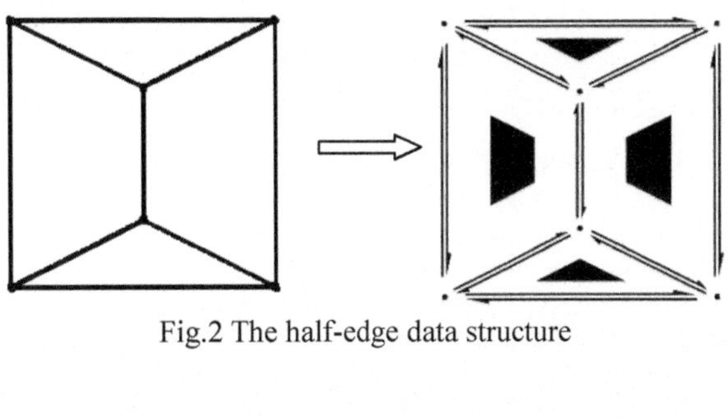

Fig.2 The half-edge data structure

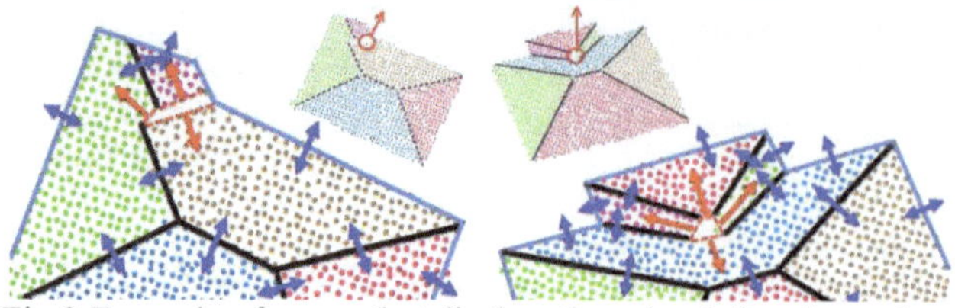

Fig.3 Example of errors that disobey the rules, left: miss a plane; right: miss a ridge (marked by red arrows and arcs).

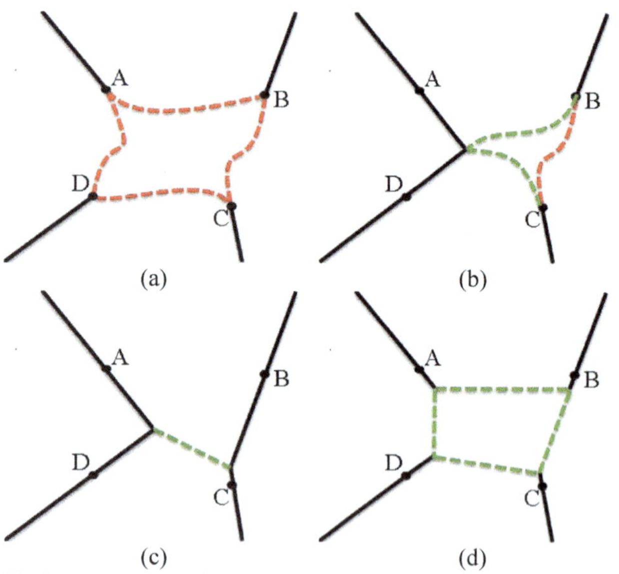


Fig.4. Examples of correct manners, (a) origin region, (b) expansion, (c) add ridges/step wall, (d) add new plane.

### 2.2.3 Errors Correction

Though the error regions can be detected by the undetermined arcs in the front step, approach to correct them may not be easy as the errors type can be various. In our problem, the correct approach is considered as the division of the error regions in 2D, and to find another side for those arcs. Three types of hypothesis with decreasing priorities are considered (Fig .4).

The expansion process (b) is used to test whether the error regions can be included into neighbour planes or the extensible ridges within the plane can be intersected. Some errors caused by neighbour competition in segmentation step can be solved. When we extend the ridges, a special check is adopted to test if the points near the ridge can be smoothly divided into two different parts. Some short ridges or inside step structure are failed to be distinguished because of data sparse, occlusion or improper segmentation results. The corresponding hypotheses are provided in (c). As for (d), if points in this area can be found, a fitted or horizontal plane will be tested. Some simple pre-defined building primaries will also be matched.

## 3. EXPERIMENTS AND EVALUATION

In this section, we experimentally explain the details of the HRTG and the completeness test, and their affects to the reconstruction results. The selected data in our experiments are from the Vaihingen data set (Germany), with an average point density of $4/m^2$.

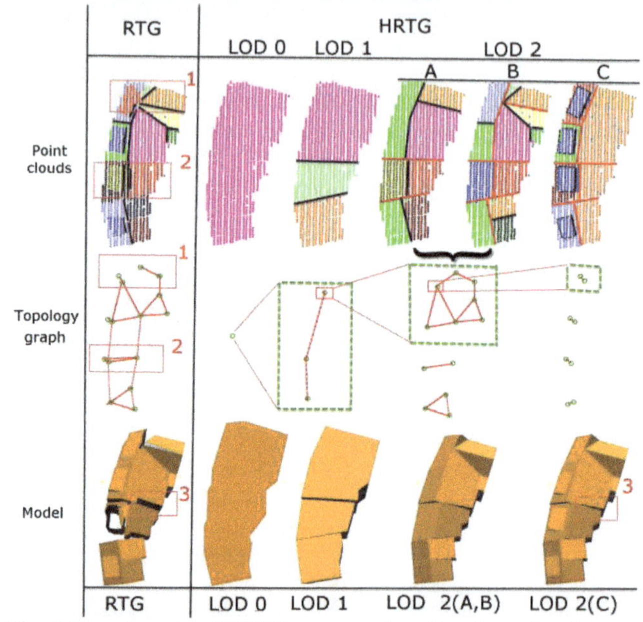


Fig .5 Experiments of HRTG construction. For the point clouds, different colours represent the point clouds to be separated in each step; Key ridges to be calculated are marked in black and the red ones are ridges extracted in previous steps.

Fig.5 gives a compare experiment of RTG and HRTG based methods. For RTG based methods, several errors occurred in the provided results, including errors caused by inaccurate segmentation or classification (error 1 and 2) and overregulation or errors in irregular boundaries (error 3). In HRTG, each level concentrates on particular property of the models which aims to avoid parts of the above errors. LOD 0 produce the outside boundary of the whole building or connective building blocks (consist of only one vertex here). In LOD 1, the cutting lines are defined and the connected building models are extended to the vertical wall thus errors 3 will be avoided.

LOD 2 discusses relations among roof planes. Unlike current RTG based methods, we divide the construction of roof topology into three sequential processes. In A, we detect the main horizontal and convex ridges which are simple but rather useful processes. In C, the roof sub-structures are detected and their points are projected into the main roof which will greatly reduce the difficulties in B, as the main roofs in B will become integral ones and the extraction of boundaries and ridges can be much easier. As the point clouds are often sparse in C, modern driven methods is adopted and several model primaries are also defined and matched. This is also very important for the understanding and storage of building models, which enhance its potential in other applications.

The difficulties of topology completeness test mainly exist in the hypothesis generation, as the building complexity makes the possible of errors ever-changing. An advantage of our method is that the HRTG divides the model of complex buildings into different parts. Some errors have been avoided and the rest ones always centre in very small regions. In our test, evident improvements are produced by using very simple hypotheses types.

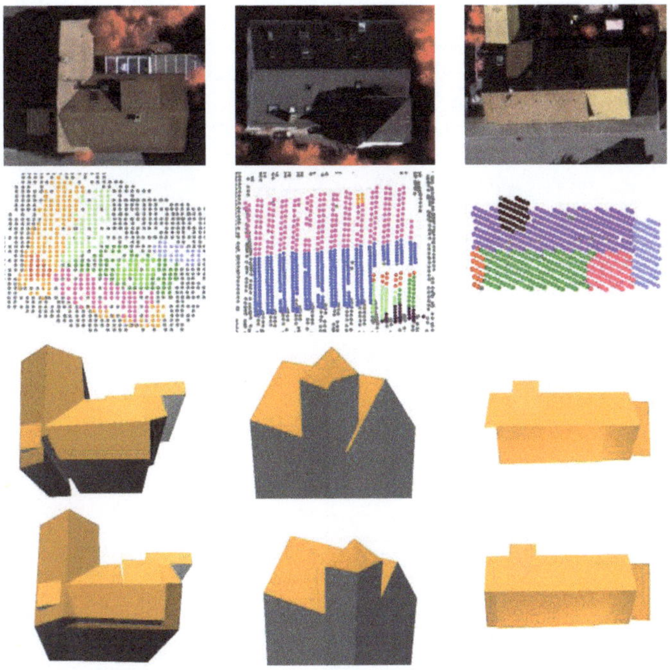

(a) (b) (c)

Fig .6 Improvements by methods using HRTG: Image, segmented points clouds, results use RTG and our final results are provided.

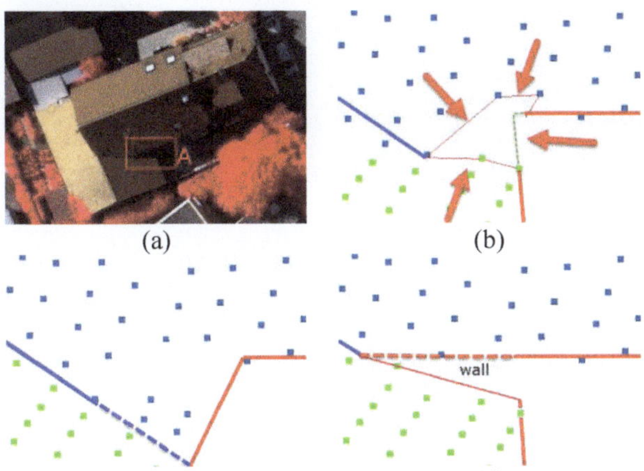


(a) (b) (c) (d)

Fig .7 Error corrections. (a): Image; (b): error detection of A in (a); (c) and (d): two hypotheses of expansion and re-segmentation.

Fig. 6 shows some buildings that are error-prone for RTG based methods. In (a), the points of right-down main roof plane is sparse and often lost. In (b), the collinear boundary of the tower and roof may also be neglected. In (c), the left-down roof may also fail to be constructed as the intersect ridge is lost. Our methods produce significant improvements in those situations.

It has to be noted that a correct RTG is not enough to ensure a right roof topology of the constructed model. In Fig. 7, there are both step edge and intersect edge between the two planes. What's more, the improper segmentation results interrupt the distinguishing of step edges. As a result, ambiguity occurs in area A. Similar situations may also occur in the data of Fig .5. Generally, low point density request looser thresholds to detect the edges or intersect ridges; therefor, while false results will also be easier produced. In our work, this problem is solved by a hypothesis-and-testing process. Take Fig .7(c) and (d) for example, different solutions are all tested and compared on the ambiguous regions to generated the most proper results based on the observed model errors. In Fig. 8, models of several connective buildings with complex structures are also provided. The HRTG shows great advantage under such situations as the buildings have very distinct hierarchy.

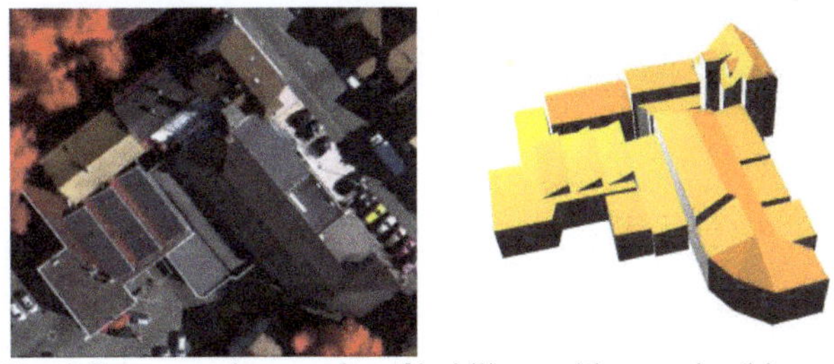

Fig. 8 Segmentation results of buidlings with complex hierarchy.

## 4. CONCLUSION

In this work, we have shown our improvements in the roof topology by taking its hierarchy and completeness into consideration. The topology hierarchy is reliazed by the HRTG in according with the CityGML LOD. It reorganizes the RTG and decompose the whole model data step by step, which can reduce the mutual interference among neighbor planes and make full use of the integral information of the child model. Models under different level of details can be produced simultaneously. The completeness of roof topology is test based on the half-edge data structure. The HRTG restricts the errors in small local regions, and then we correct the errors by a hypothesize-and-verify process. Our methods can reconstruct models under sparse data with imperfect segmentation or feature extraction results. The future works is to enlarge the hypothesis types in the correction process and find more reasonable quantitative standard to verify the hypothesis.

## ACKNOWLEDGEMENTS

This work was partially supported by the National Basic Research Program (973 Program) of China under Grant 2012CB719904.

## REFERENCE

*Boguslawski, Pawel, Gold, Christopher M, & Ledoux, Hugo. (2011). Modelling and analysing 3D buildings with a primal/dual data structure. ISPRS Journal of Photogrammetry and Remote Sensing, 66(2), 188-197.*

*Elberink, Sander Oude. (2008). Problems in automated building reconstruction based on dense airborne laser scanning data. Paper presented at the International Archives of the Photogrammetry, Remote Sensing and Spatial Information Sciences, Beijing, China.*

*Elberink, Sander Oude, & Vosselman, George. (2009). Building reconstruction by target based graph matching on incomplete laser data: analysis and limitations. Sensors, 9(8), 6101-6118.*

*Fan, Hongchao, Yao, Wei, & Fu, Qing. (2014). Segmentation of Sloped Roofs from Airborne LiDAR Point Clouds Using Ridge-Based Hierarchical Decomposition. Remote Sensing, 6(4), 3284-3301.*

*Gröger, Gerhard. (2008). OpenGIS city geography markup language (CityGML) encoding standard: Open Geospatial Consortium.*

*Haala, N., & Kada, M. (2010). An update on automatic 3D building reconstruction. Isprs Journal of Photogrammetry and Remote Sensing, 65(6), 570-580. doi: DOI 10.1016/j.isprsjprs.2010.09.006*

*Karantzalos, Konstantinos, & Paragios, Nikos. (2010). Large-scale building reconstruction through information fusion and 3-d priors. Geoscience and Remote Sensing, IEEE Transactions on, 48(5), 2283-2296.*

*Lafarge, Florent, Descombes, Xavier, Zerubia, Josiane, & Pierrot-Deseilligny, Marc. (2008). Automatic building extraction from DEMs using an object approach and application to the 3D-city modeling. ISPRS Journal of Photogrammetry and Remote Sensing, 63(3), 365-381.*

*Lafarge, Florent, & Mallet, Clement. (2011). Building large urban environments from unstructured point data. Paper presented at the Computer Vision (ICCV), 2011 IEEE International Conference on.*

*McGuire, Max. (2000). The half-edge data structure. Website: http://www. flipcode. com/articles/article halfedgepf. shtml.*

*Musialski, Przemyslaw, Wonka, Peter, Aliaga, Daniel G, Wimmer, Michael, Gool, L, & Purgathofer, Werner. (2013). A survey of urban reconstruction. Paper presented at the Computer Graphics Forum.*

*Oude Elberink, Sander Jakob. (2010). Acquisition of 3D topography: automated 3D road and building reconstruction using airborne laser scanner data and topographic maps: University of Twente.*

*Perera, Gamage Sanka Nirodha, & Maas, Hans-Gerd. (2014). Cycle graph analysis for 3D roof structure modelling: Concepts and performance. ISPRS Journal of Photogrammetry and Remote Sensing.*

*Perera, Sanka Nirodha, Nalani, Hetti Arachchige, & Maas, Hans-Gerd. (2012). An Automated Method for 3d Roof Outline Generation and Regularization in Airbone Laser Scanner Data. ISPRS Annals of Photogrammetry, Remote Sensing and Spatial, Information Sciences I-3, 281-286.*

*Rau, Jiann-Yeou, & Lin, Bo-Cheng. (2011). Automatic roof model reconstruction from ALS data and 2D ground plans based on side projection and the TMR algorithm. ISPRS Journal of Photogrammetry and Remote Sensing, 66(6), S13-S27.*

*Sampath, A., & Shan, J. (2010). Segmentation and Reconstruction of Polyhedral Building Roofs From Aerial Lidar Point Clouds. Ieee Transactions on Geoscience and Remote Sensing, 48(3), 1554-1567. doi: Doi 10.1109/Tgrs.2009.2030180*

*Sohn, G., Huang, X. F., & Tao, V. (2008). Using a Binary Space Partitioning Tree for Reconstructing Polyhedral Building Models from Airborne Lidar Data. Photogrammetric Engineering and Remote Sensing, 74(11), 1425-1438.*

*Vanegas, Carlos A., Aliaga, Daniel G., & Benes, Bedrich. (2012). Automatic Extraction of Manhattan-World Building Masses from 3D Laser Range Scans. Visualization and Computer Graphics, IEEE Transactions on, 18(10), 1627-1637. doi: 10.1109/TVCG.2012.30*

*Verdie, Yannick, Lafarge, Florent, & Alliez, Pierre. (2015). LOD Generation for Urban Scenes. ACM Transactions on Graphics.*

*Verma, Vivek, Kumar, Rakesh, & Hsu, Stephen. (2006). 3d building detection and modeling from aerial lidar data. Paper presented at the Computer Vision and Pattern Recognition, 2006 IEEE Computer Society Conference on.*

*Xiong, B., Elberink, S. O., & Vosselman, G. (2014). A graph edit dictionary for correcting errors in roof topology graphs reconstructed from point clouds. Isprs Journal of Photogrammetry and Remote Sensing, 93, 227-242. doi: DOI 10.1016/j.isprsjprs.2014.01.007*

*Zhou, Qian-Yi, & Neumann, Ulrich. (2011). 2.5 D building modeling with topology control. Paper presented at the Computer Vision and Pattern Recognition (CVPR), 2011 IEEE Conference on.*

# Permissions

All chapters in this book were first published in TIAPRSSIS, by Copernicus Publications; hereby published with permission under the Creative Commons Attribution License or equivalent. Every chapter published in this book has been scrutinized by our experts. Their significance has been extensively debated. The topics covered herein carry significant findings which will fuel the growth of the discipline. They may even be implemented as practical applications or may be referred to as a beginning point for another development.

The contributors of this book come from diverse backgrounds, making this book a truly international effort. This book will bring forth new frontiers with its revolutionizing research information and detailed analysis of the nascent developments around the world.

We would like to thank all the contributing authors for lending their expertise to make the book truly unique. They have played a crucial role in the development of this book. Without their invaluable contributions this book wouldn't have been possible. They have made vital efforts to compile up to date information on the varied aspects of this subject to make this book a valuable addition to the collection of many professionals and students.

This book was conceptualized with the vision of imparting up-to-date information and advanced data in this field. To ensure the same, a matchless editorial board was set up. Every individual on the board went through rigorous rounds of assessment to prove their worth. After which they invested a large part of their time researching and compiling the most relevant data for our readers.

The editorial board has been involved in producing this book since its inception. They have spent rigorous hours researching and exploring the diverse topics which have resulted in the successful publishing of this book. They have passed on their knowledge of decades through this book. To expedite this challenging task, the publisher supported the team at every step. A small team of assistant editors was also appointed to further simplify the editing procedure and attain best results for the readers.

Apart from the editorial board, the designing team has also invested a significant amount of their time in understanding the subject and creating the most relevant covers. They scrutinized every image to scout for the most suitable representation of the subject and create an appropriate cover for the book.

The publishing team has been an ardent support to the editorial, designing and production team. Their endless efforts to recruit the best for this project, has resulted in the accomplishment of this book. They are a veteran in the field of academics and their pool of knowledge is as vast as their experience in printing. Their expertise and guidance has proved useful at every step. Their uncompromising quality standards have made this book an exceptional effort. Their encouragement from time to time has been an inspiration for everyone.

The publisher and the editorial board hope that this book will prove to be a valuable piece of knowledge for researchers, students, practitioners and scholars across the globe.

# List of Contributors

**M. Favorskaya**
Institute of Informatics and Telecommunications, Siberian State Aerospace University, 31 Krasnoyarsky Rabochy av., Krasnoyarsk, 660014 Russian Federation

**A. Nosov**
Institute of Informatics and Telecommunications, Siberian State Aerospace University, 31 Krasnoyarsky Rabochy av., Krasnoyarsk, 660014 Russian Federation

**A. Popov**
Institute of Informatics and Telecommunications, Siberian State Aerospace University, 31 Krasnoyarsky Rabochy av., Krasnoyarsk, 660014 Russian Federation

**Qing Xiong**
State Key Laboratory of Information Engineering in Surveying Mapping and Remote Sensing, Wuhan University, P.O. Box C310, 129 Luoyu Road, Wuhan, Hubei, 430079, P.R. China

**Qing Zhu**
State Key Laboratory of Information Engineering in Surveying Mapping and Remote Sensing, Wuhan University, P.O. Box C310, 129 Luoyu Road, Wuhan, Hubei, 430079, P.R. China
Faculty of Geosciences and Environmental Engineering, Southwest Jiaotong University, Chengdu, Sichuan, P.R. China

**Sisi Zlatanova**
GISt, OTB, Delft University of Technology, Jaffalaan 9, 2628 BX Delft, The Netherlands

**Zhiqiang Du**
State Key Laboratory of Information Engineering in Surveying Mapping and Remote Sensing, Wuhan University, P.O. Box C310, 129 Luoyu Road, Wuhan, Hubei, 430079, P.R. China

**Yeting Zhanga**
State Key Laboratory of Information Engineering in Surveying Mapping and Remote Sensing, Wuhan University, P.O. Box C310, 129 Luoyu Road, Wuhan, Hubei, 430079, P.R. China

**Li Yang Zeng**
Faculty of Geosciences and Environmental Engineering, Southwest Jiaotong University, Chengdu, Sichuan, P.R. China

**R. Lacaze**
HYGEOS, Euratechnologies, 165 avenue de Bretagne, 59000 Lille, France

**Bruno Smets**
VITO, Boeretang 200, 2400 Mol, Belgium

**F. Baret**
INRA, EMMAH UMR1114, site Agroparc, 84914 Avignon, France

**M. Weiss**
INRA, EMMAH UMR1114, site Agroparc, 84914 Avignon, France

**D. Ramon**
HYGEOS, Euratechnologies, 165 avenue de Bretagne, 59000 Lille, France

**B. Montersleet**
HYGEOS, Euratechnologies, 165 avenue de Bretagne, 59000 Lille, France

**L. Wandrebeck**
HYGEOS, Euratechnologies, 165 avenue de Bretagne, 59000 Lille, France

**J.-C. Calvet**
CNRM/Météo-France, 42 avenue G. Coriolis, 31057 Toulouse Cedex, France

**J.-L. Roujean**
CNRM/Météo-France, 42 avenue G. Coriolis, 31057 Toulouse Cedex, France

**F. Camacho**
EOLAB, G.V. Marques del Turia, 20-9, 46 005 Valencia, Spain

**Luo Chengfeng**
Institute of Photogrammetry and Remote Sensing, Chinese Academy of Surveying and Mapping, Beijing 100830, P.R. China

**Yu Xiujuan**
National Calibration Center for Electro-optical Distance Meter, Chinese Academy of Surveying and Mapping, Beijing 100830, P.R. China

**Liu Caijuan**
Geomatics of College, Shandong University of Science and Technology, Qingdao 266590, P.R. China

**Du Yingkun**
Geomatics of College, Shandong University of Science and Technology, Qingdao 266590, P.R. China

**Cong Thang Pham**
Institute of Applied Mathematics and Computer Science, Tula State University-Russia

**Andrey V. Kopylov**
Institute of Applied Mathematics and Computer Science, Tula State University-Russia

**K. Kamiyaa**
Dept. of Civil Engineering, The University of Tokyo, 7-3-1 Hongo Bunkyo Tokyo 113-8656, Japan

**T. Fuseb**
Dept. of Civil Engineering, The University of Tokyo, 7-3-1 Hongo Bunkyo Tokyo 113-8656, Japan

**L. J. Martinez M**
Universidad Nacional de Colombia, Facultad de Ciencias Agrarias, Bogotá, Colombia

**A. Ramos**
Universidad Nacional de Colombia, Facultad de Ciencias Agrarias, Bogotá, Colombia

**V. Khryashchev**
P.G. Demidov Yaroslavl State University, Yaroslavl, Russia

**A. Priorov**
P.G. Demidov Yaroslavl State University, Yaroslavl, Russia

**O. Stepanova**
P.G. Demidov Yaroslavl State University, Yaroslavl, Russia

**A. Nikitin**
PicLab LLC, Yaroslavl, Russia

**M. Schönert**
BlackBridge, Dept. of Application Research, 10719 Berlin, Germany

**E. Zillmann**
BlackBridge, Dept. of Application Research, 10719 Berlin, Germany

**H. Weichelt**
BlackBridge, Dept. of Application Research, 10719 Berlin, Germany

**J. U. H. Eitel**
Geospatial Laboratory for Environmental Dynamics, University of Idaho, Moscow, ID 83844-1135, USA

**T. S. Magney**
Geospatial Laboratory for Environmental Dynamics, University of Idaho, Moscow, ID 83844-1135, USA

**H. Lilienthal**
Julius-Kühn-Institut (JKI), Federal Research Centre for Cultivated Plants, 38116 Braunschweig, Germany

**B. Siegmannd**
Institute for Geoinformatics and Remote Sensing, University of Osnabrueck, 49076 Osnabrueck, Germany

**T. Jarmerd**
Institute for Geoinformatics and Remote Sensing, University of Osnabrueck, 49076 Osnabrueck, Germany

**S. V. Savicheva**
Murom Institute (branch) or the Federal State Budget Institution of Higher Professional Education "Vladimir State University Alexander G. and Nicholai G. Stoletovs", Russia

**M. Favorskaya**
Institute of Informatics and Telecommunications, Siberian State Aerospace University, 31 Krasnoyarsky Rabochy av., Krasnoyarsk, 660014 Russian Federation

**D. Novikov**
Institute of Informatics and Telecommunications, Siberian State Aerospace University, 31 Krasnoyarsky Rabochy av., Krasnoyarsk, 660014 Russian Federation

**Y. Savitskaya**
Institute of Informatics and Telecommunications, Siberian State Aerospace University, 31 Krasnoyarsky Rabochy av., Krasnoyarsk, 660014 Russian Federation

**B. Fichtelmann**
German Aerospace Center (DLR), Earth Observation Center, German Remote Sensing Data Center, National Ground Segment, 17235 Neustrelitz, Germany

**E. Borg**
German Aerospace Center (DLR), Earth Observation Center, German Remote Sensing Data Center, National Ground Segment, 17235 Neustrelitz, Germany

**E. Schwarz**
German Aerospace Center (DLR), Earth Observation Center, German Remote Sensing Data Center, National Ground Segment, 17235 Neustrelitz, Germany

**M. Buchwitz**
Institute of Environmental Physics (IUP), University of Bremen, Bremen, Germany

**M. Reuter**
Institute of Environmental Physics (IUP), University of Bremen, Bremen, Germany

**O. Schneising**
Institute of Environmental Physics (IUP), University of Bremen, Bremen, Germany

**H. Boesch**
University of Leicester, Leicester, United Kingdom c SRON Netherlands Institute for Space Research, Utrecht, Netherlands

**I. Aben**
SRON Netherlands Institute for Space Research, Utrecht, Netherlands

**M. Alexe**
European Commission Joint Research Centre (EC-JRC), Institute for Environment and Sustainability (IES),Air and Climate Unit, Ispra, Italy

**R. Armante**
Laboratoire de Météorologie Dynamique (LMD), Palaiseau, France

**P. Bergamaschi**
European Commission Joint Research Centre (EC-JRC), Institute for Environment and Sustainability (IES),Air and Climate Unit, Ispra, Italy

**H. Bovensmann**
Institute of Environmental Physics (IUP), University of Bremen, Bremen, Germany

**D. Brunner**
Swiss Federal Laboratories for Materials Science and Technology (Empa), Dübendorf, Switzerland

**B. Buchmann**
Swiss Federal Laboratories for Materials Science and Technology (Empa), Dübendorf, Switzerland

**J. P. Burrows**
Institute of Environmental Physics (IUP), University of Bremen, Bremen, Germany

**A. Butz**
Karlsruhe Institute of Technology (KIT), Karlsruhe, Germany

**F. Chevallier**
Laboratoire des Sciences du Climat et de l'Environnement (LSCE), Gif-sur-Yvette, France

**A. Chédin**
Laboratoire de Météorologie Dynamique (LMD), Palaiseau, France

**C. D. Crevoisier**
Laboratoire de Météorologie Dynamique (LMD), Palaiseau, France

**S. Gonzi**
University of Edinburgh, Edinburgh, United Kingdom (UK)

**M. De Mazière**
Belgian Institute for Space Aeronomy (BIRA), Brussels, Belgium

**E. De Wachter**
Belgian Institute for Space Aeronomy (BIRA), Brussels, Belgium

**R. Detmers**
SRON Netherlands Institute for Space Research, Utrecht, Netherlands

**B. Dils**
Belgian Institute for Space Aeronomy (BIRA), Brussels, Belgium

**C. Frankenberg**
Jet Propulsion Laboratory (JPL), Pasadena, California, United States of America

**P. Hahne**
Karlsruhe Institute of Technology (KIT), Karlsruhe, Germany

**O. P. Hasekamp**
SRON Netherlands Institute for Space Research, Utrecht, Netherlands

**W. Hewson**
University of Leicester, Leicester, United Kingdom c SRON Netherlands Institute for Space Research, Utrecht, Netherlands

**J. Heymann**
Institute of Environmental Physics (IUP), University of Bremen, Bremen, Germany

**S. Houweling**
SRON Netherlands Institute for Space Research, Utrecht, Netherlands

**M. Hilker**
Institute of Environmental Physics (IUP), University of Bremen, Bremen, Germany

**T. Kaminski**
The Inversion Lab, Hamburg, Germany

**G. Kuhlmann**
Swiss Federal Laboratories for Materials Science and Technology (Empa), Dübendorf, Switzerland

**A. Laeng**
Karlsruhe Institute of Technology (KIT), Karlsruhe, Germany

**T. T. v. Leeuwen**
SRON Netherlands Institute for Space Research, Utrecht, Netherlands

**G. Lichtenberg**
Deutsches Zentrum für Luft- und Raumfahrt (DLR), Oberpfaffenhofen, Germany

**J. Marshall**
Max-Planck-Institute for Biogeochemistry (MPI-BGC), Jena, Germany

**S. Noël**
Institute of Environmental Physics (IUP), University of Bremen, Bremen, Germany

**J. Notholt**
Institute of Environmental Physics (IUP), University of Bremen, Bremen, Germany

**P. Palmer**
University of Edinburgh, Edinburgh, United Kingdom (UK)

**R. Parker**
University of Leicester, Leicester, United Kingdom c SRON Netherlands Institute for Space Research, Utrecht, Netherlands

**M. Scholze**
Lund University, Lund, Sweden

**G. P. Stiller**
Karlsruhe Institute of Technology (KIT), Karlsruhe, Germany

**T. Warneke**
Institute of Environmental Physics (IUP), University of Bremen, Bremen, Germany

**C. Zehner**
European Space Agency (ESA), ESRIN, Frascati, Italy

**A. Bobkov**
Saint-Petersburg State University, Institute of Earth Sciences, Department of Physical Geography and Landscape Design, 33-35 10-th line V.O., 199178 Saint-Petersburg, Russia

**E. Panidi**
Saint-Petersburg State University, Institute of Earth Sciences, Department of Cartography and Geoinformatics, 33-35 10-th line V.O., 199178 Saint-Petersburg, Russia

**N. Torlopova**
Russian Academy of Sciences, Ural Branch, Komi Scientific Center, Institute of Biology, 28 Kommunisticheskaya st., 167982 Syktyvkar, Russia

**V. Tsepelev**
Russian State Hydrometeorological University, Meteorological Department, 98 Malookhtinsky prospect, 195196 Saint-Petersburg, Russia

**A. F. Domingues**
Geophysics Centre of Évora - University of Évora (CGE-UE), Rua Romão Ramalho, 59, 7000 Évora, Portugal

**D. Bortoli**
Geophysics Centre of Évora - University of Évora (CGE-UE), Rua Romão Ramalho, 59, 7000 Évora, Portugal
Institute of Atmospheric Sciences and Climate (ISAC-CNR), Via Gobetti 101, 40129 Bologna Italy

**A. M. Silva**
Geophysics Centre of Évora - University of Évora (CGE-UE), Rua Romão Ramalho, 59, 7000 Évora, Portugal
Department of Physics, University of Évora, Rua Romão Ramalho 59, 7000 Évora, Portugal

**P. Kulkarni**
Geophysics Centre of Évora - University of Évora (CGE-UE), Rua Romão Ramalho, 59, 7000 Évora, Portugal

**R. Mendes**
Geophysics Centre of Évora - University of Évora (CGE-UE), Rua Romão Ramalho, 59, 7000 Évora, Portugal

**A. A. Zakharov**
Murom Institut Vladimir State University, CAD Department, , 602264, Orlovskaya 23, Murom, Russian Federation

**A. E. Barinov**
Murom Institut Vladimir State University, CAD Department, , 602264, Orlovskaya 23, Murom, Russian Federation

**A. L. Zhiznyakov**
Murom Institut Vladimir State University, CAD Department, , 602264, Orlovskaya 23, Murom, Russian Federation

**Martina Hodrius**
Vista Geowissenschaftliche Fernerkundung GmbH, Gabelsbergerstraße 51, 80333 München, Germany

**Silke Migdall**
Vista Geowissenschaftliche Fernerkundung GmbH, Gabelsbergerstraße 51, 80333 München, Germany

**Heike Bach**
Vista Geowissenschaftliche Fernerkundung GmbH, Gabelsbergerstraße 51, 80333 München, Germany

**Tobias Hank**
Ludwig-Maximilians-Universität München, Luisenstraße 37, 80333 München, Germany

**A. Makri**
MSc Geomatics, Faculty of Architecture and the Built Environment, Delft University of Technology, Julianalaan 134, 2628 BL, Delft, The Netherlands

**S. Zlatanova**
3D Geoinformation, Faculty of Architecture and the Built Environment, Department of Urbanism, Delft University of Technology, Julianalaan 134, 2628 BL, Delft, The Netherlands

**E. Verbree**
GIS Technology, Faculty of Architecture and the Built Environment, OTB Research Institute for the Built Environment, Delft University of Technology, Julianalaan 134, 2628 BL, Delft, The Netherlands

**E. V. Pugin**
Murom Institute (branch) Vladimir State University, Information Technologies Department, 602264 Orlovskaya str. 23, Murom, Russia

**A. L. Zhiznyakova**
Murom Institute (branch) Vladimir State University, Information Technologies Department, 602264 Orlovskaya str. 23, Murom, Russia

**V. V. Kniaz**
St. Res. Institute of Aviation Systems (GOSNIIAS), Moscow, Russia

**Z. N. Smirnova**
Gnessin Russian Academy of Music, Moscow, Russia

**J. S. J. Wijesingha**
Department of Physical Geography and Ecosystem Science, Lund University, S¨olvegatan 12, S-223 62 Lund, Sweden

**N. L. Deshapriya**
Department of Physical Geography and Ecosystem Science, Lund University, S¨olvegatan 12, S-223 62 Lund, Sweden

**L. Samarakoon**
Geoinformatics Center (GIC), Asian Institute of Technology (AIT), PO Box 04, Klong Luang, Pathumthani, Thailand

**V. S. Gorbatsevich**
The Federal State Unitary Enterprise «State Research Institute of Aviation Systems» Russia, Moscow

**D. Nguyen**
Department of Geodesy and Geoinformation, Vienna University of Technology, Austria
Department of Photogrametry and Remote Sensing, Hanoi University of Mining and Geology, Vietnam

**W. Wagner**
Department of Geodesy and Geoinformation, Vienna University of Technology, Austria

**V. Naeimi**
Department of Geodesy and Geoinformation, Vienna University of Technology, Austria

**S. Cao**
Department of Geodesy and Geoinformation, Vienna University of Technology, Austria

**Y. C. Lai**
Dept. of Aerospace and Systems Engineering, Feng Chia University, Taiwan

**C. C. Chang**
Dept. of Aerospace and Systems Engineering, Feng Chia University, Taiwan

**C. M. Tsai**
Dept. of Aerospace and Systems Engineering, Feng Chia University, Taiwan

**S. Y. Lin**
Cloud Service Technology Center, Industrial Technology Research Institute, Taiwan

**S. C. Huang**
Cloud Service Technology Center, Industrial Technology Research Institute, Taiwan

**George Sithole**
Geomatics Division, School of Architecture, Planning and Geomatics, University of Cape Town, Private Bag X3, Rondebosch, 7701, South Africa

**Langalethu Majola**
Geomatics Division, School of Architecture, Planning and Geomatics, University of Cape Town, Private Bag X3, Rondebosch, 7701, South Africa

**A. Miraliakbari**
Department of Geomatics, Computer Science and Mathematics, University of Applied Sciences Stuttgart, Schellingstraße 24, 70174 Stuttgart, Germany

**M. Hahn**
Department of Geomatics, Computer Science and Mathematics, University of Applied Sciences Stuttgart, Schellingstraße 24, 70174 Stuttgart, Germany

**S. Sok**
Department of Geomatics, Computer Science and Mathematics, University of Applied Sciences Stuttgart, Schellingstraße 24, 70174 Stuttgart, Germany

**S. Mirzaee**
University of Tehran, Dept. of Surveying and Geomatics Eng., Tehran, Iran

**M. Motagh**
GFZ German Research Centre for Geosciences, Section of Remote Sensing, 14473, Potsdam, Germany

**H. Arefi**
University of Tehran, Dept. of Surveying and Geomatics Eng., Tehran, Iran

**A. Nooryazdan**
Forest, Range and Watershed Management of Iran, Lorestan

**Xiao Tan**
State Key Laboratory for Information Engineering in Surveying, Mapping and Remote Sensing, Wuhan University, Wuhan 430079 , China
Department of Management Engineering, Naval University of Engineering, Wuhan 430033 , China

**Y. Takahashi**
Research and Develoment Center, AERO ASAHI CORPORATION, 350-1165, 3-1-1 Minamidai, Kawagoe, Saitama, Japan
Dept. of Civil and Environmental Engineering, Tokyo Denki University, 350-0394, Hatoyama, Saitama, Japan

**H. Chikatsu**
Dept. of Civil and Environmental Engineering, Tokyo Denki University, 350-0394, Hatoyama, Saitama, Japan

**Nguyen Hoang Khanh Linh**
Faculty of Land Resources and Agricultural Environment, Hue University of Agriculture and Forestry, 102 Phung Hung, Hue City, Vietnam

**Huynh Van Chuong**
Faculty of Land Resources and Agricultural Environment, Hue University of Agriculture and Forestry, 102 Phung Hung, Hue City, Vietnam

**Xiongfei Wen**
Changjiang River Scientific Research Institute, Changjiang Water Resources Commission, Wuhan 430010, PR China

**Zhe Li**
Changjiang River Scientific Research Institute, Changjiang Water Resources Commission, Wuhan 430010, PR China

**Sui Zhang**
Changjiang River Scientific Research Institute, Changjiang Water Resources Commission, Wuhan 430010, PR China

**Shaohong Shen**
Changjiang River Scientific Research Institute, Changjiang Water Resources Commission, Wuhan 430010, PR China

**Dunmei Hu**
Department of Pharmacy, Tongji Hospital Affiliated with Tongji Medical College, Huazhong University of Science and Technology, Wuhan, China 430032, PR China

**Xiao Xiao**
Changjiang River Scientific Research Institute, Changjiang Water Resources Commission, Wuhan 430010, PR China

**A. M. Helle**
IAPG, Institute for Applied Photogrammetry and Geoinformatics, Jade University of Applied Sciences, Oldenburg, Germany

**J. Pilinski**
IAPG, Institute for Applied Photogrammetry and Geoinformatics, Jade University of Applied Sciences, Oldenburg, Germany

**T. Luhmann**
IAPG, Institute for Applied Photogrammetry and Geoinformatics, Jade University of Applied Sciences, Oldenburg, Germany

**K. Yokoi**
Division of Sustainable Energy and Environmental Engineering, Osaka University, Suita, Osaka, Japan

**N. Yabuki**
Division of Sustainable Energy and Environmental Engineering, Osaka University, Suita, Osaka, Japan

**T. Fukuda**
Division of Sustainable Energy and Environmental Engineering, Osaka University, Suita, Osaka, Japan

**T. Michikawa**
Center for Environmental Innovation Design for Sustainability, Osaka University, Suita, Osaka, Japan

**A. Motamedi**
Division of Sustainable Energy and Environmental Engineering, Osaka University, Suita, Osaka, Japan

**W. Nakanishi**
Dept. of Civil Engineering, The University of Tokyo, 7-3-1 Hongo, Bunkyo-ku, Tokyo 1138656 Japan

**T. Fuse**
Dept. of Civil Engineering, The University of Tokyo, 7-3-1 Hongo, Bunkyo-ku, Tokyo 1138656 Japan

**T. Ishikawa**
Dept. of Civil Engineering, The University of Tokyo, 7-3-1 Hongo, Bunkyo-ku, Tokyo 1138656 Japan

**S. G. Artyukhina**
MSU, Faculty of Computational Mathematics and Cybernetics, Russia, 119991, Moscow, GSP-1, 1-52, Leninskiye Gory

**L. M. Mestetskiy**
MSU, Faculty of Computational Mathematics and Cybernetics, Russia, 119991, Moscow, GSP-1, 1-52, Leninskiye Gory

**Kalaivani Thangamania**
SErvice SEnsing, Assimilation, and Modeling Research Group, Human Informatics Research Institute, National Institute of Advanced Industrial Science and Technology (AIST), 1-1-1 Umezono, Tsukuba 305-8568, Japan

**Ryosuke Ichikaria**
SErvice SEnsing, Assimilation, and Modeling Research Group, Human Informatics Research Institute, National Institute of Advanced Industrial Science and Technology (AIST), 1-1-1 Umezono, Tsukuba 305-8568, Japan

**Takashi Okumaa**
SErvice SEnsing, Assimilation, and Modeling Research Group, Human Informatics Research Institute, National Institute of Advanced Industrial Science and Technology (AIST), 1-1-1 Umezono, Tsukuba 305-8568, Japan

**Tomoya Ishikawa**
Site Sensing Inc., Japan

**Takeshi Kurat**
Site Sensing Inc., Japan

**Sanghyun Yoon**
School of Civil and Environmental Engineering, Yonsei University, 134 Shinchon-dong, Seodaemun-gu, Seoul 120-749, Republic of Korea

**Jaehoon Jung**
School of Civil and Environmental Engineering, Yonsei University, 134 Shinchon-dong, Seodaemun-gu, Seoul 120-749, Republic of Korea

**Joon Heo**
School of Civil and Environmental Engineering, Yonsei University, 134 Shinchon-dong, Seodaemun-gu, Seoul 120-749, Republic of Korea

**Huanxue Zhang**
Institute of Remote Sensing and Digital Earth, Chinese Academy of Sciences, Datun Road, Chaoyang District, Beijing 100101,China

**Qiangzi Lia**
Institute of Remote Sensing and Digital Earth, Chinese Academy of Sciences, Datun Road, Chaoyang District, Beijing 100101,China

**Fan Lei**
Hunan Province Land and Resources Planning Institute, Furong Road, Yuhua District, Changsha 410007, China

**Xin Du**
Institute of Remote Sensing and Digital Earth, Chinese Academy of Sciences, Datun Road, Chaoyang District, Beijing 100101,China

**Jide Wei**
Hunan Province Land and Resources Planning Institute, Furong Road, Yuhua District, Changsha 410007, China

**V. Yu. Volkov**
Dept. of Radioengineering, The Bonch-Bruevich State Telecommunications University, Saint-Petersburg, Russia
Dept. of Radioengineering, State University of Aerospace Instrumentation, Saint-Petersburg, Russia

**T. Fuse**
Dept. of Civil Engineering, University of Tokyo, Hongo 7-3-1, Bunkyo-ku, Tokyo, 113-8656, Japan

**K. Matsumoto**
Rakuten, Inc., Higashi Shinagawa 4-12-3, Shinagawa-ku, Tokyo, 140-0002, Japan

**B. Xu**
State Key Laboratory of Information Engineering in Surveying, Mapping and Remote Sensing, Wuhan University, Wuhan, Hubei, China

**W. S. Jiang**
State Key Laboratory of Information Engineering in Surveying, Mapping and Remote Sensing, Wuhan University, Wuhan, Hubei, China

**Q. S. Zhu**
State Key Laboratory of Information Engineering in Surveying, Mapping and Remote Sensing, Wuhan University, Wuhan, Hubei, China